OLED Fundamentals

OLED Fundamentals

Materials, Devices, and Processing of Organic Light-Emitting Diodes

Edited by
Daniel J. Gaspar
Evgueni Polikarpov

CRC Press
Taylor & Francis Group
Boca Raton   London   New York

CRC Press is an imprint of the
Taylor & Francis Group, an **informa** business

Cover Image. A flexible transparent organic small molecule light emitting panels (see Chapter 6, "Small Molecule Fundamentals" by Xin Xu, Michael S. Weaver). Courtesy of Universal Display Corporation.

CRC Press
Taylor & Francis Group
6000 Broken Sound Parkway NW, Suite 300
Boca Raton, FL 33487-2742

First issued in paperback 2018

© 2015 by Taylor & Francis Group, LLC
CRC Press is an imprint of Taylor & Francis Group, an Informa business

No claim to original U.S. Government works

ISBN 13: 978-1-138-89396-2 (pbk)
ISBN 13: 978-1-4665-1518-5 (hbk)

Library of Congress Cataloging-in-Publication Data

OLED fundamentals : materials, devices, and processing of organic light-emitting diodes / editors Daniel J. Gaspar and Evgueni Polikarpov.
     pages cm
Includes bibliographical references and index.
ISBN 978-1-4665-1518-5 (alk. paper)
1. Light emitting diodes. 2. Electroluminescent devices. 3. Organic semiconductors. I. Gaspar, Daniel J., editor. II. Polikarpov, Evgueni, editor.

TK7871.89.L53O54 2015
621.3815'22--dc23
2014049674

Visit the Taylor & Francis Web site at
http://www.taylorandfrancis.com

and the CRC Press Web site at
http://www.crcpress.com

To my family and to all of the members of the Pacific Northwest National Laboratory OLED team, past and present.

Daniel J. Gaspar

To my parents Tatyana and Vladimir.

Evgueni Polikarpov

Contents

SECTION I   Materials

SECTION II   Devices and Processing

Preface

What is an organic light-emitting diode (OLED)? Why should we care? What are they made of? How are they made? What are the challenges in seeing these devices enter the marketplace in various applications? These are questions we hope to answer in this book, at a level suitable for knowledgeable non-experts, graduate students, and scientists and engineers working in the field who want to understand the broader context of their work.

At the most basic level, an OLED is a promising new technology composed of some organic materials sandwiched between two electrodes. When current is passed through the device, light is emitted. The stack of layers can be very thin and has many variations, including flexible and/or transparent materials. The organic material can be polymeric or composed of small molecules, and may include inorganic components. The electrodes may consist of metals, metal oxides, carbon nanomaterials, or other species, though of course for light to be emitted, one electrode must be transparent. OLEDs may be fabricated on glass, metal foils, or polymer sheets (though polymeric substrates must be modified to protect the organic material from moisture or oxygen). In any event, the organic material must be protected from moisture during storage and operation. A control circuit, the exact nature of which depends on the application, drives the OLED. Nevertheless, the control circuit should have very stable current control to generate uniform light emission. OLEDs can be designed to emit a single color of light, white light, or even tunable colors. The devices can be switched on and off very rapidly, which makes them suitable for displays or for general lighting.

Given the amazing complexity of the technical and design challenges for practical OLED applications, it is not surprising that applications are still somewhat limited. Although organic electroluminescence is more than 50 years old, the modern OLED field is really only about half that age—with the first high-efficiency OLED demonstrated in 1987. Thus, we expect to see exciting advances in the science, technology, and commercialization in the coming years. We hope that this book helps to advance the field in some small way.

Contributors to this book are experts from top academic institutions, industry, and national laboratories who provide comprehensive and up-to-date coverage of the rapidly evolving field of OLEDs. Furthermore, this work collects in one place, for the first time, key topics across the field of OLEDs, from fundamental chemistry and physics, to practical materials science and engineering topics, to aspects of design and manufacturing. The chapters together synthesize and put into context information scattered throughout the literature for easy review in one book. The scope reflects the necessity to focus on new technological challenges brought about by the transition to manufacturing. In the first section, all materials of construction of the OLED device are covered, from substrate to encapsulation. In the second section, for the first time, additional challenges in devices and processing are addressed.

This book is geared toward a broad audience, including materials scientists, device physicists, synthetic chemists, and electrical engineers. Furthermore, it makes a great introduction to scientists in industry and academia, as well as graduate students interested in applied aspects of photophysics and electrochemistry in organic thin films. This book is a comprehensive source for OLED R&D professionals from all backgrounds and institutions.

Editors

Daniel J. Gaspar is technical group manager of the Applied Materials Science group at Pacific Northwest National Laboratory (PNNL). Dr. Gaspar earned his PhD in physical chemistry from the University of Chicago in 1998, and his BS from Duke University in 1992. Dr. Gaspar's group concentrates on the discovery and application of new materials for energy applications, including organic light-emitting diodes (OLEDs) for energy-efficient solid-state lighting, energy storage, and separations. Dr. Gaspar leads PNNL's OLED program, setting technical direction and building partnerships with industry, academia, and other national laboratories, in order to advance the goal of accelerating adoption of energy-efficient OLED-based solid-state lighting. Dr. Gaspar's technical contributions include significant expertise in nanoscale materials characterization. Dr. Gaspar has published more than 45 papers and book chapters.

Evgueni Polikarpov is staff scientist in the Applied Materials Science group at Pacific Northwest National Laboratory. He earned his diploma in chemistry from Moscow State University in 1998, and PhD in chemistry from the University of Southern California in 2008 working on dopant materials for OLEDs. Dr. Polikarpov is a coauthor of over 20 research publications. His research interests include materials for photovoltaics and organic light-emitting devices.

Contributors

Abhinav Bhandari
PPG, Inc.
Pittsburgh, Pennsylvania

Jan Blochwitz-Nimoth
Novaled GmbH
Dresden, Germany

Damien Boesch
Samsung Cheil
San Jose, California

Curtis R. Fincher
Dupont Central Research and
    Development
Wilmington, Delaware

Daniel J. Gaspar
Pacific Northwest National
    Laboratory
Richland, Washington

David W. Gotthold
Pacific Northwest National
    Laboratory
Richland, Washington

Mark T. Greiner
Department of Materials Science and
    Engineering
University of Toronto
Toronto, Ontario, Canada

Junji Kido
Department of Organic Device
    Engineering
Yamagata University
Yamagata, Japan

Denis Y. Kondakov
Dupont Central Research and
    Development
Wilmington, Delaware

Roman Korotkov
Arkema, Inc.
King of Prussia, Pennsylvania

Valentina A. Krylova
Department of Chemistry
University of Southern
    California
Los Angeles, California

Falk Loeser
Novaled GmbH
Dresden, Germany

Min-Hao Michael Lu
OLED Lighting Design Center
Acuity Brands Lighting
Berkeley, California

Zheng-Hong Lu
Department of Materials Science and
    Engineering
University of Toronto
Toronto, Ontario, Canada

Björn Lüssem
Department of Physics
Kent State University
Kent, Ohio

Lorenza Moro
Samsung Cheil
San Jose, California

Asanga B. Padmaperuma
Pacific Northwest National
   Laboratory
Richland, Washington

Evgueni Polikarpov
Pacific Northwest National
   Laboratory
Richland, Washington

Vsevolod V. Rostovtsev
Dupont Central Research and
   Development
Wilmington, Delaware

Hisahiro Sasabe
Department of Organic Device
   Engineering
Yamagata University
Yamagata, Japan

Gary S. Silverman
Arkema, Inc.
King of Prussia, Pennsylvania

Mark E. Thompson
Department of Chemistry
University of Southern California
Los Angeles, California

Max Tietze
Institut für Angewandte Photophysik
Technische Universität Dresden
Dresden, Germany

Yuan-Sheng Tyan
First O-Lite, Inc.
Nanjing, China

Michael S. Weaver
Universal Display Corporation
Ewing, New Jersey

Xin Xu
Universal Display Corporation
Ewing, New Jersey

Xianghui Zeng
Samsung Cheil
San Jose, California

1. Introduction

Evgueni Polikarpov and Daniel J. Gaspar

Electroluminescence of organic molecules has been a well-known phenomenon since the mid-twentieth century. However, it was not until 1987, that organic light-emitting diodes (OLEDs), sometimes called organic light-emitting devices, became promising for practical applications, when Tang and van Slyke demonstrated the first high-efficiency device. Since this discovery, OLEDs have evolved from a scientific curiosity to a commercially viable technology incorporated into hand-held devices for bright, vibrant displays that offer beautiful colors and unmatched viewing angles. Furthermore, recent progress in OLED materials and manufacturing technologies suggests that significant changes such as development of larger displays, specialty lighting, and even general lighting will be possible, once current challenges to longevity and cost are overcome. These novel light sources, not only offer promise of energy efficiency, but also, may enable entirely new approaches to room lighting design due to novel form factors; and for transparent and flexible devices not accessible with other lighting technologies.

During relatively short history of OLEDs, the materials and technology have advanced rapidly. In just over 20 years, quantum efficiency (or the fraction of charge that is converted into light) has increased more than 20-fold, approaching the theoretical limit for internal quantum efficiencies. Despite these advances, the search continues for stable and efficient materials and device architectures, and for tackling increasing important challenges that have come to the fore in manufacturing technology: substrate and electrode processing, light extraction, and cost. Research and development efforts have focused on decreasing manufacturing costs while maintaining needed performance for large-area OLED devices and light sources in order to be competitive in the marketplace, while enhancing device stability and maintaining efficiency, nonetheless.

OLEDs are a unique technology, based on the use of organic molecules to conduct large amounts of charge, which recombines to emit light that is bright enough for displays or general lighting. Successful application of organic luminescence in light-emitting devices required materials and device structures that overcame the intrinsically high resistivity of the organic materials while achieving balanced charge injection from electrodes into organics. The first indication that these limitations could be addressed was cited in a paper by C. W. Tang and S. vanSlyke, which mentioned using a thin film heterostructure (stacked thin films). In a heterostructure OLED under operation, holes and electrons are injected from electrodes (usually inorganic) into organic layers, where they are then transported to the emissive layer (EML). The opposite charges form an exciton, which decays to electronic ground state while emitting a photon. This emission process can be fluorescent or phosphorescent, depending upon the emitting molecule. Individual layer thicknesses of 50 nm or less enable low drive voltages, typically <10 V, and, currently in many cases, close to theoretical limit determined by exciton energy. Furthermore, separate hole and electron conducting layers provide efficient charge injection and recombination by permitting optimization of electron, hole injection, and transport, simultaneously. The device developed by Tang and vanSlyke used an EML consisting entirely of emitting molecule. Shortly, after the introduction of thin film heterostructure-based OLEDs, two-component EMLs consisting of emitter molecules doped into an appropriate host matrix were demonstrated—leading to a great increase in device efficiency. This efficiency improvement was the result of an increased charge recombination and exciton confinement in an EML. Dispersing the emitter in a matrix also reduced self-quenching of the emitting dopants. In the late 1990s, a new family of emissive dopants was introduced that generated marked increase in OLED efficiency. These new high-efficiency emitters were based on utilization of triplet excitons (where quantum mechanical spin quantum number is 1, accounting for 75% of excitons). Efficient harvesting of triplet excitons requires a phosphorescent dopant, where both singlets and triplets lead to emissive recombination. In order to reduce degradation and ensure efficient radiative recombination relative to nonradiative recombination, phosphorescent dopant should have a radiative lifetime on a microsecond timescale. Currently, the only known way to achieve both high phosphorescence efficiency and a microsecond radiative lifetime is to incorporate a heavy metal atom into the dopant, whose spin–orbit coupling efficiently promotes, intersystem crossing between singlet and triplet states. Thus, the most efficient OLEDs involve phosphorescent precious metal complexes, particularly iridium (whose name stems from "iridescent"); small amount of these exotic materials suffice to produce extra-large areas of coating. Despite their cost, these materials are not source of current high cost of OLEDs. The challenges lie in scaling down all OLED manufacturing process steps, such as—cost of substrate manufacturing, organic film deposition, and electrode deposition. All these must be decreased for OLED-based televisions, lighting, and other large-area applications in order for OLEDs to achieve their full potential, especially in low-cost markets such as lighting.

Currently available OLED products are nearly all displays. Given the higher value and price offered for displays, they have been the main focus of OLED development, in their history. OLED-based lighting, on the other hand, has yet to overcome its unique set of challenges. For lighting applications, products must simultaneously achieve lifetime, color quality, uniformity, brightness, and efficiency acceptable to a consumer. In particular, the definitions for what constitutes "white" light are very stringent. To save energy and money for consumers, efficiency demands are also very challenging. Fortunately, resolution, pixel pitch, and ambient contrast requirements are relaxed as compared to displays. However, the biggest challenge to OLEDs for lighting from the perspective of device materials is combining high efficiency with longevity.

The two most notable features of OLED field are (1) rapid progress and (2) high degree of integration of different disciplines: OLED research encompasses expertise in electrical engineering, synthetic chemistry, optics, electrochemistry, mechanical engineering, and materials science. These two characteristics create the need for an up-to-date monograph that covers the current state of progress in OLED research with emphasis on technical aspects necessary for the development of viable OLED products, that is, from fundamentals to practical considerations of device manufacturing. The primary focus of this book is to cover this spectrum from beginning to end. To date, topics that are of increased interest to those developing OLED-based lighting solutions—comprising encapsulation, light extraction, deposition methods, and tools—have not been addressed in one monograph. However, all aspects of OLED development from organic materials design to device architecture, including light extraction, electrode design, and chemical phenomena related to materials and device degradation during operation are covered. This volume includes expanded focus on small molecule OLEDS and topics such as large-scale film processing and light extraction that are ever more important to the transition of the science behind OLEDs into commercial manufacturing.

The chapters are grouped into two sections. Section I covers OLED materials based on their role in the device, and Section II focuses on the function of devices (with a focus on OLED lighting applications), their architecture, and manufacturing.

Chapter 2 deals with rigid transparent substrates for OLEDs and discusses various types of glasses that are already in OLED production, already or have a potential to improve a device's lifetime and/or address high substrate cost issue. This is followed by a description of various flexible alternatives to glass substrates, including plastic and metal foils. Indeed, stability of OLEDs under operation is one of the most difficult problems that commercialization of this technology is facing presently. Oxygen and moisture play a big role in causing OLED degradation during its operation, with this in mind, Chapter 3 describes methods of device encapsulation and types of barrier coatings, while also providing a substantial background on the interaction of device components with ambient moisture, and mechanisms of device failures that can be mitigated by encapsulation. Next, transparent conductive oxides (TCOs) are introduced; and their chemical, electrical, and optical properties, and manufacturing challenges are described in Chapter 4. The specifics of TCO manufacturing challenges for displays and lighting applications are addressed. Emerging transparent electrode technologies that use alternatives to TCOs are also touched upon in Chapter 4. In-depth analysis of charge injection from an electrode into organic layers of a device is the subject of Chapter 5. In addition to discussing specific material combinations for charge injection at both anode- and cathode–organics interfaces, an extensive background of charge injection and methods of measuring the energy level alignment at electrode–organics interfaces. The fundamentals of small molecule organic materials for OLEDs are covered in Chapter 6. The goal of this chapter is to provide an overview of fundamental phenomena such as charge transport, energy transfer, exciton formation, electronic structure effects, and intermolecular interactions in solids that govern the behavior of small molecule OLED materials. Chapters 7 through 11 discuss organic heart of an OLED: Design, synthesis, and applications of electron (Chapter 7), hole transport (Chapter 8) layer materials, conductivity dopants (Chapter 9), hosts for emissive dopants (Chapter 10), and phosphorescent emitters (Chapter 11) are discussed in detail. Each chapter provides a description of operating principles, materials' classes, and device performance. In addition, Chapter 9 also provides a detailed description of methods to characterize conductivity of organic materials in order to better understand the relationship between structure, properties, and device performance.

The second section is dedicated to device design and manufacturing. As described in the first section of the book, it is possible to approach 100% internal quantum efficiency of emission in an OLED. However, fraction of light that exits a device and is therefore useful for a display or general illumination is considerably smaller than 100%. Therefore, optical techniques often referred to as light extraction or "outcoupling," to maximize the amount of light emitted, or external quantum efficiency, must be

Chapter 1

improved to generate higher power efficiency or lower power consumption. The fundamentals of light extraction from a planar optical device together with the current state of the art in outcoupling structures are the subject of Chapter 12. Mechanisms of device degradation in all aspects—from electrode degradation to growth of quenching centers, and chemical transformations of molecular constituents of individual organic layers—are the focus of Chapter 13. This chapter connects degradation data and mechanisms scattered throughout the research literature to provide a uniquely complete overview of OLED degradation. The most power-efficient OLEDs, to date, have been fabricated in laboratory settings using thermal evaporation techniques. These methods often prove very challenging for scaling up due to numerous factors related to large-area high-throughput processes. Also, use of large-scale vacuum techniques is often associated with increased production costs. Chapter 14 covers scale up of vapor deposition methods and issues associated with OLED manufacturing, including an overview of existing thermal evaporation technologies and industrial equipment. Although published accounts of OLEDs manufactured using solution-processing methods, have not reached power efficiencies demonstrated by their vacuum thermal evaporation counterparts, but, solution methods have already been adopted for display manufacturing. Displays have different requirements—described briefly in Chapter 15—including rather relaxed efficiency requirements, but more stringent uniformity and color quality metrics as compared to lighting. Solution processing methods also constitute a promising technology for efficient OLED lighting at a reasonable cost; this and their application to display manufacturing are covered in Chapter 15. Although, OLED displays are already commercialized but, solid-state lighting remains a precommercial application of OLEDs due to stringent cost, lifetime, and performance requirements. However, OLED lighting has the potential to transform how we bring light to our living and working spaces, offering the potential for flexible, transparent, and large-area lighting luminaires. Chapter 16 discusses the specifics of OLED applications for lighting from materials, device architecture, light outcoupling, and analyze economics perspectives.

Finally, in Chapter 17, we summarize current state of the art in OLED technology, with focus on current technical challenges, and provide a survey of current and future markets in small displays, televisions, specialty lighting, and general illumination.

Materials

2. Substrates

Abhinav Bhandari and Daniel J. Gaspar

2.1   Introduction

Organic light-emitting devices (OLEDs) offer a number of potential advantages for both lighting and displays including light weight, flexibility, transparency, color quality, and viewing angle. OLEDs are thin film devices consisting of a stack of organic layers (which may also include inorganic components or layers; see Chapters 4, 5, 7–11 for details) sandwiched between two electrodes. OLEDs operate by converting electrical current into light via an organic emitter (see Chapter 6 for details). Thus, the fabrication of an OLED involves the deposition of organic and inorganic materials onto a planar substrate—rigid or flexible—along a circuit to control the delivery of current (usually called a backplane). After deposition of the device, a top cover is deposited, laminated, or glued to the substrate to "encapsulate" or protect the device (see Chapter 3 for details on encapsulation). Together with the substrate, the encapsulation material(s) must prevent the ingress of moisture and oxygen to protect the sensitive organic materials. Furthermore, either the substrate or the encapsulation material must be transparent in order for the light to escape (and transparent OLEDs require that both surfaces must allow a high level of visible light transmission). Other materials and technologies may be introduced before or after fabrication to enhance the amount of light that escapes the device into the air; this is called light extraction or outcoupling (see Chapters 12 and 16 for more discussion). Outcoupling techniques may be incorporated into the substrate, added to the outside (air side) of the substrate, or built into the device during fabrication of the electrodes or organic layers. In all cases, the electrical, optical, and physical properties of the substrate help determine what types of outcoupling methods may be integrated into OLED.

Chapter 2

Most importantly for this chapter, OLED fabrication methods are dependent upon the properties of the substrate. Various approaches have been demonstrated for both vacuum thermal evaporation (see Chapter 14) and solution processing (see Chapter 15) of OLED materials and backplanes. The design of the fabrication process depends on the properties of the substrate. For instance, thermal cycling, as may be necessary during cleaning or deposition steps, can lead to substrate distortions, particularly if the coefficient of thermal expansion (CTE) is high and cycling leads to distortion, as for polymer substrates. Generally speaking, glass substrates are used in batch processes and polymer substrates are used in either batch or roll-to-roll fabrication. Metal foils are often demonstrated using batch-processing methods, but can conceivably be utilized in a roll-to-roll process. However, flexible glass has been introduced which can also be used in either configuration, as will be discussed later in this chapter. Finally, products may require that substrates demonstrate low weight, flexibility, thin form factor, low cost, or other characteristics important to consumers.

The rest of this chapter covers in detail performance requirements for substrates, the current status of materials used as substrates, and new developments with the potential to change the performance, cost, or properties of OLEDs.

2.2   Substrate Requirements

There are a number of requirements for OLED substrates to meet in order to ensure cost, lifetime, efficiency, and ease of processing. The specific requirements depend upon the final application. These can be summarized as low roughness, both root mean square (RMS) and R_{max}, or height of the highest asperity; high optical transparency (for bottom-emitting devices); good chemical resistance; moisture and oxygen permeability; low CTE; and low and high temperature stability.

OLEDs are thin film devices with very thin layers—the total thickness of the organic layers is typically on the order of a micron or less. The substrate, therefore, must be extremely flat to eliminate defects (see Chapter 16 for a detailed discussion) such as pinholes or shorts. Uniformity is also important in maintaining uniform emission characteristics, including color, brightness, and efficiency. Defects may also interfere with mask alignment and result in undercut patterning profiles. Planarizing layers which are compatible with OLED layers may be deposited on top of the substrate to mask the effect of defects and hence may allow for higher tolerance for this criterion. In general, the defect density and permissible defect size depends on the device architecture and application.

Second, it is obvious that a transparent OLED substrate should have high optical transmission across the visible spectrum. This translates into >90% transmission in the visible wavelength region of the spectrum, without the anode. Furthermore, it is important that the substrate have fairly uniform transmission across the visible spectrum to maintain a neutral color (i.e., not tinted by preferential absorption). Opaque substrates require a different OLED structure, described below. It may also be useful to have some control over the index of refraction (i.e., use high-index (HI) glass to match the index of refraction of the organic layers; described in more detail below) and/or ability to couple light extraction materials and methods (Chapters 12 and 16) to the substrate to maximize the amount of light that escapes into the air.

For high device yields, OLED fabrication typically requires aggressive cleaning of the substrate using ultrasound and aqueous and solvent-based chemicals. In addition, the process may involve photolithography, wet chemical etching of the anode, UV exposure and ozone treatment, and prolonged exposure to low-pressure conditions. The substrate should be able to withstand these conditions.

A major limitation on the lifetime of OLED devices is the formation and growth of nonelectroluminescent "dark spots" in the device upon exposure to atmosphere. These dark spots are formed at the interface between the low work-function metal that serves as part of the cathode and the emissive organic layer. It has been proposed that these spots are formed by the oxidation of the metal and/or organic materials by atmospheric oxygen and water, forming a highly resistive interface that prevents electroluminescence (Kolosov et al. 2001). In order to overcome this drawback, it is necessary to have a substrate with very low oxygen and moisture permeability.

The substrate must prevent oxygen and moisture ingress (and must not swell with absorption of moisture or other ubiquitous gases).

A wide thermal processing window is essential for device performance and good process economics. Compatibility with the process steps used in, for example, fabrication of the thin film transistor (TFT) backplane used for displays, has very different thermal cycling than a simple OLED, due to the need to generate high carrier mobility and low noise in the transistors. While two approaches—metal oxide and low-temperature polysilicon, or LTPS—have been pursued, they both require thermal cycling. Thus, stability with thermal cycling is a key requirement. This translates to a high thermal processing window, low CTE, and minimal distortions with thermal cycling.

To summarize, an OLED substrate should exhibit the following properties:

- Smooth surface morphology (RMS roughness < 2 nm; R_{max} < 20 nm)
- Good optical transparency for bottom-emitting devices
- Compatibility with the OLED fabrication processes, including good chemical and solvent resistance
- Low CTE and thermal cycling stability
- Low oxygen and moisture permeability

Some of these requirements have been summarized in Table 2.1. It should be noted that these properties serve as a guideline and may vary depending on the application and device fabrication methodology.

2.3 Rigid Glass Substrates

2.3.1 Borosilicate to Soda–Lime Float Glass Substrates

Glass substrates have good optical transmission, surface smoothness, and oxygen/water vapor barrier properties and provide excellent substrate materials for OLED devices. The first OLEDs were all fabricated on glass surfaces. Commercial OLED display devices have traditionally been fabricated on borosilicate glass. Borosilicate glass typically costs around \$40/m^2 and is around 5–10 times more expensive than soda-lime float glass and as the push continues to drive the costs down, soda-lime float glass is increasingly being

Table 2.1   A List of Selected Properties for OLED Substrates

Property	Requirement
Transmission	>90% in the visible spectrum; Neutral appearance
Oxygen transmission rate	$<10^{-4}$ cc/m^2.day.atm
Water vapor transmission rate	$<10^{-6}$ g/m^2 per day @ 38°C and 90% relative humidity
Roughness	$R_{RMS} < 2.0$ nm; $R_{PV} < 20$ nm
Defects	<1 pinhole of 0.5 µm/mm^2
CTE	As low as possible (<10 ppm/°C; cycling distortion minimal

Source: Adapted from Department of Energy 2011. Solid-State Lighting Research and Development: Manufacturing Roadmap, U.S. Department of Energy, July 2011; Mahon, J. K. et al., 1999. *42nd Annual Technical Conference Proceedings of Society of Vacuum Coaters*, Chicago, IL. 505: 456–459.

used as substrate for OLED devices. For example, OLEDs are considered a potential high efficiency, low cost, solid state replacement for general lighting and OLED lighting panels have been available since 2009. However, commercial offerings have been limited to expensive luminaires for decorative applications and prototyping panel kits. Widespread adoption of OLED-based solid-state lighting sources is constrained by the current high cost of these devices. By some estimates (Department of Energy 2011), the cost of the substrate needs to be driven down to $7/m^2 to make the OLED lighting cost-competitive with other commercially available products. Thus, there is a move to transition from borosilicate glass to soda-lime glass.

Borosilicate glass differs in chemical composition from soda-lime float glass, as it does not contain any alkaline earths. Boric acid and quartz sand are the main components. On the other hand, traditional soda-lime float glass or conventional float glass is a member of the alkali-alkaline-earth-silicate glass family. This glass does not contain any boron compounds. Instead it contains 12% alkaline earth content and contains a higher proportion of alkalis than borosilicate. A comparison of the chemical composition is presented in Table 2.2.

Table 2.2   Approximate Chemical Composition of Borosilicate and Soda-Lime Glass

Compound	Borosilicate Glass	Soda-Lime Glass
SiO_2	81%	73%
B_2O_3	13%	—
Na_2O	4%	14%
Al_2O_3	2%	1%
CaO	—	9%
MgO	—	3%

Source: Adapted from Schott. 2014. Available at: http://www.us.schott.com/borofloat/english/index.html. Accessed March 21, 2014.

Table 2.3 Comparison of the Approximate Physical, Mechanical, and Chemical Properties of Borosilicate and Soda-Lime Glasses

Property	Borosilicate Glass	Soda-Lime Glass
Density (g/cm³)	2.2	2.5
Modulus of elasticity (kN/mm²)	64	73
Coefficient of linear expansion at 20–300°C (K⁻¹)	3.25 ppm	9.0 ppm
Coefficient of thermal conductivity (W/mK)	1.2	1.0
Resistance to temperature difference (K)	80–110	40
Max. temperature (°C) short term/long term	500/450	300/200
Hydrolytic resistance (according to DIN 12 111)	HGB 1 (excellent)	HGB 3 (satisfactory)
Acid resistance (according to DIN 12 116)	1 (excellent)	1 (excellent)
Alkali resistance (according to DIN 52 322)	A2 (good)	A2 (good)

Source: Adapted from Schott. 2014. Available at: http://www.us.schott.com/borofloat/english/index.html. Accessed March 21, 2014; Gläser, H. J. *Large Area Glass Coating*. Dresden: Von Ardenne Anlagentechnik, 2000. ISBN3-00-004953-3.

Table 2.3 compares some of the physical, mechanical, and chemical properties of borosilicate and soda-lime glass. Compared with soda-lime glass, borosilicate glass has a lower density, lower CTE, better resistance to temperature difference, and better hydrolytic resistance. As a result of these properties, borosilicate glasses may be used in special applications with extreme thermal and/or chemical demands, where conventional float glasses of the soda-lime glass family may not be suitable.

For most OLED processes, the substrates are not subject to high temperatures and therefore the use of soda-lime float glass may be acceptable, particularly for lighting applications where a TFT backplane is not required. Soda-lime glass can offer >90% transmission in the visible wavelength region, which meets the requirement for OLED device fabrication. Figure 2.1 shows the transmission spectra for a 2.0 mm Starphire® ultra-clear float glass manufactured by PPG Industries, Inc. The transmitted color is also fairly neutral ($L^* = 96.61$, $a^* = -0.07$, $b^* = 0.29$; measured using D65 illumination source at 10°).

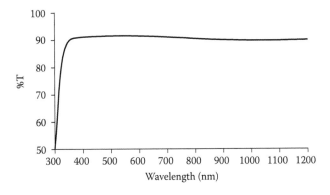

FIGURE 2.1   Transmission spectra of 2.0 mm thick Starphire ultra-clear soda-lime float glass manufactured by PPG Industries, Inc.

Chapter 2

The typical RMS roughness for soda-lime glass manufactured using float process is around 0.5 nm, which is smooth enough without further processing for use in OLED manufacturing. In a typical float glass manufacturing plant, surface defect measurements are typically achieved using a high resolution solid-state laser system and surface defects such as open bubbles, digs, rubs, scratches, top tin, and solid inclusions can be detected to a very fine resolution.

One challenge in using soda-lime glass for OLED applications is glass corrosion, which occurs upon exposure to a humid and high temperature environment. The process involves sodium ion leaching to the glass surface and formation of sodium hydroxide, and subsequently sodium carbonate and sodium bicarbonate. There are a number of ways in which this issue could be handled. The easiest way is to coat a newly manufactured glass surface (with the anode and subsequent manufacture of OLED devices) thus preserving the pristine surface. Alternatively, corrosion can be avoided by properly storing glass under controlled temperature and humidity conditions. Flat glass can also be preserved chemically by applying a suitable acidic buffer such as polymethyl methacrylate (PMMA) beads, which prevent the sodium leaching process, leading to longer shelf life. Another approach is to remove the corrosion layers by etching or polishing before coating the glass surface. Proper surface cleanliness (using ultrasonic cleaning, high-pressure jet cleaning, or brush cleaning (Gläser 2000)) is essential before coating the glass substrate as it may influence adhesion with the subsequent layers. Another concern is that sodium ions may leach from the glass substrates into the active device during fabrication and/or operation and may damage the devices. Blocking layers may be applied to the glass surface to prevent corrosion. Solarphire™ NaB is one such product available in the market that addresses sodium leaching. It is manufactured by PPG Industries.

Another potential issue with glass substrates, especially with large area devices such as 55″ televisions or 2′ × 4′ troffer lighting panels or for automotive and other applications where safety is a concern, is breakage. In these cases, tempered or laminated glass may be more suitable as a substrate. Typically, soda-lime glass thickness down to 3.0 mm can be tempered with a high surface compression in glass, which ensures the glass breaks into small diameter pieces for superior safety. Thin borosilicate glass is significantly more difficult to temper than soda-lime glass because of its low heat transfer coefficient.

2.3.2 High Index Glass Substrates

The ultimate goal of an OLED display or lighting fixture is to generate light that is seen by the observer. As described in detail in Chapter 12, the amount of light that escapes the substrate of an OLED fabricated on glass is limited by the index of refraction mismatch between the organics and the anode (typically a transparent conducting oxide such as indium tin oxide; see Chapter 4); refractive index ~1.8 in the visible) and the glass (borosilicate ~1.51, soda-lime 1.46). One novel approach to increase the light that escapes to be seen by the observer is to use a so-called HI glass. HI glass typically has a refractive index of about 1.8–2.0, eliminating reflection at the glass-organic interface. Several examples of this approach are found in the literature (Peng et al. 2004, Mladenovski et al. 2009, Reineke et al. 2009). Mladenovski and coworkers published results of an optimized green OLED with a HI glass substrate ($n \sim 1.8$) that demonstrated a record 183 lm/W (Mladenovski et al. 2009). The comparable OLED fabricated on a low-index glass substrate ($n \sim 1.5$) exhibited an efficiency of 102 lm/W, corresponding to an 80% improvement in light output. The ITO anode on the HI substrate was rougher than the commercially generated ITO on low-index glass, leading to higher leakage current. Nonetheless, this result demonstrates the potential for HI substrates. In another example, a white OLED was fabricated on a HI substrate (Schott glass: N-LAF 21; $n \sim 1.8$). This device was coupled to index matching fluid, a patterned outcoupling structure or an index-matched hemisphere to ensure minimal scattering that the HI glass–air interface. The power efficiency of this device at 1000 cd/m^2 was 102 lm/W, which was a record at the time (Reineke et al. 2009). Unfortunately, HI glasses tend to be brittle, hard to manufacture with a smooth surface, and expensive, and thus remain a novelty.

2.4 Flexible Substrates

Flexible OLED devices for displays and lighting have the potential to be used in a wide range of applications. A large number of materials have been demonstrated for use as flexible substrates. Most common is the use of a plastic substrate with an oxygen and water barrier layer (and often an integrated light extraction layer). Nonetheless, oxygen and water permeability through plastics remain a challenge and can result in low lifetime devices; see Chapter 3 for a detailed discussion of encapsulation to protect against water and oxygen. Other approaches include metal foils, flexible glass, and exotic materials such as paper. These developments present interesting opportunities for variety of flexible OLED applications, although products seem to be several years away from the marketplace. The rest of this section describes these flexible substrates and current progress toward realizing practical flexible OLEDs.

2.4.1 Polymer Substrates

Polymers were a natural choice for low-cost, flexible OLED substrates due to their optical properties, potential for low cost and mechanical properties including flexibility and toughness. In fact, it was only five years after the first modern OLED was demonstrated by Tang and VanSlyke that Heeger's group published a report detailing the fabrication of a flexible polymer-based OLED on a polyethylene terephthalate (PET) (Gustaffson et al. 1992). This publication opened up the possibility of flexible displays and lighting devices that could be made cheaply and change the way we use displays and light. In 1997, the first report detailing a first small-molecule based flexible OLED on a polymer substrate was published (Gu et al. 1997). At this point, truly flexible displays on plastic substrates remain a work in progress. Nonetheless, a great deal of progress has been made in the development of materials and methods for the incorporation of polymer substrates into flexible OLEDs. This is a very fast moving field, and materials properties continuing to improve at a rapid pace. Thus, a survey of the key requirements and properties, along with current progress is useful as only as an indicator of what suppliers and researchers can expect to achieve—not as a final status of polymer substrate technology.

Over the past three decades, a number of polymers have been explored for use as polymer substrates. An excellent review of the materials science of polymer substrates for flexible displays can be found in MacDonald et al. (2007b). The challenges for lighting are similar, with increased emphasis on enabling high power efficiency and lower cost, which in turn requires high yields and lower substrate cost. The requirements for high-resolution patterning and high temperature processing may be reduced, though, unless some types of light extraction structures are to be incorporated (e.g., metal oxide scattering structures that are deposited at high temperature, or patterned grids).

In general, the requirements for plastic OLED substrates are the same as for glass. These include thermal stability, low CTE, mechanical stability, chemical resistance, low oxygen, and moisture permeability, optical clarity in the visible wavelength region (assuming a bottom-emitting OLED is to be fabricated), and low roughness. No plastic currently meets all of these specifications, which indicates a multilayer (or other) composite structure is necessary, particularly for moisture and oxygen permeability (see Chapter 3 for a detailed discussion). Additional processing may be required to planarize the surface, but suppliers have demonstrated extremely low roughness for a variety of plastics, so this aspect should not be limiting for OLED researchers or manufacturers. Finally, a number of approaches have been demonstrated for improving the thermal and mechanical stability of the polymer, or of mitigating these properties using advanced fabrication approaches. The rest of this chapter will focus on a brief survey of polymer types that are used in OLED research and, potentially, production, along with key examples of flexible OLED displays and lighting panels fabricated on plastic substrates. For a more detailed discussion of polymer materials science relevant to plastic substrates, see MacDonald et al. (2007b).

Many classes of polymers and related materials (see Section 2.4.4 for cellulose-based materials, for instance) have been proposed for use as OLED substrates. The most successful examples of polymer substrates for flexible electronics fall into variants of five polymer families: PET (Gustaffson et al. 1992, Li et al. 2005), polyethylene naphthalate (PEN) (Pang et al. 2012), polyethersulfone (PES) (Kim et al. 2011a), polycarbonate (PC) (Wang et al. 2011), and polyimide (PI) (Kim et al. 2011b). Other classes that have been proposed include polycyclic olefins/polynorbornenes (PCO), polyetheretherketone (PEEK), and polyarlyates (PAR); these will not be discussed further as they seem to have shown less progress over the past five years. Figure 2.2 shows the structures of the first five polymers listed above, while Table 2.4 lists the properties of most of the substrate materials described throughout this chapter. PI has excellent thermal properties, but is colored and can only be used in effectively as an opaque substrate for top-emitting OLEDs (TOLEDs), competing to some extent with metal foils. PI has very good thermal properties and can enable a wider thermal processing window than other polymers, enabling, for instance, laser induced thermal imaging (LITI) pixel patterning (Lee et al. 2002).

As can be seen in Table 2.4, the various polymers exhibit a wide range of CTE values. Recent advances in polymer web fabrication processes have reduced the potential for distortion; nonetheless, all polymers may distort during thermal cycling and physical processing (i.e., roll-to-roll) (MacDonald et al. 2007a). This has led to a number of strategies to minimize distortion and ensure registry between the TFT backplane and the light-emitting region of an OLED display (Kim et al. 2014) and references therein.

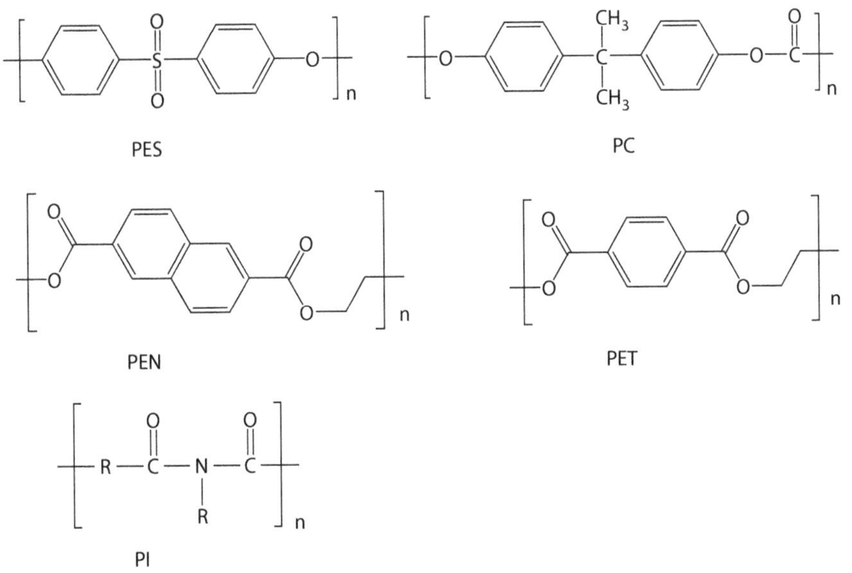

FIGURE 2.2 Structure of neutrally colored polymers most commonly used as OLED substrates. PES—polyethersulfone; PET—polyethylene terephthalate; PEN—polyethylene naphthalate; PC—polycarbonate; PI—polyimide.

Generally speaking, these have involved fixing the polymer sheet to a rigid glass (large area display or lighting) or Si wafer substrate (small displays and demonstrations) and then performing a removal step once the processing has finished. One particularly useful method has been laser release (and its close relative laser transfer) (Kim et al. 2014, Lifka et al. 2007). This method typically uses a sacrificial absorbing layer that is thermally decomposed leading to separation of the substrate from the carrier. Although there remain challenges to the use of polymeric foils as substrates for flexible OLEDs, advances in materials and processing have provided the means to overcome these, although at added cost and complexity. Continued improvements will likely lead to flexible commercial devices on plastic substrates within the next few years.

2.4.2 Metal Foils

The promise of metal foils as flexible substrates for OLEDs was recognized quite early in the history of organic electroluminescent devices. One of the primary drivers to fabricate an OLED by Wu et al. (1997) on stainless steel was the difficulty of processing a-Si TFT structures on polymer substrates due to the temperature limitations (typically <300°C); at the time, LTPS technology was still being developed. Five years after the first demonstration of a flexible OLED on a plastic substrate, a flexible OLED fabricated on stainless steel (SS) was demonstrated via a collaboration between the Wagner and Forrest groups at Princeton in 1997 (Wu et al. 1997). This work, described in more detail below, followed quickly after the development of an amorphous Si (a-Si) TFT that could be used to drive and control an OLED (Theiss and Wagner 1996). Since that time, a number of advances have been made with regard to metal foil substrate composition, properties, treatment, and processing. The rest of this section describes the challenges

Table 2.4   OLED Substrate Properties

Substrate	Max. Process Temp.°C	CTE (ppm/°C). Near 30°C	Visible Transmission.%	Chem. Resistance	Barrier Prop.	Surface	T_g or Melting Point. °C	Thermal Conduct. (W/m K)
Stainless steel	900	10–18	Opaque	Good	Good	Poor	1400	16+
Aluminum	350	~23	Opaque	Good	Good	Poor	660	>200
Glass	200–600	3–9	92	Good	Good	Excellent	>500 (T_g)	~1.0
Polyimide	200	30–60	Orange	Good	Poor	Poor	>400	0.5
Polyether-sulfone	200	15–55	90	Good	Poor	Fair	185	~0.2
Polyethylene naphthalate	150	13	87	Good	Poor	Excellent	120	~0.1
Polyethylene terephthalate	120	15	89	Good	Poor	Excellent	260	~0.2
Polycarbonate	200	60–70	90	Good	Poor	Excellent	150	~0.2
Exotic (e.g., paper)	Low	Variable	Variable	Var.	Var.	Variable	Var.	Variable

Source: Adapted from Hong, Y. T. et al. 2007. *Proceedings of the SPIE.* 6655: doi:10.1117/12.737213; With kind permission from Springer Science+Business Media: *Flexible Electronics: Materials and Applications*, Amorphous silicon: Flexible backplane, and display application, 2009, Sarma, K. R.

imposed by metal foil substrates, as well as efforts to develop SS, aluminum and exotic metal foil alloys, coatings, and processing to demonstrate high-efficiency, long-lived displays, and lighting. Table 2.3 summarizes the thermal, physical, and electrical properties of potential OLED substrate foils.

One difference between metal foils and transparent substrates (glass, polymer or other) is that the OLED must emit through the top electrode, called top emission. OLEDs fabricated in this fashion are often called TEOLEDs. Multilayer OLEDs can be fabricated in the traditional fashion, with the hole transport layer deposited onto the bottom electrode, or "inverted," with the electron transport layer deposited onto the bottom electrode. Generally, speaking inverted OLEDs are not as efficient due to poor electron injection, even with the use of an electron injection layer (see Chapter 5). The rest of this chapter will assume the OLED is fabricated in the traditional architecture. Generating a high-efficiency OLED is a substantial challenge for several reasons. First, top-emission requires what is generally called a semitransparent cathode (or anode if the device is inverted with electron transport layer on the bottom and whole transport layer on the top). The semitransparent cathode leads to a much stronger microcavity effect (see Chapter 12), which is a resonant effect and must be tuned to a particular wavelength. For white light, this ensures the microcavity is out of tune for much of the visible light range. Second, the conductivity of the semitransparent cathode is dependent on the thickness, but the transparency is inversely proportional to the thickness. A Mg:Ag electrode, developed for the first transparent OLED, maintains transparency to about 10–15 nm, so it is often used for this purpose (Gu et al. 1996). It is often used in conjunction with a metal like Mg to facilitate electron injection. In fact, Wu et al. used this cathode, together with a sputtered ITO capping layer to improve conductivity and oxidation resistance. Third, the resistance to degradation, usually oxidation of the cathode (Moon et al. 2004) leading to dark spots, is more challenging to engineer, although this challenge has been largely overcome with careful vacuum processing and encapsulation (see Chapter 3). Finally, incorporation of light extraction technologies is somewhat more challenging for this geometry. Common approaches include depositing an index-matching layer on top of the electrode, incorporating a scattering layer in the organic stack and others. See Chapter 12 for a detailed discussion of this topic.

The second challenge with metal foil substrates is roughness. Polishing approaches are capable of meeting the 2 nm RMS roughness requirement necessary for high yields. The roughness of a metal foil substrate depends upon the alloy, processing, and planarization, if any. Some alloys are first, if the foil is to be used both as a substrate and an electrode, the foil must be mechanically polished or, usually, chemomechanically or electropolished and the reverse side insulated to meet the roughness requirement described in the previous paragraph.

The third major challenge for metal foil substrates is related to electrical isolation. If the foil is polished and used as an electrode for lighting or other large area applications, the contacts on the reverse must be deposited carefully, and the rest of the panel insulated. However, if the foil is planarized with an insulating layer and a TFT backplace is to be fabricated on top of the buffer layer, capacitive coupling can take place, which reduces the achievable scan speed. In this case, Hong et al. (2007) determined that the unit-area capacitance must be maintained below ~2 nF/cm² (Hong et al. 2007); the

Chapter 2

unit-area capacitance depends upon the thickness and dielectric constant of the buffer layer. The dielectric constant of potential organic and inorganic insulating layers ranges from 2.1 (polytetrafluoroethylene) to as high as 4.5 for SiO_2. This suggests an insulating coating on the order of ~2.5 μm for organics with a lower dielectric constant down to 1 μm for SiO_2. Indeed, in Cheon et al. (2006), SiO_2 was used with a thickness of 1 μm on both sides to eliminate the capacitive coupling of the substrate to the TFT backplane. The buffer must be uniform without pinholes, even after distortion and bending. In this case, the CTE difference between SS and SiO_2 (18 ppm/°C and ~1 ppm/°C) induced bending in the substrate, which is why the authors deposited the SiO_2 on both sides in steps. Thus, the use of conductive foil substrates can lead to complicated processes, which entail coating both sides of the foil sequentially to maintain a flat substrate as in Cheon et al. (2006) and Jeong et al. (2007).

Finally, the design of a process or device architecture must take into account the thermomechanical properties of the substrate. Alloys that are to be used as OLED substrates have comparable CTE to plastics, but can be used in processes that use/involve a higher temperature. The distortions that routinely occur with plastics are much less of a problem with metal foils, but organic or polymeric planarizing layers are still subject to the same thermal restrictions as a plastic substrate. As seen (Cheon et al. 2006, Jeong et al. 2007), this large thermal expansion, compared to glass, may increase process complexity and must be accounted for during process design. One potential advantage is that the improved thermal conductivity may help with issues of OLED lifetime by better dissipating heat generated during operation, compared to plastic substrates, although a thick dielectric layer will reduce the effectiveness of the thermally conductive substrate. Note that the conductivity of aluminum is much higher than other potential substrates and would be best for this purpose. Nonetheless, the larger thermal process window and smaller amount of distortion and expansion is attractive for both display and lighting applications.

Much progress has been made toward overcoming these challenges. Wu et al. fabricated an array of a-Si TFTs directly on the Si surface using the structure shown in Figure 2.3 (Wu et al. 1997). The OLED was a simple polymer OLED using a mixed single-layer doped thin film, with poly(N-vinylcarbazole) (PVK) as the hole-transport component with the electron transport molecules 2-(4-biphenyl)-5-(4-tert-butylphenyl)-1,3,4-oxadiazole (PBD) and the fluorescent green emitter coumarin 6 (C6) doped into the mixed polymer layer. The yield of these devices was low due to the absence of a planarizing layer on the SS foil. The device performance was comparable to control devices built on glass.

Xie et al., in two reports, demonstrated the use of spin-on-glass as well as a polymer-silicon oxide two layer planarizing and insulating buffer layer. Both of these demonstrations used a semitransparent samarium layer, which is probably not practical for large-scale manufacturing due to cost (Xie et al. 2003a,b). These low efficiency Alq/α-NPD devices used an index-matching layer on top of the cathode to improve light extraction. Jeong built a fully functional RGB 5.6″ active-matrix OLED (AMOLED) display with 160 × 350 pixels on SS foil (Jeong et al. 2007). This device used an a-Si TFT backplane and maintained performance after bending. In order to fabricate a display on the originally rough SS substrate, the authors used chemomechanical polishing to reduce the RMS roughness from 81 nm to <3 nm. In the process, the thickness of the foil was reduced from 150 to 130 μm.

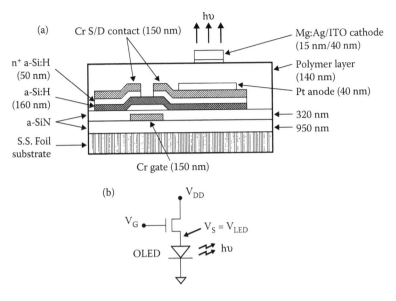

FIGURE 2.3 Structure of first OLED fabricated on stainless steel substrate. (a) Cross-sectional schematic of integrated TFT/OLED device structure. (b) Circuit diagram TFT/OLED device. (From Wu, C. C. et al. 1997. *IEEE Electron Device Letters.* 18: 609–612, with permission.)

Multiple groups have explored the use of aluminum foils for flexible substrates (Novaled 2010, Swensen et al. 2013). Aluminum is attractive from a cost, ease of processing, and thermal conductivity standpoint. However, the lower thermal processing budget due to the relatively low melting point introduces additional complications. Like SS foil, the intrinsic roughness of the foil or sheet is highly dependent on the actual alloy used, the rolling process, and any additional processing and/or coating. Two large integrated projects in Europe to advance aluminum sheet as a potential roll-to-roll processable OLED substrate, called ROLLEX and ROLLEX 2, have been undertaken by a team including a large number of companies, universities, and research institutions. Among the results of the first collaboration was the demonstration of a long-lifetime (>20,000 hr) OLED, which demonstrates the stability achievable using an Al substrate. Swensen and coworkers have demonstrated a high-efficiency OLED on Al substrates (Swensen et al. 2012). In this work, the planarization layer reduced the RMS roughness from >1 μm to ~3 nm. The device efficiency was comparable to an identical device fabricated on glass (71 lm/W compared to 67 lm/W, both at 1 mA/cm^2 and 1000 cd/m^2; shown in Figure 2.4) for a device architecture: engineered Al foil or glass substrate/Ag/Plexcore®HIL/α-NPD/4,4′,4″-tris(carbazol-9-yl)triphenylamine/4,4′-Bis(N-carbazolyl)-1,1′-biphenyl:bis(2-phenylpyridyl)iridium(iii)acetylacetonate/bathophenanthroline/Ag:Al/glass encapsulation. Glass was used as an encapsulant for demonstrations purposes only; all processes were fully compatible with lamination or thin film encapsulation.

Finally, Tan and coworkers laminated an Al sheet to a PET substrate to combine the best of both types of substrate (Tan et al. 2005). This work took a similar approach to that seen for ultrathin polymer substrates described in Section 2.4.1, but did not remove the polymer from the Al sheet that was used to stabilize the polymer. The light-emitting polymer device had a tuned microcavity which led to reasonable efficiency.

Chapter 2

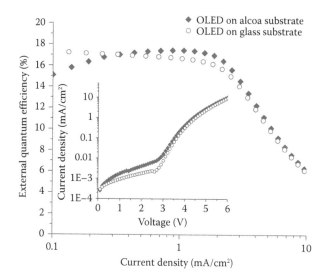

FIGURE 2.4   External quantum efficiency (EQE) versus current density (mA/cm²) for green OLEDs fabricated on an Al sheet substrate and glass with reflective bottom electrode. Device structure was engineered Al foil or glass substrate/Ag/HIL/α-NPD/TCTA/CBP:5%Ir(ppy)₂acac/Bphen/Ag:Al/glass encapsulation. (Inset) J-V curve. (From Swensen, J. S. et al. 2013. Unpublished results.)

Recently, Lee and coworkers used a surface roughness transfer approach using a Si wafer mother substrate to fabricate extremely flat Ag and INVAR, an iron–nickel alloy known generically as FeNi36 with an extremely low CTE of 1.2 ppm/°C (Lee and Kim 2012).

In this work, even though no polishing or planarization was performed, substrates with an RMS roughness of 0.57 nm (Ag) and 1.4 nm (INVAR) were obtained. No device results were presented, but this work remains very promising. It remains to be seen whether this method will work with hard, inexpensive substrates such as steels or Al.

2.4.3   Flexible Glass Substrates

Given the CTE, barrier, transparency, and surface roughness advantages of glass, it is not surprising that a number of companies and groups have explored the development of flexible glass to open up new applications in recent years. As early as 2002, groups were determining the mechanical requirements for ultrathin flexible glass. In one example, researchers at Osram, in conjunction with Infineon and other researchers in Singapore, sought to determine flexibility limits for operating devices under ambient conditions (Auch et al. 2002, Ewald et al. 2002). Using 50 μm borosilicate substrates manufactured by Schott Displayglass (Germany), buckling tests were performed to simulate the type of bending that might be experienced by a credit card or smart card with an integrated OLED. They found that the strain that leads to fracture depended upon the bend axis alignment with the draw direction, from 0.21% to 0.34%. Additional work by these groups indicated the ultrathin glass maintains the necessary moisture barrier properties to protect the OLED structure. Another approach has been to use a polymer layer to provide mechanical stability to the ultrathin glass substrate (Plichta et al. 2003).

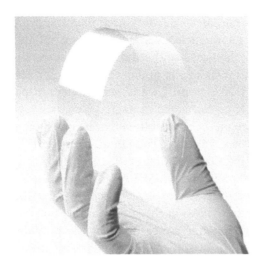

FIGURE 2.5 A picture of Corning's flexible Willow™ glass. Presented at the ACC Niche Workshop on Barrier Technologies, Arlington, VA, on September 19, 2012. (From http://www.corning.com/WorkArea/showcontent.aspx?id=51175. Accessed May 28, 2014.)

More recently, silica particles were dispersed in a PI matrix and used to demonstrate a flexible AMOLED. The film demonstrated networks between the silica particles, which led to low CTE (20 ppm/°C—comparable to metal foils and on the low end for PIs) along with good transmission in the visible wavelength range and very good flexibility.

This early work, along with developments in the glass industry going back much farther, have led to the introduction of flexible glass products into the marketplace. For example, Corning has launched Willow™ glass (Figure 2.5), which is manufactured through a proprietary draw-down fusion process.

2.4.4 Novel Flexible Substrates

This section describes the use of radically new substrates that offer dramatically new possibilities for consumers. One of the promising aspects of OLEDs is the possibility of new displays and lighting technologies that have novel form factors or are fabricated on unusual substrates. To that end, a number of groups have demonstrated OLEDs on such substrates.

One interesting approach to potential OLED products is the use of paper substrates, for low-efficiency, low-cost products. A number of groups have demonstrated simple OLEDs, typically α-NPD/Alq devices, with efficiencies a fraction of the highest efficiency seen for a device using Alq as an emitter. For instance, Legnani and coworkers (Legnani et al. 2008) demonstrated such a device on flexible membranes made of semitransparent bacterial cellulose (BC) with an indium tin oxide layer sputtered directly onto the BC membrane (or onto a silica buffer layer). The resulting substrate had a visible light transmission of about 40%, biased toward longer wavelength transmission. The optical output was a fraction of a reference OLED with the same structure. In another demonstration, cellulose paper was used as a transparent, flexible substrate. The resulting OLED had a low efficiency and brightness, but did demonstrate the concept (Min et al. 2012). Mechanical and thermal properties were not described for these paper substrates. However, Okahisa et al. fabricated a polymer-wood composite with an acceptably low CTE (15–30 ppm/°C) and very high transparency (Okahisa et al. 2009). However, no OLED performance characteristics were reported, although a picture of an OLED fabricated on the composite substrate was shown. In a slightly different approach to using paper as a substrate, Yoon et al. used a parylene-coated copy paper substrate for a top-emitting OLED (Yoon et al. 2010). In this approach, the parylene coating protected against moisture, while a Ni film was deposited onto the parylene as a reflective bottom electrode. A semi-transparent top electrode consisting of a Ca/Ag bilayer was used. The roughness of the substrate was substantial based on electron micrographs, although no roughness measurement was provided. The brightness was very low (1 cd/m² maximum). Nonetheless, these reports show the potential for generating OLEDs on extremely low cost substrates.

Finally, another novel approach is to develop light-emitting devices that go beyond planar devices. Such devices will need substrates that provide similar performance—smoothness, stability, and so on. However, the optical properties may be different. For instance, in one interesting report, O'Connor and coworkers (O'Connor et al. 2007) fabricated an Alq/α-NPD OLED on a PI fiber substrate to generate an OLED that emits in an axially symmetric fashion. Compared to identical OLEDs fabricated on planar Si and PI substrates, the fiber OLED demonstrated similar electrical performance. Roughness on the fiber substrate did lead to increased leakage current on one of two fiber OLEDs. As expected, the strong microcavity-derived angular dependence (see Chapter 12) of the planar OLED was not seen for the fiber OLED. The EQE for the fiber OLED was about half that of the planar OLEDs. Another advantage to using PI is that the mechanical and thermal properties are acceptable for OLED performance.

2.5   Summary

Substrates for OLEDs play a critical role in ensuring that OLEDs meet form, cost, efficiency, and lifetime requirements. Table 2.3 above shows a comparison of the physical, chemical, and thermal properties of various OLED substrates described in this chapter. Many of the properties shown in Table 2.4 can be found easily, but readers should be careful to distinguish between generic values for a property and the specific properties of a particular type of polymer, glass, or metal alloy.

Current transparent substrate technologies, primarily glass and polymer film, meet all necessary OLED requirements for both displays and OLED lighting to generate long-lived, high-performance devices. Top-emitting devices fabricated on opaque substrates show promise, and may have some advantages regarding thermal management and cost, but still must demonstrate the necessary efficiency and lifetime, especially with regard to the effectiveness of the top electrode (usually the cathode) and light extraction schemes. Furthermore, the additional processing required to planarize and insulate the substrate suggests higher costs, although this challenge may be overcome by new processing methods for lower cost metals such as aluminium. New materials such as paper or fibers are interesting and may someday make an impact in the commercial marketplace, but remain, at this time, a novelty. In all cases, compatibility with control circuitry has been demonstrated for a surprising number of substrates, although low temperature processing of the backplane and/or top electrodes to protect sensitive organics for lighting and displays remains challenging and costly. Fortunately, methods have been developed that work at an acceptable cost and performance so that flexible devices are starting to appear in the consumer marketplace. Thus, the market for substrates should continue to expand for the foreseeable future.

References

Auch, M. D. J., O. K. Soo, G. Ewald, and C. Soo-Jin. 2002. Ultrathin glass for flexible OLED application. *Thin Solid Films*. 417: 47–50.

Cheon, J. H., J. H. Choi, J. H. Hur, J. Jang, H. S. Shin, J. K. Jeong, Y. G. Mo, and H. K. Chung. 2006. Active-matrix OLED on bendable metal foil. *IEEE Transactions on Electron Devices*. 53: 1273–1276.

Department of Energy 2011. Solid-State Lighting Research and Development: Manufacturing Roadmap, U.S. Department of Energy, July 2011.

Ewald, G., S. K. Ramadas, F. R. Zhu, Y. L. Hong, O. K. Soo, M. D. J. Auch, K. Zhang, and C. Soo-Jin. 2002. Building blocks for ultrathin flexible organic electroluminescent devices. *Proceedings of the SPIE*. 4464: doi:10.1117/12.457476.

Gläser, H. J. *Large Area Glass Coating*. Dresden: Von Ardenne Anlagentechnik, 2000. ISBN3-00-004953-3.

Gu, G., V. Bulovic, P. E. Burrows, and S. R. Forrest.1996. Transparent organic light emitting devices. *Applied Physics Letters*. 68: 2606–2608.

Gu, G., P. E. Burrows, S. Venkatesh, S. R. Forrest, and M. E. Thompson. 1997. Vacuum-deposited, nonpolymeric flexible organic light-emitting devices. *Optics Letters*. 22: 172–174.

Gustaffson, G., Y. Cao, G. M. Treacy, F. Klavetter, N. Colaneri, and A. J. Heeger. 1992. Flexible light-emitting diodes made from soluble conducting polymers. *Nature*. 357: 477–479.

Hong, Y. T., S. J. Chung, A. Kattamis, I.-C. Cheng, and S. Wagner. 2007. Technical issues of stainless steel foil substrates for OLED display applications. *Proceedings of the SPIE*. 6655: doi:10.1117/12.737213.

Jeong, J. K., D. U. Jin, H. S. Shin, H. J. Lee, M. Kim, T. K. Ahn et al. 2007. Flexible full-color AMOLED on ultrathin metal foil. *IEEE Electron Device Letters*. 28: 389–391.

Kim, M.,Y. S. Lee, Y. C. Kim, M. S. Choi, and J. Y. Lee. 2011a. Flexible organic light-emitting diode with a conductive polymer electrode. *Synthetic Metals*. 161: 2318–2322.

Kim, S., H. J. Kwon, S. Lee, H. Shim, Y. Chun, W. Choi et al. 2011b. Low-power flexible organic light-emitting diode display device. *Advanced Materials*. 23: 3511–3516.

Kim, K., S. Y. Kim, and J.-L. Lee. 2014. Flexible organic light-emitting diodes using a laser lift-off method. *Journal of Materials Chemistry C*. 2: 2144–2149.

Kolosov, D., D. English, V. Bulovic, P. Barbara, S. R. Forrest, and M. E. Thompson. 2001. Direct observation of structural changes in organic light emitting devices during degradation. *Journal of Applied Physics*. 90: 3242.

Lee, S. T., J. Y. Lee, M. H. Kim, M. C. Suh, T. M. Kang, and Y. J. Choi et al. 2002. A new patterning method for full-color polymer light-emitting devices: Laser induced thermal imaging (LITI). *SID Symposium Digest of Technical Papers*. 33: 784–787.

Lee, J. L. and K. Kim. 2012. Metal substrates with nanometer scale surface roughness for flexible electronics. *Proceedings of the SPIE*. 8476: doi:10.1117/12.928659.

Legnani, C., C. Vilani, V. L. Calil, H. S. Barud, W. G. Quirino, C. A. Achete, S. J. L. Ribeiro, and M. Cremona. 2008. Bacterial cellulose membrane as flexible substrate for organic light emitting devices. *Thin Solid Films*. 517: 1016–1020.

Li, Y. Q., L.-W. Tan, X.-T. Hao, K. S. Ong, F. R. Zhu, and L. S. Hung. 2005. Flexible top-emitting electroluminescent devices on polyethylene terephthalate substrates. *Applied Physics Letters*. 86: 153508.

Lifka, H., C. Tanase, D. McCulloch, P. van de Weijer, and I. French. 2007. Ultra-thin flexible OLED device. *SID Symposium Digest of Technical Papers*. 38: 1599–1602.

MacDonald, W. A., R. Eveson, D. MacKerron, R. Adam, K. Rollins, R. Rustin, M. K. Looney, J. Stewart, and K. Hashimoto. 2007a. The impact of thermal stress, mechanical stress and environment on dimensional reproducibility of polyester film during flexible electronics processing. *SID Symposium Digest of Technical Papers*. 38: 373–376.

MacDonald, W. A., M. K. Looney, D. MacKerron, R. Eveson, R. Adam, K. Hashimoto et al. 2007b. Latest advances in substrates for flexible electronics. *Journal of the Society for Information Displays*. 15: 1075–1083.

Mahon, J. K., J. J. Brown, T. X. Zhou, P. E. Burrows, and S. R. Forrest. 1999. Requirements of flexible substrates for organic light emitting devices in flat panel display applications. *42nd Annual Technical Conference Proceedings of Society of Vacuum Coaters*, Chicago, IL. 505: 456–459.

Min, S. H., C. K. Kim, H. N. Lee, and D. G. Moon. 2012. An OLED using cellulose paper as a flexible substrate. *Molecular Crystals and Liquid Crystals*. 563: 159–165.

Mladenovski, S., K. Neyts, D. Pavicic, A. Werner, and C. Rothe. 2009. Exceptionally efficient organic light emitting devices using high refractive index substrates. *Optics Express*. 17: 7562–7570.

Moon, D. G., R. B. Pode, C. J. Lee, and J. I. Han. 2004. Failure of the top emission organic light-emitting device with a Ca/Ag semitransparent cathode. *Synthetic Metals*. 146: 63–68.

Novaled. 2010. Novaled demonstrates reliable OLEDs on metal substrates. Press release accessed on March 17, 2014. Available at: http://www.novaled.com/press_news/news_press_releases/newsitem/novaled_demonstrates_reliable_oleds_on_metal_substrates/.

O'Connor, B., K. H. An, Y. Zhao, K. P. Pipe, and M. Shtein. 2007. Fiber shaped organic light emitting device. *Advanced Materials*. 19: 3897–3900.

Okahisa, Y., A. Yoshida, S. Miyaguchi, and H. Yano. 2009. Optically transparent wood–cellulose nano-composite as a base substrate for flexible organic light-emitting diode displays. *Composite Science and Technology*. 69: 1958–1961.

Pang, H., P. Mandlik, P. A. Levermore, J. Silvernail, R. Ma, and J. J. Brown. 2012. Large-area high-efficiency flexible PHOLED lighting panels. *Proceedings of the SPIE*. 8476: doi:10.1117/12.928000.

Peng, H. J., Y. L. Ho, C. F. Qiu, M. Wong, and H. S. Kwok. 2004. coupling efficiency enhancement of organic light emitting devices with refractive microlens array on high index glass substrate. *SID Symposium Digest of Technical Papers*. 35: 158–161.

Plichta, A., A. Weber, A. Habeck. 2003. Ultra thin flexible glass substrates. *Materials Research Society Symposium Proceedings*. 769: 273–282.

Reineke, S., F. Lindner, G. Schwartz, N. Seidler, K. Walzer, B. Lussem, and K. Leo. 2009. White organic light-emitting diodes with fluorescent tube efficiency. *Nature*. 459:14. doi:10.1038/nature08003.

Sarma, K. R. Amorphous silicon: Flexible backplane, and display application. 2009. In: *Flexible Electronics: Materials and Applications*, eds. W. S. Wong and A. Salleo. Springer Science + Business Media, LLC, New York, NY.

Schott. 2014. Available at: http://www.us.schott.com/borofloat/english/index.html. Accessed March 21, 2014.

Swensen, J. S., E. Polikarpov, A. B. Padmaperuma, K. Shah, T. L. Levendusky, and D. J. Gaspar. 2013. Unpublished results.

Tan, L. W., X. T. Hao, K. S. Ong, Y. Q. Lai, and F. R. Zhu. 2005. An efficient top-emitting electroluminescent device on metal-laminated plastic substrate. *Materials Research Society Proceedings*. 846: 307–312.

Theiss, S. D. and S. Wagner. 1996. Amorphous silicon thin-film transistors on steel foil substrates. *IEEE Electron Device Letters*. 17: 578–580.

Wang, Z. B., M. G. Helander, J. Qiu1, D. P. Puzzo, M. T. Greiner, Z. M. Hudson et al. 2011. Unlocking the full potential of organic light-emitting diodes on flexible plastic. *Nature Photonics*. 5: 753–757.

Wu, C. C., S. D. Theiss, G. Gu, M. H. Lu, J. C. Sturm, S. Wagner, and S. R. Forrest. 1997. Integration of organic LEDs and amorphous Si TFTs onto flexible and lightweight metal foil substrates. *IEEE Electron Device Letters*. 18: 609–612.

Xie, Z. Y., L. S. Hung, and F. R. Zhu. 2003a. A flexible top-emitting organic light-emitting diode on steel foil. *Chemical Physics Letters*. 381: 691–696.

Xie, Z. Y., Y. Q. Li, F. L. Wong, and L. S. Hung. 2003b. Fabrication of flexible organic top-emitting devices on steel foil substrates. *Materials Science and Engineering B*. 106: 219–223.

Yoon, D. Y., T. Y. Kim, and D. G. Moon. 2010. Flexible top emission organic light-emitting devices using sputter-deposited Ni films on copy paper substrates. *Current Applied Physics*. 10: e135–e138.

3. OLED Encapsulation

Lorenza Moro, Damien Boesch, and Xianghui Zeng

3.1   Introduction

This chapter discusses packaging of organic light-emitting diode (OLED) devices. Packaging is a critical issue for any type of electronic device because it ultimately influences the device's reliability and service. Packaging of OLED devices is particularly critical because of their extreme sensitivity to moisture and oxygen. The related challenges become even more severe when materials other than glass are used as substrate and encapsulant, and flexibility is required for components and adhesives.

Below, we briefly review OLED failure modes related to packaging. We then present a summary of the methods used in current packaging technology, followed by a description of challenges and technologies for packaging of future thinner and flexible devices: direct

Chapter 3

thin-film encapsulation (TFE) and high-performance moisture/oxygen barrier on plastic foils. In describing the many barrier technologies proposed in the scientific/technical literature and in the market, we focus on specific advantages and drawbacks of each approach and try to distinguish between scalable and partially scaled up technologies and interesting approaches that are limited to laboratory applications. The packaging technologies we examine here are primarily for OLED displays—the only OLED application that has gained a market share. Most of the technical considerations related to device encapsulation can be extended to applications of OLEDs in solid-state lighting (SSL) and in part to organic photovoltaics.

3.1.1　Environmental Degradation of OLED Devices

OLED devices are a promising technology for displays and SSL with advanced performance, low power consumption, and new and unusual innovative form factors. Since the first commercial introduction of OLEDs in the late 1990s, amazing progress in material performance and operational lifetime has been made. Still, one of the most critical aspects to OLED deployment is the extreme sensitivity of OLED materials and devices to moisture and oxygen. The short- and long-term degradation produced by low-level exposure to these environmental species makes device encapsulation a critical and expensive step in the device fabrication process.

Moisture and oxygen can play a key role in short- and long-term failure modes of OLED devices (see Chapter 13 for a detailed discussion of degradation phenomena). The intrinsic degradation of the organic materials in an OLED is related to a series of slow phenomena triggered by the mechanisms of species diffusion and morphological changes of the materials. Among many degradation mechanisms, photodegradation of the organic emitting materials is accelerated in the presence of oxygen. Light-enhanced oxidation of the organic materials leads to a color shift or quenching of components of the light emitted by electroluminescence. In addition, in small-molecule (SM) OLED devices, oxygen diffusion into the electron and hole transport layers (ETL and HTL, respectively) may lead to less efficient carrier transport. In polymer OLED (PLED) devices, a long-term degradation phenomenon is the reduction of device lifetime in the driven devices due to the formation of a carrier-blocking layer produced by photo-oxidation of the PLED under high current operation.

Another series of degradation phenomena is induced by the interaction of oxygen and moisture with the cathode and at the interface between the cathode and the organic layer. Among these forms of degradation, the fastest and best known is the formation of black spots and pixel shrinkage due to moisture-induced oxidation of the low-work-function charge injection and/or conductivity doping layer (Ca, Ba, Mg, LiF, etc.) between the cathode and the organic material (Figure 3.1).

Black spots form where pinholes in the cathode layer allow moisture to penetrate and reach the reactive low-work-function layer at the interface. Pixel shrinkage is similarly caused by penetration of moisture from the edges of the cathode itself. Since the low-work-function layers are only a few monolayers thick, the amount of water needed to create this damage can be calculated; for complete oxidation, this number is of the order of 10^{-6} g/m^2/day for a device lifetime of 50,000 h. On a longer time scale, the reaction at the interface produces enough gaseous species to induce complete delamination of the

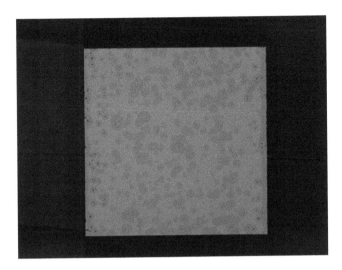

FIGURE 3.1 Optical picture of illuminated cross-electrode OLED device showing black spots and pixel shrinkage due to moisture exposure. The square pixel edge length is 1 cm. The Al electrode is vertical in the picture. Moisture is diffusing in the device through pinholes and from the edges of the Al-cathode.

Al cathode and catastrophic failure of the entire device. Oxygen and moisture can also corrode the Al cathode itself.

Some types of degradation can occur in the device during its shelf life and are not immediately related to the operation of the device, although they are accelerated upon operation. Another critical failure mode specific for operational devices is the presence of particles in the device. Particles are either present on the incoming substrate or deposited on it during the OLED fabrication process. These particles can accelerate catastrophic failure during device operation by creating shorts between layers and producing pathways for fast penetration of moisture/oxygen into the device layers that accelerate catastrophic failure by fast electrochemical reaction. These failures become more likely in devices encapsulated by TFE because of the stress induced on the device by the encapsulation film deposited directly on the device.

3.1.2 Encapsulation Requirements

The encapsulation requirements commonly accepted for OLED devices are defined in terms of water vapor transmission rate (WVTR) and oxygen transmission rate (OTR). The limits are commonly set as WVTR $< 10^{-6}$ g/m^2/day and OTR $< 10^{-4}$ cc-atm/m^2/day. The limit for WVTR was originally proposed by Burrows et al. (2001) and was calculated based on the water leak rate necessary for significantly degrading an Mg OLED cathode in a few years. The validity of the limit has been experimentally validated and extended to oxygen by many other groups with dry-box experiments and/or by leaking H$_2$O and O$_2$ in controlled vacuum environments. The leak rate specifications are suitable for leakage through glue layers in traditional encapsulation by metal can. In practice, such limits are necessary conditions for TFE of displays, but additional limits must be set on the areal density of point defects since a single point defect would lead to failure of the full display. It is commonly accepted that the maximum size for black-spot

Chapter 3

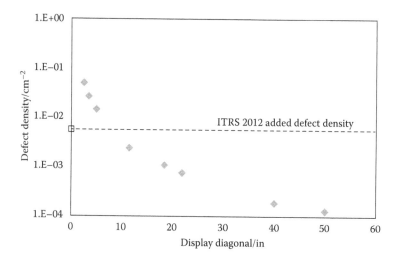

FIGURE 3.2 Defect density versus display size. Defect density above the curve leads to more than a defect per display. As a comparison the maximum acceptable added defect density for IC devices from the 2012 ITST Roadmap is also shown.

defects over the life of the displays should be less than 50 μm, or well below the detection limit of the human eye. However, for TFE or thin-film barriers deposited on plastic films, the critical defect size is of the same order of magnitude as the thickness of the layer used, typically a fraction of 1 μm and often few tens of nanometers. The maximum density depends on the size of the display to be encapsulated, as shown in Figure 3.2.

Here, we also show as a dotted line the maximum density specified for added defects by the International Technology Roadmap for Semiconductors (ITRS) for 300 mm wafers: less than 5 added per process step with size in the range 20–30 nm (ITRS 2012). By comparison, it is easy to grasp the magnitude of the challenge posed to materials, processes, and equipment by the effort of making an effective thin barrier for OLED devices: displays larger than 10″ diagonal (e.g., small tablets) have more stringent defectivity requirements than IC devices. A more detailed discussion on types and size of critical defects in thin film barrier coatings is given in Section 3.3.2.

3.1.3   Encapsulation Strategies

Encapsulation strategies are very dependent on the type of OLED display as well as size and form factors. Small bottom emission (BE), segmented, and/or passive matrix OLED displays are used as a screen for small appliances, secondary displays for cell phones, wristwatches, headsets, displays for cameras, car audio systems, and many other kinds of devices. In several of these applications, the device is a passive monochrome display, often encapsulated with metal cans, the most traditional and, at this moment, the least expensive of the encapsulation methods. Larger displays (>2″ diagonal size) are RGB active-matrix top-emission displays, which are used for A/V–MP3 players, more sophisticated smart phones, and small tablets where the market premium features lightweight, low-profile devices. The displays for these products are encapsulated by an etched-glass can. In some cases prior to encapsulation by the glass can, an inorganic passivation layer is deposited on the displays to provide more robust protection against moisture and oxygen permeation. Sometimes the encapsulated devices are chemically etched to reduce the total thickness of glass in the final product. Glass-can encapsulation is also used for microdisplays used in visors and binoculars. TFE, which is direct deposition of a hermetic moisture/oxygen barrier on the device, is being pursued for achieving higher yields and decreasing the fabrication cost of displays larger than 4″ and perhaps also those of large displays such as TVs.

OLED is the technology of choice for flexible displays. At the moment, flexible displays are mainly fabricated with active matrix technology (AMOLED) and, other than as demonstrators, they have been fabricated only for military applications. At present, the substrate is generally a metal foil, the choice being driven by the scarce compatibility of the available plastic substrates with Si-based thin film transistor (TFT) backplane fabrication. TFE and encapsulation by lamination of a plastic foil engineered with a transparent ultrabarrier coating are two competitive technologies for flexible displays. Engineered plastic foils could be used as substrates if a low-temperature, high-quality TFT technology (e.g., oxide semiconductor TFT) can be implemented.

3.2 Packaging of OLED Devices

3.2.1 Metal Can and Glass Encapsulation

The traditional and still most widely used approach to encapsulation of OLED on glass substrates is based on a metal or glass cap sealed to the rigid substrate by epoxy glue (Figure 3.3).

Because the critical amount of moisture and oxygen in proximity to the OLED is less than 0.1 ppm ($P < 10^{-4}$ Torr) and permeation through the epoxy is a few g-mil/m^2/day, desiccants are required to extend the lifetime of the encapsulated device up to the 15,000–20,000 h required for mobile applications, the most common application in use at present. Cover glass, desiccant, and an edge-seal epoxy constitute the preferred approach for top emission displays.

A schematized process flow for OLED device fabrication is shown in Figure 3.4.

Below we discuss encapsulation (lower branch of the flow diagram). As mentioned above, metal caps or cover glasses are used as the encapsulant. Metal cans were the first type of encapsulant used for the first segmented bottom-emitting OLED displays introduced to the market by Pioneer and NEC as small molecules, and by Phillips as PLED. Metal cans have been mostly substituted by cover glasses that are also compatible with top emission OLED (TOLED). Often, the cover glasses are etched to create a pocket to

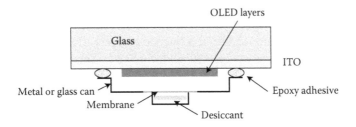

FIGURE 3.3 Schematic configuration of OLED with can/glass encapsulation.

Chapter 3

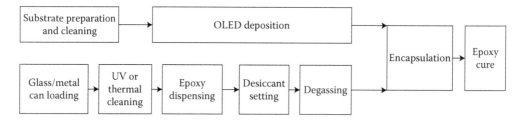

FIGURE 3.4 Simplified process flow of OLED devices fabrication and encapsulation.

house the desiccant and to avoid direct contact with the OLED device. The fabrication is then simplified by using thin adhesive desiccant patches, eliminating the need for edge posts as spacers. An additional advantage in using cover glass over a metal can is the possibility of using glass frits followed by laser treatment for edge sealing. Cover glasses with thicknesses ranging from 0.1 mm to 1 mm are produced by different companies. The transparency of glasses intersecting the emitted light path is higher than 91%–92% in the visible range. Among glass suppliers for OLED display applications, the most established are Corning (Corning 2012) and Asahi glass (Ikezaki 2011). Both companies supply glass substrates, glass covers, and sealing frits. Corning in particular has an established partnership with Samsung, the leading company in scaling up OLED production and establishing OLED displays in the global market for mobile applications. Both glass suppliers have also announced flexible/rollable glass that would be compatible as substrate and/or cover glass for flexible displays (Ikezaki 2011, VanDewoestine 2012). DuPont is another supplier that offers cover glass for OLED displays. Its technological approach, branded as Drylox™, differentiates itself from the competitors in being an integrated solution offering a single component lid with integrated getter that is activated by a thermal treatment immediately before sealing (DuPont 2011). Several smaller companies sell etched glass for glass cans in Asian markets.

The first step in the encapsulation sequence (Figure 3.4) is a treatment to remove moisture and other organic contaminants from the cover. Thermal processes are used for metal cans, and UV-ozone exposure is generally used for cover glasses. After this treatment, the glasses must be kept in vacuum or a dry atmosphere to avoid readsorption of contaminants. The next step is dispensing of the adhesive, followed by placement of the desiccant. Dispensing of the glue is done by either syringe, jet printing (which minimizes the size of the sealing line), or screen-printing. In many cases, the glue is deposited between glass ribs or intermixed with glass balls. The function of these glass features is twofold: diminish the effective side area available for gas permeation and provide a controlled spacer to prevent the active part of the device to be touched by the encapsulation cover. Different types of adhesives, mostly based on epoxy chemistry, are available from many suppliers. They are generally cured by exposure to UV at a fluence of a few thousands of mJ/m^2 and/or thermally cured at low temperature ($T < 80°C$) for 0.5–2 h. The amount of moisture and oxygen permeating the sealed volume where the OLED is packaged depends on the permeability of the adhesive, its thickness, and the edge seal width, as defined in Figure 3.5a.

Thickness is typically 20–25 μm, and width is of the order of 1–2 mm, although significantly less than 1 mm (0.3 mm) is required for the smaller devices. A typical permeation curve of a sealed system is shown in Figure 3.5b. Increasing the width of the edge

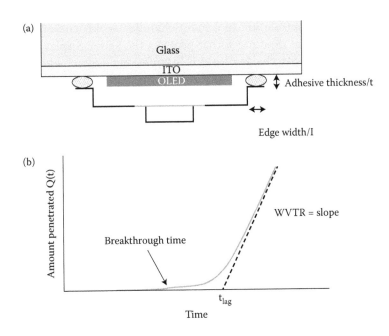

FIGURE 3.5 Definition of edge seal width and adhesive thickness with reference to scheme in (a) and schematic permeation curve through an adhesive (b).

increases the length of the moisture diffusion path toward the inside cavity as well as the lag time required for starting the filling of the volume within which the OLED is packaged. Decreasing the thickness of the adhesive reduces the cross-sectional area for moisture diffusion and therefore the total amount of permeant per unit time. However, once steady-state conditions are reached, the amount of moisture accumulating increases very fast because the typical WVTR of adhesive at room temperature is of the order of 1–4 g/m²/day. Getters for moisture and other gases that can permeate the adhesive layers are therefore necessary to increase the lifetime of the OLED. These materials, also called desiccants, can capture and hold gaseous species by chemically reacting with them. A good desiccant material is able to adsorb highly diluted gases and chemically react with them with fast kinetics at room temperature. Typical examples of chemical desiccants are oxides of the alkaline earth metals. Among the elements of Group II in the periodic table, Ba, Ca, and Sr are the most effective. For optimum performance, the purity of the materials and the effective surface area of the powders play an important role. Water capacities up to 30% in weight are possible for good desiccants. To facilitate the application of desiccant used in OLED BE displays, the desiccant powder (e.g., CaO) is packaged in a container made of metal sheet and a moisture-permeable membrane. Variations for large-scale manufacturing are layers of CaO captured in a porous silica matrix completed with an adhesive layer. The parts are supplied on a polyethylene terephthalate (PET) liner in a reel configuration for easier processing by encapsulation tools in mass manufacturing where cover glass, desiccant, and edge glue are automatically assembled. An example of such a product is DryFlex® by SAES Getter (SAES Getters Group 2008). These types of desiccant must be processed in an inert atmosphere. In encapsulation tools, the glass with the coated OLED displays is moved from the OLED

Chapter 3

fabrication tool to the encapsulation tool main chamber, which contains a UV press. The substrate is then placed on a mask that protects the active display area from UV exposure. The cover glass with the desiccant is brought into contact with the glass. The system is first pumped down to remove any air and gas contamination, and the glasses are put in contact and then equilibrated at atmospheric pressure in dry N_2. Finally, the edge-seal areas are irradiated with a UV lamp, maintaining the OLED area protected by the mask. If glass frit rather than polymeric adhesive is used, the irradiation is done by a scanning laser along the edge. The finished, encapsulated glass is then removed from the tool to the next production step. Commercial encapsulation tools for laboratory and mass manufacturing are, for example, produced by Canon-Tokki Corporation (2012) and MBraun (2012). To simplify the process and reduce the number of steps in a controlled atmosphere, alternative desiccant pastes have been developed that can be coated on the cover glass in ambient environments, and then dried and thermally cured to be activated. After curing, they must be stored in a dry environment. For TOLEDs, transparent desiccant pastes made of dispersed CaO powders in clear resins have been developed. The solutions are UV- or thermally curable. The transparency of these pastes in the visible spectrum depends on the CaO powder load, and can be higher than 95% with loads of CaO < 20%. Such a reduced load in CaO compared with the equivalent powder base proportionally shortens the display lifetime. This type of paste desiccant also improves heat sinking, enhances the light coupling from the OLED to the glass cover, and improves mechanical strength under external stress.

Metal can or glass encapsulation is an expensive and relatively slow operation, with low throughput and low yield, especially for larger displays where the probability of a faulty edge seal increases. Because of these issues, even at the start of OLED technology, TFE has been seen as a key process step to reduce the cost of OLED devices and establish the technology.

3.2.2   Encapsulation by Thin-Film Barriers

One of the driving market forces in the industry, pushing toward the widespread introduction of OLED-based displays, is the trend to make displays thinner, lighter, more rugged, and possibly flexible (bendable and even rollable). To move in this direction, it is necessary to eliminate the glass/metal cap and replace the glass substrate with a plastic substrate. Elimination of glass in displays would result in a 50% or more reduction in thickness and weight, as well as increased strength.

Figure 3.6 schematically introduces four of the methods by which thin-film barriers are used in display technology. For displays fabricated on glass substrates, the two possible approaches to encapsulation are by direct TFE (direct deposition of a thin-film barrier on the display as shown in Figure 3.6a) and lamination on the display of an engineered plastic film where a barrier has been previously deposited (Figure 3.6b).

TFE has the advantage of eliminating the need and cost of a plastic substrate. In addition, by allowing direct hermetic sealing of the critical display areas, adhesives and desiccants are avoided and seals with narrower edges are possible, thereby increasing the "active area"-to-"total glass area" ratio (Chu et al. 2009). Edge-seal reduction is particularly important for smaller displays. However, in-line processing of barrier deposition can slow down the production line, and makes the low-defectivity requirements

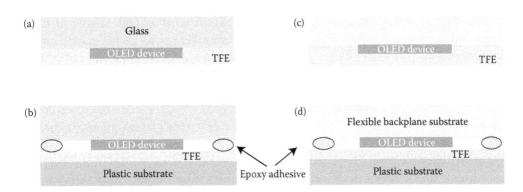

FIGURE 3.6 Possible encapsulation strategies for rigid (a, b) and flexible (c, d) OLED devices. (a) Direct TFE after deposition of OLED on glass substrate. (b) Lamination of TFE to OLED on glass with epoxy edge seal, after deposition of TFE on plastic. (c) Direct encapsulation of OLED device deposited on flexible substrate. (d) Lamination of TFE to OLED on flexible substrate with epoxy edge seal, after deposition of TFE on plastic.

for the barrier deposition process more stringent, since defects added at the back end of the process lead to scrapping of semifinished displays. Overall, the TFE approach seems more advantageous for smaller displays. Deposition of ultrabarriers on plastic film can be done at higher speed, in roll-to-roll (R2R) web systems. Preinspection of the foils to be laminated would reduce the risk of scrapping a semifinished product. However, high-quality and relatively expensive films are required as substrates for barrier deposition. These films, and the barriers deposited on them, must be highly transparent for top-emission displays. Compatible adhesive and desiccant are also required as in glass encapsulation. For this application, pressure-sensitive adhesives (PSAs) have been proposed as an alternative to clear adhesive coating. Both types must match well with the mechanical characteristics of the plastic foils. Flexible displays can also be encapsulated by TFE (Figure 3.6c) or by lamination of plastic on foil (Figure 3.6d), but in this case a plastic foil with a barrier or other type of flexible foil is the substrate of the OLED device.

Major efforts have been devoted in the past 10 years to the developments of barriers to be deposited directly on the OLED devices for TFE and/or on plastic foils with permeations low enough to substitute the glass. Barriers of this type with WVTR < 10^{-6} g/m^2/day and OTR < 10^{-4} cc/m^2-day-atm are often classified as "ultrabarriers" to distinguish them from other barrier on foil used in packaging of food and drugs or medical devices.

Achieving low permeation is not the only challenge for barriers deposited directly on the OLEDs or on plastic foil. An important concern in using TFE is avoiding damage to the electrical and optical emission properties of the devices from the deposition process and from the presence of the barrier itself. Potential sources of damage to OLEDs are heat, UV radiation, exposure to plasma in the process of deposition of inorganic and organic layers, chemical interactions between precursors and TFE materials with the layers in the device, and mechanical stresses induced by the deposited layers on the active layers. Typical solution-processed OLED materials must be processed at substrate temperatures below or equal to 100°C. If such a substrate temperature limit has to be passed in the deposition, the temperature spike should be very short (1 min or less) to minimize the total thermal budget to the OLED. Moro et al. (2004) have shown that certain types of devices are quite sensitive to plasma exposure during the sputtering deposition of an

Chapter 3

inorganic layer in an organic/inorganic multilayer barrier used as TFE. Higher turn-on voltages and, in some cases, loss in light output efficiency were measured after oxide sputtering deposition. This device damage can be minimized, but not always. Deposition of a protective, thin-film buffer layer between the device and the barrier structure is suggested as a method to avoid damage when it cannot be completely eliminated by using optimized process conditions (Moro and Visser 2007). Mandlik et al. (2012) describe how cool-down cycles are necessary for preventing damage to the device during plasma enhanced chemical vapor deposition (PECVD) deposition of a hybrid single film permeation barrier using RF plasma. The same patent application also reports that precise control of the internal stress is an important factor in the deposition of the hybrid barrier film to avoid fractures of the film and of the device (in TFE) if the film built-in tensile stress is excessive. Buckling and delamination of the device induced by the deposition of films with high compressive stress must also be avoided. Similarly, Moro and Krajewski (2010) discuss how excessive shrinkage upon cure of the decoupling organic layers in an inorganic/organic multilayer barrier must be avoided to prevent the delamination of the OLED cathode.

Stability of the deposited layer, low outgassing, and adhesion between TFE and the substrate/device must be achieved not only at the deposition, but over the lifetime of the device, including during accelerated lifetime testing at high temperature and humidity. Typical conditions are 60°C/90% RH and 85°C/85% RH passive and biased operation, as well as thermal cycling between −40°C and +80°C.

The optical properties of the encapsulation layers are also very important. Transparency higher than 90% in the visible spectrum is required for layers in the path of the OLED-emitted light. In addition, the optical properties must be tailored to avoid spectral shift and angular dependence (Moro et al. 2007) and, when possible, enhance the light out-coupling from the OLED (Visser and Moro 2010a,b, Clausen and Dabruzzi 2009).

Similar challenges, although with somewhat more relaxed requirements, must be met when the barrier is deposited on plastic foils to be used as superstrates for encapsulation by lamination or substrates. Several plastic films have been used as substrates in the production of flexible barriers on foil to be used in encapsulation by lamination or as the substrate in the fabrication of the device itself. PET and polyethylene naphthalate (PEN) are the most widely used. They are clear, have a moderate coefficient of thermal expansion (CTE), good chemical resistance, and moderate moisture absorption. The materials are generally inexpensive, although film with quality that is actually compatible with barrier deposition costs more. Display-grade PET and PEN films are produced by DuPont/Teijin Films under the brand names of Melinex® and Teonex®, respectively. Polycarbonate (PC) is another film with similar characteristics. Its main limitation is high CTE past the T_g (glass transition point) at 150°C, but this is not a drawback for using it as a superstrate encapsulant foil. Lexan® is a registered trademark for the brand of PC film produced by SABIC Innovative Plastics (formerly General Electric Plastics) and PURE-ACE® is the name for the film produced by Teijin. Polyethersulfone (PES) is a more expensive film with inferior chemical resistance. It is commercialized as Sumilite® by Sumitomo Bakelite. Polyarylate (Arylate™ by Ferrania), a polycyclic olefin, and polynorbornene, (Apparel™ by Promerus, a subsidiary of Sumitomo Bakelite) are also clear films that have been proposed for this type of applications. Extensive reviews of substrates for flexible electronics and their properties have been published by MacDonald et al. (2007) and Rutherford (2005a,b).

Most of the flexible displays demonstrated have been fabricated on metal foil (both stainless steel and aluminum), and very recently, a few on flexible glass and on plastic film (see Chapter 2 for a detailed discussion of glass, foil and plastic substrates). However, both stainless steel metal foil and flexible glass are very expensive. The CTE of metal foils is too high for fabrication of high-resolution displays. In addition, metal foils do not have the necessary smoothness and require deposition of thick dielectric layers to insulate the TFTs and create a smooth surface so to avoid shorts. Capacitive coupling between the backplane and the substrate may affect the performance in advanced devices. Flexible glass is in its infancy, and is very difficult to handle with no fractures during processing. In May 2012, Corning made available on the market its flexible Willow™ glass (Corning 2012). To make its flexible glass compatible with existing processes, Asahi Glass has presented a carrier glass technology based on a thicker and reusable glass with temporary adhesion to thin glass. The thicker glass is then removed at the end of the process. Polymeric substrates engineered with ultrabarriers on which to fabricate the backplane and deposit the display would be a viable alternative to the inorganic foils. In this case, the engineered film must sustain all the thermal and chemical process steps involved in the fabrication sequence. Also needed are resistance to chemical patterning and, following thermal treatment processes, ITO deposition, vacuum degassing, and UV ozone cleaning. Simple and small devices used in applications such as signage or displays for appliances often use a passive matrix BE configuration; therefore transparency of the plastic foil substrate is a strong requirement. Materials transparency is not required for top-emission displays. However, for AMOLED devices, a good dimensional stability of the substrate and of the barrier deposited on it is necessary at the low-temperature polysilicon (LTPS) deposition temperature ($T > 350°C$). Among the polymeric films used as substrates for these applications, the most common is polyimide (PI) (e.g., Kapton® by DuPont). Colorless versions are even commercialized, but PI use is limited by high moisture absorption and high cost. Composite substrates like fiber-reinforced plastic (FRP) by Sumitomo Bakelite have also been proposed (Ito et al. 2006). To improve the stability of these substrates to the thermal and chemical steps, barrier layers on both sides are envisioned. Liu et al. presented a flexible display fabricated on a colorless hybrid material. This plastic substrate is strong enough to sustain all the steps of the fabrication process. Its high inorganic content renders it highly thermally resistant, and its modulus, dimensional stability, and low CTE make possible the fabrication of flexible displays (Liu et al. 2011).

A key property to be considered for barriers deposited on plastic foil is flexibility of the thin layers. This is important for film production in R2R tools and for the lifetime of the device in flexible applications. Cracks in the inorganic barrier layers induced by bending and stretching are catastrophic failures for the engineered films. It is well known in thin-film mechanics that the onset of fragmentation depends on the adhesion between the film and the substrate and the cohesive strength of the films. For inorganic films that are impermeable to oxygen and moisture, thinner films crack later and faster than thicker coatings. This has been demonstrated, for example, by Leterrier et al. (1997) for SiO_2 films on PET. The optimum thickness depends on the materials and on the details of the deposition, but in general it is between 30 and 60 nm for oxide films deposited by sputtering. In general, compressive stresses improve the tensile strength and adhesion, and form better barriers that are more mechanically resistant. Carbon-rich polymer-like films, like amorphous hydrogenated carbon nitrogen (a-C:N:H) and a-Si:C:N:H, in general deposited by PECVD, have good adhesive and cohesive properties, but have low density and as a consequence poor barrier properties (Zambov et al. 2006). Therefore, for both sputtered and PECVD barrier films, multilayer coatings have been pursued to achieve both flexibility and good barrier properties. Flexible multilayer coatings made by a stack of sputtered Al_2O_3 and acrylate layers have been successfully demonstrated and introduced in the market (Rutherford et al. 2004, 3M Company 2012). It has been demonstrated that the critical failure mechanisms for these multilayer systems are also oxide cracking and oxide–inorganic delamination (Zheng et al. 2011). Improvements in the mechanical properties may be achieved by using compositionally graded inorganic oxides (Choi et al. 2010a,b, Roehrig et al. 2011) and by using a top-protective coating (Leterrier 2003, Leterrier et al. 2008, Zheng et al. 2011). The film has a discrete or graded multilayer structure, as is seen for PECVD-deposited barrier films that have been demonstrated as effective. A few more of these PECVD multilayer coatings are described as well (Schaepkens et al. 2004, Sugimoto et al. 2004, Akedo et al. 2006, Mandlik et al. 2012).

3.3   Mechanism of Permeation in Thin Film Barriers

3.3.1   Permeation in Single Layers

Permeation through single-layer barrier films can be understood using Fick's model of diffusion. The model assumes a barrier layer of thickness, l, diffusivity, D, and solubility, S, that are independent of concentration of the permeating species. The concentration varies from C_1 on the saturated side of the film and maintained at zero on the other side by a sweep gas. The flux, F, is characterized by two time regimes. At the beginning, F is low as moisture initially diffuses through the barrier. This period is called the lag time. Afterwards, F approaches a constant in the steady-state regime, which is the reported value for the WVTR or OTR when the permeant species are water vapor and oxygen, respectively. The steady-state permeation and lag time are related to D and S (Graff et al. 2004).

For a large time, t, the total mass transmitted through the layer, Q, is given by Equation 3.1:

$$Q(t \rightarrow \infty) = \int_{0}^{t \rightarrow \infty} F(x = l, t)dt = \frac{DC_1}{l}\left(t - \frac{l^2}{6D}\right) \tag{3.1}$$

Q is linear with time and offset by a lag time, L, which is given by Equation 3.2:

$$L = \frac{l^2}{6D} \tag{3.2}$$

Finally, the permeability coefficient, P, is related to the steady-state flux, F_{ss}, and the pressure gradient across the film, ΔP, by Equation 3.3:

$$P = \frac{F_{ss}l}{\Delta P} = DS \tag{3.3}$$

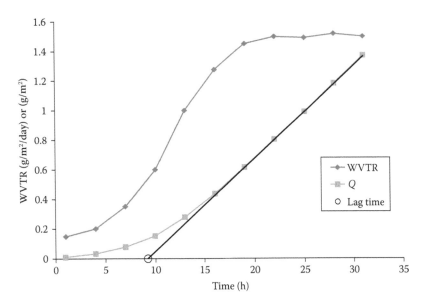

FIGURE 3.7 Typical experimental plot of permeation data. Steady state permeation is demonstrated by the linear fitting of Q at $t > 15$ h. The intercept of the fitted line defines the lag time.

Figure 3.7 provides example permeation data. Integrating the measured flux gives the total mass transmitted. For most times, the slope of Q approaches a constant, and the intercept of the extrapolated line with the x-axis gives the lag time.

To create flexible and transparent barrier layers with WVTR $< 10^{-4}$ g/m²/day, it is necessary to use thin films of inorganic materials for both TFE and the barrier on plastic foil. The materials used usually include oxides and nitrides of Al, Si, Ti, and so on. The thin inorganic barrier layer has several potential pathways for diffusion of permeating molecules. One possible pathway is diffusion through the bulk of the film. This pathway indicates a highly porous microstructure or perhaps the presence of chemical impurities. The bulk crystalline inorganic materials exhibit extremely low moisture and oxygen diffusivity, so a properly dense and stoichiometric film should behave in a similar manner. Depositing high-quality dense inorganic films on rigid substrates requires optimization of the parameters of deposition, including deposition energy, substrate temperature, and deposition rate. However, for deposition on flexible substrates or sensitive devices, these parameters are severely limited by the substrate. Due to this constraint, the optimal deposition conditions cannot be reached, which leads to lower-quality inorganic films that incorporate chemical impurities and/or microstructural flaws. The most problematic microstructural flaw for inorganic barrier films is high porosity due to low substrate temperature requirements. A porous film with many voids will allow vapor ingress through nanometer-sized channels. Thus, density of the inorganic layer is a critical factor for its barrier properties (Yamada et al. 1995). Measuring refractive index is an effective way to control the film density. A very low index indicates that a significant volume fraction of the film is composed of voids. Positron annihilation lifetime spectroscopy (PALS) has also demonstrated that barrier films with the smallest pores have the best barrier properties (Zambov et al. 2006).

Chapter 3

In certain cases, the barrier film may contain chemical impurities that promote diffusion through the film. This behavior may arise from diffusivity and solubility that are dependent on concentration, which would interfere with the models of diffusion described above. The impurities may be due to incomplete reactions during the deposition process, again due to the low temperature of the substrate during deposition. These chemical interactions may result in changes to optical and vibrational absorption characteristics measurable using FTIR or Raman spectroscopy.

For encapsulating OLED displays, barrier films are required to attain a WVTR of 10^{-6} g/m^2/day. This represents an extreme challenge to create perfect inorganic layers thin enough to maintain flexibility. One flaw in the barrier film will lead to a black spot, which will ruin the entire display. In fact, for high barrier films, permeation is dominated by defects, as we will describe in the next section.

3.3.2   Role of Defects

Measured permeation through single layers is higher than values predicted from the bulk properties. While permeation through dense inorganic layers can be slow, permeation through defects proceeds virtually unimpeded. Many researchers have reported that permeation is not inversely proportional to inorganic film thickness (Hanika et al. 2003, Graff et al. 2004). Additionally, in the case of moisture permeation, the experimentally determined activation energy is close to that of the polymer substrate (Tropsha and Harvey 1997). These findings indicate that permeation is dominated by defects in the inorganic layer.

The defects that lead to high permeation can be grouped into categories that are intrinsic and extrinsic to the inorganic deposition process. Intrinsic defects are caused by the deposition process and may include nanochannels, grain boundaries, or particles from the deposition process. The boundaries of small grains or a columnar structure may result in local areas of lower density that allow gas to transmit through the film. Grain size may be checked with atomic force microscopy (AFM), and internal structures may require transmission electron microscopy (TEM). Extrinsic defects may be caused by substrate handling or other actions unrelated to the deposition process for the barrier film. Commonly, extrinsic defects are particles that land on the substrate during handling. Other extrinsic defects may be caused by process steps prior to deposition of the barrier layer. In order to ensure flexibility of the inorganic layer, the barrier film is generally limited to less than 100 nm, so any defect or particle of that size or larger may compromise the layer. For ultrabarriers, defects in the inorganic thin film lead to localized regions where the WVTR is much higher than the acceptable limit. This eventually leads to failure of the OLED pixels in the affected region, and a scrapped display. An additional challenge for TFE is the topography of the display, which itself may lead to the formation of defects. A case of extreme topography may be the cathode separators in a PM display that are "mushroom" structures several microns tall.

For inorganic deposition of R2R barrier films, defects may be particles that become coated and subsequently fall off, or scratches from the rollers (Bishop 2011). Plastic films often incorporate antiblock particles of silicon dioxide on one surface to prevent adhesion between layers of the roll. These slip treatments may lead to particles or divots on the barrier coating, but a clean protective sheet may be used to preserve the surface upon rolling. Additionally, with unwinding, the films may become electrostatically charged and attract dust particles (Hanika et al. 2003). Finally, plastic films may contain tall spikes on the surface that originate from the production process and may induce defectivity in the barrier layer.

In the case of a rough or highly defective substrate, a smoothing layer is necessary for directional coatings. The directional coating techniques are susceptible to void formation due to shadowing effects during film growth. These effects result in a dependence of barrier properties on the roughness of the substrate surface (Low and Xu 2005). The quality of the barrier layer may be improved by using a substrate smoothing layer. Some coatings may be highly conformal, such as CVD or atomic layer deposition (ALD), and they present possibilities of creating high-quality, single-layer barrier films without a smoothing layer by overcoming rough substrates or even small particles.

3.3.3 Multiple Layers

Many researchers have reported that barrier properties of inorganic layers are improved when thin layers of alternating inorganic materials are used (Lee et al. 2010). The different materials act to disrupt the formation of voids and other defects that would grow in the direction of the film thickness. However, the total thickness is still limited by flexibility and stress considerations. Particles and defects that are large enough will remain as pathways for moisture and oxygen ingress.

It has been found that depositing a polymer smoothing layer before the inorganic layer greatly improves the barrier quality of the inorganic layer (Shaw and Langlois 1994). This pairing of smoothing and barrier layers is called a dyad. The structure may be repeated with multiple dyads and has shown to produce barrier films capable of delivering WVTR ~ 10^{-6} g/m^2/day. The smoothing layer reduces defect density in the subsequent inorganic barrier layer. Additional dyads spatially separate defects in the plane of the film. Theoretically, the steady-state permeation of water through a multiple dyad structure will be given by the series combination of the inorganic layers. This would make achieving a barrier for OLED very difficult. However, with ultrabarriers the lag time is extended to outlast the device lifetime.

Extending the lag time in multiple dyad structures is accomplished by generating films with long spacing between defects in the plane of the film in the inorganic layers. The polymer smoothing layers act to decouple defects between the inorganic films. Moisture diffusion occurs fastest through the defects, but must travel horizontally before reaching a defect in the following inorganic layer. In this way, the effective path length is given by the horizontal diffusion of water through the polymer in a "tortuous path" as shown schematically in Figure 3.8.

Greener et al. (2007) have calculated that significant barrier performance gains can be achieved for polymer thicknesses below that of the average pinhole size for multiple dyad structures. They also determined that reducing the diffusion coefficient of the organic layer or the defect density in the inorganic layer also provides improvements to the transmission rates for the barrier film. These properties indicate that the permeation in the multilayer structure is limited by diffusion through the polymer layers. With sufficiently low defect densities in the inorganic layers and low permeability in

Chapter 3

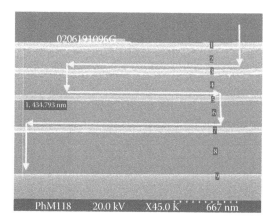

FIGURE 3.8 Pictorial description of tortuous path for moisture diffusion through defects in the inorganic layer of a multilayer barrier structure.

the smoothing layers, the diffusion lag time may be extended beyond the most rigorous accelerated aging conditions, including the standard photovoltaic test of 1000 h at 85°C and 85% relative humidity (Olsen et al. 2008).

3.3.4   Layer Adhesion and Interfaces

Adhesion of the barrier layers must be high enough to maintain the film's properties throughout its normal lifetime of use as well as during accelerated aging tests. Adhesive failure upon accelerated aging is a common failure mode for inorganic barrier materials, whether single or multilayer. As illustrated above, good adhesion between layers is important to ensure good mechanical properties and flexibility of the engineered foil in R2R production and the service life of flexible devices.

The simplest method to measure adhesion for barrier layers on foil is the tape test, ASTM D3359 (ASTM International 2009). Proper control of interfaces with substrate pretreatment may provide adequate adhesion of the film. It is recommended that adhesion be tested immediately after deposition. It may also be necessary to expose films to damp conditions before testing adhesion, commonly called wet adhesion. Finally, care must be taken not to damage the single layers during the transporting, cutting, and testing of the films. A polymer top coating may provide the structural support necessary for these real-life conditions after barrier film production.

Lateral diffusion of moisture and oxygen along the interface between the barrier layer and substrate (foil or device) leads to degradation and edge shrinkage of the encapsulated device. This preferential path for diffusion may be created by poor adhesion or by the presence at the interface of materials with polymeric characteristics. The latter phenomenon is invoked by Mandlik et al. (2008) to explain the degradation of devices encapsulated with a PECVD-deposited single-layer barrier. In a later publication the same authors claim to solve the problem by increasing the density of the layer at the interface (Mandlik et al. 2012). In a multilayer barrier made of stacks of organic/inorganic layers, lateral diffusion is avoided by using an edge seal, which allows the inorganic layers to completely cover the organic areas. When the inorganic layers are deposited by pulsed DC or AC sputtering, the adhesion to the substrate is quite good due to the ion bombardment during the deposition. In this type of multilayer, lateral diffusion is important when moisture penetrates the organic layer by a defect. Vogt et al. used a multidyad Al_2O_3/acrylic polymer model system to show the mechanism of moisture permeation from the polymer layer of the most external dyad toward the inner dyads. They also showed how adsorption of moisture at the hydrophilic oxide interface occurs and, by acting like a desiccant, increases the lag time to reach the steady-state regime of permeation and therefore the operational service life of the device (Vogt et al. 2005). Moisture diffusion to the underlayers, and eventually to the device, is reduced by the proper choice for the decoupling organic layers of highly cross-linked and stable polymeric materials (Moro and Krajewski 2010).

In the multilayer approach proposed by Terabarrier (Ramadas and Chua 2010), a similar mechanism is exploited to improve barrier performance of an engineered plastic foil by introducing oxide nanoparticles in the polymeric film that can bind moisture at their surface. However, the success of such a strategy can be realized only by compromising the transparency and clarity of the structure, because the high load of nanoparticles in the polymer leads to the formation of agglomerates acting as a center of light scattering.

3.3.5   WVTR and OTR Relationship

When the permeation of moisture and oxygen is controlled by defects, the oxygen and water permeation rates, OTR and WVTR, respectively, are in strict correlation. The relationship between the two is linearly proportional, and the activation energy is very similar. However, when the permeation decreases and the role of the extrinsic defect becomes less important, diffusion through intrinsic defects in the inorganic layer becomes the dominant mechanism. In this case, void size and void density in the film become more important. Selectivity between oxygen and water may occur due to differences in size and chemistry of the two molecules. While O_2 easily permeates as a gas, H_2O may chemically interact with porous walls and interfaces in the inorganic layers and condense there (Vogt et al. 2005). Erlat et al. (1999) report increased activation energy, E_a, for moisture permeation in a PECVD SiO_x film deposited from a hexamethyldisiloxane (HMDSO) precursor as the O_2 permeation decreases. In addition, for AlO_xN_y films deposited by sputtering, a higher activation energy for moisture permeation was measured for increasing layer thicknesses, leading to lower WVTR (Erlat et al. 2001). In contrast, in these films, OTR decreases, but the same E_a is maintained. The authors speculate that this is caused by an interaction of moisture with defects and nanopores and possible adsorption to the nanopore walls. In multilayer barrier structures, selectivity may also be introduced by faster or slower diffusion of the permeant in the polymeric layers.

It is well known that moisture adsorption in materials and films may lead to changes in the diffusion/permeation mechanisms of other species, including O_2. Normally, faster diffusion of other species, explained by non-Fickian diffusion models, is induced by swelling of polymeric layers. Therefore, in characterizing ultrabarriers, OTR should be measured in dry and wet environments.

Chapter 3

3.3.6   Test Methods for Low Levels of Permeations

ASTM F1249 (ASTM International 2011) is a standard for measuring moisture penetration through plastic films. The film is affixed to an apparatus that separates a side with moisture from a side with dry N_2 flow. The total pressure is kept the same, and the difference in partial pressure of water vapor drives moisture penetration. On the dry side, an IR sensor measures a value proportional to the water content in the gas. With this method, a flux of moisture through the film is measured with a minimum WVTR sensitivity at 5×10^{-3} g/m²/day. Makers of the apparatus for this test and similar ones are Mocon Inc. (Minneapolis, MN, USA), Brugger (Munich, Germany), and Labthink Instruments Co. Ltd. (Shandong, China).

The MOCON Aquatran uses a technique similar to ASTM F1249 with a coulometric sensor that reacts with water molecules. The coulometric sensor "increases the sensitivity by an order of magnitude over the traditional ASTM F1249" (Stevens et al. 2012). The detection limit is not capable of measuring values as low as the 1×10^{-6} g/m²-day required for OLEDs, but the testing may be performed in an accelerated aging environment.

The "tritium test" uses a technique similar to ASTM F1249 with tritiated water and a sensor to detect radioactivity. The sensor detects radioactivity from HTO and possibly atomic tritium. The detection limit is ~1×10^{-6} g/m²/day, determined by the background radiation (Moriconie et al. 2009). A second detection method involving tritium uses a salt to collect the HTO. Periodically, the salt is removed and the tritium content is detected with a scintillator. The reported detection limit for this technique is below 1×10^{-6} g/m²/day (Park et al. 2011).

The Ca test is a commonly used technique for measuring WVTR in the 10^{-6} g/m²/day range and below. In this test, a thin film of Ca is encapsulated by barrier layers. Encapsulation may be direct, simulating the direct TFE of devices. Alternatively, the Ca coupon may be laminated to a barrier film. If moisture penetrates the barrier, the water hydrolyzes the metallic Ca according to the reaction:

$$2H_2O + Ca \rightarrow Ca(OH_2) + H_2 \tag{3.4}$$

The $Ca(OH)_2$ is transparent and nonconductive, whereas the metallic Ca is opaque and electrically conductive. The Ca test is based on measuring the change in thickness of the metallic Ca and relating that information to the number of Ca atoms reacted. Over a period of damp heat testing, the number of Ca atoms reacted gives the WVTR. Using oxygen isotopes to distinguish oxygen originating from O_2 from oxygen originating from water, it has been determined that Ca does not react with molecular oxygen (Cros et al. 2006). Other researchers conducting experiments at higher temperatures established that O_2 oxidation is possible, but is a secondary effect (Reese et al. 2011). The two methods for measuring the variation of thickness of the metallic Ca are via optical transmission and electrical resistance. For the former, the transmittance of the Ca is related to the thickness through the Beer-Lambert law. The measurement may detect a few nanometers change in thickness of the Ca over many hours of testing. For the electrical Ca test, the resistance of the Ca film is measured in the plane of the substrate. As the metallic Ca is consumed and becomes thinner, the resistance increases (Reese et al.

2011). The hydrolyzed Ca does not provide a barrier to further penetration due to the morphology, dimension change, and release of H_2 gas. This is evidenced by the successful use of Ca films as desiccants in a variety of applications.

The Ca test is also very useful in pinpointing defects with localized WVTR much higher than 10^{-6} g/m²/day. In areas where localized water permeation is fast due to defects, transparent holes are formed in the metallic Ca. Investigation with optical microscopy, SEM, and other methods may reveal a particle responsible for the defect. Figure 3.9 shows Ca squares backlit after accelerated aging at 85°C and 85% RH. The edges of the Ca squares are faint due to the shadow mask used. The degradation evident in the right square of Figure 3.9a was due to a large defect. The degradation in the left square of Figure 3.9b was caused by defects in the laminated barrier film. The other Ca square shows slight hydrolysis of the upper right corner, which was due to ingress from the edge of the laminated structure. In the literature, there are many different Ca test experimental procedures as there is no standard method. Researchers at the National Renewable Energy Laboratory (NREL) are developing a standard test for high barrier films using the electrical method.

Another method capable of measuring very low WVTR is through mass spectrometry in a UHV test chamber. A barrier film separates a region with controlled water vapor from a dry chamber. The dry chamber is allowed to accumulate water. A mass spectrometer measures the baseline water content and the water content after accumulation. The reported detection limit is ~10^{-7} g/m²/day, but this method has a very low sample throughput for testing (Moriconie et al. 2009).

Currently, measuring oxygen transmission is mainly limited to the ASTM D3985 (ASTM International 2010) technique, which is similar to ASTM F1249 with an oxygen sensor. For example, the MOCON Oxtran has permeation sensitivity at 5×10^{-3} cc/m²/day. Another standard test is ASTM F1927 (ASTM International 2007), which measures oxygen permeation in the presence of humidity in case moisture affects the barrier properties of the film. It is possible that methods similar to the Ca test may be developed with sensitivity to oxygen permeation.

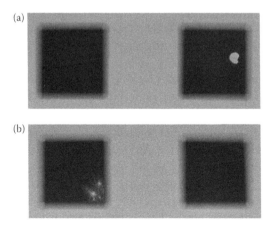

FIGURE 3.9 Examples of deterioration of metallic-Ca due to failure of the encapsulation after exposure to moisture for 1000 h in accelerated aging at 85°C and 85% RH. The 2 cm edge length Ca-squares were deposited on glass and encapsulated with TFE (a) and laminated barrier film (b).

Chapter 3

For accelerated aging, a conversion factor is needed to arrive at the room temperature WVTR. Permeation is given by an Arrhenius equation with activation energy, E_a. Activation energies can be found experimentally by measuring permeation at different temperatures. Values between 15 and 55 kJ/mole have been reported. The spread comes from the range of barrier structures tested, as well as the method used to measure the water and/or oxygen permeation through the barrier (ASTM methods, Ca-tests, OLED degradation). In fact, different degradation mechanisms have their own different activation energies. Another source of confusion is that some groups do not correct for the water concentration change at different temperatures, which is the dominant contribution to E_a (41 kJ/molK is the temperature dependence of the water concentration with temperature). The order of magnitude of the acceleration factor is generally agreed upon despite differences in specific values. It is also important to consider that the accelerated aging tests are standards established by industrial bodies, so it is necessary to pass the appropriate tests for the application of the barrier film. In general it is accepted that 1000 h at 60°C/90% RH is equivalent to a lifetime of a few years (<5) at 20°C/50% RH, and 1000 h at 85°C/85% RH is required for applications with lifetimes longer than 10 years.

3.4   Thin-Film Barrier Technologies

3.4.1   Inorganic Thin-Film Deposition

Barrier layers with quality sufficient to encapsulate OLEDs may be deposited using DC or AC reactive sputtering of inorganic thin films in a vacuum chamber. Sputtering is capable of depositing very dense layers with controlled stoichiometry, uniformity, and low substrate temperature. The quality of the layers is achieved by the high energy of the impinging particles that eventually coalesce into the inorganic film. However, the drawback of this energy is heating and possible damage to the substrate. Reactive sputtering requires low base pressures (~1×10^{-6} torr) to eliminate background gas, especially water vapor. Additionally, sputtering plasmas require pressures at a few millitorr to operate, which limits the mean free path of sputtered atoms in vacuum. These requirements limit the throughput for sputter-deposited films; nonetheless, the quality of layers deposited at low substrate temperature makes sputtering the primary technique historically used for high-barrier thin films.

The most common material deposited by sputtering for barrier films is Al_2O_3 using a metallic aluminum target. The target undergoes the normal hysteresis characteristic of reactive sputtering, wherein the target surface changes from metal to oxide with increased oxygen gas flow. A sputtering feedback loop is required to control the oxygen flow and thereby the stoichiometry of the deposited film. The feedback loop may monitor the target voltage, residual oxygen gas, or plasma emission line of the metal.

Reactive sputtering of Al_2O_3 is possible using pulsed DC power supplies that reverse the applied voltage repeatedly at a frequency of the order of tens of kilohertz. The voltage reversals discharge oxidized regions of the target and prevent arcing. Current arcs may damage the substrate, eject particles from the target, and possibly send the reactive sputtering process out of control. During the sputter deposition of barrier oxides via pulsed DC sputtering, as the chamber is coated with oxide, the anode becomes coated and gradually becomes insulated from the sputtering plasma. The plasma must spread to uncoated regions, which is called the disappearing anode effect. This is a problem with DC sputtering of insulating materials, but can be mitigated through specialized anode design. Dual-cathode AC sputtering may be employed using two targets. The configuration of dual-cathode sputtering uses one target as anode and the other target as cathode with a sinusoidal sputtering voltage output from the power supply. When the polarity is reversed, the targets switch roles and the anode is clear of oxide, which eliminates the disappearing anode problem. However, in contrast to pulsed DC sputtering, the duty cycle is reduced and two cathodes are required compared to pulsed DC sputtering. Additionally, AC power causes higher substrate ion bombardment, which may damage the substrate or encapsulated device (Szczyrbowski et al. 1997).

An important factor with sputtering is the substrate heating or damage induced by energetic species. These energetic particles may be ions from the plasma and neutrals sputtered from the target. Substrate damage is a key problem for depositing barrier films on foil and directly on OLED devices. For R2R sputtering, substrate heating induces thermal expansion of the plastic foil. A chilling drum is used to cool the plastic during sputter deposition. However, if the heat load is too high, the induced stresses may overcome the tension at the roller surface, leading to wrinkles and lost thermal contact with the drum. For direct encapsulation of OLEDs, excessive substrate damage will result in areas with diminished brightness. A technique to reduce damage to the OLED is to use a thin film deposited with milder conditions as a protection layer for the device (Moro et al. 2007, Moro and Chu 2010). Outgassing is another complication resulting from substrate heating. For barrier films deposited on foil, the outgassing of water vapor from plastic substrates into the vacuum chamber may be enough to reduce the quality of the deposited oxide or overwhelm the reactive sputtering process. This problem is overcome by separate pumping of unwind and pretreatment zones and with in-between a zone with high gas separation ratio.

Organizations employing reactive sputtering for barrier layer deposition on plastic foils include 3 M (St Paul, MN) and Materion (Westford, MA). Organizations using reactive sputtering in the past included Symmorphix and Vitex Systems. Symmorphix reported single layers deposited by a proprietary PVD process with WVTR in the 10^{-5}–10^{-6} g/m^2/day range (Narasimhan 2005). Vitex Systems reported multilayer polymer and oxide barrier films with WVTR <10^{-6} g/m^2/day for TFE (Moro et al. 2004, Chu et al. 2007) and barrier on film (Rutherford 2005a,b, Kapoor et al. 2006, Ramos 2006).

PECVD has also been used for thin-film barrier deposition at low temperature. A typical setup with parallel plate electrodes is shown in Figure 3.10.

PECVD has several advantages: high deposition rate, good step coverage, easy control of thin-film composition by changing precursor gas ratio, the ability to perform continuous deposition, and the ability to deposit a wide range of oxides and other compounds (Rossnagel et al. 1990). However, these advantages do not directly translate to deposition conditions compatible with TFE or barrier deposition on plastic because of the restriction in substrate temperature, plasma exposure, and, in TFE, compatibility of process gases with OLEDs. SiN_x, SiO_2, and SiC have been chosen as PECVD barriers. A WVTR of 10^{-4} g/m^2/day can be realized with a single-layer PECVD barrier film. Use of remote-plasma PECVD has successfully avoided overheating of substrates or devices by plasma and damage of devices by radiation. In this configuration, the substrate is moved

Chapter 3

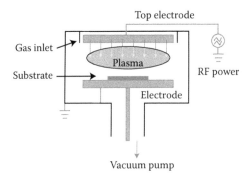

FIGURE 3.10   Schematic configuration of PECVD reactor.

out of the discharge region of the plasma (TU/e 2012). The remote plasma is produced by microwave, radiofrequency coil, or DC arcing; it can be designed in the same chamber as the substrate or in a separate chamber (Figure 3.11).

Compared with conventional PECVD, remote PECVD involves less bombardment on film surfaces and substrates by energetic particles during deposition, and so induces less damage on the underlying device and substrate and on the surface of the already grown film. Remote PECVD can reduce the unreacted hydrogen bond density of precursors (Alexandrov et al. 1995). Due to reduced heating by plasma, remote PECVD can realize deposition at room temperature. Drawbacks of this approach are the relatively low deposition rate and the low density of the film deposited (van Assche et al. 2004).

In considering PECVD for barrier application, a concern is that the particles produced in the gas-phase reaction are possible origins of nanometer-sized defects in PECVD films. Larger particles can also be generated by the unwanted coating deposited on chamber walls, shower heads, and mobile parts flaking off. The solution used in microelectronics applications of PECVD is the introduction of cleaning cycles between runs. The same approach has been extended to batch coating TFE (Hemerik et al. 2006), but clever engineering may be necessary to overcome the problem in R2R production

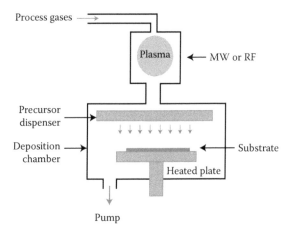

FIGURE 3.11   Schematic configuration of remote PECVD reactor.

systems. Another concern is that the incomplete reaction of precursors at low temperature PECVD leads to certain bonds in the film that are susceptible to oxidation and hydrolysis when exposed to humid environments, and so the integrity of the film may be compromised. For example, Chiang et al. (1989) reported N–H bond hydrolysis in SiN_x films by PECVD at saturated water vapor pressure.

Several schemes have been explored to enhance the barrier properties of PECVD films, including multilayer structures as well as organic and inorganic graded structures. Most of the authors identify high porosity/void fraction in PECVD films deposited at low temperature as the reason why multilayer structures are required (van Assche et al. 2004, Rosink et al. 2005, Zambov et al. 2006).

Significant progress has been made toward the real application of PECVD barriers. A summary of published barrier results is listed in Table 3.1. By using hexamethyl disiloxane (HMDSO) and oxygen as precursors, Mandlik et al. (2008) demonstrated single-layer barrier encapsulation of OLED with a lifetime of about 1 year at 65°C and 85% RH, far exceeding the industrial requirement of 1000 h. The company OLED Technologies B.V. (OTB) has developed and deployed commercial OLED production lines with integrated TFE based on a remote plasma source with high plasma density, realized in a single chamber (van Assche et al. 2004, Hemerik et al. 2006).

ALD has also been proposed for deposition of inorganic barrier layers in TFE and on film. ALD was developed in the 1960s and 1970s; it is able to deposit a wide range of materials and has been successfully used in the semiconductor industry for the deposition of high-quality dielectric and diffusion barrier materials. Several review articles on ALD have been published (Suntola 1992, 1994, George 2010). ALD is very similar to CVD except that in ALD the precursors are separated and introduced sequentially, and the film grows step-wise instead of continuously. Most ALD processes are based on

Table 3.1   Summary of Performance of PECVD Barrier Films

Coating Material	Structure	Substrate	Deposition Temperature	WVTR (g/m²/d)	Reference
SiO_xN_y/ SiO_xC_y	Graded layer	Polycarbonate	<100°C	5×10^{-6} (23°C/50% RH)	Schaepkens et al. (2004)
SiO_x/SiN_x	Multilayer	PET	120°C	2.5×10^{-7} (23°C/40% RH)	Chen et al. (2007)
SiC_x	Single layer	PET	25°C–50°C	10^{-4}	Zambov et al. (2006)
SiO_2/ pp-HMDSO	Single layer	OLED on glass	<80°C	Not reported	Mandlik et al. (2008)
Nitride	Multilayer	OLED on glass	80°C	Not reported	van Assche et al. (2004); Hemerik et al. (2006)
NONON	Nitride/ oxide multilayer	OLED on glass; PET	<130°C	3.6×10^{-6}	Rosink et al. (2005); Van Assche et al. (2008)

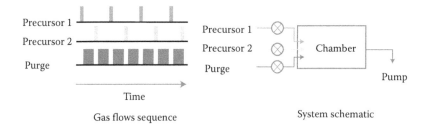

FIGURE 3.12 Schematic time sequence for reactant introduction in ALD deposition (left) and system schematic (right).

binary reaction sequences in which two surface reactions occur and deposit a binary compound film. A typical ALD deposition sequence and a schematic of the system are shown in Figure 3.12.

There is an optimal temperature window for ALD where a perfect and completely reacted monolayer is deposited on the surface. Figure 3.13 shows a scheme of the processes involved.

Below a certain temperature, T1, at low precursor flows there is incomplete reaction of the surface available reactive bonds, while at higher flows the precursors tend to condense, resulting in CVD-like deposition. Above a certain temperature, T2, the precursors desorb at low flows or decompose on the surface at high flows leading again to CVD-like deposition. The phenomena outside the process window lead to low deposition rate and/or the presence of unreacted species in the film. When operating at low temperature to remove condensed and unreacted material, the purge time must be increased, so the time per cycle increases and deposition rate decreases.

For barrier film deposition, the ALD Al_2O_3 is a typical model of the process, in which trimethylaluminum (TMA) and H_2O are used. The surface reaction process is given in Equations 3.4 and 3.5:

$$AlOH^* + Al(CH_3)_3 \rightarrow AlOAl(CH_3)_2^* + CH_4 \qquad (3.5)$$

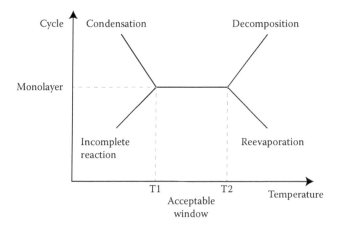

FIGURE 3.13 Growth regimes in an ALD system (see description in text).

$$AlCH_3^* + H_2O \rightarrow AlOH^* + CH_4 \tag{3.6}$$

ALD has two advantages: atomic layer thickness control and perfect conformality. Also, ALD can deposit thin films at low temperatures. These properties make it suitable for gas diffusion barrier deposition on plastic substrates or directly on OLED devices, although concern about the long-term effects of exposing OLEDs to water in the first stage of the deposition has been raised. Table 3.2 summarizes several ALD barrier processes and performance data. Single layer ALD Al_2O_3 on PEN and Kapton substrates have a WVTR of 1.0×10^{-3} g/m²/day (Groner et al. 2006, Langereis et al. 2006). Carcia et al., also using single-layer 25-nm ALD Al_2O_3 on PEN substrate and Ca test, reported a WVTR of 1.5×10^{-5} g/m²/day at 38°C and 85% RH (Carcia et al. 2006). By using a multilayer structure of Al_2O_3/SiO_2, Dameron achieved a WVTR of 5.0×10^{-5} g/m²/day on PEN at room temperature and 100% RH (Dameron et al. 2008). In a barrier with PECVD SiO_x or SiN_x as the bottom layer and ALD Al_2O_3 as the top passivation layer, Kim et al. (2009) obtained WVTR of 2.0×10^{-5} g/m²/day at 20°C and 50% RH on glass. It has been argued that the sealing of pinholes by the top ALD layer is the primary mechanism leading to improvement of barrier performance. By depositing a Al_2O_3/ZrO_2 nanolaminate structure on Si, Meyer et al. (2010) obtained a WVTR of 3.2×10^{-4} g/m²/day at 85°C and 85% RH (corresponding to WVTR $< 10^{-6}$ g/m²/day at 20°C/50% RH), and found that chemical bond formation at the interface is responsible for the improvement of barrier properties and that the nanometer layer structure prevents the formation of grains and grain boundaries.

TEM analysis shows that low-temperature ALD Al_2O_3, SiO_x, and ZrO_2 have amorphous structures, which is favorable for gas barriers. The ALD process avoids the bombardment of energetic particles to the underlying device or polymer layer in the multilayer structure.

The conventional ALD process is suitable for batch processing, but usually has low deposition rate and low precursor use efficiency. These are major concerns in using

Table 3.2 Summary of Process and Performance of ALD Barrier Films

Material	Method	Substrate, Temp.	Deposited Structure	WVTR (g/m²/d)	OTR (cc/m²/day)	Reference
Al_2O_3 25 nm	ALD	PEN 120°C	Single layer	1.5×10^{-5} (38°C/85% RH)	Not reported	Carcia et al. (2006)
Al_2O_3 1–26 nm	ALD	PEN, Kapton 100°C–175°C	Single layer	1×10^{-3} (RT)	5×10^{-3} (23°C/ 50% RH)	Groner et al. (2006)
Al_2O_3 40 nm	PEALD	PEN 100°C	Single layer	5×10^{-3} (21°C/60% RH)	Not reported	Langereis et al. (2006)
Al_2O_3/ZrO_2 100 nm	ALD	Si/SiO₂ 80°C	Nano- laminate	3.2×10^{-4} (85°C/85% RH)	Not reported	Meyer et al. (2010)
Al_2O_3/SiO_2 26 nm/60 nm	ALD	PEN, Kapton 175°C	Multilayer	5×10^{-5} (RT/100% RH)	Not reported	Dameron et al. (2008)
$(SiO_x/SiN_x)/$ Al_2O_3 100 nm/50 nm	ALD + PECVD	Glass 110°C	Multilayer	2×10^{-5} (20°C/50% RH)	Not reported	Kim et al. (2005)

Chapter 3

ALD for mass barrier production. However, great progress has been made in hardware design and process development toward batch-to-batch mass production and R2R production, and several prototypes have been demonstrated. For example, by using the synchronously modulated flow and draw (SMFD) process, a short (~1 s) cycle time can be realized (Sundew 2012), and the use efficiency of precursors increased to between 30% and 90%.

Another approach is to use a coating head where the two precursors are spatially separated. By the relative movement of the coating head to the substrate, the substrate is exposed to both precursors separately, thus realizing ALD film growth (Levy et al. 2008) (Figure 3.14).

This spatial-ALD can operate at pressures close to atmosphere, and a high deposition rate has been reported (Beneq 2010). The spatial ALD can be adapted for batch-to-batch and R2R processes. The quality of the barrier needs to be investigated.

R2R ALD with shared zones is under development by Lotus (Dickey 2012). In this concept the web is transported back and forth between precursor zones in serpentine path. In the simplest configuration three total zones are defined: one zone for each of the two precursors (e.g., TMA and water) separated by one zone with high pressure of N_2 for purging. This technology is currently targeted toward ultrabarrier films. Further development by the same group considers the substitution of an oxygen-rich plasma for water with the prospective of better separation between the two precursors, higher deposition rate and better film quality by reduction of unreacted precursors.

Groner et al. (2006) reported that even a thin layer of 10-nm ALD Al_2O_3 film can form an effective barrier and suggested that the thinner film needed for ALD than for other deposition techniques improves the throughput and reduces cost. However, Carcia et al. (2010) show how thicker layers (~25 nm) grown at higher temperatures (100°C) are necessary to create a barrier on polymer able to prevent the visible oxidation of Ca films upon exposure at 60°C/85% RH for 500 h. More recent data reported by Söderlund with a Beneq system have shown that even thicker layers (~100 nm) were required in a roll-to-roll system run in batch mode (Söderlund 2012a,b).

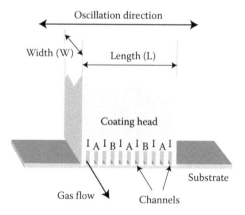

FIGURE 3.14 Schematic configuration of spatial-ALD. (Reproduced with permission from Levy D. H. F. et al. 2008. *Appl. Phys. Lett.* 92: 192102–192104.)

A few other issues related to ALD need to be pointed out. Because the ALD process is based on a surface chemical reaction, the initial nucleation of the ALD film is extremely important for continuous and pinhole-free thin-film barriers. Some plastic substrates such as PEN and PET do not have ideal surface functional groups for ALD growth; the initial growth will be in the island mode. This could lead to increased surface roughness and defect density as well as reduced adhesion to substrates. The relatively high WVTR of single-layer ALD Al_2O_3 with PEN (Groner et al. 2006) could be an indication of such defects. Proper surface treatment on substrates is necessary to ensure high-quality barrier growth.

Incomplete reaction of precursors is an important issue for low-temperature ALD films; they may function as defects and provide channels for gas diffusion. Proper choice of precursors and process parameters is very important. Kim et al. studied the influence of using O_3 and H_2O as oxidants on the impurity concentration in ALD Al_2O_3 films and discovered a higher concentration of carbon when using O_3 as oxidant (Kim et al. 2006) (Figure 3.15).

Also, the partially reacted precursors can continue to react when they are exposed to air or humidity, which may lead to a volume change in the barrier layer, thereby producing stress and degrading the performance of the barrier.

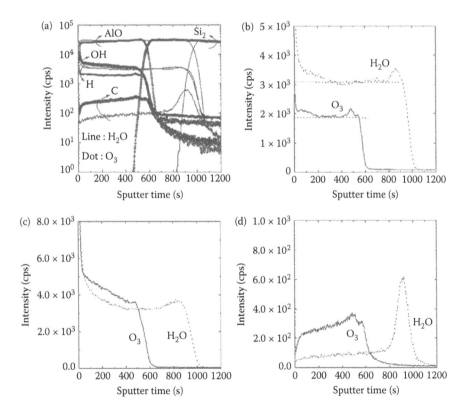

FIGURE 3.15 ToF-SIMS depth profile results of a sample grown at 100°C using O_3 or H_2O as reactant (a). Profiles in linear scale: (b) H-profile, (c) OH-profile, (d) C-profile. (Reproduced with permission from Kim S. K. et al. 2006. *J. Electrochem. Soc.* 153: F69–F77.)

Other inorganic deposition techniques have been suggested in the literature for the fabrication of barrier layers on OLED or plastic substrates. Inorganic barrier layers may be deposited using RF sputtering of ceramic targets. Some researchers have reported low WVTR properties of RF-sputtered thin films, especially when two oxide materials are combined in alternating layers (Choi et al. 2010a,b). RF sputtering is beneficial in terms of charge buildup and arcing at the target, and the film density is low. However, the deposition rate is very low, and additional equipment is necessary to match the RF supply to the target. Another configuration of RF sputtering presented in the literature is called neutral beam assisted sputtering (NBAS), which uses RF sputtering with an additional source of neutral atoms impacting the substrate to contribute additional kinetic energy to the depositing film and, therefore, increase the film density and its performance as a barrier (Yoo et al. 2008, Lee et al. 2011).

An alternative low-temperature CVD is catalytic chemical vapor deposition (Cat-CVD). Its thin-film barrier in TFE and R2R applications been developed by Prof. Matsumura's group at the Japanese Advanced Institute of Science and Technology (JAIST), and major Japanese equipment manufacturers like ULVAC and Canon-Anelva have considered it for industrial applications. It uses a filament made of a catalytically active metal (usually tungsten) to decompose source gases such as silane, ammonia, and hydrogen, and realize film deposition. Cat-CVD operates at ultra-high vacuum; high-quality films of SiN_x have been prepared at substrate temperatures between 20°C and 100°C. A high deposition rate of 100 nm/min and low hydrogen bond concentration have been reported (Heya et al. 2008). Gas barrier films using organic silicon compounds and Cat-CVD have also been reported (Matsumura and Ohdaira 2008), overcoming the problem of carbonization of the W catalyst. Single-layer barriers have modest performance not suitable for OLED encapsulation, but Matsumura's group has demonstrated SiN_x/SiO_xN_y stacked thin films (7 layers) with each layer thickness of 300 nm that fully protected OLED on glass for 1000 h at 60°C/90% RH (Ogawa et al. 2008). Barrier film on PEN produced in R2R configurations gave a more modest performance with measured WVTR of 0.01 g/m^2/day due to particle defects and cracking (Heya et al. 2008). An OLED on glass encapsulated by lamination using this barrier survived with no black spots or shrinkage for 400 h at 60°C/90% RH. Unfortunately, in spite of the good barrier performance, absence of plasma damage, high deposition rate, high use of reactant gases (10 X compared to PECVD), and easy scalability, lingering problems such as the short life of the catalyst wire and difficulty in implementing *in situ* chamber cleaning are obstacles to the scale up and commercialization of the technology (Asari et al. 2008, Matsumura 2008).

3.4.2   Organic Layer Deposition

Organic layers have been used to improve the performance of thin-film ultrabarriers. Organic layers are utilized as "decoupling" layers between inorganic layers that constitute the real barrier to permeation. In addition to improving the permeation as discussed in Section 3.3.3, organic layers make the structure more robust and flexible. Individual inorganic layers in the organic/inorganic multilayer structure can be kept thinner, so both the inorganic layer and the total structure are intrinsically more flexible. Furthermore, the brittle ceramic layers are sandwiched between compliant organic layers, so the total structure is more robust toward fragmentation and cracking.

While it is generally claimed that the chemistry of the organic layer is irrelevant to the overall permeation performance, Moro et al. (2010) reported how the ultimate performance of the barrier depends on its chemistry. For example, high cross-linking slows the diffusion of moisture and O_2, and as a consequence, the lag time increases, as does the time to reach the steady-state flow. Also, swelling of the polymer layers upon moisture absorption can result in cracking of the thin inorganic layer.

Padiyath et al. showed how the matching of T_g between the interlayer polymeric material and the substrate is important for achieving lower WVTR and OTR (Padiyath and Roehrig 2009). The organic materials proposed by Vitex Systems and 3 M as organic interlayers in organic/inorganic ultrabarriers are blends of vacuum-deposited polyacrylate. Vacuum polymer deposition (Affinito et al. 1999) offers a flexible technique to overcome the effects of substrate nonuniformities detrimental to a vacuum-deposited ceramic vapor barrier. The technique involves the flash evaporation of an acrylic monomer onto a cooled substrate in roll-coating configuration. By designing proprietary hardware, Vitex Systems extended the use of polymer vacuum deposition to batch-to-batch encapsulation of OLED by TFE for R&D applications (Moro et al. 2004) and mass manufacturing (Moro et al. 2006). Subsequent to condensation, the monomer is immediately cured, yielding a highly cross-linked polyacrylate film. Traditionally, plasma irradiation, e-beam curing and Hg-bulb UV lamps have been used in R2R systems for the deposition of barriers on plastic foils. More recently, for direct TFE of OLEDs, LED lamps have been implemented with the significant advantage of reducing the temperature during curing, making the curing less harmful to the sensitive organic OLED materials, and to reduce size, complexity, and cost of the curing stations (Rosenblum et al. 2009). The organic thin film vacuum evaporation technique enables extremely high deposition rates (up to 1 μm/s), which is important for R2R coating applications. More importantly, for vapor barrier applications, the deposition and curing of the liquid monomer results in a flexible, transparent polymer film that is nonconformal to the substrate, effectively decoupling substrate nonuniformities and morphologies from a barrier layer that can be deposited in the same vacuum system. AFM measurements of a PET substrate show a reduction from an initial surface roughness (R_a) > 15 nm to less than 1 nm after deposition of a 0.25-μm-thick layer of polyacrylate. A layer of polymer up to ~5 μm thick can be applied to a substrate in one pass. Using multiple passes in a vacuum system or roll coater, films of arbitrary thickness can be built up with similar surface roughness. The acrylate and methacrylate used in vacuum evaporation are mostly carbon–based monomers with molecular weight (MW) in the range 150–600 amu. The evaporator temperature is kept at or around 200°C, although higher temperatures up to 350°C have been reported to evaporate heavier compounds. Silicon-containing monomers (Roehrig et al. 2012) and sulfonyl-containing monomers (Agata et al. 2008) have also been proposed. Mikhael et al. of Sigma Technologies (Mikhael and Yializis 2002) expanded the polymerization approaches beyond free-radical polymerization of acrylates by suggesting that epoxy compounds can be coated by vacuum evaporation. They demonstrated cation-polymerization by using acid created *in situ* by electron or UV radiation.

Nonvacuum deposition techniques have been proposed for the coating of the organic layers. These include gravure roll-coating, spray-coating (Padiyath and David 2009), and ink-jet printing (Lin et al. 2010). By avoiding the evaporation step, these methods enable the coating of materials with larger MW.

Chapter 3

Other vacuum deposition techniques of polymers are those using parylene (Liao et al. 2006), initiated CVD or iCVD, a variation of Cat-CVD (Coclite et al. 2010, Spee et al. 2012), and more traditional CVD approaches (Iwanaga 2010). Many of these approaches have been applied only at the level of basic research, and the reports do not discuss the most important issues related to deposition of organics in vacuum (e.g., contamination of vacuum chambers and introduction of particles by heterogeneous condensation).

A few authors propose the use of polymeric layers with an active role. For example, Ramadas and Chua (2010) introduce nanoparticles of reactive oxides and carbon nanotubes as desiccants in the organic layer. However, a significant gettering effect may be achieved at relatively high loads of nanoparticles that reduce the transparency of the film and increase haze. The same researchers suggest a healing effect by migration of particles into the defects and cracks, although the mechanistic details of such migration are not hypothesized or discussed. In a clever experiment, Aresta et al. (2012) show how for low-density PECVD barriers, the liquid precursor of the organic layer actually condenses into the nanodefects, filling them and stuffing the permeation path. After polymerization, the hybrid composite material at the interface constitutes an impermeable barrier to moisture and oxygen. While the model may be not easily extendable to other systems with more dense inorganic layers (e.g., sputtered layers), the results well explain and agree with many prior publications reporting on PECVD barriers. Scientists at the Fraunhofer-Institut für Silicatforschung (ISC) in Germany have been proposing ORMOCER®, a wet coating of vacuum-deposited inorganic layers using an inorganic/organic hybrid monomer (Amberg-Schwab and Weber 2009). After wet-coating with ORMOCER, the barrier performance of single-layer inorganic films improves by two orders of magnitude. The authors suggest that sealing of defects is achieved by the reaction of the moiety in the wet coating with hydroxyl groups at the surface of the vacuum-deposited films. The best performance reported of these types of barriers in a stack composed by sputtered-Al_2O_3/ORMOCER/sputtered-A_2O_3 on PET (WVTR < 2. eq. < 10^{-4} g/m^2/day at 20°C, 50% RH) is not enough for OLED applications. According to the authors the limitation comes from added particles in the process of coating the wet and dry layer on different R2R systems.

3.4.3 Multilayer Barriers

Among the several multilayer technologies described in the literature, the first to be introduced and advanced toward commercial applications is the Barix™ technology proposed initially by Battelle Pacific Northwest Division and developed by Vitex Systems for application as TFE or as a barrier on plastic foil. A Barix multilayer stack is shown in Figure 3.16.

In the Barix process, alumina layers with typical thickness between 10 and several hundred nanometers are deposited by DC reactive sputtering of a metallic Al (99.999%) target. Polymer layers are deposited by flash evaporation of liquid monomers in vacuum as described above. The monomer vapors are condensed on the sample surface in the liquid phase and cross-linked by UV radiation to form a solid film. A proprietary blend of acrylate monomers (Barix Resin System) optimized for encapsulation of OLED devices is used to give a highly cross-linked acrylic organic film (Moro and Krajewski 2010). Typical thicknesses of the organic films vary between 0.25 and several micrometers.

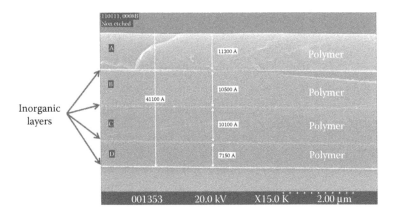

FIGURE 3.16   SEM cross section of organic/inorganic multilayer barrier film.

What is really unique about this process is that the organic phase is deposited as a liquid: the film is very smooth (<0.2 nm variation) locally and also has extremely good planarizing properties over high topographical structures like "cathode separators," "ink jet wells," and "active matrix pixel" structures (Figure 3.17).

So, while the resulting local flatness creates an ideal surface for growing an almost defect-free inorganic layer, the liquid covers all of the topographic features. It should also be mentioned that while even nonconformal methods to deposit oxides like CVD have difficulty covering cathode separators without creating voids, they also struggle to coat structures higher than 4 microns in an acceptable process time. The Barix barrier properties depend on the number of organic/inorganic layers deposited. The first reports by Vitex showed cross sections of the film with four dyads or more (as many as 6), but with technology development the number of layers was reduced to three dyads or fewer with >95% yield (Figure 3.18).

The Vitex Barix process for TFE has been shown to meet telecommunication application specifications for a wide variety of OLED displays: passive and active matrix displays, bottom, top, and transparent displays, and it works equally well for SM, polymer, and phosphorescent OLEDs. The Barix process has also been applied to encapsulation

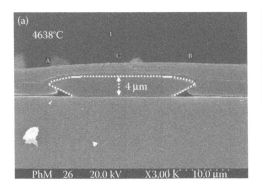

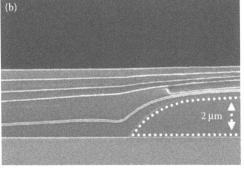

FIGURE 3.17   Examples of planarization by vacuum deposition of polymer in an organic/inorganic multilayer barrier: Cathode separator in a passive-matrix OLED display (a), active-matrix pixel structures in an AMOLED display (b).

Chapter 3

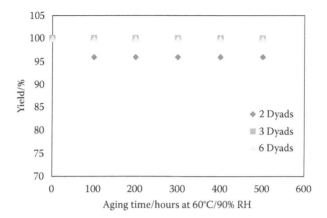

FIGURE 3.18 OLED devices yield versus time for TFE Barix structures with different number of dyads.

by both the TFE and lamination technologies of organic and inorganic solar cells. For inorganic solar cells, the Barix encapsulation technology was able to pass the lifetime test at 1000 h at 85°C/85% RH (Olsen et al. 2008), proving both the remarkable thermal stability of the barrier structure and the exceptionally low permeability to H_2O and O_2. For the TFE process of OLED displays, Vitex also introduced and implemented an edge seal by designing the encapsulation tool with mask-exchange capability. Initially the minimum edge width (distance between the edge of the active area and the edge of the barrier layer) was 3 mm, but later published results demonstrated a 1.5-mm edge seal with mechanical alignment and 0.5 mm with CCD alignment (Figure 3.19) (Chu et al. 2007).

Some types of OLED devices are sensitive to the plasma exposure in the TFE process. The detrimental effects of plasma exposure can be avoided by using a Vitex proprietary protective layer. The protection is also effective for deposition energetic conditions compatible with mass manufacturing (Moro et al. 2006).

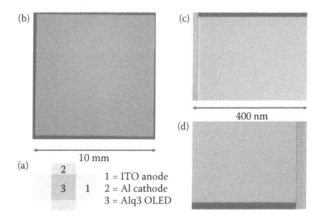

FIGURE 3.19 TFE encapsulated cross-electrode OLED pixel with 1.5 mm edge seal after 500 h 60°C/90% RH accelerated aging: schematic of the device (a); low magnification image of illuminated pixel (b); magnified picture of top-left (c); and bottom-right corner (d) of the illuminated pixel with back light. There is no evidence of pixel shrinkage due to moisture diffusion.

Flexible displays encapsulated with Barix TFE have been shown at numerous shows and trade fairs (Chwang et al. 2006, Jin et al. 2006, Fischer et al. 2011). The barrier has been shown to be mechanically robust to impact both on Ca (XC95131 2009) and for TFE encapsulated flexible displays (TTTiagoTTT 2010) by repeated hammering. The excellent performance of the Barix barrier has also been demonstrated for barrier on plastic foil. Several flexible displays on plastic foil have been demonstrated. Excellent barrier performance has been shown by using Ca lifetime with two-sided thin film encapsulation (PEN/barrier/Ca/TFE barrier): only 16% transmission change after 570 h, 60°C/90% RH, and no pinholes, equivalent to WVTR ~ 8×10^6 g/m²/day for encapsulant + barrier substrate at 21°C, and compatibility with basic process steps such as photo-patterning, ITO deposition, thermal treatments, and flex test (Rutherford 2005a,b, Ramos 2006). Recent modifications and improvements to the Barix multilayer approach have demonstrated excellent barriers on foil with a reduced number of dyads after testing at 85°C/85% RH for 1000 h (Moro et al. 2011). WVTR $< 10^{-6}$ g/m²/day has been demonstrated using the Ca test for a single dyad barrier on glass with yields higher than 50%. Using a batch process, a single dyad barrier on PEN has also been demonstrated with Ca test and aging at 85°C/85% RH. The change in transmission of samples like the one shown in Figure 3.20 with a structure PEN/1dyad barrier/Ca/3dyad barrier is 2–6 × 10^{-6} g/m²/day, corresponding to $<1 \times 10^{-7}$ g/m²/day at 20°C/50% RH.

The encapsulation process proposed by OTB is also multilayer (Hemerik et al. 2006). The TFE consists of six alternating silicon nitride and organic polymer layers. The silicon nitride (SiN) is deposited through a remote inductively coupled plasma source. The

FIGURE 3.20 Encapsulated transparent and flexible Ca samples with a TFE barrier PEN/1dyad barrier/ Ca/3dyad barrier with WVTR of 2–6 × 10^{-6} g/m²/day at 85°C/85% RH, corresponding to $<1 \times 10^{-7}$ g/m²/ day at 20°C/50% RH.

Chapter 3

NH_3 gas is injected through a showerhead directly into the plasma volume. The resulting radicals diffuse to the substrate and react with the SiH_4 gas that is injected right in front of the substrate, thus forming a SiN layer on the substrate. The polymer layers are printed by inkjet. The different polymer layers have layer thicknesses varying between 0.3 and 12 µm. The authors comment that multiple layers are necessary because the SiN layers can only be processed at a temperature of around 100°C, so the density of the SiN layer is low. Layer properties and uniformity were tuned by changing process parameters and deposition rates of 1–10 nm/s were achieved. After optimization of the different layers, the complete encapsulation stack was applied to 20″ substrates with 90 RGB printed PM OLED. The devices passed a 500-h, 60°C/90% RH aging test with no black spot growth. No damage related to the deposition was detected.

Barrier multilayers on R2R were also demonstrated by General Electric using a PECVD described by Schaepkens et al. (2004). The barrier process was integrated in an R2R demonstration line for the production of OLED for SSL. The line, including the R2R deposition of the barrier on foil, was developed in the frame of a NIST Advanced Technology Program (ATP Program 2003–2007) in partnership with Energy Conversion Devices (http://www.atp.nist.gov/index.html). The goal of the 4-year project was to prove that continuous R2R OLED fabrication is possible (Duggal 2009). Finally a multilayer approach similar to Barix is the strategy chosen by 3 M in the development of its ultrabarrier on foil (McCormick 2011).

3.5 Production Systems

As illustrated in the previous sections, the academic, scientific, and technical literature regarding ultrabarriers is very rich. Most of the publications focus on the chemical/physical properties of films and on barrier performance on small-size devices or substrates. Many of the proposed solutions, while obtaining some success on small areas, basically fail on large surface area (>few cm^2) and real devices because of the presence of particles, topography, cracks and defects in the layer, and residual stress. In addition, the technical goal must be achieved in an economically feasible way. The process characteristics to be considered for deployment in production are defect density, yield of encapsulated devices or barrier film, tool throughput, maintenance requirements, and tool uptime. For these reasons, only a few technologies for TFE encapsulation or deposition of ultrabarriers on foil have been deployed for R&D/pilot line. All of them are based on a multilayer approach (Section 3.4.3).

For the Barix technology, some of the issues related to mass manufacturing have been addressed both for TFE (Moro et al. 2006, Chu et al. 2007) and for R2R (Kapoor et al. 2006). A few companies under license by Vitex Systems have commercialized batch-to-batch TFE systems based on the Barix technology (Vitex System, Inc. 2008). According to experts in the sector, this technology is being introduced on commercial OLED displays by Samsung SMD (OLED Association 2012).

OTB has been commercializing a similar multilayer technology where the SiN_x inorganic layer is deposited by remote plasma CVD and the organic layer is based on acrylic resins deposited by inkjet and cured by UV (OTBv 2009). The tool is now commercialized by Roth and Rau (2012). Cannon-Tokki has instead licensed (Tokki 2007) the graded barrier PECVD technology developed by GE.

The Barix technology has also been licensed for R2R production of ultrabarriers on foil substrates (Vitex Systems, Inc. 2009). Commercial availability of ultrabarriers on foil with WVTR < 10^{-6} g/m²/d has been announced by FujiFilm (FujiFilm Corporation 2009) and DNP (IdTechEx 2009). Mitsubishi (Mitsubishi 2012) has announced a transparent barrier for PV application and e-readers with WVTR in the 10^{-4} g/m²/day that may be extended to OLED applications.

In May 2012, 3M announced on its website that its flexible, optically clear barrier films for water vapor and oxygen are commercially available. Previously, these films had been sold primarily under limited R&D agreements. The film has a water vapor transfer rate of less than 1×10^{-3} g/m²/day at 20°C with surface roughness of about 1 nm. The barrier film consists of a base polyester layer (50–125 microns thick) with a very thin (less than 2.0 microns) barrier coating made up of layers of polymer and oxide, a very similar technology to Barix technology (3M Company 2012). 3 M is developing a range of barrier films using different substrates, additional film layers, and even higher barrier performance for applications ranging from solar panels to electrophoretic, electrochromic, and OLED displays.

Other companies have been publishing their work on R2R deposition tools of barrier by PECVD (Zambov et al. 2011) and ALD, but close observation and discussions indicate that the efforts are still at the preliminary R&D stage. For example, Soderlund describes an R&D R2R system for flexible substrates (Soderlund, 2012). In this system, the flexible substrate is fixed on a rotating cylinder within the reaction chamber and is surrounded by a number of linear nozzles, each creating an isolated gas region over the full width of the substrate. As the cylinder is rotated, the substrate passes through different gas regions and is coated. This experimental system reached WVTR ≤ 10^{-3} g/m²/day at 25 nm Al_2O_3 at a linear speed of 6 m/min, and WVTR of 10^{-6} g/m²/day at 100 nm (Soderlund, 2012).

3.6 Summary

Due to the sensitivity of OLED devices to water and oxygen, encapsulation is a key production step for successful and widespread acceptance in display and solid state lighting applications. Traditional can encapsulation is currently the most common technique used in commercial devices. Much effort, documented in a rich literature, has been spent at the R&D level to move towards TFE and encapsulation by engineered plastic foils. These techniques may allow new form factors that revolutionize the use of displays and lighting. Only a few of the numerous various approaches described in literature have been developed to pilot and production lines, as work remains to solve technical and marketing challenges for these new technologies.

Compatibility of materials and deposition techniques with devices, and their fabrication process should be considered in proposing new encapsulation approaches. Depending on the type of OLED device (AMOLED vs. passive matrix devices, bottom or top emission, etc.) and role of the encapsulant (substrate or top encapsulation) the requirements for the barrier are different, as are the most suitable deposition techniques, the materials used in the films, and the substrates. Depending on the type of encapsulation additional functionality may be included in the design of the barrier (e.g., light extraction, heat dissipation, heat shielding).

Chapter 3

Rather than new deposition techniques, ingenuity and creativity is required in the engineering of the deposition tools to be used in pilot and production lines to avoid extrinsic defects deleterious to barrier performance and to meet the productivity required. For example, tools for defect detection with adequate sensitivity are necessary to improve film quality on a large scale. These considerations allow for the coexistence of multiple technologies and the further application of exciting science and technology in materials science, optics, device physics, deposition techniques, and vacuum science.

References

3M Company. 2012. 3M Announces Commercial Availability of 3M™ FTB3 Barrier Film. Press Release May 22, 2012. Accessed July 8, 2012, http://news.3m.com/pressrelease/company/3m-announces-commercial-availability-3m-ftb3-barrier-film.

Advanced Technology Program (ATP), National Institute of Standards and Technology (NIST), US Department of Commerce, 100 Bureau Drive, STOP 4701 Gaithersburg, MD 20899-4701. Accessed July 8, 2012, http://www.atp.nist.gov/index.html.

Affinito J. D., Gross M. E., Mournier P. A., Shi M.-K., Graff G. L. 1999. Ultrahigh rate, wide area, plasma polymerized film from high molecular weight/low vapor pressure liquid or solid monomer precursors. *J. Vac. Sci. Technol. A* 17: 1974–1981.

Agata Y., Tsukamoto N., Ishikawa S.-I. 2008. Gas barrier laminated film and image display device comprising same. US Patent Application 2008/0081205.

Akedo K., Miura A., Fujikawa H., Taga Y. 2006. Flexible OLEDs for automobiles using SiN_x/CN_x:H multi-layer barrier films and epoxy substrates. *J. Photopolym. Sci. Technol.* 19: 203–208.

Alexandrov S. E., Hitchman M. L., Shamlian S. H. 1995. Remote plasma-enhanced chemical vapor deposition of silicon nitride films: The effect of diluting nitrogen with helium. *J. Mater. Chem.* 5: 457–460, doi:10.1039/JM9950500457.

Amberg-Schwab S., Weber U. 2009. Inorganic-organic polymers in combination with vapor deposited inorganic thin layers an approach to ultra barrier films for technical applications. Presented at *LOPE-C, International Conference and Exhibition for the Organic and Printed Electronics Industry*, Frankfurt, Germany, June 23–25.

Aresta G., Palmans J., van de Sanden M. C. M., Creatore M. 2012. Evidence of the filling of nano-porosity in SiO_2-like layers by an initiated-CVD monomer. *Microporous Mesoporous Mater.* 151: 434–439.

Asari S., Fujinaga T., Takagi M. et al. 2008. ULVAC research and development of Cat-CVD applications. *Thin Solid Films* 516: 541–544.

ASTM International. 2007. ASTM F1927-07 Standard Test Method for Determination of Oxygen Gas Transmission Rate, Permeability and Permeance at Controlled Relative Humidity through Barrier Materials Using a Coulometric Detector. Accessed July 21, 2012, 1996–2012 ASTM, http://www.astm.org/Standards/F1927.htm.

ASTM International. 2009. ASTM D3359-09e2 Standard Test Methods for Measuring Adhesion by Tape Test. Accessed July 14, 2012, 1996–2012 ASTM, http://www.astm.org/Standards/D3359.htm.

ASTM International. 2010. ASTM D3985-05(2010)e1 Standard Test Method for Oxygen Gas Transmission Rate Through Plastic Film and Sheeting Using a Coulometric Sensor. Accessed July 21, 2012, 1996–2012 ASTM, http://www.astm.org/Standards/D3985.htm.

ASTM International. 2011. ASTM F1249-06(2011) Standard Test Method for Water Vapor Transmission Rate Through Plastic Film and Sheeting Using a Modulated Infrared Sensor. Accessed July 21, 2012, 1996–2012 ASTM, http://www.astm.org/Standards/F1249.htm.

Atomic Layer Deposition Beneq Oy, Vantaa, Finland. 2010. Accessed July 15, 2012, http://www.beneq.com/atomic-layer-deposition.html.

Barix Barrier Film Survives the Hammer test! YouTube, uploaded by xc95131 on June 19, 2009. Accessed July 15, 2012, http://www.youtube.com/watch?v=a0iTg1pufdE.

Bishop C. 2011. How does the substrate affect barrier performance of vacuum-deposited coatings? *Converting Quart.* 1: 14.

Burrows P. E., Graff G. L., Gross M. E. et al. 2001. Gas permeation and lifetime tests on polymer-based barrier coatings. In *Proceedings of SPIE-The International Society for Optical Engineering*, 4105 (Organic

Light-Emitting Materials and Devices IV), ed. Z. H. Kafafi, 75–83. SPIE-The International Society for Optical Engineering, Bellingham, WA.

Carcia P. F., McLean R. S., Reily M. H. et al. 2006. Ca test of Al_2O_3 gas diffusion barriers grown by atomic layer deposition on polymers. *Appl. Phys. Lett.* 89: 031915–031917.

Carcia P. F., McLean R. S., Reilly M. H. 2010. Permeation measurements and modeling of highly defective A_2O_3 thin films grown by atomic layer deposition on polymers. *Appl. Phys. Lett.* 97: 221901-1-3.

Chen T.-N., Wuu D. S., Wu C.-C. et al. 2007. Improvements of permeation barrier coatings using encapsulated parylene interlayers for flexible electronic applications. *Plasma Processes Polym.* 4: 180–185.

Chiang J. N., Ghanayem S. G., Hess D. W. 1989. Low-temperature hydrolysis (oxidation) of plasma-deposited silicon nitride films. *Chem. Mater.* 1: 194–198.

Choi J.-H., Kim Y.-M., Park Y.-W. et al. 2010a. Highly conformal SiO_2/Al_2O_3 nanolaminate gas-diffusion barriers for large-area flexible electronics applications. *Nanotechnology* 21: 475203, doi:10.1088/0957-4484/21/47/475203.

Choi J.-H., Bulliard X., Benayad A. et al. 2010b. Design and fabrication of compositionally graded inorganic oxide thin films: Mechanical, optical and permeation characteristics. *Acta Mater.* 58: 6495–6503.

Chu X., Burrows P. E., Mast E. S. et al. 2009. Method for edge sealing barrier films. U.S. Patent Application 2009/0191342, filed December 30, 2008, and published July 30, 2009.

Chu X., Moro L., Rosenblum P.M. et al. 2007. Barix thin film encapsulation of OLEDs. Manufacturing aspects and high temperature performance. Paper presented at *IMID 2007 Meeting*, Daegu, South Korea, August 27–31.

Chwang A., Hewitt R., Urbanik K. et al. 2006. Full color 100 dpi AMOLED display on flexible stainless steel substrate. In *Society for Information Display (SID) 2006 International Symposium–Digest of Technical Papers*, Vol. XXXVII: 1858–1861. San Francisco, California, 4–6 June 2006.

Clausen T., Dabruzzi C. 2009. Substrates & encapsulation & light outcoupling components to enable OLED SSL. Paper presented at *DOE Solid-State Lighting Manufacturing Workshop*, Vancouver, WA, June 24–25, http://apps1.eere.energy.gov/buildings/publications/pdfs/ssl/clausen_vancouver09.pdf.

Coclite A. M., Ozaydin-Ince G., Palumbo F. 2010. Single-chamber deposition of multilayer barriers by plasma enhanced and initiated chemical vapor deposition of organosilicones. *Plasma Processes Polym.* 7: 561–570.

Corning Display Technologies. 2012. Corning Incorporated. Accessed July 15, 2012, http://www.corning.com/WorkArea/showcontent.aspx?id=48969.

Corning® WillowTM Glass—Fact Sheet, *Corning Incorporated*, 2012. Accessed July 15, 2012, http://www.corning.com/WorkArea/showcontent.aspx?id=48969.

Cros S, Firon M., Lenfant S. et al. 2006. Study of thin calcium electrode degradation by ion beam analysis. *Nucl. Instrum. Methods Phys. Res. B* 251: 257–260.

Dameron A. A., Davidson S. D., Burton B. B. et al. 2008. Gas diffusion barrier on polymers using multilayers fabricated by Al_2O_3 and rapid SiO_2 atomic layer deposition. *J. Phys. Chem. C* 112: 4573–4580.

Dickey E. 2012. Progress in roll-to-roll atomic layer deposition. Paper presented at the *New Industrial Chemistry and Engineering (NIChE) "Workshop on Barrier Technologies"* organized by the Council for Chemical Research, Arlington, Virginia, September 10.

DryFlex Product Line-up, SAES Getters Group. 2008. Accessed May 28, 2012, flat_panels@saes-group.com.

Duggal A. 2009. Roll-to-roll OLEDs. Paper presented at *DOE (US Department of Energy) Solid State Lighting Manufacturing Workshop*, Fairfax, VA, April 21–22, http://www1.eere.energy.gov/buildings/ssl/fairfax09_materials.html.

DuPont. 2011. OLED Product Technology Drylox Coverglass, DuPont, 2011. Accessed May 28, 2012, http://www2.dupont.com/OLED/en_US/products/drylox_coverglass.html.

Erlat A. G., Henry B. M., Ingram J. J. et al. 2001. Characterization of aluminium oxynitrade barrier films. *Thin Solid Films* 388: 78–86.

Erlat A. G., Spontak R. J., Clarke R. P. et al. 1999. SiO_x gas barrier coatings on polymer substrates: Morphology and gas transport considerations. *J. Phys. Chem. B* 103: 6047–6055.

Fischer H., Tietke M., Fritze F. 2011. Electronic passports with AMOLED displays. *J. SID* 19: 163–169.

Fujifilm Corporation. 2009. Next-generation material with extremely high moisture barrier property, essential for flexible electronic devices. Press Release February 17, 2009. Accessed July 8, 2012, http://www.fujifilm.com/news/n090224.html.

George S. M. 2010. Atomic layer deposition: An overview. *Chem. Rev.* 110: 111–131.

Graff G. L., Williford R. E., Burrows P. E. 2004. Mechanisms of vapor permeation through multilayer barrier films: Lag time versus equilibrium permeation. *J. Appl. Phys.* 96: 1840–1849.

Chapter 3

Greener J., Ng K. C., Vaeth K. M. et al. 2007. Moisture permeability through multilayered barrier films as applied to flexible OLED display. *J. Appl. Polym. Sci.* 106: 3534–3542.

Groner M. D., George S. M., McLean R. S. et al. 2006. Gas diffusion barriers on polymers using Al_2O_3 atomic layer deposition. *Appl. Phys. Lett.* 88: 051907–051909.

Hanika M., Langowski H.-C., Moosheimer U., W. et al. 2003. Inorganic layers on polymeric films—influence of defects and morphology on barrier properties. *Chem. Eng. Technol.* 26: 605–614. Available at: http://onlinelibrary.wiley.com/doi/10.1002/ceat.200390093/abstract.

Hemerik M., Van Erven R., Vangheluwe R., Yang J., Van Rijswijk T., Winters R., Van Rens B. 2006. Lifetime of thin-film encapsulation and its impact on OLED device performance. *SID Symposium Digest of Technical Papers* 37(1): 1571–1574, doi:10.1889/1.2433297.

Heya A., Minamikawa T., Niki T. et al. 2008. Cat-CVD SiN passivation films for OLEDs and packing. *Thin Solid Films* 515: 553–557.

IDTechEx. 2009. Ultimate flexible barrier film. Posted on Printed Electronics web site on December 28, 2009, *1999–2012 IDTechEx*. Accessed July 15, 2012, http://www.printedelectronicsworld.com/articles/ultimate-flexible-barrier-film-00001946.asp?sessionid=1.

Ikezaki T. 2011. AGC new damage-resistance speciality glass: Dragontail. Paper presented at *SID Display Week*, Los Angeles, May 18, 2011.

Ito H., Oka W., Goto H. et al. 2006. Plastic substrates for flexible displays. *J. J. Appl. Phys.* 45: 4325–4329.

ITRS 2012. International Technology Roadmap for Semiconductors, last updated on March 9, 2012. Accessed June 3, 2012, http://www.itrs.net/2009, 2010.

Iwanaga H. 2010. Gas-barrier film, substrate film, and organic electroluminescent device. US Patent, Publication number 7,815,983, filed March 26, 2007, and issued October 19, 2010.

Jin D. U., Jeong J. K., Shin H. S., Kim M. K., Ahn T. K., Kwon S. Y., Kwack J. H., Kim T. W., Mo Y. G., Chung H. K.. 2006. 5.6-inch flexible full color top emission AMOLED display on stainless steel foil. *SID Symposium Digest of Technical Papers* 37(1): 1855–1857, doi:10.1889/1.2433405.

Kapoor S., Seta M. E., Visser R. J. et al. 2006. Pilot production of ultrabarrier substrate for flexible displays. *Society of Vacuum Coaters 49th Annual Technical Conference Proceedings*, 625–631. Washington, USA, April 22–27, 2006.

Kim N., Potscavage W. J., Domercq B. et al. 2009. A hybrid encapsulation method for organic electronics. *Appl. Phys. Lett.* 94: 163308–163310.

Kim S. K., Lee S. W., Hwang C.S. et al. 2006. Low temperature (<100°C) deposition of aluminum oxide thin films by ALD with O_3 as oxidant. *J. Electrochem. Soc.* 153: F69–F77.

Kim T. W., Yan M, Erlat A. G. et al. 2005. Transparent hybrid inorganic/organic barrier coatings for plastic organic light-emitting diode substrates. *J. Vac. Sci. Technol.* 23: 971–977.

Langereis E., Creatore M., Heil S. B. S. et al. 2006. Plasma-assisted atomic layer deposition of Al_2O_3 moisture permeation barriers on polymers. *Appl. Phys. Lett.* 89: 081915–081917.

Lee G. H., Yun J., Lee S. et al. 2010. Investigation of brittle failure in transparent conductive oxide and permeation barrier oxide multilayers on flexible polymers. *Thin Solid Films* 518: 3075.

Lee J. Y., Lee Y. J., Jang Y. S. et al. 2011. Improvement of room temperature processed high quality TCO thin film using the DC/RF superimposed magnetic field shielded sputtering (MFSS). Paper presented at *MRS Fall Meeting & Exhibit*, Hynes Convention Center, Boston, MA, November 28–December 2, http://www.mrs.org/f11-abstracts-z/Z1.

Leterrier Y. 2003. Durability of nanosized oxygen-barrier coatings on polymers. *Prog. Mater. Sci.* 48: 1–55.

Leterrier Y., Andersons J., Pitton Y. et al. 1997. Adhesion of silicon oxide layers on poly (ethylene terephthalate). II: Effect of coating thickness on adhesive and cohesive strength. *J. Polym. Sci. B Polym. Phys.* 35: 1463–1472.

Leterrier Y., Singh B., Bouchet J. et al. 2008. Diffusion-barrier coating for protection of moisture and oxygen sensitive devices. International Patent Application WO 2008/122292 filed April 4, 2007, published October 16, 2008.

Levy D. H., Freeman D., Nelson S. F. et al. 2008. Stable ZnO thin film transistors by fast open air atomic layer deposition. *Appl. Phys. Lett.* 92: 192102–192104.

Liao J.-Y., Liu P.-C., Hsiao C.-H. 2006. The characterization of the gas barrier comprising Parylene-C thin films for flexible OLED application. Paper presented at *IDW '06 The 13th International Display Workshop*, Otsu, Japan, December 6–8.

Lin S., Chu X., Rosenblum P. M. 2010. Ultra-barrier coatings enabled by inkjet printing. Paper presented at the *Spring 2010 Meeting of American Chemical Society-Division of Polymeric Materials: Science and Engineering*, San Francisco, California, USA, March 21–25. *PMSE Preprints of Volume* 102: 855.

Liu J.-M, Lee T.-M., Wen C. H. et al. 2011. High-performance organic–inorganic hybrid plastic substrate for flexible displays and electronics. *J. SID* 19: 63–69.

Low H. Y., Xu Y. 2005. Moisture barrier of Al_xO_y coating on poly(ethylene terephthalate), poly(ethylene naphthalate) and poly(carbonate) substrates. *Appl. Surf. Sci.* 250: 135.

MacDonald W. A., Looney M. K., MacKerron D. et al. 2007. Latest advances in substrates for flexible electronics. *J. SID* 15: 1075–1083.

Mandlik P., Gartside J., Han L. et al. 2008. A single-layer permeation barrier for organic light-emitting displays. *Appl. Phys. Lett.* 92: 103309–103311.

Mandlik P., Silvernail J., Mq Ruiquing. 2012. Permeation barrier for encapsulation of devices and substrates. US Patent Application 2012/0068162, filed September 10, 2010, published March 22, 2012.

Matsumura H., Ohdaira K. 2008. Recent situation of industrial implementation of Cat-CVD technology in Japan. *Thin Solid Films* 516: 537–540.

McCormick F. 2011. Barrier films and adhesives for display applications. Paper presented at *OLED Materials for Lighting and Displays, A New Industrial Chemistry and Engineering (NIChE) Workshop by the Council for Chemical Research*, Minneapolis, Minnesota, June 6–8, http://www.ccrhq.org/publications_docs/OLED_McCormick.pdf.

Meyer J., Schmidt H., Kowalsky W. T. et al. 2010. The origin of low water vapor transmission rates through Al_2O_3/ZrO_2 nanolaminate gas-diffusion barrier grown by atomic layer deposition. *Appl. Phys. Lett.* 96: 243308–243310.

Mikhael M. G., Yializis A. 2002. Vacuum deposition of cationic polymer systems. US Patent, Publication number 6,468,595, filed February 13, 2001, and issued October 22, 2002.

Moriconie T. J., Reese M. O., Dameron A. A. et al. 2009. Understanding moisture ingress and packaging requirements for photovoltaic modules. *Photovoltaics International*, 5th Edition, 30, 2012 *Solar Media Limited*. Accessed July 10, 2012, http://www.photovoltaicsinternational.com/technical_papers/list/category/thin_film/P30/.

Moro L., Chu X. 2010. Multilayer barrier stacks and methods of making multilayer barrier stacks, Publication number 7,648,925, filed July 12, 2007, and issued January 19, 2010.

Moro L., Krajewski T. A. 2010. Encapsulated devices and method of making. U.S. Patent 7,767,498, filed August 24, 2006, and issued August 3, 2010.

Moro L., Visser R.-J. 2007. Displays & lighting: Substrate and encapsulation challenges. Paper presented at *Printed Electronics USA, Masterclass 2: Displays & Lighting*, San Francisco, California, November 12–15.

Moro L., Boesch D., Maghsoodi S. et al. 2011. Barrier and encapsulants. Paper presented at the *FlexTech Alliance Hybrid Nanocomposites and Interfaces for Printed Electronics Workshop*, Atlanta, GA, September 13–13.

Moro L., Chu X., Hirayama H. et al. 2006. A mass manufacturing process for Barix encapsulation of OLED displays: A reduced number of dyads, higher throughput and 1.5 mm edge seal, in *IMID/IDMC '06 Digest*, Daegu, South Korea, August 22–25, 754–758.

Moro L., Chu X., Rosenblum M. P. et al. 2007. Method of making an encapsulated plasma sensitive device. Publication number WO/2007/139643, filed August 24, 2007, and published June 12, 2007.

Moro L., Krajewski T. A., Rutherford N. M. et al. 2004. Process and design of a multilayer thin film encapsulation of passive matrix OLED displays. In *Proceedings of SPIE-The International Society for Optical Engineering 5214(Organic Light-Emitting Materials and Devices VII)*, eds. Z. H. Kafafi and P. A. Lane, 83–93. SPIE-The International Society for Optical Engineering, Bellingham, WA.

Narasimhan M. 2005. Progress on barrier on plastic substrates. Paper presented at the *Fourth Annual Flexible Displays and Microelectronics Conference of the U.S. Display Consortium (USDC)*, Phoenix, Arizona, January 31–February 3.

Ogawa Y., Ohdaira K., Oyaidu T. et al. 2008. Protection of organic light emitting diodes over 50,000 h by Cat-CVD SiO_x/SiO_xN_y stacked films. *Thin Solid Films* 515: 611–614.

OLED display manufacturing equipment, Canon Tokki Corporation. 2012. Accessed July 8, 2012, http://www.canon-tokki.co.jp/eng/product/el/index.html.

OLED PLED Technology. 2012. *MBraun Incorporated*. Accessed July 8, 2012, http://www.mbraunusa.com/OLEDhome.htm.

Our Products. 2009. *OTBv-OLED Technologies B.V.* Accessed July 15, 2012, http://www.oled-ts.com.

OLED Association (OLED-A) web page. 2008. Samsung Mobile Display Reaches New Levels of Profitability and Introduces the YOUM Brand for Flexible Displays, posted April 10, 2012. Accessed July 01, 2012, http://www.oled-a.org/news_details.cfm?ID=760.

Chapter 3

Olsen L. C., Gross M. E., Graff G. L. et al. 2008. Approaches to encapsulation of flexible CIGS cells. In *Reliability of Photovoltaic Cells, Modules, Components, and Systems*, ed. N. G. Dhere, *Proceedings of SPIE*, Vol. 7048. San Jose, California, USA, doi:10.1117/12.796104.

Padiyath R., David M. M. 2009. Moisture barrier coating. US Patent Application, Publication number 2009/0169770, filed March 9, 2009, and issued July 2, 2009.

Padiyath R., Roehrig. 2009. Flexible high-temperature ultrabarrier. US Patent Publication number 7,486,019, filed November 14, 2005, and issued February 3, 2009.

Park J. S., Chae H., Chung H. K. et al. 2011. Thin film encapsulation for flexible AM-OLED: A review. *Semicond. Sci. Technol.* 26: 034001/1-8.

Ramadas S. K., Chua S. J. 2010. Nanoparticulate encapsulation barrier stack. U.S. Patent Application US 2010/0089636 filed December 18, 2009, and published April 15, 2010.

Ramos T. 2006. Progress toward early-to-market flexible OLEDs: Packaging of low-resolution plastic and high-resolution foil-based. Paper presented at *USDC Fifth Flexible Displays & Microelectronics Conference*, Phoenix, AZ, February 6–9.

Reese M. O., Dameron A. A., Kempe M. D. 2011. Quantitative calcium resistivity based method for accurate and scalable water vapor transmission rate measurement. *Rev. Sci. Instrum.* 82: 085101–085110. Available at: http://rsi.aip.org/resource/1/rsinak/v82/i8/p085101_s1.

Roehrig M. A., Nachtigal A. K., McCormick F. B. R. J. 2011. Gradient composition barrier. U.S. Patent Application 2011/0223434, filed May 11, 2011, and published September 15, 2011.

Roehrig M. A., Nachtigal A. K., Weigel M. D. 2012. Moisture resistance coating for barrier films. US Patent Application, Publication number 2012/0003484, filed July 2, 2010, and issued January 5, 2012.

Rosenblum M. P., Lin S., Chu X. 2009. UV-polymerized films enable ultra-barrier coatings for thin-film photovoltaics and flexible electronics. *Radtech Report* July/August, 21–26.

Rosink J. J. W. M., Lifka H., Rietjens G. H. et al. 2005. Ultra-thin encapsulation for large-area OLED displays. In *Society for Information Display (SID) 2005 International Symposium—Digest of Technical Papers*, Vol. XXXVI: 1272–1275. Boston, USA, May 22–27, 2005.

Rossnagel S. M., Cuomo J. J., Willion Dickson Westwood. 1990. *Handbook of Plasma Processing Technology*. New York: Noyes Publications/William Andrew Publishing.

Roth and Rau B.V. PiXDRO Systems Applications: Accurate multi-layer and multi-function printing. Press Release on June 12, 2008. Accessed December 08, 2012, http://www.roth-rau.com/pixdro.

Rutherford N. M. 2005a. Flexible displays—A low cost substrate/encapsulation packaging solution. Paper presented at *USDC Fourth Flexible Displays & Microelectronics Conference*, Phoenix, AZ, January 31–February 3.

Rutherford N. M. 2005b. Flexible substrates and packaging for organic displays and electronics. *Information Display (SID Magazine)* 21: 20–25. Available at: http://www.informationdisplay.org/article.cfm?year=2005&issue=11&file=art5.

Rutherford N. M., Moro L., Chu X. et al. 2004. Plastic barrier substrate and thin film encapsulation: Progress toward manufacturability. Paper presented at the *Third Annual Flexible Displays and Microelectronics Conference of the U.S. Display Consortium (USDC)*, Phoenix, Arizona, February 10–12.

Samsung Flexible OLED, YouTube, uploaded by TTTiagoTTT on February 19, 2010. Accessed July 15, 2012, http://www.youtube.com/watch?v=JGNs2gtWDdo&feature=related.

Schaepkens M., Kim T. W., Erlat A. G. et al. 2004. Ultrahigh barrier coating deposition on polycarbonate substrates. *J. Vac. Sci. Technol A* 22:1716–1722.

Shaw D. G., Langlois M. G. 1994. Use of vapor deposited acrylate coatings to improve the barrier properties of metallized film. In *Society of Vacuum Coaters 37th Annual Technical Conference Proceedings*, 240. Albuquerque, NM.

Spee D., van der Werf K., Rath J. et al. 2012. Excellent organic/inorganic transparent thin film moisture barrier entirely made by hot wire CVD at 100C. *Phys. Status Solidi RRL* 6: 151–153, doi:10.102/pssr.20120635.

SMFD-ALD™: Overcoming Obstacles to ALD Productivity Sundew Technologies Inc., Broomfield, CO, USA. 2002. Accessed July 14, 2012, http://www.sundewtech.com/smfdald.htm.

Söderlund M. 2012a. Atom by atom. *OPE J.* 1: 26–27. Available at: www.ope-journal.com.

Söderlund M. 2012b. Roll-to-roll atomic layer deposition. Paper presented at *Functional Materials Summer Festival 2012 Annual Seminar of the Tekes Functional Materials Programme*, Helsinki, Finladia Hall, May 29–30. Accessed July 15, 2012, http://tapahtumat.tekes.fi/uploads/44f6862/soderlund_jakoon-3271.pdf.

Stevens M., Tuomela S., Mayer D. Water vapor permeation testing of ultra-barriers: Limitations of current methods and advancements resulting in increased sensitivity. MOCON Corporation. Accessed July 14, 2012, http://www.mocon.com/permeation.php.

Sugimoto A., Ochi H., Fujimura S. et al. 2004. Flexible OLED displays using plastic substrates. *IEEE J. Sel. Top. Quantum Electron.* 10 (1): 107–114.

Suntola T. 1992. Atomic layer epitaxy. *Thin Solid Films* 216: 84–89.

Suntola T. 1994. Atomic layer epitaxy. in *Handbook of Crystal Growth*, ed. D. T. J. Hurle, Vol. 3, Part B, Chapter 14: Growth mechanisms and dynamics. Amsterdam: Elsevier.

Super High Gas Barrier Film/Sheet X-Barrier. Accessed July 15, 2012, *Mitsubishi Plastics, Inc.*, http://www.x-barrier.com/en/product/xbarrier/index.html.

Szczyrbowski J., Brauer G., Ruske M. et al. 1997. Properties of TiO_2—Layers prepared by medium frequency and DC reactive sputtering. *Society of Vacuum Coaters 40th Annual Technical Conference Proceedings*, 237–242. New Orleans, LA.

The Expanding Thermal Plasma (ETP) technique. Accessed July 14, 2012, Eindhoven University of Technology (TU/e), Department of Applied Physics Plasma, Physics and Radiation Technology Research Cluster. Available at: http://web.phys.tue.nl/nl/de_faculteit/capaciteitsgroepen/plasmafysica_en_stralingstechnologie/plasma_amp_materials_processing/scientific_projects/the_expanding_thermal_plasma/.

Tokki, GE to Codevelop EL Display Production Equipment January 26, 2007, 2007 *Jiji Press English News Service*. Accessed July 15, 2012, http://www.redorbit.com/news/science/816068/tokki_ge_to_codevelop_el_display_production_equipment/.

Tropsha Y. G., Harvey N. G. 1997. Activated rate theory treatment of oxygen and water transport through silicon oxide/poly(ethylene terephthalate) composite barrier structures. *J. Phys. Chem. B* 101: 2259.

van Assche F., Rooms H., Young E. et al. 2008. Thin-film barrier on film for organic LED lamps. Paper presented at *AIMCAL Fall Technical Conference*, Myrtle Beach, South Carolina, October 19–22, 2008.

van Assche F. J. H., Vangheluwe R. T., Maes J. W. C. et al. 2004. A thin film encapsulation stack for PLED and OLED displays. In *Society for Information Display (SID) 2004 International Symposium—Digest of Technical Papers*, Vol. XXXV: 695–697. Seattle, Washington, USA, May 23–28.

VanDewoestine J. 2012. Flexible glass: An emerging materials and process revolution. Presented at *2012 FlexTech*, Tucson, February 6–9.

Visser R. J., Moro L. 2010a. Encapsulated RGB OLEDs having enhanced optical output. U.S. Patent Application 2010/0156277, filed December 22, 2008, and published June 24, 2010.

Visser R. J., Moro L. 2010b. Encapsulated white OLEDs having enhanced optical output. U.S. Patent Application 2010/0159792, filed December 22, 2008, and published June 24, 2010.

Vitex Systems, Inc. Vitex systems achieves lifetime record on flexible copper indium gallium selenide solar cells. Press Release on June 12, 2008. Accessed July 15, 2012, http://www.prnewswire.com/news-releases/vitex-systems-achieves-lifetime-record-on-flexible-copper-indium-gallium-selenide-solar-cells-57451137.html.

Vitex Systems, Inc. Vitex systems signs licensing agreement for its flexible glass technology with leading optical films manufacturer. Press Release on September 13, 2009. Accessed July 15, 2012, http://www.prnewswire.com/news-releases/vitex-systems-signs-licensing-agreement-for-its-flexible-glass-technology-with-leading-optical-films-manufacturer-56767432.html.

Vitex Systems, Inc. Vitex systems expands license agreement with advanced neotech systems. Press release on January 27, 2009. Accessed July 15, 2012, https://www.edn.com/electronics-products/electronic-product-releases/other/4366175/Vitex-Systems-Expands-License-Agreement-with-Advanced-Neotech-Systems.

Vogt B. D., Lee H.-J, Prabhu V. M. et al. 2005. X-ray and neutron reflectivity measurements of moisture transport through model multilayered barrier films for OLED applications. *J. Appl. Phys.* 97: 114509.

Yamada Y. Iseki K., Okuyama T. et al. 1995. The properties of a new transparent and colorless barrier film. *Society of Vacuum Coaters 38th Annual Technical Conference Proceedings*, 28–31.

Yoo S. J., Kim D. C., Joung M. et al. 2008. Hyperthermal neutral beam sources for material processing. *Review Scientific Instruments* 79: 02C301–102C301-5. Available at: http://dx.doi.org/10.1063/1.2801343.

Zambov L., Shamamiam V., Weider K. et al. 2011. Advanced roll-to-roll plasma-enhanced CVD silicon carbide barrier technology for protection from detrimental gases. *Chem. Vap. Deposition* 17: 253–260, doi:10.1002/cvde.201106923.

Zambov L., Weidner K., Shamamian V. et al. 2006. Advanced chemical vapor deposition silicon carbide barrier technology for ultralow permeability applications. *J. Vac. Sci. Technol. A* 24(5): 1706–1713.

Zheng J., Tucker M. B., Teng L. 2011. Failure mechanics of organic-inorganic multilayer permeation barrier in flexible electronics. *Composites Sci. Technol.* 71: 365–372.

Chapter 3

4. Transparent Electrodes for OLED Devices

Roman Korotkov and Gary S. Silverman

4.1 Introduction to Transparent Conducting Materials

Transparent conducting materials (TCM) are a broad class of materials that can be used in a variety of optoelectronic devices such as low emissivity windows, touch screens, thin film transistors, thin film photovoltaics, and organic light-emitting diodes (OLEDs) to simultaneously provide transparency and conductivity. For current commercial OLEDs, transparent conducting oxides (TCO) have been predominately used with indium tin oxide (ITO) or indium zinc oxide (IZO) being the TCO of choice. This chapter will discuss conducting materials at a high level, since there are recent books that delve into great depth in many areas inclusive of characterization techniques, alternative TCOs, structure, and theory (Ginley 2010, Luque and Hegedus 2011). Subsequently, this chapter will focus on specific TCO manufacturing and application issues that OLEDs face

Chapter 4

with an emphasis on OLED lighting applications. The last section of this chapter will discuss potentially new TCMs that may eventually compete with the current TCOs in OLED devices.

Conductivity is defined as $\sigma = 1/\rho = \mu n e$, where ρ is resistivity, μ is charge mobility, n is carrier concentration, and e is elementary charge. Resistivity is also defined as sheet resistance (SR, expressed in Ω/sq) multiplied by thickness (expressed in cm). Therefore, a scientist simply needs to use readily available equipment to measure film thickness and sheet resistance to determine the resistivity (or conductivity as defined above) of a film. While it is common to compare sheet resistance at a defined thickness for conductive substrates, it is more efficient and appropriate to use resistivity (or conductivity) as the comparison tool. Using resistivity will eliminate confusion when comparing different materials by mathematically taking the two previously mentioned parameters into account. More elaborate testing using Hall equipment allows one to determine mobility (μ) and carrier concentration (n). For TCMs used in OLED devices the conductivity needs to be at the higher end of semiconducting or metallic conducting properties ($\sigma > 2000\ \Omega^{-1}\ cm^{-1}$), while maintaining high transmission across the visible spectrum (>85%).

4.1.1 Differences in TCM Requirements for Display and Lighting Applications

The two emerging applications of OLED devices are displays and lighting with OLED displays enjoying the faster growth rate due to smart phone and tablet applications. Clearly, these applications have significantly different inherent value. OLED displays improve the quality of entertainment, communication, and instant information access such that consumers are willing to pay a significant premium for them, relative to general lighting, resulting in a high performance requirement versus cost constraints. However, for OLED lighting the economic pressure is more severe and the reliance on expensive materials such as ITO or IZO as anode material is one of the economic product development barriers to the successful commercialization of OLED devices for the general illumination market (Gasman and Nolan 2010, Markowitz and Gasman 2010). In fact, development of TCOs to replace ITO/IZO with more economical electrodes was voted as one of the top five needs for OLEDs at the United States Department of Energy Solid State Lighting (DOE SSL, 2010) 2005 R&D workshop and again at the 2007–2011

Table 4.1   TCM Parameter Guidelines for OLED Lighting

Resistivity ($10^{-4}\ \Omega$ cm)	Transmission (% Visible Light Spectrum)	Surface Roughness (nm) RMS Z_{max}		Patterning (μ)	Homogeneity	Stability at 60rh%: 80°C
<4	>85%	<4	<30	100	±3% of measurement	<3% variation in T% and ρ at 504 h

workshops (United States Department of Energy 2005, 2007, 2008, 2009, 2010, 2011). Table 4.1 lists the desired TCM characteristics for OLED lighting.

4.1.2 Critical Properties

Critical properties and issues are difficult to generalize over the entire range of potential OLED products since there are multiple applications in both lighting and displays that have different requirements. For example, flexibility of the TCM is important for flexible OLED devices, but not for rigid OLED devices. Flexible OLED displays are projected to have market penetration within the next 5 years due to their inherent economic value and since they can afford the current state-of-the-art barrier coatings. However, flexibility adds to the material requirements for the TCM, which is currently a development area for organic conducting materials and new inorganic stack structures, such as TCO/metal/TCO. A similar approach has been the focus of several lighting manufacturers for flexible OLED lighting that may be important in the future; however, this approach has economic barriers that require a significant discovery for general illumination. Thus, we will focus on rigid glass substrates for OLED lighting.

Typical critical parameters for the TCMs are the resistivity (typically expressed in ρ) and visible transmission, shown in Table 4.1. Clearly, 100% transmission is ideal, but substrate absorption and stack reflection have led to the current specification of >85%. Having a smooth surface is important to prevent shorts in the OLED device. While manufacturers want surface roughness as low as possible there are other layers (hole injection layer [HIL], hole transport layer [HTL], or charge injection) that can planarize the TCM. Note, the type of roughness is also important, that is, the peak to valley (Z_{max}) measurement is more important from a shorting perspective than the root mean square (RMS) measurement. Patterning of the TCM is important and is certainly less complicated in lighting applications versus displays where each pixel ($\sim 30 \times 30~\mu m^2$) needs to be addressed. Patterning can be done by photolithography, but due to the simplicity of the pattern for lighting devices many groups are moving to laser ablation that results in decreased process time and eliminates the steps that are required in photolithography. The homogeneity is critical for obtaining uniform emission color from the device. Therefore, efforts are focused on TCO film thickness uniformity since it has a high impact on color variability in the OLED device.

4.2 Transparent Conducting Oxides

In this section, we will discuss some of the most commonly used TCO types, their processing, optical and electronic performance, dispersion parameters that enable curve fitting for surface mapping, and crude economics based on the information available.

4.2.1 Transparent Conducting Oxides General Survey

TCOs are a class of metal oxide semiconductors that, when doped by a heteroatom, are conductive while still retaining high transparency in the visible region. This combination of seemingly contradictory properties makes TCO films extremely useful in a variety of applications that use transparent electrodes (displays, OLEDs, touch screens,

Chapter 4

static dissipation, thermal insulation, low emissivity glass, and oven/refrigerator doors). The properties reported for selected TCO materials are presented in Table 4.2. The materials in the selected list of TCOs do not have significant health and environmental risks, meet the performance needs of the application, and potentially can be made in an economical commercial process.

Table 4.2   Electrical and Optical Properties of Selected TCOs

TCO	Deposition Method[a]	E_g (eV)	Resistivity ($10^{-4}\,\Omega$ cm)	Mobility (cm²/Vs)	n (10^{20} cm⁻³)	T_{vis} (%)	References
SnO₂:F	CVD	3.8	6	20	5	85	Russo and Lindner (1986)
	PLD	4.3	5	15	8	87	Kim et al. (2008)
SnO₂:Sb	CVD	3.8	10	20	3	70	Russo and Lindner (1986)
	SP	3.8	8.4	12.3	6	80–83	Lee and Park (2006)
Sn:In₂O₃ (ITO)	Sputter	2.8	1.6	60	6	85	Tominaga et al. (1996)
	Sputter	3.2	1.5–2.5			85	Commercially available
Mo:In₂O₃	TRE	3.2	1.7	80–130	2–3	75	Meng et al. (2001)
	Sputter	4.3	5.9	20.2	5.2	90	Miao et al. (2006)
F: In₂O₃	CVD	3.5	2.9			>85	Maruyama and Fukui (1990)
Nb:TiO₂	PLD	3.0	2–4		0.1–20	80	Furubayashi et al. (2005)
B:ZnO	CVD	3.2	0.4–4	2.5–35	3.5–5.5	90	Hu and Gordon (1992)
Al:ZnO	FCVA	3.37	9.6	7.2	9	80–85	Lee et al. (2004)
	PLD		0.9	47	15	83	Agura et al. (2003)
	APCVD	3.2	1.4	34	13	91	Stricker et al. (2010)
Ga:ZnO	Sputter	3.2	2.6	18	13	80–90	Fortunato et al. (2004)
	APCVD	3.2	1.5	29	14	89–92	Stricker et al. (2012)
In:ZnO	Sputter	3.2	2.9	31	7	95[b]	Minami et al. (1996)

[a] CVD = chemical vapor deposition, FCVA = filtered cathodic vacuum arc, PLD = pulsed laser deposition, TRE = thermal reactive evaporation, SP = spray pyrolysis, Arkema results are on 15 × 15 cm² substrates.
[b] Corrected for substrate.

4.2.2 Indium Tin Oxide/Indium Zinc Oxide

ITO and more recently developed amorphous IZO (Minami et al. 1996) are the current TCOs of choice for OLED applications. The optoelectronic properties of ITO/IZO are "good enough" and meet the minimum technical requirements for OLED devices. TCO and IZO are made by physical vapor deposition (PVD) and then annealed at 300°C–400°C to obtain the desired optoelectronic properties. Amorphous IZO has lower curing temperatures and is a potential material for flexible substrates. While IZO can be made by PVD, some manufacturers are developing sol–gel processing with the promise of lower manufacturing cost (Zhi-hua et al. 2008).

The barrier for hole injection from the anode into the organic stack is defined by the work functions and electron affinities of the TCM and organics, as well as a formation of the dipole charges at the inorganic–organic interface (Hwang et al. 2009, Ishii et al. 1999). Highest occupied molecular orbital (HOMO) and lowest unoccupied molecular orbital (LUMO) can be schematically represented using the simplified energy diagram shown in Figure 4.1. In this diagram, HOMOs and LUMOs for a series of charge injection and transport moieties are offset with respect to the vacuum level. Hole injection barriers can be estimated from the differences between conduction (n-type) or valence band (VB) (p-type) offsets of the TCM, shown with thin bars, and the LUMO of the

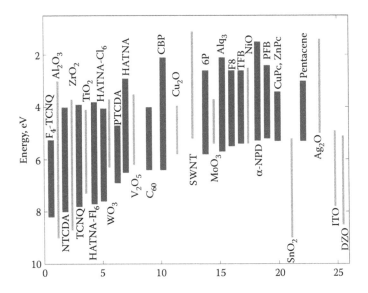

FIGURE 4.1 Schematic energy band diagram for different OLED materials, including HOMO and LUMO, and VB and conduction band of some inorganic materials measured with respect to the vacuum level. 2,3,5,6-Tetrafluoro-7,7,8,8-tetracyanoquinodimethane (F$_4$-TCNQ); naphthalenetetracarboxylic dianhydride (NTCDA); 7,7,8,8-tetracyanoquinodimethane perylenetetracarboxylic dianhydride (TCNQ); 2,3,8,9,14,15 hexafluoro diquinoxalino[2,3-a,2′,3′-c]phenazine (HATNA-Fl$_6$); 2,3,8,9,14,15 hexachlorodiquinoxalino[2,3-a,2′,3′-c]phenazine (HATNA-Cl$_6$); (PTCDA); diquinoxalino[2,3-a,2′,3′-c] phenazine (HATNA); dicabazolyl-biphenyl (CBP); para-sexiphenyl (6P); tris-(8-hydoxyquionoline) aluminum (Alq$_3$); poly(9,9′-dioctylfluorene) (F8); poly(9′9′-dioctylfluorene-co-bis-N,N′-(4-butylphenyl) diphenylamine) (TFB); α-N,N′-diphenyl-N,N′-bis(1-napthyl)-1,10-biphenyl-4,4″-diamine (α-NPD); (poly-9,9′-dioctylfluorene-co-bisN,N′-phenyl-1,4-phenylenediamie, PFB); per-fluorinated copper phthalocyanine (F$_{16}$-CuPc); copper phthalocyanine (CuPc); zinc phthalocyanine (ZnPc).

Chapter 4

HIL. The work function for ITO/IZO is relatively shallow (reported in the range of 4.6–5.1 eV) (Ishii et al. 1999) and requires modification to work for blue- and green-emitting triplet (phosphorescent) OLED systems, that is, by the use of a deeper work function HIL. For example, most of the blue emitters will have very low injection barriers for the TCM with the work function between 7 and 8 eV (Ishii et al. 1999). Unfortunately, most of the conductive oxides have work function between 4 and 5.2 eV. Doped zinc oxide (DZO) work function was found to be 5.1 eV by x-ray photoelectron spectroscopy (XPS). Thus, utilization of a thin (<10 nm) semi-insulating inorganic layer is required. The work functions of some of the potential inorganic injection layers vary from 5.6 through 5.7 eV for MoO_3 (Kanai et al. 2009), 6.2 eV for WO_3 (Li et al. 2005), to 9 eV for Al_2O_3 (Heo et al. 2009). It needs to be noted that utilization of the semi-insulators may inherently increase the serial resistance of the OLED devices. Therefore, the use of very thin HIL or p-type TCM, such as NiO or Cu_2O on top of the semi-insulating material may be required (Irwin et al. 2008).

ITO/IZO films are unstable in reducing atmosphere and degrade under high electric fields, leading to the formation of active indium and oxygen species that can migrate into organic dielectric layers resulting in clouding, degradation, and ultimately loss in device performance (Chao et al. 1996, Lee et al. 1999). Due to this instability these TCOs are typically activated just before patterning and stack deposition primarily by ultraviolet (UV) ozone or plasma treatment.

Indium is a rare element, accounting for only about 0.1 ppm of the earth's crust. Due to its scarcity and rapid growth of demand in nonlighting markets, the price of indium has increased from \$97/kg in 2002 to a September 2011 price of \$775/kg (Jansseune 2005, Kitco Base Metals, United States Geological Survey 2005, 2007). In fact, Nanomarkets projected that indium is heading to over \$1000/kg and higher pricing has been suggested in the Chinese press, to as high as \$3000/kg. This potential instability is in part due to indium falling under the China Export policy which when enacted in 2011 resulted in an abrupt 68% increase in price in February 2011 (\$550/kg to \$925/kg), followed by a leveling price near \$700/kg. Calculations presented at the 2010 DOE SSL R&D manufacturing meeting indicates that OLED lighting could consume up to 50% of the indium market and would require significant expansion of mining and recovery capital. Another concern is that indium is a heavy metal and several indium compounds are currently under Environmental Protection Agency (EPA) consent orders due to toxicity concerns (OSHA Chemical Sampling Information).

4.2.3   Doped ZnO

Zinc oxide-based TCOs have garnered significant interest as an alternative to ITO (Agura et al. 2003, Hu and Gordon 1992, Lee et al. 2004, Ozgar et al. 2005). When ZnO films are doped with Group 3 elements such as boron, aluminum, or gallium, the resulting TCO can have both electrical conductivity and high visible transparency comparable to or exceeding that of ITO. Interest in DZO has been largely cost driven with zinc approximately three orders of magnitude more abundant than indium, which results in significantly lower raw material costs (less than \$3/kg) (United States [US] Geological Survey 2010, OSHA Health Guidelines). Additionally, the environmental, health, and safety effects of zinc and aluminum (the most common dopant for ZnO) are well understood due to their extensive industrial use (OSHA Health Guidelines). Therefore, the source materials (chemical precursors and sputtering targets) necessary for producing commercial scale ZnO coatings are economically and environmentally favorable relative to the corresponding indium materials. DZO is air stable at ambient temperature and does not require the UV ozone activation that is necessary for ITO anodes. Doped ZnO can be readily etched using dilute acid solutions in a photolithography process in addition to laser ablation. Based on economics and performance, DZO is already ubiquitous in the photovoltaic industry and is an ideal candidate to replace ITO/IZO as an OLED anode for lighting applications. Thus, this specific TCO is included in this section versus potential future TCMs for OLED devices that will be discussed in the last section.

4.3   Methods of Deposition

For the production of commercial quantities of large-area coated glass, the primary deposition methods employed are magnetron sputtering and atmospheric pressure chemical vapor deposition (APCVD). Both methods may be used to deposit similar materials based on chemical stoichiometry; however, there are distinct differences in certain material properties such as morphology, and in process economics of the respective deposition methods. In the following sections we will discuss the pros and cons of these two methods of TCO deposition.

4.3.1   Physical Vapor Deposition

Though many variations are employed, sputtering is a PVD method in which high-energy ions bombard a solid target(s) composed of the desired coating material causing the target to eject particles into the gas phase (Ohring 2002). The ejected particles then contact a substrate to form a film. A wide variety of materials can be deposited using sputtering, with the only material limitation being the availability of high purity targets. Reactive sputtering (common for ITO and DZO) often leads to sub-oxide stoichiometry that necessitates UV–ozone treatment of the film to generate the desired optoelectronic properties. Because sputtering is a physical deposition method, individual particles are deposited with essentially no chemical bonding to either the substrate or to other grains within the coating. Thus, resulting films have low shear strength and physical adhesion relative to APCVD, which can lead to pathways through which humidity and/or cleaning chemicals can adversely affect the coatings. In OLEDs, this may lead to water reactivity with the emissive layer and poor physical adhesion of the TCO with the substrate. These shortfalls are well-known in both the research and the commercial literature of manufacturers selling PVD systems (Siegel and Culp 2002). In addition, because sputtering relies on well-controlled plasma, the process is conducted under relatively high vacuum, which adds significant operational and working capital expenses. There are complications associated with the preparation of DZO by reactive sputtering. If the oxide partial pressure is too high, the films will be insulating, but if the oxide pressure is too low, the films will exhibit considerable absorption in the visible region. Unfortunately, the transition between the regimes is particularly abrupt, making it difficult to control the process on large substrates, resulting in nonuniformity in film thickness, sheet resistance, and optical transmission as well

Chapter 4

as poor reproducibility. To overcome these limitations, even more complicated and expensive approaches generally need to be undertaken (Hao et al. 2006, Jeong et al. 2006, Hovsepian 2000, Silverman et al. 2009).

4.3.2 Atmospheric Pressure Chemical Vapor Deposition

APCVD is a process in which gaseous precursor molecules are introduced at atmospheric pressure to a substrate where they undergo a chemical reaction to produce a film with chemical conversion yields of 10%–40% (Ohring 2002). Typically, the substrate is heated to a temperature sufficient to initiate the chemical reaction. In the case of large-area online coating, film deposition can be performed in the bath of a float glass line (600°C–700°C), in the physical gap between the bath and the lehr (~550°C), or in an annealing lehr (100°C–500°C). Most primary float glass manufacturers coat in-bath or in the lehr gap. For in-bath coating the glass is above the glass transition temperature, T_g, (~550°C, dependent on composition) with one side floating on a molten tin bath (>600°C) and the other side experiencing a cooling effect from the impinging coating precursor vapor stream (160°C–225°C) and the coater (heat sink at 160°C–225°C) 3–8 mm above the glass surface. However, because the temperature is above T_g, the glass is relaxed and these cooling effects impart no stress. Lehr coating is similar, but the coating is performed at a temperature closer to the T_g of glass. Coating well below the T_g (375°C–425°C) can result in glass cupping/curling or sagging if the ΔT between surfaces is large, especially with thin glass. All three potential online APCVD process-coating zones utilize the "free" thermal energy provided by the hot glass to initiate the chemical vapor deposition (CVD) reaction leading to lower operational costs relative to PVD techniques. However, two critical deposition parameters, deposition temperature and rate, are constrained by the float glass process. Thus, depending upon the float bath design and thickness of the glass being produced, the residence time of the glass underneath the coater (and hence the time available for film deposition) is in the order of 2–8 s. With a TCO target thickness in the range of 100–150 nm, the deposition rate for a single slot, online process may need to be as high as 70 nm/s in order to match the flat-glass production line speed. Multiple slots or a larger deposition zone can be employed to decrease the required deposition rate.

A commercial off-line process requires designation of a manufacturing site, design, fabrication, and installation of off-line coating equipment. In spite of this need for capital investment, there are many benefits to producing OLED anodes in an off-line process. The effects of glass yield online versus off-line, TCO dopant choice, and chemical efficiency of the APCVD process has significant impact on the processing cost. While off-line chemical efficiency is higher due to glass yield (essentially quantitative) the volume of chemical use and throughput makes online processing significantly more economically favorable. Note that the variable cost for the DZO components is dwarfed by the glass variable cost so the overall impact is less than 20% of total variable cost.

4.4 Factors Affecting Choice of TCOs (ITO, IZO, and DZO)

For flat panel displays the optoelectrical performance of the TCO material is the key factor; however for OLED lighting these performance properties must be couched with economic parameters too. Section 4.4.1 discusses optical performance, and Section 4.4.3 discusses dispersion parameters in order to understand the transmission and reflection properties. Discussion of multilayered films is provided in Section 4.4.2 with economics discussed in Section 4.4.4.

4.4.1   Optical Performance of ITO and DZO

Optical transmission of conductive oxides consists of a broad transparency window situated between a sharp rise in absorption due to the material band gap in UV and increase in reflection due to the interaction of light with free electrons (Figure 4.2). The band gap for the unintentionally doped ITO was recently measured at 2.67 eV (molecular beam epitaxy) and 2.9 eV (r.f.) deposited films (Bourlange et al. 2008, Walsh et al. 2008a). The ITO band diagram is characterized by the upper VB states having the wrong symmetry for the allowed optical transitions, resulting in the first allowed optical transitions taking place 0.8 eV below VB top (Ginley 2010). Slightly wider band gaps of 3.4 eV are usually reported for undoped ZnO (Ginley 2010). Some variation in the band gap for doped ZnO is due mostly to Burstein–Moss shift (Burns 1990) to the blue and many-body interaction effects to the red part of the energy spectrum (band-gap renormalization) (Childs 1989, Walsh et al. 2008b). For the samples considered in this discussion (doping levels > 1.3×10^{21} cm^{-3}) the absorption edge of the DZO is shifted by 0.6–0.7 eV to the higher energies relative to undoped ZnO, opening up the transmission window toward UV. In the infrared (IR), optical properties of the DZO and ITO samples are defined by a strong reflection >70% at 2500 nm that reduces overall transmittance starting at around 1200 nm. The plasma wavelength depends strongly on doping levels and lies between 1.0 and 1.4 μm for the studied DZO and ITO samples. The IR transparency window may be further extended under optimized doping conditions. The transmittance of ITO, shown in Figure 4.2, is somewhat lower in the visible than that of the doped ZnO. This evidence was confirmed by previous experimental studies (Berry et al. 2008). In part, this can be ascribed to a smaller band gap of ITO relative to ZnO, resulting in relaxation of the

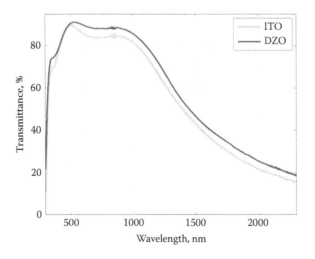

FIGURE 4.2   Optical transmittance of ITO and DZO layers of similar thickness.

Chapter 4

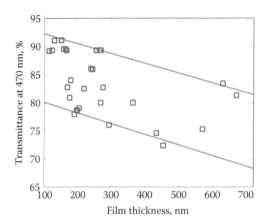

FIGURE 4.3   Effect of DZO thickness on transmittance (at 470 nm).

selection rules for the forbidden transition in real-world thin films due to inherent film stress, and lattice deformation due to heavy doping.

The selection of the optimum thickness of the conductive anode depends on the required sheet resistance that is directly connected to the metal grid spacing of the OLED. The presence of the Fabry–Perot interference fringes applies its own limitation to the transparency window of the conductive electrode. Transmission at a maximum wavelength of 470 nm is a modulation function of the film thickness (Figure 4.3). For example, DZO films with thickness of 120–150 nm have SR between 16 and 20 Ω/sq and transmittance including the substrate of 90%–92.8%. The next maximum in the transmission curve comes at 89%–90% (470 nm) for the 250–260 nm thick films, having sheet resistance of 8–10 Ω/sq. In contrast, very low transmittance minima of 79% and 76% are observed for the films having thickness 195 and 325 nm, respectively. Therefore, depending on the required wavelength range, the structure can be chosen to satisfy both transmission and conductivity targets. This approach allows maximizing transmission within a narrow wavelength range. If improvement in transmission is required in a wider range, a multilayered structure needs to be implemented.

4.4.2   Optical Performance of Multilayer Structures

Optical performance of the TCO layers can be further improved by the incorporation of the refractive index matching undercoats. This approach will not only improve transmission in certain region of the spectrum, but can also reduce the observed reflected color of the TCM coatings. The optical transmittance of 145 nm thick DZO deposited on glass (Figure 4.4) was compared with similar thickness DZO deposited on Al_2O_3 undercoats (Korotkov et al. 2011, Padmaperuma and Gaspar 2011). One may observe considerable flattening of the reflectance and transmittance curves in the visible range (390–690 nm). Total transmittance above 90% was measured between 440 and 700 nm. Note that the blue reflected color of the coating was also eliminated using this technique. Further improvements in visible transmission may require use of the antireflection coatings. Since these coatings are deposited on top of the DZO, one needs to select

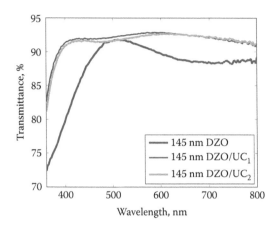

FIGURE 4.4 Optical transmittance of DZO (inclusive of glass substrate) with and without undercoats.

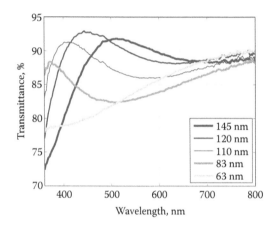

FIGURE 4.5 Optical transmittance of the DZO (inclusive of glass substrate) with and without undercoats.

these layers carefully limiting the effect of these layers on the overall electrical properties of the stack. One potential approach is to reduce the TCM layer thickness. When the thickness of the DZO layer is reduced below 100 nm, the peak of the reflected color is shifted in UV (Figure 4.5). Despite considerable reduction in color, the reflectivity of these coatings is still high. Visible transmission of 60–80 nm coatings is between 80% and 90%. Further reduction in thickness below 60 nm improves visible transmission somewhat. However, this improvement is accompanied by considerable increase in sheet resistance and resistivity (Table 4.3).

Table 4.3 Properties of Thin DZO Samples

DZO Thickness (nm)	64	75	80	110	120	130	144
Sheet resistance (Ω/sq)	58.3	43.4	38	23.8	20.2	18.5	16.8
Resistivity $\times 10^{-4}$ (Ω cm)	3.73	3.2	3.04	2.62	2.42	2.4	2.4

Chapter 4

4.4.3   Dispersion Parameters of the ITO and DZO

Development of nondestructive characterization techniques for determining electrical properties of doped ZnO can greatly enhance optimization studies. Commonly the electrical properties of thin film materials are evaluated using Hall effect measurements. This technique consists of several steps, including wafer dicing and deposition of Ohmic contacts, which render the substrate unusable in the subsequent manufacturing steps during OLED production. One of the advantages of spectroscopic ellipsometry lies in its nondestructive nature. The disadvantage is attributed to the tedious nature of modeling optical properties of the material. Several sample preparation steps can make the model development easier. For example, the surface roughness of the films may affect the calculations and is often modeled using the Bruggeman approximation. In addition, modeling of the considerable back reflection, present in transparent substrates, is complicated and is usually treated as an additional parameter.

In recent work by Fujiwara and Kondo (2005), thin films were polished to the thickness of 170–180 nm as determined by surface profilometry. The polishing step eliminates the need for the Bruggeman correction during simulation and reduces the effect of variations in the oscillator parameters as a function of film depth. Calculation of the dispersion parameters to derive the electrical properties for the studied coatings has complications from precise definition of the effective mass, m^*, and the high-frequency dielectric constant, ε_∞. The effective mass in doped ZnO shows a large variation as a function of electron concentration (Fujiwara and Kondo 2005, Singh et al. 2004, Young et al. 2000); this was corroborated with a large sample set of different electron concentration values (0.7–1.54×10^{21} cm^{-3}) using spectroscopic ellipsometry, Hall effect measurements, and optical measurements, shown in Figure 4.6, that deviated from the expected $m^* \sim n^{1/3}$ dependence (shown with a solid line) for both DZO and ITO (Fujiwara and Kondo 2005).

In this work, two independent ellipsometry parameters (Ψ and Δ) were measured for each point of the 11×11 point matrix using SE at a scattering angle of $70°$. The dispersion of ZnO was modeled using a double oscillator, Tauc–Lorentz (TL) model (von

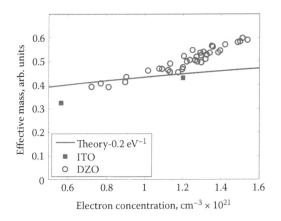

FIGURE 4.6   Variation of the effective mass with doping level in ITO and DZO.

Blanckenhagen et al. 2002) for the UV–visible and Drude (D) model for the IR region resulting in the dielectric function being expressed by (Korotkov et al. 2012)

$$\varepsilon(E) = \varepsilon_{TL}(E) + \varepsilon_D(E) \tag{4.1}$$

The imaginary part of the dielectric function in the TL model is given by

$$\varepsilon_i(E) = \frac{AE_0 C(E - E_G)^2}{[(E^2 - E_0)^2 + C^2 E^2]E}, \quad E > E_G$$

$$\varepsilon_i(E) = 0, \quad E \le E_G \tag{4.2}$$

where A is an amplitude, C is a broadening parameter, E_0 is a peak transition energy, and E_G is a Tauc optical gap. An additional parameter, E_{inf} represents the contribution of the optical transitions at higher energies and is found in the Kramers–Kronig relationship for the real part of the TC dielectric constant. The near IR free-carrier absorption is commonly modeled using the Drude formula, involving two main parameters, Γ_D (oscillator damping term) and $A_D = \varepsilon_\infty E_p^2$ (a plasma energy-dependent term), along with the polarization, P. The sum of P and E_{inf} is equal to the high-frequency dielectric function ε (see Equation 4.1). In these calculations, $E_{inf} = 0.0$ was fixed, and the dielectric constant was derived directly as P from function minimization calculations. The values of the dielectric function, in which the real part of ε_1 is plotted as a function of $1/[E^2 + \Gamma_D^2]$, were similar to the values calculated above (Korotkov et al. 2011). High-frequency dielectric function values reported in the literature vary from 3 through 4 (Korotkov et al. 2008). The summation of the two oscillators may introduce some difficulties in correct estimate of A_D and E_p. The nonzero component of the TL ε_1 around E_p can lead to modification of the A_D and E_p values by increasing or reducing the real part of the Drude dielectric function. Thus, great care must be taken to arrive at the proper value of the dielectric constant at infinity as a summation of both oscillators (Korotkov et al. 2012).

Equations 4.1 and 4.2 were used to derive the dispersion dependencies of DZO and ITO (Figures 4.7 and 4.8). The refractive index and extinction coefficient were affected by the electron concentrations in these samples as measured by the Hall effect. With increasing electron concentration, the plasma wavelength shifts toward the visible bringing the refractive index curve down and extinction coefficient curve up. The maximum changes are normally observed with doping in the IR. However, the refractive index of DZO at 500 nm decreased from 2 in the undoped films to 1.8 in heavily doped films. This reduction leads to better refractive index matching of the anode to the organic stack as well as the glass substrate. The increase of the extinction coefficient with doping has a major affect in the IR and a relatively small effect in the visible portion of the spectra.

Automated four-point probe and spectroscopic ellipsometry mapping tools can be used to characterize the spatial distribution of small variations in the electrical and optical properties in transparent conductive films. For instance, in the work of Korotkov et al. (2012), measurements of doped ZnO films measured the thickness variations across a 6″ × 6″ wafer. In these measurements, the variation does not exceed σ/χ ratio of 2%, for example, the average thickness of the selected coating is 220 ± 4 nm as measured

Chapter 4

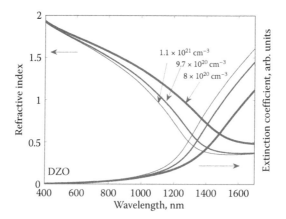

FIGURE 4.7 Dispersion parameters of DZO as a function of wavelength for different level of doping.

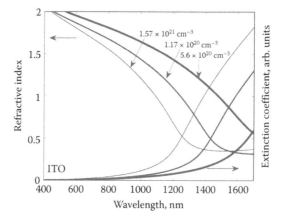

FIGURE 4.8 Dispersion parameters of ITO as a function of wavelength for different levels of doping.

over 142×142 mm² surface (Figure 4.9). Small thickness variations observed at the edge of the wafer were attributed to a temperature gradient at the perimeter of the substrate during thin film deposition process. Taking the inside 130×130 mm² area, the average thickness for the sample is 222 ± 2 nm (Figure 4.9). Sheet resistance values averaged over 142×142 and 130×130 mm² areas are 9.8 ± 0.4 and 9.6 ± 0.3 Ω/sq, respectively. The sheet resistance mapping demonstrates highly homogeneous films with σ/χ ratio of approximately 4% for the $6'' \times 6''$ wafers (Figure 4.10). Resistivity maps were calculated from the sheet resistance and thickness values ($2.26 \pm 0.07 \times 10^{-4}$ Ω cm) (Figure 4.11).

Fitting the SE measurements to obtain dielectric constants indicated that first oscillator parameters were similar across many doped ZnO films. Therefore, these parameters were fixed and Drude model parameters were varied during function minimization calculations when calculating electron concentration and electron mobility maps (Figure 4.12a and b).

The near IR free-carrier absorption is commonly modeled using the second oscillator. In this formalism, the damping term as previously discussed is inversely proportional to the electron mobility. The effective mass is included in the calculation of electron

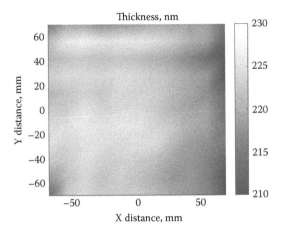

FIGURE 4.9   DZO thickness surface map.

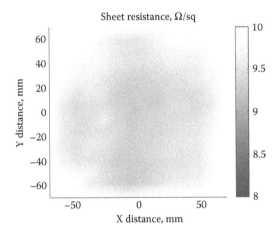

FIGURE 4.10   DZO sheet resistance surface map.

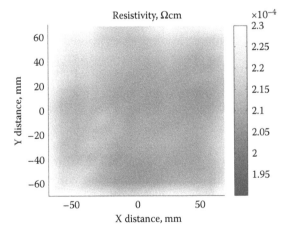

FIGURE 4.11   DZO resistivity surface map.

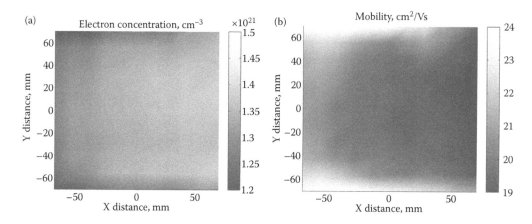

FIGURE 4.12   (a) DZO electron concentration surface map, (b) DZO electron mobility surface map.

concentration and mobility (Figure 4.6). In these measurements, excellent homogeneity was observed for optically calculated electrical properties. For example, electron concentration and mobility of $1.34 \pm 0.026 \times 10^{21}$ cm^{-3} and 20.6 ± 0.76 cm^2/Vs were measured over a 130×130 mm^2 area of the DZO wafer.

4.4.4   Economics

Economic benchmarking indicates that processing costs for 3–6 mm thick low emissivity glass (fluorine-doped tin oxide) produced using an online APCVD FTO process are approximately 20%–40% lower than typical costs for manufacturing low emissivity Ag multilayer stack coatings by sputtering processes. Note that this comparison does not include raw material costs. Films produced by APCVD are chemically bound to the glass surface, making them more mechanically durable than films produced by sputtering. This improved durability reduces the number of flaws introduced into the film during postdeposition handling and limits the potential for forming fatal defects that can lead to OLED device failures, both immediately and over time. From a logistics perspective, the postdeposition handling and shipping of APCVD coated glass is significantly simplified versus sputtered coatings, leading to additional cost advantages to the end user (Silverman et al. 2009). This advantage was confirmed in the Arkema: Philips DOE SSL project (Silverman et al. 2010) where sputtered and APCVD DZO were compared in a standard 60:80 test (60°C and 80% relative humidity for 504 h) and the APCVD DZO retained its original properties while sputtered samples showed significant discoloration and loss of conductivity.

Consensus pricing on commercial large volume ITO on FPD glass was $60/m^2 with the ITO component approximately $20/m^2 at the 2010 DOE SSL Manufacturing Workshop. At the 2011 DOE SSL R&D Manufacturing Workshop there was a push to raise this price; however, captive ITO production should keep pricing near 2010 level. At the 2012 DOE SSL R&D Manufacturing Workshop it was disclosed by a glass manufacturer that a 3 m ITO target is approximately $120,000 with most processes using five of these targets for homogeneity and faster throughput. Using the above values and some assumptions on yield, results in a raw material cost estimate in the range of $8–12/m^2 or a 40%–60% variable margin that is consistent with a large volume price of $20/m^2. Note that the near term DOE SSL volume (3 year horizon) estimate for OLED lighting is less than 100,000 m^2, which is not enough volume for most primary float glass manufacturers to invest capital for OLED lighting alone so this value will be significantly larger for the emerging market. In fact it was stated at the 2012 DOE SSL R&D Manufacturing Workshop that a fully patterned substrate with metallization for the interconnects was currently at $1000/m^2. While there are several extra steps included in the manufacturing of the structured TCO, it is clear that the processing and raw material costs need to be addressed. The market for DZO (photovoltaics) has already demonstrated approximately 25% cost savings relative to ITO and it is expected on scale (greater than 5 million m^2) to approach the commodity FTO coated window glass pricing that is considerably lower than the raw material cost of ITO.

4.5 Emerging Transparent Materials with Metallic Conductivity

The resistivity limit of conventional TCOs is reached at about $8–9 \times 10^{-5}$ and 9×10^{-5} Ω cm demonstrated for ZnO:Ga (Berry et al. 2008) and ITO (Tominaga et al. 1996), respectively, in the lab environment on small-scale substrates using pulsed laser deposition. This resistivity barrier leads to restrictions on the drivable pixel size of flat panel displays and OLED lighting devices. In particular, at this resistivity, there is a voltage drop in the middle of the pixel, which is detectable by the human eye as a decrease in brightness. This is commonly known as "current spreading"—a problem which is currently commonly resolved by the deposition of the metallic grids on the surface of the TCOs. The development of materials with bulk metal resistivities—1.52×10^{-6} Ω cm for Ag and 1.67×10^{-6} Ω cm for Cu—in transparent electrodes could dramatically reduce the cost of these electrodes by eliminating the expensive photolithographic steps utilized in grid production. However, metals commonly possess very high visible absorption and are highly reflective due the location of the plasma wavelength in the UV. Nonetheless, work is progressing in finding ways to use thin film stacks or nanostructured materials to overcome the absorption and reflection limitation of metals. In the following sections we will discuss SMS structures, Ag nanomaterials, and graphene/single wall carbon nanotubes (SWCNTs).

4.5.1 Semiconductor–Metal–Semiconductor Structures

The concept of induced transmission, as applied to semiconductor–metal–semiconductor (SMS) structures, has been known for many years (Berning and Turner 1957). It relies on the vacuum deposition of a series of thin films, with a low reflective index metal layer, sandwiched between two transparent layers possessing high refractive index. There have been many semiconductor materials studied in SMS structures including ZnO, TiO$_2$, CeO$_2$, and ZnS. At the same time, only a few metals, one of which is Ag, have low enough visible absorption and high enough conductivity to be used in the transparent electrode applications.

Chapter 4

One of the common uses that had been originally envisioned for SMS structures was as heat reflectors for transparent low-e windows (Fan et al. 1976). Transparent heat mirrors work as band pass filters, allowing visible light to pass through the window while reflecting IR radiation. In a theory of optical filters the amplitude and phase transformations of the reflected and transmitted light passing through layer *j* is governed by the characteristic matrix (Macleod 1986)

$$M_j = \begin{pmatrix} \cos(\delta) & \dfrac{i}{n}\sin(\delta) \\ in\sin(\delta) & \cos(\delta) \end{pmatrix} \tag{4.3}$$

where *n* is a complex refractive index for the *j* layer and δ is a phase thickness. If more layers are added to the stack, the characteristic matrix takes a form of $\prod(M_i)$. Metals with their low refractive index are inherently suited for this induced transmission purpose. For a metal with extinction coefficient of $k_M = 3.4$ and semiconductor with $n = 2$–2.4, the antireflection film (S) thickness that causes the maximum phase lag must be greater than $d_S > \lambda/[8n_S] \sim 29$ nm. Absorption losses of the SMS construction are mostly due to the nonvanishing real part of the refractive index and maximum transmission is expected for metal layers considerably thinner than $d_M < (\lambda/4\pi k_M)\log_e[2k_M/n_M] \sim 60$ nm (Kostlin and Frank 1982).

Using the method of Macleod (1986), the dispersion parameters of the ITO and bulk Ag can be used to calculate the visible reflectance for SMS structures (Equation 4.3, Figure 4.13a and b). For simplicity, an infinite substrate can be assumed. In most cases, introduction of the finite thickness substrate shifts reflectance curves up. If the ITO layer thickness is kept at 45 nm and the silver layer thickness was varied from 5 through 30 nm, the reflectance curves shown in Figure 4.13a are obtained. In these stacks, a well-developed minimum at ~500 nm is observed (Figure 4.13a). The minimum reflectance increased from <1% through 26% with increasing Ag layer thickness. For 15 nm silver layers the reflectance increased from its minimum at 518 nm to 25% and 20% at

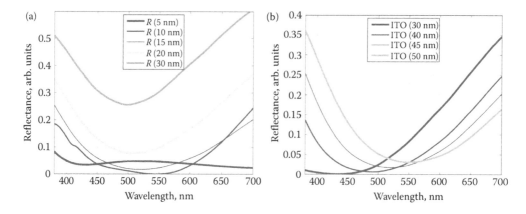

FIGURE 4.13 Calculated reflectances of the ITO/Ag/ITO/glass (∞) stacks (a) variable Ag layer thickness with fixed ITO of 45 nm and (b) variable ITO thickness with fixed Ag layer thickness of 15 nm.

380 and 700 nm, respectively. Maintaining metal layers as thin as possible keeps the visible reflectance low. In practice, the thickness of the evaporated silver layer is kept in a very narrow range (10–30 nm) for two reasons. First, silver coatings with thicknesses <10 nm tend to form incomplete coverage on the semiconductor surface. Second, very low transmission, >50%, is observed in SMS stacks with Ag layer thickness >30 nm due to high absorption. The absorption is a consequence of an extinction coefficient between 1.24 and 6.8 and refractive index of silver are between 1.07 and 0.21 in the range of 200–1000 nm. This leads to an absorption coefficient between 8.5 and 7.8×10^5 cm^{-1} (200–1000 nm) (Weber 2003).

The variation of the semiconductor layer thickness affects the shape of the reflectance curve (Figure 4.13b). If the Ag layer thickness is fixed at 15 nm and both ITO layer thicknesses are varied symmetrically from 30 through 50 nm, the reflectance curves shift toward the IR, thus reducing the blue transparency window. For these calculations, the optimum optical properties are achieved for an ITO layer thickness of 45 nm, where R (518 nm) = 2%. Some further improvements in the blue transmittance window may be achieved by using asymmetric stacks with R < 20% calculated at 380 and 700 nm.

The high transparency of the SMS structure is only one prerequisite for the transparent electrode—low conductivity is also required. The resistivity of the silver coatings in this thickness range is somewhat higher ($3.3–9 \times 10^{-6}$ Ω cm) than that of the bulk silver layers 1.52×10^{-6} Ω cm (Klöppel et al. 2000). Generally, the resistivity of Ag is reduced with increasing thickness of the coatings (Figure 4.14) (Bender et al. 1998, Choi et al. 1999, Klöppel et al. 2000) due to incomplete coalescence (Choi et al. 1999) and grain boundary scattering (Jung et al. 2003) mechanisms. A comprehensive literature survey indicates that the coalescence thickness occurs between 11 and 16 nm for sputtered silver coatings (Table 4.4). At this point, the percolation threshold for the current conduction is reached in these layers and the measured SR of the SMS stack improves dramatically (Klöppel et al. 2000).

To characterize the properties of the SMS structures, a figure of merit defining the ratio of luminous transmittance (as related to the sensitivity of the human eye) to the

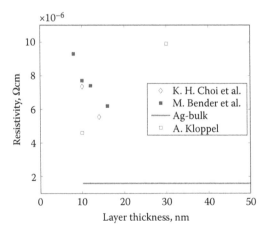

FIGURE 4.14 Resistivity of Ag thin film of varying thickness (Bender 1998) and calculated as part of a SMS structure as a function of the thickness of the Ag layer.

Table 4.4 Optoelectronic Properties of Selected SMS Structures

Stack Construction	T(500 nm), (%)	Coalescence Thickness (nm)	SR (Ω/sq)	F_{TC}	References
ZnO–Ag–ZnO	60–80 (4–13 nm)[a]	11	4–10	14.9–24.4	Sahu and Huang (2009)
ITO–AgCu–ITO	70–82	>8	4–18	8.7–69.7	Bender et al. (1998)
ZnS–Ag–ZnS	88	>12	3	40–70	Liu et al. (2001)
ITO–Ag–ITO	90	>14	4	—	Choi et al. (1999)
ITO–Ag–ITO	88	—	4.5		Klöppel et al. (2000)
ZnO (150 nm)	86–92[a]	—	13–15	59–68	Arkema (2012, Unpublished results)

[a] Total transmittance: Optical coating stack plus the substrate.

sheet resistance, is useful to compare various stack structures with varying properties (Liu et al. 2001)

$$F_{TC} = \frac{T_{lum}^{10}}{R_{Sheet}} \tag{4.4}$$

where T_{lum} is the luminous transmittance of the film stack (Haacke 1973)

$$T_{lum}^{10} = \frac{\int f(\lambda)T(\lambda)d\lambda}{\int f(\lambda)d\lambda} \tag{4.5}$$

where $f(\lambda)$ is relative photopic sensitivity of the human eye.

One can use Equations 4.4 and 4.5 for SMS structures shown in Table 4.4 to determine the value of F_{TC}. Most of the reports in the literature do not mention the substrate properties of the SMS structure, suggesting that the reported transmission and luminous transmittance values are for the thin film only; glass substrates, depending on thickness, reduce visible transmittance by 8%–12%. In Table 4.4, the studies reporting total transmission including the substrate are labeled with an asterisk (*). Some spread in the reported F_{TC} data is seen—from 8 through 70×10^{-3} Ω^{-1}. The best F_{TC} values are seen in the high contrast ZnS/Ag/ZnS system due to high refractive index of ZnS. A comparison of the F_{TC} values between SMS structures and single DZO layers of comparable thickness suggests superior optical properties of ZnO with total transmission of 91% for 150 nm thick coatings. In addition, deposition of a color matching undercoat reduces the Fabry–Perrot interference fringes across a very broad wavelength range (382–916 nm) keeping transmittance above 90% (Figure 4.15, top trace). At the same time, due to high reflectance losses, transmission may be reduced near the IR. In any case, the SR of a 150 nm thick DZO coating (~13–15 Ω/sq) cannot match that of an SMS structure (~3–5 Ω/sq).

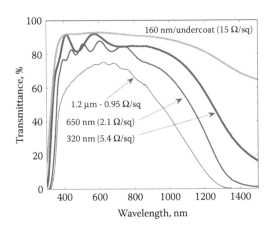

FIGURE 4.15 Optical transmittance for different thickness of DZO layers.

It is likely that gridless display constructions will require SR values of <1 Ω/sq. Such low sheet resistances are within reach of the SMS structures. However, their total transmission will be well below 50%. At the same time, the optical transmission of 1.2 μm thick ZnO having a SR of 0.95 Ω/sq is still above 70% at 500 nm (Figure 4.15). With a SR of 3–4 Ω/sq, metal grid line spacings can be increased, but not eliminated. Development of high throughput screenprinting techniques for the deposition of the grid lines may help reduce the total production cost further.

High volume production of the OLEDs for lighting will require low total accumulated cycle (TAC) times. Current deposition rates for just the silver layers are ~0.2–0.5 nm/s (Choi et al. 1999), suggesting a 100 s window for the deposition of the SMS structures, assuming that Ag deposition is the rate limiting step. In a continuous process, this TAC time is within the limits of the other steps, such as small molecule deposition.

The low RMS roughness of the SMS structures typically obtained in the sputtering process, <1 nm (Ohring 2002), is one key advantage compared to other materials. In addition, the fact that these coatings can be deposited at temperatures below 200°C allows them to be used on temperature sensitive substrates, such as polymers.

There have been some stability issues with SMS structures. In particular, stability issues with respect to moisture, scratching and temperature have been reported (Jung et al. 2003, Sahu and Huang 2009, Siegel and Culp 2002). Multiple strategies for improving the temperature stability of SMS structures have included introduction of additional metal layers such as Cu and Al, with considerable success. In certain instances, a compromise has been reached between improving thermal stability and maximizing optical transmission. However, the low mechanical stability of the SMS coatings is inherent to the deposition technique; these structures cannot be handled easily without incorporation of sophisticated capping layers that may also affect transparency and SR (Siegel and Culp 2002).

4.5.2 Silver Nanowires and Nanomeshes

While SMS structures have improved flexibility relative to some of the conventional TCOs, improvement is still needed for them to be used in flexible displays. This is easily

Chapter 4

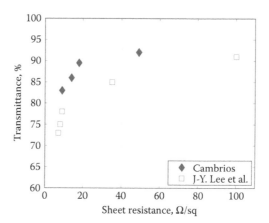

FIGURE 4.16   Transmission as a function of sheet resistance for a series of Ag nanowire thin films.

resolved using transparent mesh electrodes. These electrodes are solution processed from precursor inks, enabling faster production speeds consistent with roll-to-roll processing. The nanomesh samples deposited on flexible substrates can be bent to the radius of 4 mm. Additionally, this solution process does not cause damage, as is often associated with sputtering deposition on top of the organic active layers (Gu et al. 1996). The ink production is based on the reduction of the $AgNO_3$ in the presence of the poly(vinyl pyrrolidene) in the ethylene glycol. The resulting nanowires are 9–20 nm long (with a diameter ~100 nm). After the nanowire ink is deposited on a substrate, the resulting film is a random mesh of Ag nanowires. Again, it is useful to simultaneously compare the sheet resistance and transmission (Figure 4.16). From Figure 4.16, it is clear that one can achieve an excellent sheet resistance of 7 Ω/sq at 73% transmission, or achieve a transmission of 92% with a sheet resistance of 50 Ω/sq. One of the drawbacks of the nanowire thin films is their high roughness, >20 nm (Lee et al. 2008), that leads to significant number of shorts in the OLED devices or reduces process yields. However, high roughness also provides strong light scattering: most of the light (~80%) in nanomesh assemblies is scattered at angles above 10° (Lee et al. 2008). The application of the nanowire thin films in combination with a top hole injection and planarization layer helps resolve some of these issues, but may also add cost (Pschenitzka 2011).

4.5.3   Graphene and Carbon Nanotube Thin Films

Most of the TCOs and SMSs discussed in the previous sections are n-type conductors, with a few exceptions such as NiO and Cu_2O. This imposes limitations to the applications of these materials for hole injection into the hole conduction side of an OLED. However, p-type charge transport is a key signature of carbon-based nanomaterials such as SWCNT and graphene thin films (Wu et al. 2004). P-type charge transfer within these materials permits new applications of alternative photonic coupling schemes, though the measured work function of these materials, ~5.2 eV, may still impose limitations on efficient charge injection (Kim 2005, Wu et al. 2004). In addition, common TCOs and SMS behave as optical filters with a limited transmission window (Macleod

1986). The transmission of a SWCNT is predicted to be high (90%) in the 2–5 μm spectral range (Wu et al. 2004).

Deposition methods to generate transparent films of carbon nanomaterials involve a variety of techniques: drop casting from solvent, airbrushing, Langmur–Blodgett deposition, multiwall carbon nanotube (MWCNT) drawing and aerogel compression process (Zhang et al. 2005), and vacuum filtration (Wu et al. 2004). Interestingly, a vacuum filtration process that uses dilute, surfactant-based suspension of purified SWCNT to generate free-standing films seems to be limited only by the size of the filtration membranes. The last step—transferring the film to the substrate—is the most complex step and still requires more process development for commercial feasibility. Doping of these materials is accomplished using a nitric acid treatment during SWNT purification that produces a series of doped and undoped tubes. 50 nm thick doped SWCNT films deposited using filtration have demonstrated a transmission of 70% in the visible wavelength range with a sheet resistance of 30 Ω/sq (Figure 4.17) (Wu et al. 2004). A large range of sheet resistance, from 80 through 300 Ω/sq has been demonstrated for SWCNT arrays (De et al. 2009). Considerably higher sheet resistances of 700 Ω/sq are typical for free-standing bundles of MWCNT films (Zhang et al. 2005). However, a maximum transmission of 78% has been demonstrated for coated nanotubes with a SR below 100 Ω/sq (Eigler 2009).

Due to its honeycomb crystal lattice, graphene has inherited flexibility to deformation and is a promising candidate for flexible transparent electrode applications and has multiple advantages, such as good chemical and mechanical stability, acceptable electrical properties, and satisfactory optical transmission (De et al. 2009, Eigler 2009, Feng et al. 2012). There is considerable variation in optical and electrical properties of this material depending on the process parameters (Figure 4.17). In general, transmission of these layers decreases with decreasing SR. Each single layer of graphene absorbs 2.3% of visible light (Eigler 2009). Sheet resistances of graphene on Ni surfaces varied from 30 through 3000 Ω/sq, dramatically affected by doping to obtain lower resistance, with the transmission between 75% and 93% for graphene layers only (Feng et al. 2012). Higher SRs were obtained for graphene oxide (GO). Despite constant

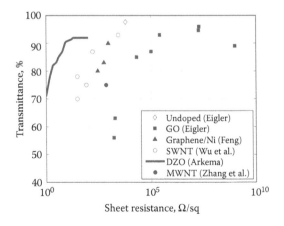

FIGURE 4.17 Transmission as a function of sheet resistance for undoped SWCNT, MWCNT, GO = graphene oxide, graphene, graphene/Ni, c-nanotubes, and DZO thin films.

Table 4.5 Opto-Electronic Properties Materials Relevant to Transparent Conductors

Material	ρ (10^{-4} Ω/sq)	T(%)	RMS (nm)	Work Function (eV)	Thermal Stability (°C)	Process Method	Stability to Deformation
SMS	0.3–1	50–85	<1	—	<300	PVD	+
Graphene	6–100	2.3/layer	<1	4.5	Stable	Solution	++
SWCNT	1.5–10^6	70–90	>30	5.2	Stable	Solution	++
MWCNT	30–10^6	70–90	>30	5.2	Stable	Solution	++
TCOs	0.7–5	70–92	3–6	4.8–5.3	<400–700	CVD, PVD	–
Nanomesh	1–3	60–90	>20	4.7	<200	Solution	++

improvements in this field, conventional TCMs such as DZO display better electrical and optical properties (solid line, Figure 4.17). With constantly improving optoelectrical properties there is potential for high volume applications of carbon-based thin films as electrode materials in the future. High deposition temperature ($T > 1000$°C) limits its application. While some novel transfer techniques are being developed, commercial manufacturing techniques need to come to fruition before this can become a commercially viable TCM.

Collecting the properties and key parameters of the reviewed materials in this chapter (depicted in Table 4.5), it is clear that each technology has its pro and cons. Subsequently, we expect there will continue to be opportunities for alternative approaches and discoveries to meet the needs of specific applications.

References

Agura, H., Suzuki, A., Matsushita, T., Aoki, T., Okuda, M. 2003. Low resistivity transparent conducting Al-doped ZnO films prepared by pulsed laser deposition. *Thin Solid Films* 445: 263–267.

Bender, M., Seeling, W., Daube, C., Frankenberger, H., Ocker, B., Stollenwerk, J. 1998. Dependence of film composition and thickness on optical and electrical properties of ITO-metal-ITO multilayers. *Thin Solid Films* 326: 67–71.

Berning, P. H., Turner, A. F. 1957. Induced transmission in absorbing films applied to band pass filter design. *Journal of the Optical Society of America* 47: 230–239.

Berry, J. J., Ginley, D. S., Burrows, P. E. 2008. Organic light emitting diodes using a Ga:ZnO anode. *Applied Physics Letters* 92: 193304–193307.

Bourlange, A., Payne, D. J., Egdell, R. G., Foord, J. S., Edwards, P. P., Jones, M. O., Schertel, A., Dobson, P. J., Hutchison, J. L. 2008. Growth of In$_2$O$_3$(100) on Y-stabilized ZrO$_2$(100) by O-plasma assisted molecular beam epitaxy. *Applied Physics Letters* 92: 092117–092120.

Burns, G. 1990. *Solid State Physics*. London: Academic press.

Chao, C. I., Chuang, K. R., Chen, S. A. 1996. Failure phenomena and mechanisms of polymeric light-emitting diodes: Indium–tin–oxide damage. *Applied Physics Letters* 69: 2894–2897.

Childs, G. N. 1989. Electron-impurity collision broadening of energy states in heavily doped n-type indium oxide. *Solar Energy Materials* 19: 403–410.

Choi, K. H., Kim, J. Y., Lee, Y. S., Kim, H. J. 1999. ITO/Ag/ITO multilayer films for the application of a very low resistance transparent electrode. *Thin Solid Films* 341: 152–155.

De, S., Lyons, P. E., Sorel, S. et al. 2009. Transparent, flexible, and highly conductive thin films based on polymer-nanotube composites. *ACS Nano* 3(3): 714–717.

Eigler, S. 2009. A new parameter based on graphene for characterizing transparent, conductive materials. *Carbon* 47: 2933–2937.

Fan, J. C. C., Frank, J., Bachner, F. J. 1976. Transparent heat mirrors for solar-energy applications. *Applied Optics* 15: 1012.

Feng, T., Xie, D., Tian, H., Peng, P., Zhang, D., Fu, D., Ren, R., Li, X., Zhu, H., Jing, Y. 2012. Multi-layer graphene treated by O_2 plasma for transparent conductive electrode applications. *Materials Letters* 73: 187.

Fortunato, E., Assuncao, V., Goncalves, A., Marques, A., Aguas, H., Pereira, L., Ferreira, I., Vilarinho, P., Martins, R. 2004. High quality conductive gallium-doped zinc oxide films deposited at room temperature. *Thin Solid Films* 451–452: 443–447.

Fujiwara, H., Kondo, M. 2005. Effects of carrier concentration on the dielectric function of ZnO: Ga and In_2O_3:Sn studied by spectroscopic ellipsometry: Analysis of free-carrier and band-edge absorption. *Physical Review B* 71: 075109–075114.

Furubayashi, Y., Hitosugi, T., Yamamoto, Y., Inaba, K., Kinoda, G., Hirose, Y., Shimada, T., Hasegawa, T. 2005. A transparent metal: Nb-doped anatase TiO_2. *Applied Physics Letters* 86: 252101–252103.

Gasman, L., Nolan, R. 2010. OLED lighting market. Webinar by Nanomarket, LLC, www.nanomarkets.net.

Ginley, D. (Ed.). 2010. *Handbook of Transparent Conductors*. Associate Editors Hosono, H. and Pine, D. New York: Springer Science + Business Media, LLC.

Gu, G., Bulovic, V., Burrows, P. E., Forrest, S. R. Thompson, M. E. 1996. Transparent organic light emitting devices. *Applied Physics Letters* 68: 2606–2608.

Haacke, G. 1973. Transparent conducting coatings. *Annual Review of Materials Science Journal of Applied Physics* 44: 4618–4621.

Hao, X. T., Tan, L. W., Ong, K. S., Zhu, F. 2006. High-performance low-temperature transparent conducting aluminum-doped ZnO thin films and applications. *Journal of Crystal Growth* 287: 44–47.

Heo, G.-S., Park, J.-W., Choi, S.-E. 2009. Modification of optical and electrical properties of ITO using a thin Al capping layer. *Thin Solid Films* 518: 1160–1163.

Hovsepian, P. E. 2000. PVD/CVD Hybrid Coatings Craft IV: Project No. BE97–5318, Nanotechnology Centre for PVD Research, Sheffield Hallam University, 2000 and Richter Precision Inc. literature, http://www.richterprecision.com/DCD_coatings.htm.

Hu, J., Gordon, R. 1992. Deposition of boron doped zinc oxide films and their electrical and optical properties. *Journal of the Electrochemical Society* 139: 2014–2022.

Hwang, J., Wan, A., Kahn, A. 2009. Energetics of metal-organic interfaces: New experiments and assessment of the field. *Material Science and Engineering* R 64: 1–31.

Irwin, M. D., Buchholz, D. B., Hains, A. W., Chang, R. P. H., Marks, T. J. 2008. P-type semiconducting nickel oxide as an efficiency-enhancing anode interfacial layer in polymer bulk-heterojunction solar cells. *Proceedings of the National Academy of Sciences of the United States of America* 105: 2783–2787.

Ishii, H., Sugiyama, K. Ito, K. Seki, K. 1999. Energy level alignment and interfacial electronic structure at organic/metal and organic/organic interfaces. *Advanced Materials* 11: 605–625.

Jansseune, T. 2005. Indium price soars as demand for displays continues to grow. *Compound Semiconductor* May 2005. Available at: http://www.compoundsemiconductor.net/articles magazine /11/5/5/1#Csind3_05–05.

Jeong, W. J., Kim, S. K., Park, G. C. 2006. Preparation and characteristic of ZnO thin film with high and low resistivity for an application of solar cell. *Thin Solid Films* 506–507: 180–183.

Jung, Y. S., Choi, Y. W., Lee, H. C. 2003. Effect of thermal treatment on the electrical properties of silver based indium tin/metal/indium tin oxide structures. *Thin Solid Films* 440: 278–284.

Kanai, Y., Matsushima, T., Murata, H. 2009. Improvement of the stability for organic solar cells by using molybdenum trioxide buffer layer. *Thin Solid Films* 518: 537–540.

Kim, H., Auyeung, R. C. Y., Pique, A. 2008. Transparent conducting F-doped SnO_2 thin films grown by pulsed laser deposition. *Thin Solid Films* 516: 5052–5056.

Kim, W., Javey, A., Tu, R., Cao, J., Wang, Q. 2005. Electrical contacts to carbon nanotubes down to 1 nm in diameter. *Applied Physics Letters* 87, 173101–173103.

Kitco Base Metals. 2012. Available at: http://www.kitcometals.com/charts.

Klöppel, A., Kriegseis, W., Meyer, B. K., Scharmann, A., Daube, C., Stollenwerk, J., Trube, J. 2000. Dependence of the electrical and optical behavior of ITO–silver–ITO multilayers on the silver properties. *Thin Solid Films* 365: 139–146.

Korotkov, R. Y., Ricou, P., Fang, L., Coffey, J., Silverman, G. S., Ruske, M., Schwab, H., Elmer, K., Klein, A., Rech, B. 2008. *Transparent Conductive Zinc Oxide*. New York: Springer.

Korotkov, R. Y., Ricou, P., Fang, L., Silverman, G. S. 2012. Properties of TCO anodes deposited by atmospheric pressure chemical vapor deposition and their application to OLED lighting. *Proceedings of SPIE* 8263: 826308-1.

Korotkov, R., Smith, R., Silverman, G. S., Stricker, J., Carson, S. 2011. Patent application: OLED substrate consisting of transparent conducting oxide (TCO) and anti-iridescent undercoat. WO 2011/005639 A1.

Kostlin, H., Frank, G. 1982. Optimization of the transparent heat mirrors based on a thin silver film between antireflection films. *Thin Solid Films* 89: 287–293.

Lee, H. W., Lau, S. P., Wang, Y. G., Tse, K. Y., Hng, H. H., Tay, B. K. 2004. Structural, electrical and optical properties of Al-doped ZnO thin films prepared by filtered cathodic vacuum arc technique. *Journal of Crystal Growth* 268: 596–601.

Lee, J.-Y., Connor, S. T., Cui, Y., Peumans, P. 2008. Solution-processed metal nanowire mesh transparent electrodes. *Nano Letters* 8(2): 689–692.

Lee, S. T., Gao, Z. Q., Hung, L. S. 1999. Metal diffusion from electrodes in organic light-emitting diodes. *Applied Physics Letters* 75: 1404–1407.

Lee, S. Y., Park, B. O. 2006. Structural, electrical and optical characteristics of SnO_2:Sb thin films by ultrasonic spray pyrolysis. *Thin Solid Films* 510: 154–158.

Li, J., Yashiro, M., Ishida, K., Yamada, H., Matsushige, K. 2005. Enhanced performance of organic light emitting device by insertion of conductive/insulating WO_3 anodic buffer layer. *Synthetic Metals* 151: 141–146.

Liu, X., Cai, X., Mao, J., Jin, C. 2001. ZnS/Ag/ZnS nano-multilayer films for transparent electrodes in flat display application. *Applied Surface Science* 183: 103–110.

Luque, A., Hegedus, S. 2011. Transparent conductive oxides for photovoltaics. In: Delahoy, A. and Guo, S (Eds.), *Handbook of Photovoltaic Science and Engineering*, Second Edition, United Kingdom: John Wiley & Sons, pp. 716–796.

Macleod, H. A. 1986. *Thin-Film Optical Filters*. New York: Macmillan.

Markowitz, P., Gasman, L. 2010. Transparent conductors market. Webinar by Nanomarket, LLC, www.nano-markets.net.

Maruyama, T., Fukui, K. 1990. Fluorine-doped indium oxide thin films prepared by chemical vapor deposition. *Japanese Journal of Applied Physics Part 2-Letters* 29: L1705–L1707.

Meng, Y., Yang, X. L., Chen, H. X., Shen, J., Jiang, Y., Zhang, Z. 2001. A new transparent conductive thin film In_2O_3:Mo. *Thin Solid Films* 394: 218–222.

Miao, W. N., Li, X. F., Zhang, Q. 2006. Transparent conductive In_2O_3:Mo thin films prepared by reactive direct current magnetron sputtering at room temperature. *Thin Solid Films* 500: 70–73.

Minami, T., Kakumu, T., Takeda, Y., Takata, S. 1996. Highly transparent and conductive ZnO-In_2O_3 thin films prepared by d.c. magnetron sputtering. *Thin Solid Films* 290–291: 1–5.

Ohring, M. 2002. *Materials Science of Thin Films*. San Diego, CA: Academic Press.

OSHA Chemical Sampling Information (Indium and Compounds), 2012. http://www.osha.gov/dts/chemicalsamplingdata/CH_247101.html.

OSHA Health Guidelines (Zinc Oxide), 2012. http://www.osha.gov/SLTC/healthguidelines/zincoxide/recognition.html.

Ozgar, U., Alivov, Y. I., Liu, C., Teke, A., Reshchikov, M., Dogan, S., Avrutin, V., Cho, S. J., Morkoc, H. 2005. A comprehensive review of ZnO materials and devices. *Journal of Applied Physics* 98: 0413011–0412114.

Padmaperuma, A., Gaspar, D. 2011. Properties of TCO anodes deposited by APCVD and their applications to OLEDs. *Proceedings of SPIE* 7939: 793919–793921.

Pschenitzka, F. 2011. Solution-processable transparent conductive hole-injection electrode for OLED SSL. DOE Solid State Lighting Workshop, San Diego, February 2.

Russo, D. A., Lindner, G. H. 1986. Liquid coating composition for producing high quality, high performance fluorine-doped tin oxide coatings, US4601917.

Sahu, D. R., Huang, J.-L. 2009. Development of ZnO-based transparent conductive coatings. *Solar Energy Materials and Solar Cells* 93: 1923–1927.

Siegel, J., Culp, T. 2002. The MSVD Low E 'Premium Performance' myth—Actual energy conservation performance of different types of Low E glazings in residential windows. *International Glass Review* 1, 55.

Silverman, G., Bluhm, M., Korotkov, R., Salemi, A., Smith, R., Stricker, J., Xu, C., Carson, S. 2009. Arkema: Philips DOE SSL EERE DE-FC26-08NT01576. Application of developed atmospheric pressure chemical vapor deposition (APCVD) transparent conducting oxides and undercoat technologies for economical OLED lighting. Presented at the DOE SSL 2009 R&D Workshop Poster Session, San Francisco, CA.

Silverman, G., Bluhm, M., Korotkov, R., Salemi, A., Smith, R., Stricker, J., Xu, C., Carson, S. 2010. Progressing toward commercially viable OLED devices. Presented at the DOE SSL R&D Workshop, Raleigh, NC, February 3.

Singh, A. V., Mehra, R. M., Yoshida, A., Wakahara, A. J. 2004. Doping mechanism in aluminum doped zinc oxide films. *Applied Physics* 95: 3640–3643.

Stricker, J., Abrams, M., Korotkov, R., Silverman, G. S., Smith, R. 2010. Method of coating a zinc coated article, US 7,732,013 B2.

Stricker, J., Smith, R., Abrams, M., Korotkov, R., Silverman, G. S., Sanderson, K., Ye, L., Benito, G. 2012. Method of making a low resistivity doped zinc oxide coating and the articles formed thereby, US 8,163,342 B2.

Tominaga, K., Ueda, T., Ao, T., Kataoka, M., Mori, I. 1996. ITO films prepared by facing target sputtering system. *Thin Solid Films* 281–282: 194–197.

United States Department of Energy, 2005. *Solid-State Lighting Program Planning.* Workshop Report, February 3–4, San Diego, California.

United States Department of Energy, 2007. *Solid-State Lighting.* Workshop Report—Getting SSL to Market, January 31–February 2, Phoenix, Arizona.

United States Department of Energy, 2010. *Multi-Year Program Plan FY10-FY15 Solid State Lighting Research and Development.*

United States Department of Energy, 2010. *Solid-State Lighting Research and Development: Manufacturing Roadmap.*

United States Department of Energy, 2011. *Solid-State Lighting Research and Development: Manufacturing Roadmap.*

United States Department of Energy, 2008. *Solid-State Lighting.* Workshop Report—Transformations in Lighting, January 29–31, Atlanta, Georgia.

United States Department of Energy, 2009. *Solid-State Lighting.* Workshop Report—Getting SSL to Market, February 3–5, San Francisco, California.

United States Geological Survey, 2005. Minerals commodity summary (indium), http://minerals.usgs.gov/minerals/pubs/commodity/indium/indiumyb05.pdf.

United States Geological Survey, 2007. Minerals commodity summary (indium), http://minerals.usgs.gov/minerals/pubs/commodity/indium/indiumcs07.pdf.

US Geological Survey, 2010. Minerals commodity summary (zinc), http://minerals.usgs.gov/minerals/pubs/commodity/zinc/mcs-2010-zinc.pdf.

von Blanckenhagen, B., Tonova, D., Ullmann, J. 2002. Application of the Tauc–Lorentz formulation to the interband absorption of optical coating materials. *Applied Optics* 16: 3137–3141.

Walsh, A., Da Silva, J. L. F., Wei, S.-H. 2008a. Origins of band-gap renormalization in degenerately doped semiconductors. *Physical Review B* 78: 0752110–0752113.

Walsh, A., Da Silva, J. L. F., Wei, S.-H., Korber, C., Klein, A., Piper, L. F. J. 2008b. Nature of the band gap of In_2O_3 revealed by first-principles calculations and X-ray spectroscopy. *Physical Review Letters* 100: 167402–167406.

Weber, M. J. 2003. *Handbook of Optical Materials.* Boca Raton, FL: CRC Press LLC.

Wu, Z., Chen, Z., Du, X. et al. 2004. Transparent, conductive carbon nanotube films. *Science* 305: 1273–1276.

Young, D. L., Coutts, T. J., Kaydanov, V. I., Gilmore, A. S., Mulligan, W. P. 2000. Direct measurement of density-of-states effective mass and scattering parameter in transparent conducting oxides using second-order transport phenomena. *Journal of Vacuum Science and Technology A* 18: 2978.

Zhang, M., Fang, S. Zakhidov, A. A. et al. 2005. Strong, transparent, multifunctional, carbon nanotube sheets. *Science* 309: 1215–1219.

Zhi-hua, Li, Yu-peng, Ke, Dong-yan R. 2008. Effects of heat treatment on morphological, optical and electrical properties of ITO films by sol-gel technique. *Transactions of Nonferrous Metals Society of China* 18: 366–371.

Chapter 4

5. Charge-Injection Layers

Mark T. Greiner and Zheng-Hong Lu

5.1 Importance of Charge-Injection Layers

High-efficiency organic light-emitting diodes (OLEDs) rely heavily on good-quality interfaces. The word *interface* here refers to the contact area between two different solid materials. As OLEDs are typically composed of multiple layers of different materials, several interfaces are present in an OLED. There are generally numerous organic semiconductor layers sandwiched between two inorganic electrode materials, as illustrated in Figure 5.1.

During OLED operation, electrons are injected from the cathode, and are transported through multiple organic layers to the anode. Along their path, electrons traverse several

Chapter 5

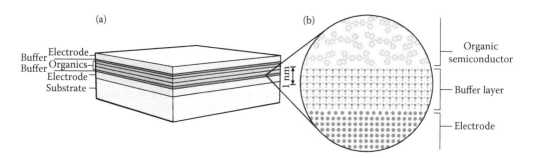

FIGURE 5.1　Schematic of an OLED showing (a) location of buffer layers in the stack and (b) expanded view of interface.

interfaces. At each interface, there is the possibility of an *energy barrier* that the electron (or hole) must overcome in order to enter the next material. Energy barriers increase the driving voltages of OLEDs, and also result in charge-accumulation regions that increase the probability of nonradiative recombination (Scott 2003; Hughes and Bryce 2005; Kampen 2006). Consequently, interfacial energy barriers give rise to high OLED driving voltages and low device efficiencies.

While organic–organic interfaces can present energy barriers, organic semiconductors can be engineered to eliminate these barriers (Ishii, Sugiyama et al. 1999; Kampen 2006; Xu, Yu et al. 2006; Koch 2007, 2008). Electrode–organic interfaces have proven to be much more problematic. Electrodes are generally made of a transparent conducting oxide (TCO)—such as tin-doped indium oxide (ITO), aluminum-doped zinc oxide (AZO), or fluorine-doped tin oxide (FTO)—or of some sort of metal—such as Au, Al, Ag, or the like. The interfacial energy barriers between the electrode materials and common organic semiconductor tend to be large.

In order to mitigate the problems with charge-injection energy barriers at electrodes, it is often necessary to engineer the electrode surface by incorporating a *buffer layer* of some sort between the electrode and the organic layer, as illustrated in Figure 5.1a and b. Buffer layers are generally very thin (ca. 1–10 nm), and are used to control the *energy-level alignment* across an interface (Ishii, Sugiyama et al. 1999; Kampen 2006; Xu, Yu et al. 2006; Koch 2007, 2008). Buffer layers reduce charge-injection energy barriers by aligning the charge-transport levels of organic semiconductors and electrodes.

5.2　Meaning of "Energy-Level Alignment"

The magnitude of interfacial energy barriers depends on the relative positions of donor and acceptor energy levels on either side of an interface (Ishii, Sugiyama et al. 1999; Blyth, Duschek et al. 2001; Scott 2003; Koch 2007). For an organic–organic interface, the relevant energy levels are the *highest occupied molecular orbital* (HOMO) and the *lowest unoccupied molecular orbital* (LUMO).

To illustrate the concept of energy-level alignment at an organic–organic interface, we will conceptually follow an electron as it passes through an OLED. Figure 5.2 illustrates a schematic energy-level diagram of a two-layer OLED. Electrons are injected into an OLED from the Fermi level of the cathode, by hopping from the cathode's Fermi level into the LUMO of the electron-transporting organic layer (labeled A in Figure 5.2).

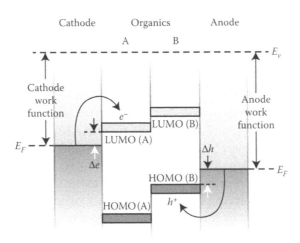

FIGURE 5.2 Schematic energy-level diagram of a simple, two-layer OLED. The arrows indicate the movement of electrons (e^-) and holes (h^+). Electrons and holes move from the Fermi level (E_F) to the LUMO and HOMO levels, respectively. The electron-injection barrier (Δe) and hole-injection barrier (Δh) are indicated. E_v represents the vacuum level.

The electron then travels through the bulk of an organic semiconductor by hopping from the LUMO of one molecule to the LUMO of a neighboring molecule. Note: This electron motion is not entirely barrier-less because molecules polarize when they are negatively charged, resulting in a polaron binding energy.

When an electron approaches an interface to a different organic material (call it material B), having a different LUMO level, then the electron can either hop into the LUMO of B or recombine with holes in the HOMO of material B. Electrons and holes recombine either radiatively (i.e., emitting light) or nonradiatively (i.e., generating heat). In order for holes to be available in the HOMO of material B, they must be provided from the anode. Holes hop from the Fermi level of the anode into the HOMO level of the hole-transporting material (labeled B in Figure 5.2). The energy barrier for hole-injection is the difference between the Fermi level and the HOMO energy, labeled Δh in Figure 5.2.

The electron- and hole-injection barriers can be overcome through an external applied voltage. The minimum injection barriers are achieved when the organic semiconductor frontier orbitals (i.e., HOMO and LUMO levels) are aligned with the Fermi levels of the respective electrodes. At the cathode interface, the LUMO of the electron-transporting material should be aligned with the cathode Fermi level, and at the anode interface, the HOMO level of the hole-transporting material should be aligned with the anode Fermi level. This is what is referred to as *energy-level alignment*.

Metallic electrodes are often coated with a semiconducting buffer layer to improve injection. In these cases—so long as the buffer layer is sufficiently thin—the Fermi level is still the relevant energy level for charge-injection. However, if one of the organic frontier orbitals (i.e., HOMO or LUMO levels) is aligned with the semiconductor's valence or conduction bands, then the interface can become more conductive. For example, if an organic material's HOMO level is aligned to the unoccupied conduction band of a semiconductor buffer layer, or if an organic material's LUMO is aligned to the occupied valence band of the semiconductor buffer layer, then charge transfer can occur from the organic frontier orbitals into the semiconductor valence or conduction band (Matsushima, Jin et al. 2011).

5.3 Types of Charge-Injection Buffer Layers

Many strategies have been employed to engineer barrier-less interfaces. Here, we will briefly review some of the types of buffer layers that have been used. Buffer layers can first be divided into two types: (1) anode buffer layers and (2) cathode buffer layers. Anode buffer layers are designed to enhance injection of *holes* from the anode to the organic layers by aligning the Fermi level of the anode to the HOMO level of the organic material. Cathode buffer layers are designed to enhance injection of *electrons* from the cathode to the organic layers by aligning the Fermi level of the cathode to the LUMO level of the organic material.

5.3.1 Anode Buffer Layers

Here, we will survey the various types of anode buffer layers that are used. Each type of buffer layer has its benefits. Sometimes, certain restrictions or limitations make one type of buffer layer preferable over another.

In the typical "bottom-emission" OLED, the anode is a TCO such as ITO, FTO, or AZO. Thin metal films can also be used as anodes for bottom-emission OLEDs if the metal film is thin enough to remain semitransparent. Thick metal films can be used as anodes in "top-emission" OLEDs. Both TCOs and metals tend to have large hole-injection barriers with most commonly used hole-transporting organic semiconductors. Most emissive organic materials have relatively deep HOMO levels—deeper than the Fermi levels of TCOs and high-work-function metals—thus, energy barriers must be overcome in order to move holes into the HOMO of the emissive material. Most of the research on anode buffer layers has been focused on modifying metal and TCO surfaces.

5.3.1.1 Small-Molecule Buffer Layers

Hole-injection from an anode to the hole-transporting layer can be improved by incorporating an intermediate "buffer" organic layer between the anode and the hole-transporting layer. The buffer organic should have a lower hole-injection barrier with the electrode than the hole-transporting layer does. In this case, hole-injection occurs from the anode, directly into the HOMO of the hole-injection organic buffer material, and then holes hop from the buffer material to the hole-transporting material. Unmodified TCOs and metals have work functions no higher than ~5.3 eV. Thus, such charge-injection materials tend to have relatively shallow HOMO levels, and may not be perfectly matched to the HOMO level of the emissive material. Nonetheless, the intermediate layer allows for a step-wise charge-injection mechanism that improves overall charge-injection.

Small-molecule organic semiconductors can be easily sublimed in vacuum at mild temperatures (usually between 200°C and 400°C). Commonly used small-molecule buffer layers include copper phthalocyanine (CuPc), (VanSlyke, Chen et al. 1996) triphenyldiamine (TPD), naphthyl-phenyl-amino-biphenyl (α-NPD), tricarbizoletriamine

(TCTA), and a family of triphenyl-amines, such as *m*-MTDATA (Shirota, Kuwabara et al. 1994) and 2-TNATA.

5.3.1.2 Spin-Cast Polymer Buffer Layers

Hole-injection from the ITO can also be improved by coating the ITO with a layer of high-work-function polymer blend. The most commonly used polymer blend is poly(3,4-ethylene-dioxythiophene):poly(4-styrenesulfonate) (PEDOT:PSS) (Brown, Kim et al. 1999). This material has a work function of ~5.2–5.6 eV, is highly conductive, and is transparent. It can be easily spin-cast from solution onto TCOs or other electrode materials. Other commonly used polymer blends include polyaniline:camphorsulfonic acid (PANI:CSA).

5.3.1.3 Chemically Attached Polymer Layers

Methods have been employed to chemically attach various hole-injection organic materials to the surfaces of TCOs. Chemically attaching the hole-injection organic material provides an intimate contact between the anode and the injection layer that is mechanically and chemically stable. Hole-injection layers can be covalently bound to ITO through siloxane linkages (Hatton, Day et al. 2001; Lee, Jung et al. 2002; Huang, Evmenenko et al. 2005) and phosphonic acid linkages (Bardecker, Ma et al. 2008). In these strategies, hole-transporting materials—such as triarylamines—are functionalized with silanol or phosphonic acid binding groups. The binding groups undergo a condensation reaction with hydroxyl groups on the TCO surface, giving rise to chemically attached hole-transporting molecules. Thiol-based chemical linkages have also been used to attach hole-injection polymers to Au electrodes to provide improved charge-injection (Marmont, Battaglini et al. 2008).

It is also possible to attach molecules with permanent dipoles to an ITO surface (Ganzorig, Kwak et al. 2001; Khodabakhsh, Poplavskyy et al. 2004). The dipoles effectively reduce the interfacial dipole barrier between the electrode and the hole-transporting material.

5.3.1.4 Metal-Oxide Layers

Various metal oxides have been used as buffer layers for cathodes (Tokito, Noda et al. 1996; Qiu, Xie et al. 2003), including MoO_3 (Tokito, Noda et al. 1996; Matsushima, Kinoshita et al. 2007; Lee, Cho et al. 2008; Matsushima, Jin et al. 2008; Wang, Qiao et al. 2008; Xie, Meng et al. 2008; Hamwi, Meyer et al. 2009; Kroger, Hamwi et al. 2009; Matsushima and Murata 2009; Yi, Jeon et al. 2009; Yook and Lee 2009), WO_3 (Meyer, Winkler et al. 2008; Wang, Helander et al. 2009), NiO (Chan, Hsu et al. 2002; Park, Choi et al. 2005; Im, Choo et al. 2007; Wei, Yamamoto et al. 2007; Irwin, Buchholz et al. 2008; Chun, Han et al. 2009; Woo, Kim et al. 2009; Steirer, Chesin et al. 2010), CuO (Wang, Osasa et al. 2006; Murdoch, Greiner et al. 2008), RuO_2 (Tokito, Noda et al. 1996), V_2O_5 (Tokito, Noda et al. 1996; Zhu, Sun et al. 2007; Zhang and Choy 2008; Meyer, Zilberberg et al. 2011), Fe_3O_4 (Zhang, Feng et al. 2009), Ag_2O (Xiao, Shang et al. 2005), and ZnO. These oxides tend to have high work functions, and thus increase the work function of the anode (Greiner, Helander et al. 2012). Some oxides can be evaporated at relatively mild temperatures—such as MoO_3, WO_3, and V_2O_5—and are often used for devices that are fabricated in vacuum. Some oxides have also been solution deposited—such as

NiO (Steirer, Chesin et al. 2010), V_2O_5 (Zilberberg, Trost et al. 2011), and MoO_3 (Meyer, Khalandovsky et al. 2011)—and are suitable for solution processed OLEDs (Meyer, Khalandovsky et al. 2011).

Metal oxides can exhibit a broad range of electronic properties, making rectifying contacts possible. Their electronic properties can be tuned by changing the metal cation oxidation state or by introducing defects. Metal oxides can also have extremely high work functions, with values greater than 7 eV quoted for some oxides prepared and measured in ultra-high vacuum.

Perhaps the most commonly used oxide in OLEDs is MoO_3. This is mainly because it is easy to prepare—with a low sublimation temperature of approximately 560°C—and it has a very high work function of ca. 6.9 eV (Kroger, Hamwi et al. 2009; Meyer, Kroger et al. 2010). MoO_3 also has a very low-lying conduction band, making it suitable for use as a p-type dopant in organic blends (Hamwi, Meyer et al. 2009; Meyer, Kroger et al. 2010).

5.3.1.5 Halogenation

Electrodes can also be modified by chemically attaching halogen atoms to the surface. For example, ITO surfaces can be covered with a monolayer of Cl atoms (Helander, Wang et al. 2011). Cl radicals are formed with UV light, and the radicals then chemically bond to the ITO surface. The monolayer of electronegative Cl atoms adds a dipole to the ITO surface that results in an increased work function. Other halogens (I, F, and Br) can also be attached to ITO in a similar fashion, and have a similar effect.

5.3.1.6 Oxygen Plasma and Ozone Treatments

Surface treatments, such as exposure to oxygen plasma or ozone, have been used to improve the surface properties of ITO (Ding, Hung et al. 2000; Sugiyama, Ishii et al. 2000; Kim, Lee et al. 2004). These treatments clean the ITO surface by removing surface hydrocarbons and leave the surface oxygen-terminated to provide a surface dipole that increases ITO's work function. Ozone has also been used to treat Au electrode surfaces (Helander, Wang et al. 2010). The highly oxidizing ozone atmosphere causes an Au-oxide monolayer to form that increases the Au work function and improves hole-injection.

5.3.1.7 Air Exposure

Some metals—such as Au—interact strongly with organic overlayers. This interaction results in a so-called interfacial dipole that results in a charge-injection barrier. The interfacial dipole interaction can be weakened by coating the Au surface with a thin layer of atmospheric contaminants (e.g., O_2, H_2O, and CO_2) (Wan, Hwang et al. 2005). The contaminated Au exhibits lower hole-injection barriers to hole-transporting organic materials. Air contamination was also performed on oxides to decrease the interfacial dipole barrier (Cheung, Song et al. 2010; Meyer, Shu et al. 2010). In some cases, the contamination had no effect on energy-level alignment or device performance (Meyer, Shu et al. 2010), while in other cases, improved energy-level alignment was observed (Cheung, Song et al. 2010; Meyer, Shu et al. 2010). However, it has also been shown that too much atmospheric contamination may eventually worsen energy-level alignment (Greiner, Helander et al. 2010).

5.3.2   Cathode Buffer Layers

Materials suitable for electron injection are somewhat more difficult to come by. In order to have a small electron-injection barrier into a high-lying LUMO level, an electrode or buffer layer must have a low work function. This necessarily means that it must be relatively easy to remove electrons from the electrode. Consequently, most low-work-function materials are very easy to oxidize and are thus not chemically stable. For example, the alkali and alkaline earth metals, Ca, Mg, Li, and Na, all have very low work functions, but they react chemically when in contact with organic molecules, or otherwise react with air and water to form an oxide film (Koch 2007). The challenge is to find a low-work-function material that is also chemically stable.

5.3.2.1   Inorganic Salts

Several alkali-metal halides—such as LiF (Hung, Tang et al. 1997), CsF (Jabbour, Kippelen et al. 1998), NaF (Lee, Park et al. 2003), MgF_2 (Park, Lee et al. 2001), and CaF_2 (Lee, Park et al. 2002)—have been used as cathode buffer layers. The most commonly used is LiF. It can be thermally evaporated onto organic materials without damaging the organic materials and acts as a barrier to protect the organic materials when the low-work-function metal electrode is deposited. Cesium carbonate (Cs_2CO_3) has also been used to form low-work-function electrodes (Chen and Wu 2008). Cs_2CO_3 can be thermally evaporated and solution deposited (Huang, Xu et al. 2007).

5.3.2.2   Metal Oxides

Some transition metal oxides have also demonstrated the capability of electron injection—such as TiO_2, ZnO, and ZrO_2 (Bolink, Coronado et al. 2007; Bolink, Coronado et al. 2008; Tokmoldin, Griffiths et al. 2009). However, these oxides tend to have very high sublimation temperatures, making processing difficult. Some methods have been developed for solution deposition of cathode oxides (Waldauf, Morana et al. 2006; de Bruyn, Moet et al. 2012).

5.3.2.3   Organic Molecules and Polymers

Various electron-transporting organic materials—such as Alq_3, and various oxadiazole-, triazole-, pyridine-, triazine-, and quinoline-based molecules (Hughes and Bryce 2005)—can be used as electron-injection materials as a consequence of their high LUMO levels. However, the shallow LUMO level of some emissive materials may prevent many of these materials from being efficient electron injectors. Alternatively, one can alter the cathode's work function by inducing a surface dipole.

Several polymers have been used as cathode buffer layers, where they introduce an interfacial dipole to the electrode surface that lowers the electrode's work function. An insulating polymer called PEIE has been used as a surface modifier to lower the work functions of a number of metals and TCOs to as low as ~3.2 eV (Zhou, Fuentes-Hernandez et al. 2012). This polymer introduces a surface dipole to improve electron injection, and was shown to be a versatile method of adapting electrodes for good electron injection. Low-work-function polymer layers are solution deposited from hydrophobic or hydrophilic solvents. Several conductive low-work-function polymers have also been effectively used as cathode buffer layers, including poly[9,9-bis(6′-(diethanolamino) hexyl) fluorine] (PFN-OH) (Huang, Niu et al. 2007) and various water-soluble polymers with surfactant-like sidechains (Ma, Iyer et al. 2005; Oh, Vak et al. 2008; Ma, Yip et al. 2010).

5.4 Theories of Energy-Level Alignment

Understanding the energy-level alignment of organic–electrode interfaces has been the central focus of many research groups for about two decades. The relevant energy levels for organic materials are the HOMO—which conducts holes—and the LUMO—which conducts electrons. For electrodes, the relevant energy levels are the Fermi level, the valence band, and the conduction band.

These energy levels are measured as ionization energy (IE), which represents the position of the HOMO level relative to the vacuum level, the electron affinity (EA), which represents the position of the LUMO level relative to the vacuum level, and the work function (ϕ), which represents the position of the Fermi level relative to the vacuum level.

5.4.1 Work Function, Fermi Level, and Local Vacuum Level

Before delving into the theories of energy-level alignment, it is helpful to review the concept of work function. Work function is often considered to be one of the most important factors governing the energy-level alignment between molecular adsorbates and solid substrates. The importance of work function is clearly demonstrated in numerous reviews (Steinberger and Wandelt 1987; Jablonski and Wandelt 1991; Ishii, Sugiyama et al. 1999; Cahen and Kahn 2003; Koch 2007; Braun, Salaneck et al. 2009). The reason why it is so important is that it represents the energetic requirements of removing or adding an electron to or from a substrate material. Thus, the energetic barrier to charge exchange at a metal–molecule interface is a function of the work function and the molecule's donor or acceptor levels (Cahen and Kahn 2003). Differences in work function are also believed to determine which way electrons will flow when two materials are placed in contact.

The work function of a material (measured via photoelectron spectroscopy or Kelvin probe measurements) has two contributions: (1) electron chemical potential and (2) surface dipole (Wandelt 1997; Ishii, Sugiyama et al. 1999; Cahen and Kahn 2003). As illustrated in Figure 5.3, the electron chemical potential (μ_e) represents the Fermi energy (E_F) relative to the absolute vacuum level ($E_{V\infty}$). The surface dipole (δ) represents an additional electrostatic barrier to removing an electron from a solid surface.

The Fermi energy can be considered the chemical potential of electrons in a solid, relative to the absolute vacuum level. Chemical potential is defined as the partial derivative of Gibbs free energy with respect to the number of particles. The chemical potential of a particle—denoted by i—is given by

$$\mu_i \equiv \left(\frac{\partial G}{\partial N_i} \right)_{T,P,N_{j \neq i}} \tag{5.1}$$

where G is the Gibbs free energy, N is the particle number, and all other variables (T, P, and $N_{j \neq i}$) are held constant. Thus, the Fermi energy represents the energy of adding (or removing) an electron to (or from) a solid.

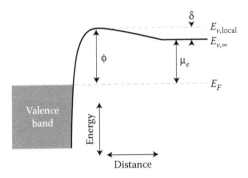

FIGURE 5.3 Illustration of electrostatic energies near the surface of a solid, indicating work function (ϕ), Fermi level (E_F), surface dipole (δ), electron chemical potential (μ_e), local vacuum level ($E_{v,local}$), and absolute vacuum level ($E_{v,\infty}$).

When an electrically conductive solid is grounded (i.e., equilibrated with "earth"), the solid's electron chemical potential will equilibrate with that of *earth*, and their Fermi levels will align via charge transfer. Thus, a solid with a lower electron chemical potential than earth will acquire additional electrons (i.e., become charged), which creates a potential at the solid's surface that gives rise to a so-called local vacuum level. Note that the local vacuum level is different from the absolute vacuum level. The local vacuum level represents the electrostatic energy of a point charge just outside the surface of a solid, while the absolute vacuum level is a quantity that represents the energy of a point charge at rest, infinitely far from any other object or charged particle, and under no external force or field (Cahen and Kahn 2003).

Figure 5.4a shows a schematic energy-level diagram of an electrically conductive, solid sample that is electrically isolated. The vertical scale represents energy and the horizontal scale represents distance. This is a hypothetical situation in which a solid material is completely isolated from any other material—that is, floating in infinite space with no external electric fields. There is an energy level associated with "ground"

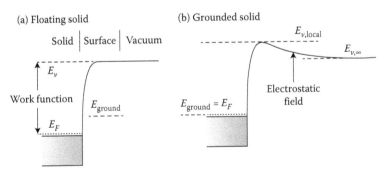

FIGURE 5.4 (a) Schematic energy-level diagram of the hypothetical situation for a solid material, floating in infinite space, with no external electric fields. (b) Schematic energy-level diagram of the solid material after being grounded, where the sample's Fermi level equilibrates with the "ground" energy level. The vertical scale is energy and the horizontal scale is distance. $E_{v,local}$ represents the local vacuum level, $E_{v,\infty}$ represents the absolute vacuum level, E_{ground} represents the electron chemical potential of "ground," and E_F represents the solid's Fermi level.

(E_{ground}). "Ground" is the electrochemical potential of an infinite reservoir of electrons. Experiments have positioned this energy level at ~4.09 eV below the absolute vacuum level (Anderson and Nyberg 1990). Figure 5.4b shows the scenario where the solid becomes grounded. Assuming the solid has sufficient conductivity for electron chemical potentials to equilibrate, the solid's Fermi level (E_F) aligns with that of ground. This results in an electrostatic field just outside the surface of the solid—as indicated in Figure 5.4b. This electrostatic field gives rise to the so-called local vacuum level.

The local vacuum level of a material and absolute vacuum level are generally not equal. Unfortunately, the absolute vacuum level is not a particularly useful reference level. It is the local vacuum level and Fermi level that govern energy-level alignment and charge transfer.

When one measures the work function of a material, the measured value represents the difference between the Fermi level and the local vacuum level. However, in addition to the surface potential due to grounding, there is another contribution to the local vacuum level: the *surface dipole*. The surface dipole represents an additional electrostatic barrier at the surface of a solid. This barrier can be caused by an electrostatic field formed by electron density spilling from the solid's surface, but it can also be altered by adsorption of polar molecules to a surface. The surface dipole depends on crystal orientation (Smoluchowski 1941), surface roughness (Smoluchowski 1941; Li and Li 2005), a material's dielectric properties, and the presence of adsorbed molecules (Steinberger and Wandelt 1987; Wandelt 1997).

Unfortunately, there is no known method to experimentally determine what portion of the measured work function comes from the surface dipole and what portion comes from the solid's electron chemical potential (Ishii, Sugiyama et al. 1999). As a result of the many contributing factors that can alter the surface dipole and the many effects that can alter the electron chemical potential—such as doping concentration, chemical changes, oxidation state—reported work function values quoted for materials can fluctuate wildly. In fact, work function should not be considered to be a material constant.

However, under appropriate preparation conditions, reproducible work function values can be obtained. For example, polycrystalline sputter-cleaned Au can be reproducibly prepared to yield a work function of 5.35 eV. Polycrystalline Cu will have a work function of 4.66 eV. Polycrystalline MoO_3 prepared by vacuum sublimation of MoO_3 and without exposure to adsorbates will have a work function between 6.8 and 6.9 eV.

So, while at times it may seem like there is an excessive amount of variation in reported work function values for a given material, it is likely because work function is so sensitive to many factors. However, there have been numerous experiments showing that work function affects the energy-level alignment of adsorbed molecules. This is a result of the effect of the electrostatic field outside the surface of a grounded sample on an adsorbed molecule's orbital energies.

Consider what happens when a molecule adsorbs to the sample surface. Figure 5.5a shows an energy-level diagram for a molecule separated by infinite distance from a solid's surface. The solid is grounded, so there is a surface electrostatic field. The molecule's vacuum level aligns with the absolute vacuum level (Ishii, Sugiyama et al. 1999).

As the molecule comes close to the solid's surface—depicted in Figure 5.5b—the molecule interacts with the surface electrostatic field. This field shifts all of the

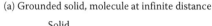

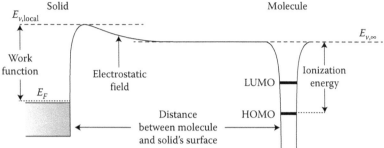

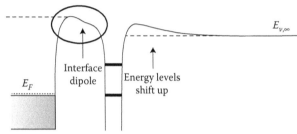

FIGURE 5.5 (a) Energy-level diagram of a grounded solid (right) and an isolated molecule at infinite distance away from the solid. (b) Scenario where the molecule comes close to the solid's surface. The surface electrostatic field pushes the molecule's energy levels upwards.

molecular orbital energies upward (Ishii, Sugiyama et al. 1999). It was originally thought that one could predict the energy-level alignment between a molecule and a solid by comparing the gas-phase IE of the molecule (i.e., the HOMO energy relative to the absolute vacuum level) with the work function of the solid. This is known as the "vacuum-level alignment rule." However, it was later realized that molecular adsorption causes a change to the solid's surface dipole that results in a drop in electrostatic potential across the solid–molecule interface, known as the interfacial dipole (Ishii and Seki 1997; Ivanco, Netzer et al. 2007). The physical origin of this interfacial dipole, and how its magnitude could be predicted, was the focus of much research over the past two decades. It is generally now believed that the interfacial dipole arises due to several possible mechanisms involving charge transfer or electron repulsion (Ishii and Seki 1997; Ishii, Sugiyama et al. 1999; Braun, Salaneck et al. 2009; Flores, Ortega et al. 2009; Hwang, Wan et al. 2009).

5.4.2 Weakly Interacting Interfaces, Fermi–Level Pinning, and the ICT Model

Several theories have developed to explain energy-level alignment of organic semi-conductors. While the prominent theories are mutually consistent with one another, the model that best describes a given interface depends on the strength of the

Chapter 5

organic–electrode interactions. Here, we address three different scenarios: (1) weakly interacting interfaces, (2) moderately interacting interfaces, and (3) strongly interacting interfaces.

When the incoming material is only physisorbed at an interface, there is a convenient relationship between energy-level alignment, organic IE/EA, and electrode work function. The trend shows that when an electrode's work function is within a certain range, the hole-injection barrier and electron-injection barriers are linearly dependent on the electrode work function, as illustrated in Figure 5.6. This is referred to as the "vacuum-level alignment" regime (even though vacuum levels of electrodes and organic materials often do not really align perfectly here). *Increasing* the electrode work function will *decrease* the hole-injection barrier and *increase* the electron-injection barrier. Conversely, *decreasing* the electrode's work function will *increase* the hole-injection barrier and *decrease* the electron-injection barrier.

If an electrode's work function is so high that it exceeds a certain threshold value (the magnitude of the threshold value depends on the organic molecule), then the HOMO level becomes "pinned" to the Fermi level. "Pinned" means that the alignment between the HOMO and the Fermi level stays constant with further increases in electrode work function. Conversely, if an electrode's work function is lower than another threshold value, the Fermi level becomes "pinned" to the LUMO level, meaning the alignment between the Fermi level and LUMO level does not change with further decreases in electrode work function.

This trend was first observed for polymer–polymer interfaces by Tengstedt et al. in 2006, for which they proposed the integer charge-transfer model (ICT model) (Tengstedt, Osikowicz et al. 2006). They proposed that the threshold values for which pinning occurs are the positive polaronic state (for HOMO-level pinning) and the negative polaronic state (for LUMO-level pinning). Polaronic levels are related to IE and

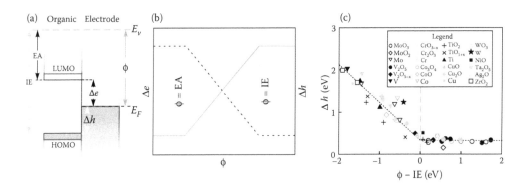

FIGURE 5.6 (a) Schematic energy-level diagram of an organic–electrode interface, where IE and EA are the organic's ionization energy and electron affinity, respectively. Δh and Δe are the hole- and electron-injection barriers, respectively. E_F is the Fermi level, E_v is the vacuum level, and ϕ is the electrode's work function. (b) An illustration of the energy-alignment trend for weakly interacting systems, showing how hole-injection barrier and electron injection barrier depends on work function, ionization energy, and electron affinity. (c) Experimental photoemission measurements of energy-level alignment at the anode interface for several organic semiconductors on numerous metals and metal oxides. (Adapted from Greiner, M. T. et al. 2012. *Nature Materials* 11(1): 76–81.)

EA except that there is an additional relaxation component caused by electron-density distortions on highly polarizable molecules. The positive polaronic level represents the energy of removing an electron from the polymer's HOMO level; thus, it is the sum of the polymer's IE and an additional relaxation energy caused by molecular polarization. Likewise, the negative polaronic level represents the sum of the EA and a relaxation energy.

The energy-alignment trend associated with the ICT model has now been confirmed on several occasions (Crispin, Crispin et al. 2006; Fukagawa, Kera et al. 2007; Osikowicz, de Jong et al. 2007; Lange, Blakesley et al. 2011; Greiner, Helander et al. 2012). While the original study used photoemission spectroscopy (PES) to observe the energy-alignment trend—which introduces concerns regarding the surface photo-voltage caused by the photoemission probes—Lange et al. have confirmed the same trend using Kelvin probe methods (Lange, Blakesley et al. 2011).

While Tengstedt's original study demonstrated the ICT energy-alignment trend using *ex situ*-prepared electrodes and spin-cast polymers, recently, Greiner et al. demonstrated the same behavior using a broad range of *in situ*-prepared metals and metal oxides (Greiner, Helander et al. 2012). In this study, it was found that the organic molecules' IEs and EAs—rather than their polaronic levels—were the relevant critical pinning levels. The discrepancy between this study and Tengstedt et al.'s finding is likely due to the smaller relaxation stabilizing energy of small molecules than large polymers. Thus, in the case of small molecules, the positive and negative polaronic energies approach the IE and EA values, respectively.

In the HOMO-pinning regime, hole-injection barriers appear to be molecule-independent (Greiner, Helander et al. 2012). The HOMO binding energies found in Greiner et al.'s study were pinned at ca. 0.3 eV below the Fermi level. It is yet uncertain how general this finding is, as the molecules examined in Greiner et al.'s study were all rather similar to one another.

Here, we will discuss the physical origin behind the ICT trend by considering what happens at the anode interface as the anode work function is progressively increased. In the linear region of Figure 5.6b—the so-called vacuum-level alignment region—the relation between HOMO binding energy (Δh) and electrode work function occurs due to the linear increase in surface potential as an electrode's work function increases. This is illustrated in Figure 5.7a and b. In this region, the molecules adsorbed to the electrode's surface remain neutral and no charge transfer occurs across the interface.

The onset of the *pinning region* occurs when the electrode work function equals or exceeds the organic's IE/positive-polaronic energy. At this point, owing to the coupling of molecular ionization processes with the electrode's Fermi level, electrons will hop from the molecule's HOMO to the electrode's Fermi level. This results in ionized molecules at the interface, as illustrated in Figure 5.7c.

As the electrode's work function continues to increase, the number of ionized molecules increases, as illustrated in Figure 5.7d. This corresponds to the pinning region in the HOMO versus ϕ relationship. Electron transfer, from the molecules to the substrate's Fermi level, results in high concentrations of ionized molecules at the interface. The electrostatic field generated from the ionized molecules keeps the HOMO level pinned to the Fermi level.

Chapter 5

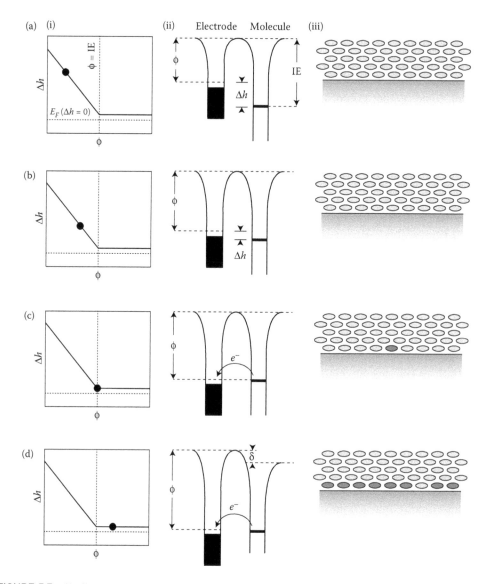

FIGURE 5.7 (i) Illustration of the observed relation between HOMO binding energy and oxide work function, (ii) schematic energy-level diagram, and (iii) illustration of the microscopic picture of an oxide–molecule interface for four scenarios: (a) $\phi_{ox.} < IE_{mol.}$, (b) $\phi_{ox.} = IE_{mol.}$, (c) $\phi_{ox.} > IE_{mol.}$, and (d) $\phi_{ox.} \gg IE_{mol.}$.

5.4.3   Moderately Interacting Interfaces and the IDIS Model

While the ICT model works well to describe interfaces in which the molecules do not interact strongly with the electrode, when stronger interactions occur—such as significant wavefunction overlap between the molecular orbitals and the electrode's surface electron bands—deviations from the ICT behavior occur.

In particular, for many metal–organic interfaces, under the conditions for which vacuum-level alignment should be observed (i.e., $IE > \phi > EA$), it is found that the slope that describes the linear correlation between Δh (or Δe) and ϕ is not unity (Ishii and

Seki 1997; Hill, Rajagopal et al. 1998; Koch and Vollmer 2006). In fact, the slope of this relationship varies from molecule to molecule. This observation gives rise to the *slope parameter*, defined in

$$S = -\frac{\Delta h}{\Delta \phi} \tag{5.2}$$

This parameter describes the relationship between the electrode's Fermi-level position relative to the HOMO level (i.e., the HOMO binding energy, or Δh) and the electrode work function. In the case of vacuum-level alignment, $S = 1$. In the case of Fermi-level pinning, $S = 0$. For the moderately interacting interfaces, S is somewhere between these two extremes.

These findings eventually gave rise to the *induced-density-of-interface-states* (IDIS) model being applied to organic interfaces by Vázquez et al. (Vazquez, Flores et al. 2004; Vazquez, Gao et al. 2005; Vazquez, Dappe et al. 2007; Vazquez, Flores et al. 2007). In this model, wavefunction overlap between the metal electrode electron bands and the adsorbed organic semiconductor molecular orbitals gives rise to interface states. However, the wavefunction overlap is *moderate* and—instead of giving rise to local-ized chemical bonds—it gives rise to a Lorentzian-type broadening of the molecular orbital wavefunctions at the interface. This broadening results in a continuum of elec-tronic states within the HOMO–LUMO energy gap. The continuum of interface states tends to be filled with electrons up to the charge-neutrality level (CNL) of the organic semiconductor.

It was proposed that the driving force for energy-level alignment is the difference between the electrode Fermi level and the organic CNL. The slope parameter represents the degree by which the difference ($\phi - E_{\mathrm{CNL}}$) is screened, so that the difference between the Fermi level and the CNL when the molecule and metal are in contact is given by

$$E_F - E_{\mathrm{CNL}} = S(\phi - E_{\mathrm{CNL}}) \tag{5.3}$$

where E_F is the Fermi level, E_{CNL} is the CNL, S is the slope parameter, and ϕ is the work function. This scenario is illustrated in Figure 5.8. Figure 5.8a shows the energy lev-els of a metal and organic molecule at infinite separation. The difference ($\phi - E_{\mathrm{CNL}}$) is indicated by the arrows. Figure 5.8b shows the energy levels of the metal and organic molecule once they are in contact. The shaded region in between the metal and organic energy-level diagrams represents the continuum of gap states. Once contact is made, charge transfer occurs to move the organic CNL closer to the metal Fermi level. This process induces a vacuum-level shift. The final position of the organic CNL, relative to the Fermi level, depends on the original metal work function, the energy of the CNL, and the screening parameter. The screening parameter varies from one material combi-nation to another. It was proposed that the screening parameter depends on the density of interface states.

The IDIS model has been successfully applied to predict the energy-level alignment of many metal–organic and organic–organic interfaces.

Chapter 5

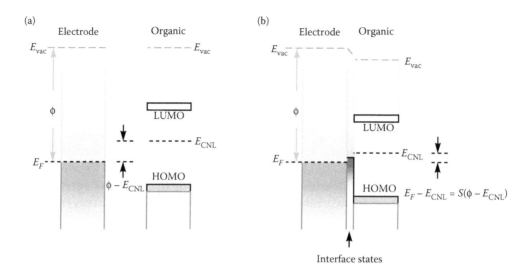

FIGURE 5.8 (a) Energy-level diagrams of a metal and molecular organic semiconductor at infinite separation. (b) Energy-level diagrams of a metal in contact with an organic semiconductor, where the contact gives rise to a continuum of gap states.

5.4.4 Strongly Interacting Interfaces and Chemical Reactivity

When there is a strong localized overlap between wavefunctions at an interface, it is possible to form localized chemical bonds between the electrode and organic molecule. This scenario has been shown numerous times experimentally (Greczynski, Fahlman et al. 2000; Schlaf, Merritt et al. 2001; Zou, Kilian et al. 2006; Tautz 2007; Duhm, Gerlach et al. 2008; Koch 2008). An example is shown in Figure 5.9.

In this example of PTCDA on various coinage metals, the *s*- and *d*-bands of the metals overlap strongly with the PCTDA HOMO and LUMO levels. The result is the formation of distinct electronic states within the PTCDA HOMO–LUMO gap. These states can clearly be seen in the ultraviolet photoemission spectroscopy (UPS) spectra shown in Figure 5.9a.

The energy-level alignment in these scenarios is not trivial to predict, at least using simple models. Instead, it is necessary to calculate the chemical bond formation using density functional theory (DFT) calculations, or at least molecular orbital arguments. Such calculations are material specific, as the formation of chemical bonds depends on the precise energy and special distribution of molecular orbital and solid-state-band wavefunctions.

5.5 Measuring Energy-Level Alignment

5.5.1 Photoemission Spectroscopy

Interfacial energy-level alignment is most commonly measured using PES. PES makes use of the photoelectric effect to generate a spectrum that represents electronic energy levels. In the photoemission process, incident photons collide with bound electrons,

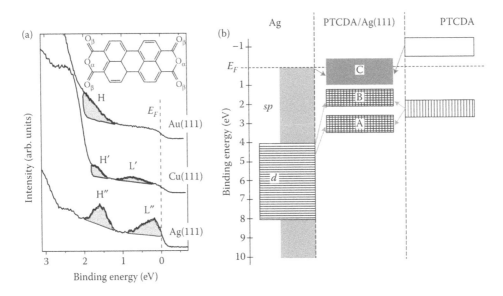

FIGURE 5.9 (a) UPS spectra of PTCDA on Au(111), Cu(111), and Ag(111) surfaces. Shaded regions indicate interface states. (Adapted from Koch, N. 2008. *Journal of Physics—Condensed Matter* 20(18): 184008.) (b) Schematic energy-level diagram, illustrating how interaction between molecular orbitals and metal bands gives rise to interface states. (Adapted from Zou, Y. et al. 2006. *Surface Science* 600(6): 1240–1251.)

transferring energy to the electrons. If the amount of energy transferred is greater than the potential energy keeping the electrons bound to the sample, then the electrons will emit from the sample. The potential energy that keeps an electron bound to a sample is the sum of the electron's binding energy (E_B) and the sample's work function (ϕ).

An electron's binding energy represents the electrostatic attraction between the electron—in a given quantum mechanical electronic state (i.e., 1s, 2s, 2p, etc.)—and the effective charge of the nucleus. The work function represents an additional potential well that is associated with condensed materials.

To illustrate the photoemission process, a schematic of the electron configuration of copper is shown in Figure 5.10a. Here, $h\nu$ is the photon energy, E_F is the Fermi level, E_V is the vacuum level, ϕ is the work function, $E_K^{(2p)}$ and $E_K^{(3p)}$ are the kinetic energies of electrons photoemitted from the Cu 2p and Cu 3p electron orbitals, respectively, and $E_B^{(2p)}$ and $E_B^{(3p)}$ are the binding energies of electrons photoemitted from the Cu 2p and Cu 3p electron orbitals, respectively.

Photoemission occurs when $h\nu > E_B + \phi$. When this is the case, photoemitted electrons leave the sample surface carrying excess energy in the form of kinetic energy (E_k). The amount of kinetic energy it carries is: $E_k = h\nu - E_B - \phi$. By determining the kinetic energy of the photoemitted electrons, one can determine their binding energies, giving rise to a spectrum as shown in Figure 5.10b. Each electron orbital gives rise to a peak in the photoemission spectrum. The intensities of the peaks in the spectrum are proportional to the number density of electrons with a given binding energy. There are

Chapter 5

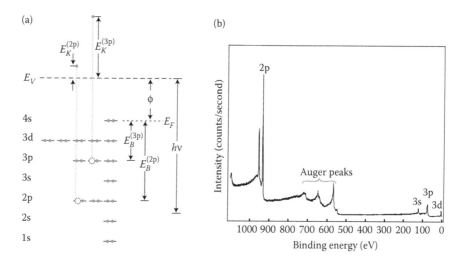

FIGURE 5.10   (a) Schematic energy diagram showing the electron configuration of copper and the photoemission process. E_V is the vacuum level, E_F is the Fermi level, ϕ is the work function, $h\nu$ is the photon energy, $E_K^{(2p)}$ and $E_K^{(3p)}$ are the kinetic energies of photoemitted Cu2p and Cu3p electrons, and $E_B^{(2p)}$ and $E_B^{(3p)}$ are the binding energies of Cu2p and Cu3p electrons. (b) XPS spectrum of copper metal, with core-level peaks and Auger peaks indicated.

also features from multielectron emission processes, such as Auger electrons, and from inelastically scattered electrons.

This example uses x-rays for photoemission (in this case, Al K_α photons, $h\nu = 1486.7$ eV). With such high photon energies, core-level electrons can be ionized. However, in energy-level alignment measurements, it is more common to use photons in the ultraviolet energy range (~5–100 eV). This type of spectroscopy is called UPS. UPS has been widely used in catalysis and adsorption studies over the past 40 years. It has become an extremely popular technique for studying organic semiconductor interfaces in the past 15 years. UPS is primarily used to determine work function, IE, and the position of the HOMO level relative to the Fermi level (i.e., the HOMO binding energy).

Low-energy UV photons are incapable of photoemitting from most core levels, but they are sensitive to valence electrons. In laboratory spectrometers, the commonly chosen photon energy is 21.22 eV, which is the He I_α emission line from a helium plasma discharge lamp. In synchrotron facilities, one can choose from a continuum of photon energies.

To a first approximation, UPS spectra represent the valence density of states of a material. Thus, UPS is primarily used for measuring a material's valence-band electronic structure. In contrast to core levels—where electron orbital binding energies can be separated by tens to hundreds of eV—there are often dozens of valence levels within only a few eV of one another. Consequently, most UPS spectra are a dense convolution of peaks.

UPS spectra are like a fingerprint, whose details can only be fully understood with the aid of molecular orbital calculations, and often require more sophisticated calculation methods such as DFT. For instance, in extended solids, valence states are delocalized and thus do not take on a Gaussian shape. In these cases, valence band structures can only be accurately interpreted using complex calculations of the density of states. Furthermore, in charge-transfer insulators and Mott–Hubbard insulators, electron correlation effects, and multielectron excitation processes are involved in the photoemission process. Consequently, the valence spectra of some materials are not conducive to an interpretation using band theory or "density-of-states."

In spite of the complexities of valence spectra, many simple interpretations are often possible, as is the case with many metals, such as Au. The UPS spectrum of sputter-cleaned polycrystalline Au is shown in Figure 5.11. This spectrum was measured using nonmonochromated He I_α radiation ($h\nu = 21.22$ eV). Thus, only the electronic states with $E_B + \phi < 21.22$ eV are measured. The zero on the binding energy scale represents Fermi level of the sample; in this case, the sample is grounded to the spectrometer, which means the sample and spectrometer Fermi levels are aligned.

The panel on the left shows the general features of a UPS spectrum: the secondary-electron background, the secondary-electron cut-off, the valence states, and the Fermi level. The panel on the right shows an expanded view of the valence features. The contributions from the Au s-band and d-band are highlighted. In this case, the UPS spectrum is a good representation of the valence band density of states. The spectrum matches well with theoretical density-of-states calculations that show the Au valence band consists of a narrow d-band and a broad s-band. The s-band crosses the Fermi level. One can also see that the electron density drops off at the Fermi level, where the spectrum takes on the shape of the Fermi–Dirac distribution.

The secondary-electron cut-off is also a valuable feature of UPS spectra. It is used to determine the work function. The secondary-electron signal arises from a cascade of inelastically scattered electrons. These electrons have lost much of their kinetic energy while passing through the sample so their intensities no longer represent valence states.

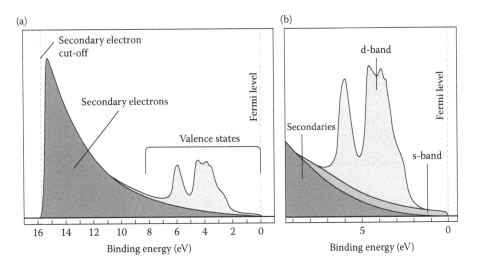

FIGURE 5.11   Illustration of the He I_α UPS spectrum of sputter-cleaned gold using binding energy as the x-axis. (a) Spectrum showing the entire valence band as well as the secondary-electron cut-off. (b) Expanded view of the shallow valence region, illustrating the contributions from various electron bands.

Chapter 5

The secondary-electron intensity increases rapidly toward low kinetic energy, and then cuts off abruptly at zero kinetic energy. This cut-off indicates that electrons have lost so much of their kinetic energy that they are no longer able to escape the surface of the sample.

To illustrate how the work function is determined from the secondary-electron cut-off, Figure 5.12 is replotted using various energy scales for the *x*-axis. On the *binding energy* scale, the Fermi level is zero. On the *ionization energy* scale, the local vacuum level is zero. And on the *kinetic energy* scale, the secondary-electron cut-off is zero. One can determine the vacuum level by drawing a line—whose length is equal to the photon energy—from the secondary-electron cut-off, toward the high-kinetic-energy end of the spectrum. The work function is then the difference between the photon energy and the kinetic energy of electrons coming from the Fermi level.

While the valence states of metals are delocalized and form bands that do not take the form of Gaussian-shaped peaks, organic semiconducting molecules are isolated entities with closed electron shells. Their valence states are more localized and are well represented by Gaussian-shaped peaks. The UPS spectrum of a typical molecular organic semiconductor (α-NPD) is shown in Figure 5.13. One can see that there is still a secondary-electron cut-off, and thus one can still define the vacuum level for condensed molecular films.

While the UPS spectra of organic molecules consist of many peaks, the most important feature in energy-level alignment is the HOMO level. The HOMO is always the valence feature that is closest to the Fermi level. It is the occupied state that requires the

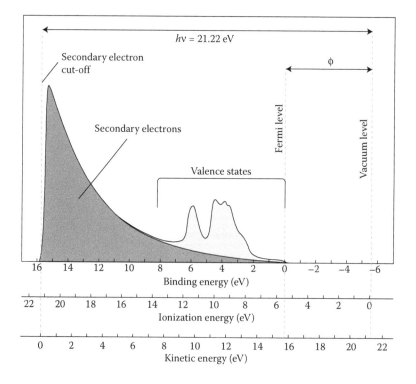

FIGURE 5.12 Illustration of a He Iα spectrum of sputter-cleaned gold, with the *x*-axis plotted using various scales: binding energy, ionization energy, and kinetic energy.

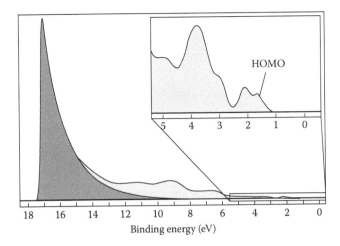

Binding energy (eV)

FIGURE 5.13 Illustration of the He $I\alpha$ UPS spectrum of an α-NPD film. The inset shows an expanded view of the HOMO level and shallow valence features.

least energy to ionize and therefore it is the state that is involved in electron transfer at an interface.

The binding energy of the HOMO level is generally taken to represent the hole-injection barrier, although the quantitative accuracy of this interpretation is under debate (Krause, Casu et al. 2008). As discussed in Section 5.4, the HOMO binding energy depends on the substrate onto which the organic film is coated. Thus, HOMO binding energy is not a constant. However, the IE of the HOMO (i.e., the sum of HOMO binding energy and work function) is independent of the substrate to which an organic material is adsorbed (except perhaps very near to the interface, where substrate polarization effects take effect) (Amy, Chan et al. 2005).

A compilation of UPS spectra for several common organic semiconductors is shown in Figure 5.14. The horizontal axes of Figure 5.14a and b are plotted in electron kinetic energy instead of binding energy because binding energy is substrate-dependent. Figure 5.14c shows an expanded view of the HOMO levels for each of these organic materials. The horizontal axis for Figure 5.14c is converted to IE (by subtracting the photon energy, 21.22 eV, from the kinetic energy scale). The HOMO ionization energies are shown for an each organic material.

5.5.2   Layer-by-Layer UPS to Measure Energy–Level Alignment

In energy-level alignment studies of organic molecules, one is primarily interested in the substrate work function (ϕ), HOMO binding energy (ΔE_H), and organic ionization energy (IE_{org}), as shown in the energy-level diagram in Figure 5.15. Figure 5.15b and c show UPS spectra and how interface parameters are extracted from UPS measurements (note that in Figure 5.15b and c, the secondary-electron cut-off is replotted 21.22 eV toward lower binding energy of the Fermi level because it represents the vacuum level, and plotting in this way makes the spectra easier to compare to the energy-level diagram).

Owing to the very shallow probe depth of UPS (<1 nm), it is common to characterize an interface using the layer-by-layer method. This method involves sequentially

Chapter 5

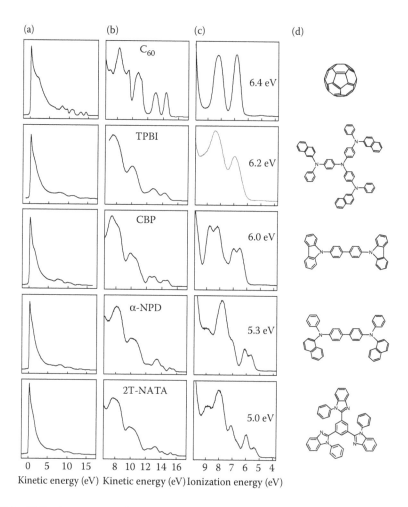

FIGURE 5.14 UPS valence spectra for several organic semiconductors. (a) Full spectrum, plotted in electron kinetic energy, (b) expanded view of valence features, plotted in kinetic energy, and (c) expanded view of HOMO features, plotted in ionization energy. The numbers shown in (c) are the HOMO ionization energies. Molecular structures are shown in (d). (Adapted from Greiner, M. T. 2012. *Transition Metal Oxides in Organic Electronics*. Ph.D., University of Toronto.)

depositing ultra-thin layers of organic molecules onto a substrate. UPS measurements are taken after each layer is deposited, as illustrated in Figure 5.16.

The layer-by-layer measurement routine generates a stack of valence and secondary cut-off spectra. The purpose for measuring HOMO binding energy and work function of the organic film in several steps, as opposed to using a single-point measurement is to generate a binding-energy-versus-thickness or work-function-versus-thickness profile. Examples of work-function, HOMO-binding-energy, and ionization-energy profiles are shown in Figure 5.17.

Figure 5.17 shows profiles from a common hole-transporting material, CBP, on various metal and metal-oxide electrodes, determined from layer-by-layer UPS measurements. One can see that the HOMO and work-function profiles are off-set from one another, depending on the electrode, while the ionization-energy profiles are essentially independent of the electrode.

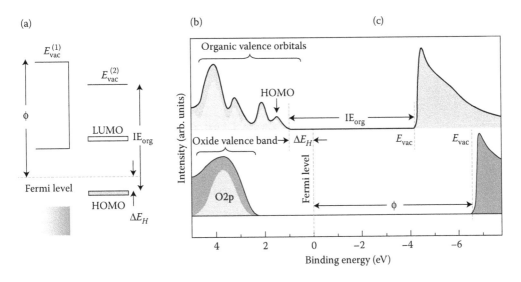

FIGURE 5.15 (a) Schematic energy-level diagram of an oxide–organic interface, with the substrate work function (Φ), vacuum level (E_{vac}), ionization energy (IE), and HOMO off-set (ΔE_H) indicated. (b) Valence photoemission spectrum of 2T-NATA (top) and MoO_3 (bottom). (c) Secondary-electron spectrum of 2T-NATA (top) and MoO_3 (bottom). The arrows indicate how the parameters in (a) are determined from the spectra in (b) and (c). (From Greiner, M. T. et al. 2012. *Nature Materials* 11(1): 76–81.)

UPS and layer-by-layer characterization methods have been extensively utilized to understand organic–electrode interfaces. They have been used to measure interfacial dipoles (Schlaf, Schroeder et al. 2000), band bending in organic films (Schlaf, Parkinson et al. 1998; Schlaf, Schroeder et al. 1999; Schroeder, France et al. 2002a,b), interfacial reactivity (Schlaf, Merritt et al. 2001; Schwieger, Peisert et al. 2004), molecular polarization near an interface (Peisert, Knupfer et al. 2002; Amy, Chan et al. 2005), Fermi-level pinning (Tengstedt, Osikowicz et al. 2006; Greiner, Helander et al. 2012), and trends in energy-level alignment (Weiler, Mayer et al. 2004).

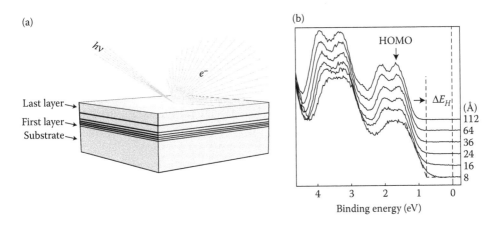

FIGURE 5.16 (a) Illustration of the layer-by-layer photoemission technique of interface characterization. (b) Stack of UPS spectra of an organic molecule, 4-4′-bis(carbazol-9-yl)-biphenyl (CBP), sequentially deposited onto a substrate. Note: A substrate reference spectrum was subtracted from these spectra for the very thin films (8–16 Å). (Adapted from Greiner, M. T. et al. 2012. *Nature Materials* 11(1): 76–81.)

Chapter 5

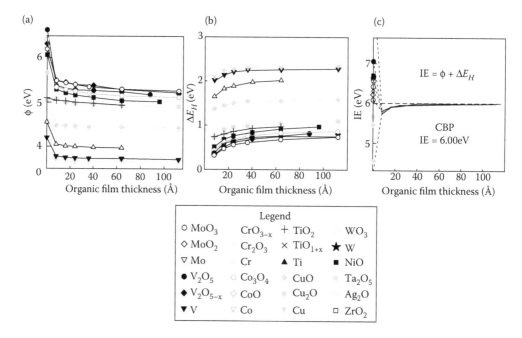

FIGURE 5.17   (a) Work-function profiles, (b) HOMO-binding-energy profiles, and (c) ionization-energy profiles determined from layer-by-layer UPS characterizations of CBP deposited onto various metal and metal-oxide electrodes. (Adapted from Greiner, M. T. et al. 2012. *Nature Materials* 11(1): 76–81.)

5.5.3   Transition Metal Oxide Buffer Layers

Transition metal oxides buffer layers are particularly versatile due to their diverse electronic properties and wide range of work functions. Transition metal oxides can have electronic structures that include n-type wide band gap semiconductors, defective semiconducting oxides, p-type semiconductors, p-type Mott–Hubbard insulators, and metallic conductors. This broad diversity of electronic structures allows oxides to be used, not only as charge-injection layers but also as rectifying charge-blocking layers. Schematic energy-level diagrams of various transition metal oxides and several common organic semiconductors are shown in Figure 5.18.

In addition to their diverse electronic properties, transition metal oxide work functions are also tunable by changing the cation oxidation and by changing the defect concentration. In general, there is a correlation between cation oxidation state and oxide work function, as shown in Figure 5.19a. Metal-oxide work functions tend to decrease with decreasing cation oxidation state.

Oxidation state and work function are correlated because, as one decreases the oxidation state of a cation, one decreases the cation's electronegativity (Li and Xue 2006). The overall electronegativity of a compound can be calculated using the concept of *group electronegativity*, which is the geometric mean of the electronegativity values of the individual components in a compound as shown in (Nethercott 1974)

$$\chi_{AB} = \left(\chi_A^m \chi_B^n\right)^{1/(m+n)} \tag{5.4}$$

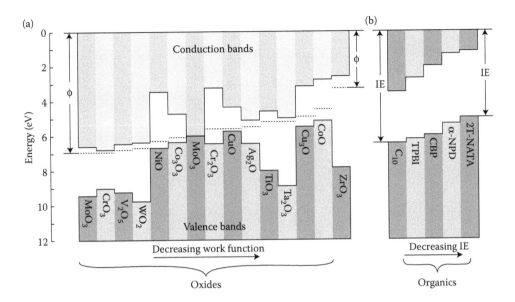

FIGURE 5.18 Schematic energy-level diagrams of (a) several transition metal oxides and (b) several organic semiconductors. The lower shaded regions represent the valence bands and the upper shaded regions represent the conduction bands. The dashed lines indicate the position of the Fermi level for each oxide. Oxides with their Fermi levels close to the valence band are p-type semiconductors and oxides with their Fermi levels close to the conduction band are n-type semiconductors. (Adapted from Greiner, M. T. et al. 2012. *Nature Materials* 11(1): 76–81.)

where χ_{AB} is the group electronegativity of compound A_mB_n, where m and n represent stoichiometric coefficients. χ_A and χ_B are the electronegativities of components A and B, respectively. Consequently, the overall "group electronegativity" of the oxide is decreased when low-oxidation-state cation concentrations are increased. Electronegativity is closely related with Fermi level. In fact, in Mulliken's definition, electronegativity is midway between the IE and EA (i.e., midgap) (Mulliken 1934). Likewise, the Fermi level in undoped semiconductors is midgap (Matar, Campet et al. 2011). Thus, there is a direct relationship between electronegativity and Fermi level (Nethercott 1974).

Using Equation 5.4, it is possible to calculate the electronegativity of an oxide. A plot of calculated Fermi level (based on group electronegativity) versus measured midgap position is shown in Figure 5.19b for various transition metal oxides.

Beyond compositional tuning, the work function of an oxide can be continually tuned by inducing defects (Greiner, Chai et al. 2012). For example, oxygen vacancy defects cause low-oxidation-state cation defects to arise in oxides. By increasing the oxygen-vacancy concentration, one can increase the concentration of low-electronegativity cations. The work function of an oxide can then be continually tuned using this method, as demonstrated for MoO_3 in Figure 5.20.

The effect of increasing the concentration of low-electronegativity cations on work function—in this case—can be calculated using the group electronegativity formulation, leading to

$$E_F = \left((\chi_{Mo^{6+}})^{1-2x}(\chi_{Mo^{5+}})^{2x}(\chi_{O^{2-}})^{3-x} \right)^{1/(4-x)} \tag{5.5}$$

Chapter 5

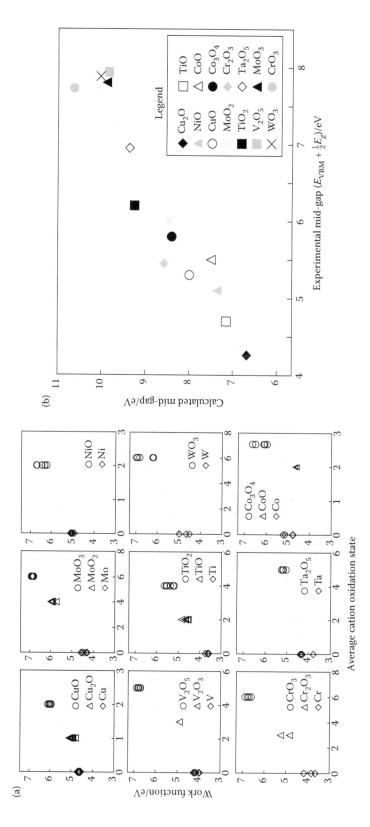

FIGURE 5.19 (a) Plots of work function versus nominal average oxidation state of metal atoms in several transition metals and transition metal oxides. (b) Calculated midgap positions (using ionic electronegativities) versus the experimental midgap position (determined from UPS measurements and literature band gaps) for several transition metal oxides. The dashed line shows the linear regression. (Adapted from Greiner, M. T. et al. 2012. *Advanced Functional Materials* 22(21):4557–4568.)

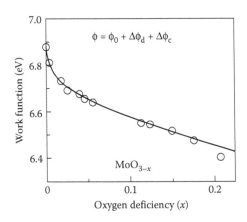

FIGURE 5.20 Plot of work function versus oxygen deficiency for MoO_3. (Adapted from Greiner, M. T. et al. 2012. *Advanced Functional Materials* 22(21): 4557–4568.)

where $\chi_{Mo^{6+}}$, $\chi_{Mo^{5+}}$, and $\chi_{O^{2-}}$ are the electronegativities of Mo^{6+}, Mo^{5+}, and O^{2-}, respectively, and x is the degree of nonstoichiometry in MoO_{3-x}.

Therefore, in order to have a high-work-function oxide, one would want to use a highly stoichiometric oxide with a cation that has a high electronegativity—such as V_2O_5, MoO_3, or CrO_3. Conversely, for a cation buffer layer, one would want to use an oxygen-deficient oxide and a cation with a low electronegativity, such as TiO_{2-x} or ZrO_{2-x}. Given the broad range of transition-metal-oxide work functions—as listed in Table 5.1—and the possibility of tuning their work functions continually, it is no surprise that transition metal oxides have found so much use in organic electronic devices.

Table 5.1 Summary of Work Function Values for Metals and Metal Oxides Measured by Photoemission Spectroscopy

Material	ϕ (eV)	Material	ϕ (eV)	Material	ϕ (eV)
TiO_2	5.4 ± 0.2	TiO	4.7 ± 0.2	Ti	3.7 ± 0.2
MoO_3	6.82 ± 0.05	MoO_2	5.9 ± 0.2	Mo	4.4 ± 0.2
CuO	5.9 ± 0.1	Cu_2O	4.9 ± 0.1	Cu	4.63 ± 0.03
NiO	6.3 ± 0.2	—	—	Ni	4.99 ± 0.07
WO_3	6.8 ± 0.4	$W_{18}O_{49}$	6.4 ± 0.1	W	4.8 ± 0.3
V_2O_5	6.8 ± 0.1	V_2O_3	4.9	V	4.0 ± 0.2
CrO_{3-x}	6.75 ± 0.2	Cr_2O_3	5.0 ± 0.5	Cr	4.0 ± 0.3
Ta_2O_5	5.2 ± 0.2	—	—	Ta	4.1 ± 0.4
Co_3O_4	6.3 ± 0.3	CoO	4.6 ± 0.2	Co	4.8 ± 0.2

Source: Adapted from Greiner, M. T. et al. 2012. *Advanced Functional Materials* 22(21): 4557–4568.

Note: Error ranges represent the 95% confidence interval; however, oxide work functions are biased toward low values, and metal work functions are biased toward high values.

Chapter 5

References

Amy, F., C. Chan et al. 2005. Polarization at the gold/pentacene interface. *Organic Electronics* 6(2): 85–91.

Anderson, S. E., G. L. Nyberg. 1990. Binding-energies and reference levels in photoelectron-spectroscopy. *Journal of Electron Spectroscopy and Related Phenomena* 52: 293–302.

Bardecker, J. A., H. Ma et al. 2008. Self-assembled electroactive phosphonic acids on ITO: Maximizing hole-injection in polymer light-emitting diodes. *Advanced Functional Materials* 18(24): 3964–3971.

Blyth, R. I. R., R. Duschek et al. 2001. Band alignment in organic devices: Photoemission studies of model oligomers on In_2O_3. *Journal of Applied Physics* 90(1): 270–275.

Bolink, H. J., E. Coronado et al. 2007. Air stable hybrid organic-inorganic light emitting diodes using ZnO as the cathode. *Applied Physics Letters* 91(22): 223501.

Bolink, H. J., E. Coronado et al. 2008. Inverted solution processable OLEDs using a metal oxide as an electron injection contact. *Advanced Functional Materials* 18(1): 145–150.

Braun, S., W. R. Salaneck et al. 2009. Energy-level alignment at organic/metal and organic/organic interfaces. *Advanced Materials* 21(14–15): 1450–1472.

Brown, T. M., J. S. Kim et al. 1999. Built-in field electroabsorption spectroscopy of polymer light-emitting diodes incorporating a doped poly(3,4-ethylene dioxythiophene) hole injection layer. *Applied Physics Letters* 75(12): 1679–1681.

Cahen, D., A. Kahn. 2003. Electron energetics at surfaces and interfaces: Concepts and experiments. *Advanced Materials* 15(4): 271–277.

Chan, I. M., T. Y. Hsu et al. 2002. Enhanced hole injections in organic light-emitting devices by depositing nickel oxide on indium tin oxide anode. *Applied Physics Letters* 81(10): 1899–1901.

Chen, M. H., C. I. Wu. 2008. The roles of thermally evaporated cesium carbonate to enhance the electron injection in organic light emitting devices. *Journal of Applied Physics* 104(11): 113713.

Cheung, C. H., W. J. Song et al. 2010. Role of air exposure in the improvement of injection efficiency of transition metal oxide/organic contact. *Organic Electronics* 11(1): 89–94.

Chun, J. Y., J. W. Han et al. 2009. Application of high work function anode for organic light emitting diode. *Molecular Crystals and Liquid Crystals* 514: 445–451.

Crispin, A., X. Crispin et al. 2006. Transition between energy level alignment regimes at a low band gap polymer-electrode interfaces. *Applied Physics Letters* 89(21): 213503.

de Bruyn, P., D. J. D. Moet et al. 2012. All-solution processed polymer light-emitting diodes with air stable metal-oxide electrodes. *Organic Electronics* 13(6): 1023–1030.

Ding, X. M., L. M. Hung et al. 2000. Modification of the hole injection barrier in organic light-emitting devices studied by ultraviolet photoelectron spectroscopy. *Applied Physics Letters* 76(19): 2704–2706.

Duhm, S., A. Gerlach et al. 2008. PTCDA on Au(111), Ag(111) and Cu(111): Correlation of interface charge transfer to bonding distance. *Organic Electronics* 9(1): 111–118.

Flores, F., J. Ortega et al. 2009. Modelling energy level alignment at organic interfaces and density functional theory. *Physical Chemistry Chemical Physics* 11(39): 8658–8675.

Fukagawa, H., S. Kera et al. 2007. The role of the ionization potential in vacuum-level alignment at organic semiconductor interfaces. *Advanced Materials* 19(5): 665–668.

Ganzorig, C., K. J. Kwak et al. 2001. Fine tuning work function of indium tin oxide by surface molecular design: Enhanced hole injection in organic electroluminescent devices. *Applied Physics Letters* 79(2): 272–274.

Greczynski, G., M. Fahlman et al. 2000. Polymer interfaces studied by photoelectron spectroscopy: Li on polydioctylfluorene and Alq(3). *Thin Solid Films* 363(1–2): 322–326.

Greiner, M. T. 2012. *Transition Metal Oxides in Organic Electronics*. Ph.D., University of Toronto.

Greiner, M. T., L. Chai et al. 2012. Transition metal oxide work functions: The influence of cation oxidation state and oxygen vacancies. *Advanced Functional Materials* 22(21): 4557–4568.

Greiner, M. T., M. G. Helander et al. 2010. Effects of processing conditions on the work function and energy-level alignment of NiO thin films. *The Journal of Physical Chemistry C* 114(46): 19777–19781.

Greiner, M. T., M. G. Helander et al. 2012. Universal energy-level alignment of molecules on metal oxides. *Nature Materials* 11(1): 76–81.

Hamwi, S., J. Meyer et al. 2009. p-Type doping efficiency of MoO_3 in organic hole transport materials. *Applied Physics Letters* 94(25): 253307.

Hatton, R. A., S. R. Day et al. 2001. Organic electroluminescent devices: Enhanced carrier injection using an organosilane self assembled monolayer (SAM) derivatized ITO electrode. *Thin Solid Films* 394(1–2): 292–297.

Helander, M. G., Z. B. Wang et al. 2010. Oxidized gold thin films: An effective material for high-performance flexible organic optoelectronics. *Advanced Materials* 22(18): 2037–2040.

Helander, M. G., Z. B. Wang et al. 2011. Chlorinated indium tin oxide electrodes with high work function for organic device compatibility. *Science* 3326032: 944–947.

Hill, I. G., A. Rajagopal et al. 1998. Molecular level alignment at organic semiconductor-metal interfaces. *Applied Physics Letters* 73(5): 662–664.

Huang, F., Y. H. Niu et al. 2007. A conjugated, neutral surfactant as electron-injection material for high-efficiency polymer light-emitting diodes. *Advanced Materials* 19(15): 2010–2014.

Huang, J., Z. Xu et al. 2007. Low-work-function surface formed by solution-processed and thermally deposited nanoscale layers of cesium carbonate. *Advanced Functional Materials* 17(12): 1966–1973.

Huang, Q. L., G. A. Evmenenko et al. 2005. Covalently bound hole-injecting nanostructures. Systematics of molecular architecture, thickness, saturation, and electron-blocking characteristics on organic light-emitting diode luminance, turn-on voltage, and quantum efficiency. *Journal of the American Chemical Society* 127(29): 10227–10242.

Hughes, G., M. R. Bryce. 2005. Electron-transporting materials for organic electroluminescent and electrophosphorescent devices. *Journal of Materials Chemistry* 15(1): 94–107.

Hung, L. S., C. W. Tang et al. 1997. Enhanced electron injection in organic electroluminescence devices using an Al/LiF electrode. *Applied Physics Letters* 70(2): 152–154.

Hwang, J., A. Wan et al. 2009. Energetics of metal-organic interfaces: New experiments and assessment of the field. *Materials Science and Engineering R-Reports* 64(1–2): 1–31.

Im, H. C., D. C. Choo et al. 2007. Highly efficient organic light-emitting diodes fabricated utilizing nickel-oxide buffer layers between the anodes and the hole transport layers. *Thin Solid Films* 515(12): 5099–5102.

Irwin, M. D., B. Buchholz et al. 2008. p-Type semiconducting nickel oxide as an efficiency-enhancing anode interfacial layer in polymer bulk-heterojunction solar cells. *Proceedings of the National Academy of Sciences of the United States of America* 105(8): 2783–2787.

Ishii, H., K. Seki. 1997. Energy level alignment at organic/metal interfaces studied by UV photoemission: Breakdown of traditional assumption of a common vacuum level at the interface. *IEEE Transactions on Electron Devices* 44(8): 1295–1301.

Ishii, H., K. Sugiyama et al. 1999. Energy level alignment and interfacial electronic structures at organic metal and organic organic interfaces. *Advanced Materials* 11(8): 605–625.

Ivanco, J., F. P. Netzer et al. 2007. On validity of the Schottky-Mott rule in organic semiconductors: Sexithiophene on various substrates. *Journal of Applied Physics* 101(10): 103712.

Jabbour, G. E., B. Kippelen et al. 1998. Aluminum based cathode structure for enhanced electron injection in electroluminescent organic devices. *Applied Physics Letters* 73(9): 1185–1187.

Jablonski, A., K. Wandelt. 1991. Quantitative aspects of ultraviolet photoemission of adsorbed xenon—A review. *Surface and Interface Analysis* 17(9): 611–627.

Kampen, T. U. 2006. Electronic structure of organic interfaces—A case study on perylene derivatives. *Applied Physics A—Materials Science and Processing* 82(3): 457–470.

Khodabakhsh, S., D. Poplavskyy et al. 2004. Using self-assembling dipole molecules to improve hole injection in conjugated polymers. *Advanced Functional Materials* 14(12): 1205–1210.

Kim, S. Y., J. L. Lee et al. 2004. Effect of ultraviolet-ozone treatment of indium-tin-oxide on electrical properties of organic light emitting diodes. *Journal of Applied Physics* 95(5): 2560–2563.

Koch, N. 2007. Organic electronic devices and their functional interfaces. *ChemPhysChem* 8(10): 1438–1455.

Koch, N. 2008. Energy levels at interfaces between metals and conjugated organic molecules. *Journal of Physics—Condensed Matter* 20(18): 184008.

Koch, N., A. Vollmer. 2006. Electrode-molecular semiconductor contacts: Work-function-dependent hole injection barriers versus Fermi-level pinning. *Applied Physics Letters* 89(16): 162107.

Krause, S., M. B. Casu et al. 2008. Determination of transport levels of organic semiconductors by UPS and IPS. *New Journal of Physics* 10: 085001.

Kroger, M., S. Hamwi et al. 2009. Role of the deep-lying electronic states of MoO(3) in the enhancement of hole-injection in organic thin films. *Applied Physics Letters* 95(12): 123301.

Lange, I., J. C. Blakesley et al. 2011. Band bending in conjugated polymer layers. *Physical Review Letters* 106(21): 216402.

Lee, H., S. W. Cho et al. 2008. The origin of the hole injection improvements at indium tin oxide/molybdenum trioxide/N,N'-bis(1-naphthyl)-N,N'-diphenyl-1,1'-biphenyl-4,4'-diamine interfaces. *Applied Physics Letters* 93(4): 043308.

Chapter 5

Lee, J., B. J. Jung et al. 2002. Modification of an ITO anode with a hole-transporting SAM for improved OLED device characteristics. *Journal of Materials Chemistry* 12(12): 3494–3498.

Lee, J., Y. Park et al. 2002. Tris-(8-hydroxyquinoline)aluminum-based organic light-emitting devices with Al/CaF$_2$ cathode: Performance enhancement and interface electronic structures. *Applied Physics Letters* 80(17): 3123–3125.

Lee, J., Y. Park et al. 2003. High efficiency organic light-emitting devices with Al/NaF cathode. *Applied Physics Letters* 82(2): 173–175.

Li, K. Y., D. F. Xue. 2006. Estimation of electronegativity values of elements in different valence states. *Journal of Physical Chemistry A* 110(39): 11332–11337.

Li, W., D. Y. Li. 2005. On the correlation between surface roughness and work function in copper. *Journal of Chemical Physics* 122(6): 064708.

Ma, H., H. L. Yip et al. 2010. Interface engineering for organic electronics. *Advanced Functional Materials* 20(9): 1371–1388.

Ma, W. L., P. K. Iyer et al. 2005. Water/methanol-soluble conjugated copolymer as an electron-transport layer in polymer light-emitting diodes. *Advanced Materials* 17(3): 274–277.

Marmont, P., N. Battaglini et al. 2008. Improving charge injection in organic thin-film transistors with thiol-based self-assembled monolayers. *Organic Electronics* 9(4): 419–424.

Matar, S. F., G. Campet et al. 2011. Electronic properties of oxides: Chemical and theoretical approaches. *Progress in Solid State Chemistry* 39(2): 70–95.

Matsushima, T., G. H. Jin et al. 2008. Marked improvement in electroluminescence characteristics of organic light-emitting diodes using an ultrathin hole-injection layer of molybdenum oxide. *Journal of Applied Physics* 104(5): 054501.

Matsushima, T., G. H. Jin et al. 2011. Interfacial charge transfer and charge generation in organic electronic devices. *Organic Electronics* 12(3): 520–528.

Matsushima, T., Y. Kinoshita et al. 2007. Formation of Ohmic hole injection by inserting an ultrathin layer of molybdenum trioxide between indium tin oxide and organic hole-transporting layers. *Applied Physics Letters* 91(25): 253504.

Matsushima, T., H. Murata. 2009. Observation of space-charge-limited current due to charge generation at interface of molybdenum dioxide and organic layer. *Applied Physics Letters* 95(20): 203306.

Meyer, J., R. Khalandovsky et al. 2011. MoO$_3$ films spin-coated from a nanoparticle suspension for efficient hole-injection in organic electronics. *Advanced Materials* 23(1): 70–73.

Meyer, J., M. Kroger et al. 2010. Charge generation layers comprising transition metal-oxide/organic interfaces: Electronic structure and charge generation mechanism. *Applied Physics Letters* 96(19): 193302.

Meyer, J., A. Shu et al. 2010. Effect of contamination on the electronic structure and hole-injection properties of MoO$_3$/organic semiconductor interfaces. *Applied Physics Letters* 96(13): 133308.

Meyer, J., T. Winkler et al. 2008. Transparent inverted organic light-emitting diodes with a tungsten oxide buffer layer. *Advanced Materials* 20(20): 3839–3843.

Meyer, J., K. Zilberberg et al. 2011. Electronic structure of vanadium pentoxide: An efficient hole injector for organic electronic materials. *Journal of Applied Physics* 110(3): 033710.

Mulliken, R. S. 1934. A new electroaffinity scale; Together with data on valence states and on valence ionization potentials and electron affinities. *The Journal of Chemical Physics* 2(11): 782–793.

Murdoch, G. B., M. Greiner et al. 2008. A comparison of CuO and Cu$_2$O hole-injection layers for low voltage organic devices. *Applied Physics Letters* 93(8): 083309.

Nethercott, A. 1974. Prediction of Fermi energies and photoelectric thresholds based on electronegativity concepts. *Physical Review Letters* 33(18): 1088–1091.

Oh, S.-H., D. Vak et al. 2008. Water-soluble polyfluorenes as an electron injecting layer in PLEDs for extremely high quantum efficiency. *Advanced Materials* 20(9): 1624–1629.

Osikowicz, W., M. P. de Jong et al. 2007. Formation of the interfacial dipole at organic-organic interfaces: C-60/polymer interfaces. *Advanced Materials* 19(23): 4213–4217.

Park, S. W., J. M. Choi et al. 2005. Inverted top-emitting organic light-emitting diodes using transparent conductive NiO electrode. *Applied Surface Science* 244(1–4): 439–443.

Park, Y., J. Lee et al. 2001. Photoelectron spectroscopy study of the electronic structures of Al/MgF$_2$/tris-(8-hydroxyquinoline) aluminum interfaces. *Applied Physics Letters* 79(1): 105–107.

Peisert, H., M. Knupfer et al. 2002. Full characterization of the interface between the organic semiconductor copper phthalocyanine and gold. *Journal of Applied Physics* 91(8): 4872–4878.

Qiu, C. F., Z. L. Xie et al. 2003. Comparative study of metal or oxide capped indium-tin oxide anodes for organic light-emitting diodes. *Journal of Applied Physics* 93(6): 3253–3258.

Schlaf, R., C. D. Merritt et al. 2001. Determination of the orbital lineup at reactive organic semiconductor interfaces using photoemission spectroscopy. *Journal of Applied Physics* 90(4): 1903–1910.

Schlaf, R., B. A. Parkinson et al. 1998. Determination of frontier orbital alignment and band bending at an organic semiconductor heterointerface by combined x-ray and ultraviolet photoemission measurements. *Applied Physics Letters* 73(8): 1026–1028.

Schlaf, R., P. G. Schroeder et al. 1999. Observation of strong band bending in perylene tetracarboxylic dianhydride thin films grown on SnS_2. *Journal of Applied Physics* 86(3): 1499–1509.

Schlaf, R., P. G. Schroeder et al. 2000. Determination of interface dipole and band bending at the Ag/tris (8-hydroxyquinolinato) gallium organic Schottky contact by ultraviolet photoemission spectroscopy. *Surface Science* 450(1–2): 142–152.

Schroeder, P. G., C. B. France et al. 2002a. Energy level alignment and two-dimensional structure of pentacene on Au(111) surfaces. *Journal of Applied Physics* 91(5): 3010–3014.

Schroeder, P. G., C. B. France et al. 2002b. Orbital alignment at p-sexiphenyl and coronene/layered materials interfaces measured with photoemission spectroscopy. *Journal of Applied Physics* 91(11): 9095–9107.

Schwieger, T., H. Peisert et al. 2004. Direct observation of interfacial charge transfer from silver to organic semiconductors. *Chemical Physics Letters* 384(4–6): 197–202.

Scott, J. C. 2003. Metal-organic interface and charge injection in organic electronic devices. *Journal of Vacuum Science and Technology A* 21(3): 521–531.

Shirota, Y., Y. Kuwabara et al. 1994. Multilayered organic electroluminescent device using a novel starburst molecule, 4,4′,4″-tris(3-methylphenylphenylamino)triphenylamine, as a hole transport material. *Applied Physics Letters* 65(7): 807–809.

Smoluchowski, R. 1941. Anisotropy of the electronic work function of metals. *Physical Review* 60(9): 661–674.

Steinberger, I. T., K. Wandelt. 1987. Ionization energies of valence levels in physisorbed rare-gas multilayers. *Physical Review Letters* 58(23): 2494–2497.

Steirer, K. X., J. P. Chesin et al. 2010. Solution deposited NiO thin-films as hole transport layers in organic photovoltaics. *Organic Electronics* 11(8): 1414–1418.

Sugiyama, K., H. Ishii et al. 2000. Dependence of indium-tin-oxide work function on surface cleaning method as studied by ultraviolet and x-ray photoemission spectroscopies. *Journal of Applied Physics* 87(1): 295–298.

Tautz, F. S. 2007. Structure and bonding of large aromatic molecules on noble metal surfaces: The example of PTCDA. *Progress in Surface Science* 82(9–12): 479–520.

Tengstedt, C., W. Osikowicz et al. 2006. Fermi-level pinning at conjugated polymer interfaces. *Applied Physics Letters* 88(5): 053502.

Tokito, S., K. Noda et al. 1996. Metal oxides as a hole-injecting layer for an organic electroluminescent device. *Journal of Physics D—Applied Physics* 29(11): 2750–2753.

Tokmoldin, N., N. Griffiths et al. 2009. A hybrid inorganic-organic semiconductor light-emitting diode using ZrO2 as an electron-injection layer. *Advanced Materials* 21(34): 3475–3478.

VanSlyke, S. A., C. H. Chen et al. 1996. Organic electroluminescent devices with improved stability. *Applied Physics Letters* 69(15): 2160–2162.

Vazquez, H., Y. J. Dappe et al. 2007. A unified model for metal/organic interfaces: IDIS, "pillow" effect and molecular permanent dipoles. *Applied Surface Science* 254: 378–382.

Vazquez, H., F. Flores et al. 2004. Barrier formation at metal-organic interfaces: Dipole formation and the charge neutrality level. *Applied Surface Science* 234(1–4): 107–112.

Vazquez, H., F. Flores et al. 2007. Induced density of states model for weakly-interacting organic semiconductor interfaces. *Organic Electronics* 8(2–3): 241–248.

Vazquez, H., W. Gao et al. 2005. Energy level alignment at organic heterojunctions: Role of the charge neutrality level. *Physical Review B* 71(4): 041306.

Waldauf, C., M. Morana et al. 2006. Highly efficient inverted organic photovoltaics using solution based titanium oxide as electron selective contact. *Applied Physics Letters* 89(23): 233517.

Wan, A., J. Hwang et al. 2005. Impact of electrode contamination on the alpha-NPD/Au hole injection barrier. *Organic Electronics* 6(1): 47–54.

Wandelt, K. 1997. The local work function: Concept and implications. *Applied Surface Science* 111: 1–10.

Wang, F. X., X. F. Qiao et al. 2008. The role of molybdenum oxide as anode interfacial modification in the improvement of efficiency and stability in organic light-emitting diodes. *Organic Electronics* 9(6): 985–993.

Wang, S. A., T. Osasa et al. 2006. CuOx films as anodes for organic light-emitting devices. *Japanese Journal of Applied Physics Part 1—Regular Papers Brief Communications and Review Papers* 45(11): 8894–8896.

Wang, Z. B., M. G. Helander et al. 2009. Analysis of charge-injection characteristics at electrode-organic interfaces: Case study of transition-metal oxides. *Physical Review B* 80(23): 235325.

Wei, B., S. Yamamoto et al. 2007. High-efficiency transparent organic light-emitting diode with one thin layer of nickel oxide on a transparent anode for see-through-display application. *Semiconductor Science and Technology* 22(7): 788–792.

Weiler, U., T. Mayer et al. 2004. Electronic energy levels of organic dyes on silicon: A photoelectron spectroscopy study of ZnPc, F16ZnPc, and ZnTPP on p-Si(111): H. *Journal of Physical Chemistry B* 108(50): 19398–19403.

Woo, S., J. Kim et al. 2009. Influence of nickel oxide nanolayer and doping in organic light-emitting devices. *Journal of Industrial and Engineering Chemistry* 15(5): 716–718.

Xiao, B. W., Y. F. Shang et al. 2005. Enhancement of hole injection with an ultra-thin Ag_2O modified anode in organic light-emitting diodes. *Microelectronics Journal* 36(2): 105–108.

Xie, G. H., Y. L. Meng et al. 2008. Very low turn-on voltage and high brightness tris-(8-hydroxyquinoline) aluminum-based organic light-emitting diodes with a MoOx p-doping layer. *Applied Physics Letters* 92(9): 093305.

Xu, X. J., G. Yu et al. 2006. Electrode modification in organic light-emitting diodes. *Displays* 27(1): 24–34.

Yi, Y., P. E. Jeon et al. 2009. The interface state assisted charge transport at the MoO(3)/metal interface. *Journal of Chemical Physics* 130(9): 094704.

Yook, K. S., J. Y. Lee. 2009. Low driving voltage in organic light-emitting diodes using MoO3 as an interlayer in hole transport layer. *Synthetic Metals* 159(1–2): 69–71.

Zhang, D. D., J. Feng et al. 2009. Enhanced hole injection in organic light-emitting devices by using Fe_3O_4 as an anodic buffer layer. *Applied Physics Letters* 94(22): 223306.

Zhang, H. M., W. C. H. Choy. 2008. Highly efficient organic light-emitting devices with surface-modified metal anode by vanadium pentoxide. *Journal of Physics D—Applied Physics* 41(6): 062003.

Zhou, Y. H., C. Fuentes-Hernandez et al. 2012. A universal method to produce low-work function electrodes for organic electronics. *Science* 336(6079): 327–332.

Zhu, X. L., J. X. Sun et al. 2007. Investigation of Al- and Ag-based top-emitting organic light-emitting diodes with metal oxides as hole-injection layer. *Japanese Journal of Applied Physics Part 1—Regular Papers Brief Communications and Review Papers* 46(3A): 1033–1036.

Zilberberg, K., S. Trost et al. 2011. Solution processed vanadium pentoxide as charge extraction layer for organic solar cells. *Advanced Energy Materials* 1(3): 377–381.

Zou, Y., L. Kilian et al. 2006. Chemical bonding of PTCDA on Ag surfaces and the formation of interface states. *Surface Science* 600(6): 1240–1251.

6. Small Molecule Fundamentals

Xin Xu and Michael S. Weaver

6.1 Introduction to Organic Compounds

Organic compounds are typically carbon-rich compounds that may also contain hydrogen, oxygen, nitrogen or sulfur atoms. They are generally referred to as "soft" materials with a melting point typically lower than 350°C (Gutmann et al. 1981, Pope and Swenberg 1999). Therefore, they are compatible with flexible or nonplanar substrates such as metal foils, plastic substrates and even paper sheets, in contrast to inorganic materials, which are generally hard and brittle materials. Electronic devices or circuits based on organic

Chapter 6

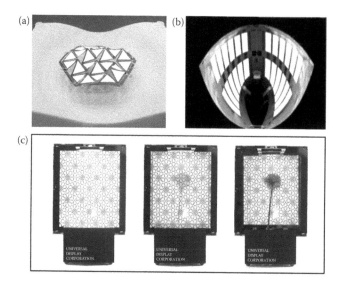

FIGURE 6.1   Demonstrations of (a) flexible, (b) rollable, and (c) transparent organic small molecule light-emitting panels.

materials have the potential to realize novel form factors such as lightweight, transparent, flexible, or rollable options as shown in Figure 6.1.

The first organic light-emitting device (OLED), reported by Tang and Vanslyke in 1987 (Tang and Vanslyke 1987), used small molecule organic compounds. This is a type of organic compound with a defined structure and a fixed molecular weight. To better understand small molecule materials, we will start with an introduction of the fundamental properties of organic compounds in this chapter.

6.1.1   Classification and Main Properties of Organic Compounds

There are four main types of chemical bonds: ionic, metallic, covalent, and intermolecular bonds, which classify solid compounds into four groups. Organic compounds are typically bound by the van der Waals interaction between discrete molecules (Gutmann et al. 1981, Pope and Swenberg 1999).

With an increasing molecular weight and complexity, organic compounds can also be classified into three types: small molecules, polymers, and biological molecules. Small molecules, typically with a molecular weight less than 1000 atomic mass units (amu), are composed of a fixed number of atoms and have a well-defined structure. By contrast, polymers consist of chains of repeating structural units where the size is determined by an average of the molecular weights. Biological molecules, which will not be the focus of this chapter, are characterized by their synthetic origin within living organisms and have the most complex molecular structures.

The van der Waals interaction energy is typically a few meV. Unlike covalent bonds (bonding energy ~eV) when electrons are shared between atoms, the van der Waals force mainly originates from the emergence of a dipole from a momentary fluctuation.

Compared with a conventional inorganic semiconductor, an organic semiconductor in general has lower carrier mobility (typically lower than 1 cm^2/V·s) due to weak van der Waals bonding. A thick layer of organic semiconductor film will then have high sheet resistance due to the low electrical conductivity of the film. However, organic semiconductor electronic devices can be built in a thin form factor (e.g., 200 nm). Thus the overall resistance of the devices can remain low (e.g., an operating voltage of a few volts).

Since the electrons are highly delocalized, organic compounds tend to have high oscillator strengths. Thus the absorption coefficient of organic materials in the visible wavelength range can exceed 10^5 cm^{-1} in many cases, which represents an absorption length <100 nm. It then becomes possible to build an ultra-thin yet highly efficient organic optoelectronic device (Peumans et al. 2000).

6.1.2 Small Molecule Organic Compounds

Unlike polymers, small molecule organic materials have defined structures and molecular weight. Thin films of small molecule solids may have various morphologies which significantly affect film properties such as density, hardness, carrier transport, transparency, and so on.

While organic single crystals have the highest structural order, they are very fragile and therefore are a challenge to use in electronic devices. They can be described by a periodic wavefunction. The charge transport in organic single crystals at low temperature is dominated by a band-like model (Emin and Holstein 1969, Marcus 1964). Single-crystalline devices are generally not easy to produce with small molecule organic materials. The formation of a crystal requires that the energy gained during bonding is greater than the loss in entropy (Atkins 1994).

In contrast, polycrystalline films are more robust and practical for making thin film small molecule organic devices. Polycrystalline films can be fabricated by using a quasi-epitaxial deposition method (Forrest 1997, Lunt et al. 2007). For example, a planar molecule, such as perylene-3,4,9,10-tetracarboxylic dianhydride (PTCDA), can be grown on a substrate where it can stack parallel to the surface and form an underlying crystalline lattice for subsequent layers to grow and crystallize on top.

Amorphous organic films have the least structural order and are the most practical and easiest organic films to form. Lack of crystallinity offers amorphous films a more stable configuration, which allows the codeposition of two or more materials, that is, doping, of the film. This becomes very important for device applications since the optical and electrical characteristics of the film, such as emission and conductivity, can then be altered by doping (Bulovic et al. 2001, Pfeiffera et al. 2003, Sze 1981).

6.1.3 Intermolecular Forces

As mentioned previously, many characteristics of an organic compound are fundamentally determined by its molecular bonding, which is a sum of attractive and repulsive interactions between neighboring molecules. The intermolecular force mainly contains four parts:

- Pauli repulsive force
- Electrostatic force between two permanent diploes

Chapter 6

- Induction between one permanent dipole and one induced dipole
- Dispersion between two induced dipoles

The Pauli repulsive force, originating from the Pauli exclusion principle, increases rapidly when two molecules approach each other and orbitals overlap. The electrostatic force is a dipole–dipole interaction and is also known as the Keesom force. The induction force and the dispersion force are both attractive forces between dipoles of opposite polarities. The induction force, known as the polarization or Debye force, describes an interaction between a dipole and an induced dipole. The dispersion force, or London dispersion force, is a type of polarization resulting from momentary dipole fluctuations. Therefore, it is an interaction between the instantaneous dipole and the induced dipole.

For a nonpolar molecule, the van der Waals force only arises from the dispersion force among the three attractive forces and Pauli repulsive force. For a polar molecule, dipole–dipole interactions and dipole-induced dipole interactions are added to the van der Waals force. Each of the dipole–dipole interactions scale with the sixth power of intermolecular distance, r, and can be expressed as r^{-6}. Therefore, the total intermolecular force $U_{L-J}(r)$ can be represented by the sum of the attractive van der Waals force at long range ($U_{attr}(r)$) and the Pauli repulsive force at short range ($U_{rep}(r)$). A simplified empirical model, known as the Lennard–Jones 6–12 potential (Atkins 1991), is often used to describe the intermolecular force to a good approximation:

$$U_{L-J}(r) = U_{rep}(r) + U_{attr}(r) = 4\varepsilon\left[\left(\frac{\sigma}{r}\right)^{12} - \left(\frac{\sigma}{r}\right)^{6}\right] \tag{6.1}$$

where ε reflects the depth of the molecular potential well and σ reflects the distance of zero intermolecular potential.

6.1.4 Electron Configuration

To help understand the electronic structure of organic compounds, we discuss here the electronic configuration of a carbon atom. Carbon has an atomic number of 6. For a single carbon atom in the ground state, there are two electrons in an inner core shell and the other four electrons are in an outer shell. The outer electron shell is composed of four electron orbitals. According to electron spin, each orbital may accommodate up to two electrons. Thus two of the four outer shell electrons in the carbon occupy the lower energy "s" orbital and another two unpaired electrons occupy two of the three higher energy "p" orbitals; this state is commonly expressed as an s^2p^2 configuration.

According to valence bond theory, the two electrons in the p orbitals of the carbon atom comprise its "valence" electrons. Thus, a carbon atom should only form two covalent bonds with other atoms such as hydrogen. However, the CH_2 fragment formed through such a pairing is unstable and highly reactive, whereas CH_4 is a ubiquitous stable compound containing four C–H bonds of equal energy. Explaining this phenomenon requires the concept of hybridization. In general, hybridization describes the mixing of existing orbitals into new orbitals with energies intermediate between those of their original constituents. Again with the example of the carbon atom, when the

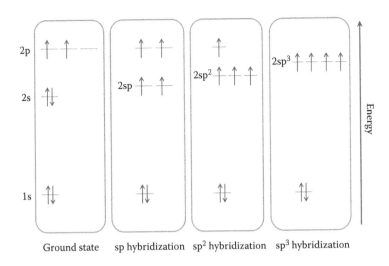

Ground state   sp hybridization   sp² hybridization   sp³ hybridization

FIGURE 6.2 The electronic configurations of the carbon atom at ground state and three hybridized configurations between 2s and 2p orbitals: sp hybridization, sp² hybridization, and sp³ hybridization (from left to right).

s orbital and three p orbitals mix, their linear combination results in four tetrahedrally oriented "sp³" orbitals whose energies are equivalent and intermediate between the formal s and p orbital energies (Figure 6.2). This is called sp³ hybridization. Each sp³ orbital contains 25% s and 75% p character. When overlapped with the 1s orbital of a hydrogen atom, a σ bond is formed which refers to a head-on overlapping configuration. Due to the symmetry, all four bonds have the same strength and bond length; thus it requires the same energy to remove any hydrogen atom from the CH_4 molecule.

In the case of sp² hybridization, the 2s orbital and two of three 2p orbitals are mixed, leading to three sp² orbitals formed in the same plane separated by 120°. The remaining unhybridized 2p orbital is oriented perpendicular to the plane defined by the three sp² orbitals (Figure 6.3). When two sp² hybridized carbon atoms approach, two sp² orbitals combine to form a head-on bond (the σ bond), and the two perpendicular p orbitals are able to form a more diffusely overlapping bond (the π bond) (Figure 6.3). In general, a σ bond is much stronger than a π bond given that head-on overlap is more direct and effective. This combination of a strong σ and weaker π bond describes the ubiquitous C=C double bond.

When two identical atoms approach and start to form a molecule such as in the simple case of forming H_2, the original energy level of the hydrogen atom splits into two molecular orbitals (MOs) according to the Pauli exclusion principle. One orbital is higher in energy than the original energy level. This is the antibonding MO (σ*). The other orbital has a lower energy and is thus more stable than the original level. This is the bonding MO (σ). Therefore, when two carbon atoms form a C=C bond, the σ MO is more stable and has a lower energy compared to the π MO due to the fact that a σ bond is stronger than the π bond. The corresponding antibonding σ* MO has a higher energy than the antibonding π* MO (Figure 6.4). The two pairs of electrons of the C=C bond will then occupy the MOs with lowest energy which in this case are σ MOs and π MOs. Thus, the π MO becomes the highest occupied molecular orbital (HOMO). Similarly, antibonding

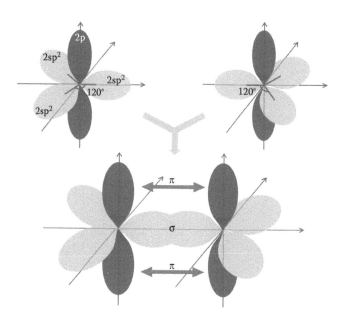

FIGURE 6.3 Illustration of sp² hybridization with three sp² hybridized orbitals within the same plane with a bond angle of 120 . The unhybridized p orbital remains perpendicular to the plane. When one atom approaches the other, σ bond is form within the plan with two sp² orbitals and π bond is formed between two p orbitals.

π* MO is the lowest unoccupied molecular orbital (LUMO). In organic semiconductors, the π bond, or more specifically the delocalized electron density of the π bond, plays an important role in charge transport (Pope and Swenberg 1999). The HOMO and LUMO levels in organic semiconductors are similar concepts to valence and conduction bands in inorganic semiconductors in the sense that they are states through which holes and

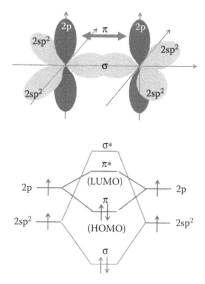

FIGURE 6.4 Formation of MOs and antibonding MOs between 2p and 2sp² orbitals of two adjacent carbon atoms. The π MO filled with paired electrons is the HOMO. The π* antibonding MO is the LUMO.

electrons can be transported, respectively. The energy gap between HOMO and LUMO levels is the minimum energy required to generate a free electron and hole in an organic semiconductor.

If we consider an extended case from the previous example, for example, a hexagonal benzene ring (C_6H_6), every two neighboring carbon atoms should have an alternating C=C double bond and C—C single bond. Since the C=C double bond is stronger than the single bond, the bond distance of the C=C double bond is shorter. However, all six bonds of the benzene ring are found to be of the same length, longer than the C=C double bond but shorter than the C—C single bond. The reason for that is the delocalized electron density of the π bond allowing sharing of π electrons between the six carbons of the benzene ring. Such a system is described to be "conjugated."

In an OLED device, increased conjugation of the light-emitting organic molecules typically results in a red shift of the emission peak wavelength and the energy gap between the HOMO and LUMO is reduced. When the emitter compound is less conjugated, then the peak wavelength of its emission spectrum is usually bluer, with a larger HUMO–LUMO energy bandgap. In general, a planar organic compound is more conjugated than a nonplanar compound.

6.2   Electronic Structure and Excitons

6.2.1   One Electron Wavefunction

In quantum mechanics, the state of a system (e.g., a molecule) is described by its wavefuntion Ψ. For an electron, the wavefunction is determined by the following quantum numbers:

- The principal quantum number, n
- The angular quantum number, ℓ
- The magnetic quantum number, m_ℓ
- The spin quantum number, s
- The spin projection quantum number, m_s

The principal quantum number, n, describes the shell of an electron. The angular quantum number, ℓ (or an orbital quantum number), describes the subshell. Thus ℓ represents the integers ranging from 0 to $n - 1$. As an example, for the third shell ($n = 3$), there are three subshells ($\ell = 0, 1, 2$) which are 3s, 3p, 3d. The magnetic quantum number m_ℓ describes the orbitals projection of ℓ on the z axis, and m_ℓ can be any integer ranging from $-\ell$ through $+\ell$. For example, when $\ell = 2$, it represents the d subshell which has 5 orbitals since $m_\ell = -2, -1, 0, +1, +2$. The fourth quantum number is the spin quantum number which has a magnitude of half integer, or $s = 1/2$. The orientation of spin is defined by the projection of spin on the z axis, m_s. When $m_s = +1/2$, it represents a "spin-up" orientation denoted by $| \uparrow \rangle$ using Dirac notation. When $m_s = -1/2$, it is a "spin-down" state, denoted by $| \downarrow \rangle$. The first three quantum numbers define the electron spatial wavefunction ψ. The latter two quantum numbers define the spin wavefunction χ. Thus, the electron wavefunction Ψ_e can be rewritten as the product of two components:

$$\Psi_e = \psi\chi \tag{6.2}$$

6.2.2 Wavefunction of an Isolated Molecule

The wavefunction Ψ describes the state of a molecule and can be constructed using MO theory. Under the Born–Oppenheimer approximation, which assumes that the motion of the electrons and the nuclei are independent given the significant difference in their masses ($m_e \ll m_N$), the molecular wavefunction can be treated as the product of the nuclear wavefunction and the electron wavefunction. Since the spin–orbit coupling is typically weak in most organic molecules (Turro 1991), the spin can also be treated independently of nuclear and electron motion. Thus the molecule wavefunction can be expressed as

$$\Psi = \psi_N \psi \chi \tag{6.3}$$

where ψ_N is the nuclear wavefunction, ψ is the electron spatial wavefunction, and χ is the spin wavefunction.

Further approximation can be made by assuming the electron–electron interactions are negligible. Thus the electron spatial wavefunction ψ may be simplified as a linear combination of atomic orbitals (LCAO) (Daudel et al. 1983, Levine 1991, Lowe 1993)

$$\psi = \sum_{i=0}^{n} a_i \phi_i \tag{6.4}$$

where a_i are linear coefficients and ϕ_i is the wavefunction of a single electron.

Then the energy levels of the one-electron MOs can be determined by

$$<E> = \frac{<\psi|\mathcal{H}|\psi>}{<\psi|\psi>} \tag{6.5}$$

where $\mathcal{H}$ is the Hamiltonian operator. According to the Pauli principle, each MO can only be occupied by two electrons with opposite spins. Thus the electrons fill up MOs starting from the one with lowest energy. The highest level filled with electrons is the HOMO level and the next lever higher than the HOMO level is the LUMO.

6.2.3 Classification of Excitons

In an organic semiconductor, when an electron is excited from HOMO to LUMO in the case of photon absorption, it leaves behind a localized hole in the HOMO. This pair of electron and hole forms a bound energy state which is called an exciton. An exciton is a neutral quasi-particle with zero net charge. The exciton binding energy is defined as the energy needed for separating the bound electron–hole pair into discrete charge carriers. Thus it equals the difference between the HOMO–LUMO energy gap and the energy of the exciton.

Excitons can be classified into three groups based upon the intermolecular distance between the bound electron–hole pair, as shown in Figure 6.5. The excitons with the shortest intermolecular distance are Frenkel excitons, which are localized on a single molecule.

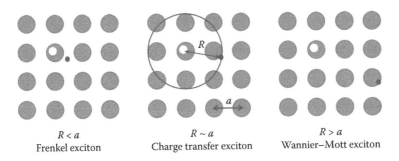

$R < a$
Frenkel exciton

$R \sim a$
Charge transfer exciton

$R > a$
Wannier–Mott exciton

FIGURE 6.5 Schematic representation of Frenkel, CT, and Wannier–Mott excitons with increased delocalization from left to right. R is the distance between the electron component and hole component of the exciton and a represents lattice constant.

Thus the binding energy of Frenkel excitons is strong. With increased delocalization, charge transfer (CT) excitons are spaced by two or more lattice constants, with the electron component residing on one molecule or the donor group of the molecule, and the hole component residing on a neighboring molecule or the acceptor group of the molecule. The binding energy of Frenkel and CT excitons are typically on the order of 0.1–1 eV (Alvarado et al. 1998, Barth and Bassler 1997, Hill et al. 2000, Knupfer et al. 1998). Due to the weak van der Waals intermolecular force, the excitons generated in organic small molecule thin films are localized, and can be either Frenkel or CT excitons (Bulovic et al. 2001). The third type of exciton—the Wannier–Mott exciton—is highly delocalized, and has a much reduced exciton binding energy of a few millielectron volt. Wannier–Mott excitions appear more often in inorganic covalent semiconductors where dielectric constants are high enough to screen the Coulomb attraction (Ashcroft and Mermin 1976).

6.2.4 Singlets and Triplets

In a two-electron system, the total spin S is either 0 or 1 given that spin of a single electron $s = 1/2$. When $S = 0$, m_s has only one allowed value, 0. This is a singlet state of the system. If $S = 1$, then there are three allowed values of m_s, −1, 0, +1. Therefore, $S = 1$ is called a triplet state of the system due to the triple degeneracy.

$$|S = 0> = \frac{1}{\sqrt{2}}\left\{|\uparrow\downarrow> - |\downarrow\uparrow>\right\} \tag{6.6}$$

$$|S = 1> = |\uparrow\uparrow> \tag{6.7}$$

$$|S = 1> = |\downarrow\downarrow> \tag{6.8}$$

$$|S = 1> = \frac{1}{\sqrt{2}}\left\{|\uparrow\downarrow> + |\downarrow\uparrow>\right\} \tag{6.9}$$

The singlet state with $S = 0$ is antisymmetric with respect to electron exchange. The triplet state with $S = 1$ is symmetric under electron exchange.

According to the Pauli exclusion principle, the wavefunction of the system should be antisymmetric under the exchange of indistinguishable electrons. This means that the two components of the wavefunction Ψ should have opposite symmetries. When spatial wavefunction ψ is symmetric, the spin state χ has to be antisymmetric. When spatial wavefunction ψ is antisymmetic, then the spin state χ is symmetric.

A stable closed-shell molecule with valence shell completely filled usually has a symmetric spatial wavefunction. Thus the spin state is antisymmetric which means the ground state S_0 is a singlet state. The first excited singlet state above ground state is labeled as S_1. The first triplet state above the ground state is labeled as T_1. A triplet state typically has lower energy than a singlet state due to its relatively low electron–electron repulsion. The nonradiative transition from a singlet to a triplet exciton is called intersystem crossing (ISC). This transition is forbidden as we describe in the next section, and as a consequence, the ISC rate is in general very slow.

6.2.5 Fermi's Golden Rule

In quantum mechanics, Fermi's golden rule is the selection rule for an electronic transition between states by providing the transition rate (s^{-1}) from the initial state to the final state under a weak perturbation. Since the transition probability, $T_{i \to f}$, is proportional to the coupling between initial and final states and the density of final states, it can be calculated as

$$T_{i \to f} = \frac{2\pi}{\hbar} |M_{if}|^2 \, \rho_f \tag{6.10}$$

where ρ_f is the density of final states and M_{if} corresponds to the matrix element of the perturbation between initial and final states. Thus, the matrix element can also be written as

$$M_{if} = \langle \Psi_f | \mu | \Psi_i \rangle = \int \Psi_f^* \mu \Psi_i d\tau \tag{6.11}$$

where Ψ_i and Ψ_f represent the wavefunctions of the initial and final states involved in the transition, respectively, and μ is the dipole operator of the molecule.

Thus we can determine if the transition is forbidden by finding out when the transition rate becomes zero. The matrix element is nonzero only if $\Psi_f^* \mu \Psi_i$ is symmetric. Since the dipole operator is symmetric, thus the final state and initial state must have the same symmetry, which means the only allowed transitions are from triplet to triplet state or from singlet to singlet state. The emission from an excited singlet state to the singlet ground state is a fast decay process that occurs on a nanosecond time scale, known as fluorescence. Since the ground state of a closed-shell molecule is a singlet state, the emission resulting from the relaxation of an excited triplet to a ground state singlet has a low radiative rate and long excited state decay; this process is called phosphorescence. However, we will discuss in detail later how molecular spin–orbit interactions can increase the radiative rate for triplet to singlet transitions.

6.3   Exciton Energy Transfer

Exciton energy transfer between molecules can be classified as nonradiative or radiative transfer. Radiative energy transfer can usually happen at long ranges (>10 nm) since the two molecules involved in the transfer do not have a direct interaction. Nonradiative transfer, which typically dominates at a shorter range (<10 nm), can be further divided into two types: Förster energy transfer and Dexter energy transfer.

6.3.1   Radiative Energy Transfer

Radiative energy transfer is a photon reabsorption process, which also may be further classified as trivial energy transfer or cascade transfer (Pope and Swenberg 1999). In principle, it may happen at an unlimited range as long as the emission spectrum of the excited donor molecule and the absorption spectrum of the acceptor molecule overlap. The energy transfer is initiated by the exciton recombination of a donor molecule, which emits a photon. Due to the spectral overlap, the photon may be absorbed by the acceptor. The energy transfer rate is determined by the quantum yield of emission from the donor molecule and the absorption coefficient of the acceptor.

6.3.2   Förster Energy Transfer

Förster energy transfer is a resonant energy transfer mechanism which occurs via dipole–dipole coupling between molecules. Initially, the donor molecule is in an excited state and the acceptor is in the ground state. Transition of a photon from the donor to the acceptor occurs when the two wavefunctions interact causing the wavefunction of the acceptor to oscillate at the same frequency. After the transition, the donor molecule stays in the ground state and the acceptor molecule moves to the excited state (Figure 6.6). This energy transfer also requires an overlap between the emission spectrum of the

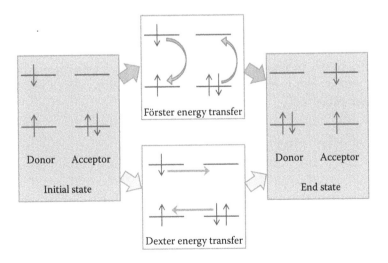

FIGURE 6.6   Schematic of energy transfer from identical initial states to identical end states by two different energy transfer mechanisms: Förster energy transfer (top) and Dexter energy transfer (bottom).

Chapter 6

donor molecule and the absorption spectrum of the acceptor molecule. Förster energy transfer can be thought of as radiative transfer but without the transferred photon being observed. Integrating over all possible (random) orientations of the donor and acceptor, the transfer rate, $\Gamma_{D \to A}$ is given by (Pope and Swenberg 1999)

$$\Gamma_{D \to A} = \frac{1}{\tau_D} \frac{1}{R^6} \left(\frac{3}{4\pi} \int \frac{c^4}{\omega^4 n^4} F_D(\omega) \sigma_A(\omega) d\omega \right) \tag{6.12}$$

where τ_D is the natural lifetime of the donor, R is the distance between donor and acceptor, c is the speed of light, ω is the frequency, n is the refraction index of the surrounding medium, F_D is the normalized emission spectrum of the donor, and σ_A is the absorption cross section of the acceptor.

The transfer rate can be expressed as a simplified expression

$$\Gamma_{D \to A} = \frac{1}{\tau_D} \left(\frac{R_0}{R} \right)^6 \tag{6.13}$$

where R_0 is the critical transfer distance, known as the Förster radius. It represents the oscillation strength between donor and acceptor. Since the range of Förster transfer is limited by the strength of the dipole–dipole interaction, it is on the same order as the Förster radius, which is typically a few nanometers (D'Andrade et al. 2001).

6.3.3 Dexter Energy Transfer

Unlike Förster energy transfer which occurs through coherent oscillation, Dexter transfer occurs with the physical exchange of electrons from donor to acceptor molecules. An electron from the initial excited state of the donor molecule moves to the acceptor. At the same time, an electron from the initial ground state of the acceptor molecule moves to the donor (Figure 6.6). In order to achieve these electron transfers, the separation between donor and acceptor molecules must be shorter than the Förster radius, that is, typically within a couple of nanometers. A significant overlap between wavefunctions of the neighboring molecules is required for the actual diffusion of excitons. The transfer rate of the Dexter process, $\Gamma_{D \to A}$, is expressed as

$$\Gamma_{D \to A} = \frac{2\pi}{\hbar} |\beta_{ex}|^2 \int F_D(E) F_A(E) dE \tag{6.14}$$

where β_{ex} represents the exchange interaction between donor and acceptor, E is the energy, $F_D(E)$ and $F_A(E)$ corresponds to the normalized emission spectrum of donor and normalized absorption spectrum of acceptor, respectively. Although it is difficult to quantitatively estimate the value of β_{ex}, it may be simplified to an exponential decay term with respect to the spatial separation between donor and acceptor molecules. This is based on the assumption that the electronic distribution of both molecules decreases exponentially when the spatial separation increases. Thus, the Dexter transfer rate can be simplified to (Turro 1991)

$$\Gamma_{D \to A} = AJ \exp(-2R_{DA}/L)   \tag{6.15}$$

where A represents the orbital interactions, J corresponds to the spectral overlap integral normalized for the extinction coefficient of the acceptor, and R_{DA} is the spatial separation between donor and acceptor relative to their van der Waals radii L (Turro 1991).

Although Dexter energy transfer may also be applied to singlet–singlet transfer, it is not a dominant mechanism due to its significantly smaller rate with respect to the Förster energy transfer rate. In contrast, Dexter energy transfer is the predominant transfer mechanism in triplet exciton transfer. In triplet to triplet transfer the absorptivity (extinction coefficient) of the triplet state is very low so Förster energy transfer is not efficient. Triplet states are also very long lived allowing time for an exchange to take place.

6.3.4   Annihilation

Energy transfer not only occurs between a donor molecule in an excited state and an acceptor molecule in the ground state as previously mentioned, but transitions may also occur which lead from one excited state to another excited state, resulting in a ground state and a higher energy excited state. This energy transfer process becomes more significant when the excitation density is high and is known as annihilation.

Annihilation can occur between an exciton and a polaron (Gutmann et al. 1981), or just between excitons. We will focus on the latter case. Annihilation between excitons can be classified into three types, singlet–singlet annihilation (SSA), singlet–triplet annihilation (STA), and triplet–triplet annihilation (TTA).

SSA can be expressed as

$$S_1 + S_1 \to S_0 + S_n^* \to S_0 + S_1 + \text{heat}   \tag{6.16}$$

where S_0 represents the singlet ground state, S_1 represents the first excited state and S_n^* represents a higher (nth) excited state.

Similarly, STA is expressed as

$$S_1 + T_1 \to S_0 + T_n^* \to S_0 + T_1 + \text{heat}   \tag{6.17}$$

The energy transfer mechanism for an annihilation process involving a singlet tends to occur via Förster energy transfer. One singlet transfers its energy to another exciton and promotes it to an upper excited state. Then the higher excited state relaxes back nonradiatively to a lower excited state with an excess amount of energy generated in the form of heat; this process is called internal conversion.

TTA may have two outcomes. The first one results in a triplet excited state. This process is also called triplet quenching:

$$T_1 + T_1 \to S_0 + T_n^* \to S_0 + T_1 + \text{heat}   \tag{6.18}$$

Chapter 6

The other outcome of TTA results in a singlet exciton, known as "delayed fluorescence:"

$$T_1 + T_1 \rightarrow S_0 + S_n^* \rightarrow S_0 + S_1 + \text{heat} \tag{6.19}$$

Since a triplet exciton is longer lived than a singlet exciton, the corresponding fluorescence derived from the TTA process thus has a longer lifetime, hence the term delayed fluorescence. The TTA process is based on Dexter energy transfer.

The rate equation of the annihilation process is given by

$$\gamma = 4\pi D_{TOT} R \tag{6.20}$$

where D_{TOT} corresponds to the sum of reactant diffusivities and R represents the radius of the interaction. For the TTA process, the interaction radius R is the spatial separation between neighboring molecules according to the Dexter transfer mechanism. For annihilation involving a singlet, based on Förster transfer, R is then the Förster radius, which can be derived from Equation 6.13.

6.4 Charge Carrier Transport in Organic Semiconductors

There are two extreme cases that are usually applied to study charge carrier transport in organic semiconductors. One is a delocalized system which assumes that the mean free paths of the charge carriers are dramatically longer than the lattice constant. The motion of the charge carrier is then expressed in the form of a Bloch wave. In this case, the band model is applied. The second case describes a system where carriers are localized due to weak intermolecular interactions and strong local polarization. Here charge transport is considered to be a hopping process where the charge carrier "jumps" from site to site as a particle without conservation of the wavevector.

6.4.1 The Band Model

The band model is applied only to a highly delocalized system in which the mean free paths of charge carriers significantly exceed the lattice constant. The delocalization of charge carriers depends on the intermolecular bonding and the overlap integral between molecules. When the surrounding medium has a small polarization and the overlap integral is large, the charge carriers are delocalized and the band model can be applied.

In a highly delocalized system, the weak intermolecular interaction leads to narrow bandwidth (Wright 1995). In narrow band theory, the charge mobility, μ_b, is related to the overlap integral, lattice constant, and temperature, and expressed as

$$\mu_b = \frac{q\lambda}{kT} \frac{Ja}{\hbar} \tag{6.21}$$

where q is the electronic charge, λ is the electron mean free path, k is Boltzmann's constant, T is temperature, J is overlap integral between molecules, and a is the lattice constant. The temperature dependence of carrier mobility is expected due to phonon scattering, and is observed in a few organic molecules (Wright 1995). For example, the mobility of the small molecule anthracene is $\mu \sim 10^{-4} \, m^2/V \cdot s$, giving $J = 0.01$ eV, $a = 0.6$ nm, and $\lambda = 0.29$ nm (Wright 1995).

6.4.2 The Polaron Hopping Model

In organic semiconductors, when the charge carriers are localized due to the weak intermolecular van der Waals force, the mean free path is not longer than the lattice constant (Kepler 1962, Pope and Swenberg 1999). Therefore, the band model is not applicable. Instead, the dominant transport mechanism for such a system is polaron hopping. This section describes the model explaining polaron hopping.

In this model, the intramolecular coupling between electrons and phonons is much stronger than the intermolecular interaction, leading to a polarization or distortion of the surrounding lattice induced by the motion of a free charge. When the charge carrier (i.e., electron or hole) moves slowly in the organic solid, it carries the distortion of the lattice along with it. The combination of a charge carrier and the resulting lattice polarization accompanying the charge forms a quasiparticle, called a polaron (Silinsh and Capek 1994, Wright 1995).

The lattice relaxes to a new equilibrium state due to the polarization. This new configuration becomes a potential well for the polaron. Thus, the movement of the polaron is confined and the mobility of the polaron is reduced (Pope and Swenberg 1999). The transport of the charge carrier in this case requires the polaron to "hop" from one molecule to a neighboring molecule.

The hopping probability depends on the polaron binding energy and temperature. Thus the carrier mobility, μ_h, is given by

$$\mu_h \propto \left(\frac{qa^2}{\hbar kT} \right) T^{-m} \exp\left(-\frac{E}{2kT} \right) \tag{6.22}$$

where q is the electronic charge, a is the lattice constant, k is Boltzmann's constant, T is temperature, E is the polaron binding energy, m power has a value of either 0 or 0.5 based on various conditions (Wright 1995). When the charge carriers are not relaxed, it corresponds to a dispersive transport. In this case, the value of m is 0. After the relaxation process is completed, the charge carriers have reached thermal equilibrium. Here the m value corresponds to 0.5. As the mobility is temperature dependent, it is dominated by the exponential term at low temperature. When the temperature is high (i.e., $kT \sim E/2$), the power law, $T^{-(m+1)}$, dominates.

In organic solids, charge carrier transport is determined by the competition between delocalization and localization energies. In organic molecular crystals, such as anthracene, charge carriers are highly delocalized. Thus the mobility is high ($\mu \sim 1 \, cm^2/V \cdot s$). However, the charge carriers in organic amorphous materials are localized due to the weak intermolecular interactions. In these cases, the hopping process dominates charge transport and the mobility is usually much lower ($\mu \ll 1 \, cm^2/V \cdot s$) (Pope and Swenberg 1999).

Chapter 6

6.5 Spin–Orbit Coupling and Phosphorescence

6.5.1 Decay Pathways

An exciton may decay radiatively or nonradiatively. Radiative decay is accompanied by photon emission. As discussed earlier, the radiative transition from a triplet exciton to the singlet ground state is usually forbidden since the spin is not conserved during this process. The allowed radiative transition from a singlet excited state to the ground state is known as fluorescence. The emission rate of fluorescence is typically on the order of 10^9 s^{-1}, which corresponds to a short lifetime of nanoseconds.

On the other hand, an exciton can also decay through internal conversion which is a nonradiative process. The nonradiative decay may result from an exciton colliding with other atoms or molecules which is called dynamic quenching. The energy released during the decay is in the form of heat, or phonons. Nonradiative decay is more commonly seen in small-bandgap materials (Turro 1991).

There are also other mechanisms that can cause excitation loss such as annihilation and ISC. Thus, the quantum yield (Φ) which measures the ratio of emitted photons versus excitons is given by

$$f = \frac{\Gamma_r}{\Gamma_r + \Gamma_{nr} + k} \tag{6.23}$$

where Γ_r, Γ_{nr}, and κ are decay rates corresponding to radiative, nonradiative, or other mechanisms, respectively.

6.5.2 Spin–Orbit Coupling

Since electronic spins are not correlated when generated electrically (unlike optical excitation), the formation of singlet and triplet excitons through electrical excitation has a statistical ratio of approximately 1:3. Thus fluorescence via electrical excitation is an inefficient transition since electrical excitation of a fluorescent emitter uses 25% of the electrically generated excited states. The remaining 75% triplet excitons generally do not contribute to fluorescent emission during electroluminescence.

However, triplet radiative transition can occur through spin–orbit coupling. In general, spin–orbit coupling in organic molecules is very weak which leads to a weak interaction between electron spin and the orbit-induced magnetic field. However, spin–orbit coupling can be significantly enhanced when a heavy atom, preferably a lower d-block metal atom such as Os, Ir, Pt, or Au, is introduced into the molecule. The heavy atom increases the magnetic field by accelerating the electron during its motion. The enhanced magnetic field induces angular momentum which allows the spins to flip. Thus the strong spin, orbit coupling can result in a significant mixing of singlet and triplet spin configurations and promote ISC. Consequently, the radiative rate of decay from a triplet exciton to the singlet ground state has been greatly improved. Phosphorescence from a heavy atom organometallic phosphor, for example, *fac*-tris(2-phenylpyridine) iridium (Ir(ppy)$_3$), has a much lower decay rate (~10^6 s^{-1}) and a longer lifetime (~microseconds) compared to fluorescence (Baldo et al. 2000, Kalinowski et al. 2002).

6.5.3 Phosphorescent Organic Light-Emitting Devices

An OLED contains a layer of one or more organic compounds as the emissive layer and transporting layers sandwiched between anodes and cathodes. Electrons and holes are injected from the cathodes and anodes, respectively. The holes and electrons then combine in the emissive layer to form excitons. The excitons can relax to the ground state radiatively with light emission.

With fluorescent emitters, the maximum internal quantum efficiency (IQE) from singlet emission (fluorescence) is ~25% (in practice, TTA resulting in delayed fluorescence may lead to as much as 40% IQE). When phosphorescent emitters are used, the theoretical IQE is close to 100% (Adachi et al. 2001), a factor of 4 increase compared with fluorescence. Thus phosphorescent materials are very attractive in building efficient OLEDs (Baldo et al. 1998, Cleave et al. 1999, Hoshino and Suzuki 1996, Kido et al. 1990, Ma et al. 1998) because they can harvest both singlet and triplet excitons.

The phosphorescent dyes that are studied and used in OLEDs are typically organometallic compounds or transition metal compounds with organic ligands (Adachi et al. 2000, Baldo et al. 1998, Chen et al. 2006, Rayabarapu et al. 2005, Sajoto et al. 2005, Tokito et al. 2003). In organometallic compounds, the two most common heavy metals incorporated are platinum (Pt) and iridium (Ir). As an example, the pioneering work of Baldo et al. (1998) demonstrated an OLED with efficient energy transfer from both singlets and triplets by doping a phosphorescent dye, 2,3,7,8,12,13,17,18-octaethyl-21H, 23H-porphine platinum(II) (PtOEP), into a host organic compound (Figure 6.7). Baldo et al. (1999) also demonstrated another OLED incorporating an iridium compound, *fac*-tris(2-phenylpyridine) iridium (Ir(ppy)$_3$), which is now a well-known green emitter (Figure 6.7). Other transition metals such as osmium (Os) and ruthenium (Ru) may also be used (Carlson et al. 2002, Tung et al. 2005). With all of these types of organometallic compounds, the presence of a heavy metal results in a metal-to-ligand charge-transfer (MLCT) triplet state, which mixes with singlet states through strong spin–orbit coupling and leads to efficient phosphorescent emission. The ISC rate can approach ~100%.

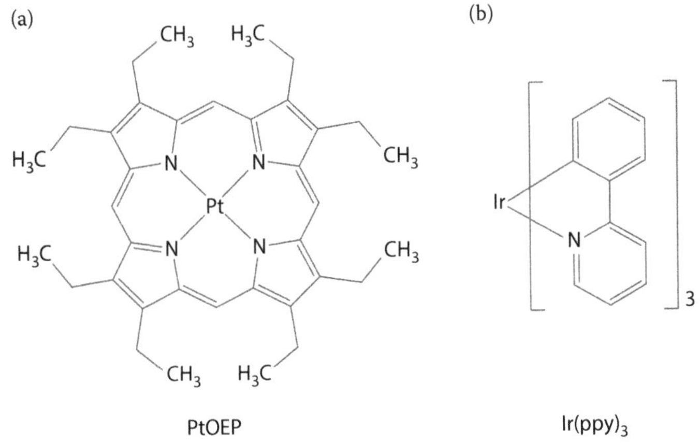

FIGURE 6.7 Molecular structures of two examples of organometallic compounds: PtOEP (a) and Ir(PPY)$_3$ (b).

Chapter 6

Table 6.1 Device Performance (Including Color, Efficacy, and Lifetime) Update in 2012 on a Variety of OLEDs Made of Organic Small Molecule Phosphorescent Materials from Universal Display Corporation

PHOLED Performance (at 1000 cd/m²)	1931 CIE Color Coordinates	Luminous Efficacy (cd/A)	Operating Lifetime [hrs]	
			LT 95%	LT 50%
Deep red	(0.69, 0.31)	17	14,000	250,000
Red	(0.66, 0.34)	29	23,000	600,000
Red	(0.64, 0.36)	30	50,000	900,000
Yellow	(0.44, 0.54)	81	85,000	1,450,000
Green	(0.31, 0.63)	85	18,000	400,000
Light blue	(0.18, 0.42)	50	700	20,000

Note: All results are for bottom-emitting structures (with no cavities) fabricated by vacuum thermal evaporation. Lifetime data are based on accelerated current drive conditions at room temperature without any initial burn-in.

Thus both singlet and triplet excitons in the system eventually result in phosphorescence and close to 100% IQE can be achieved.

OLED displays were first commercialized in the 1990s, with the first displays based on fluorescent materials. Subsequently, due to their high-efficiency characteristics, OLEDs based on phosphorescent materials have attracted a great deal of interest and attention from academia and industry. They were first used in commercial displays in 2003 by Pioneer Corporation. Since this time, many OLED products ranging from cell phones with OLED displays, OLED TVs to OLED lighting products, have been commercialized. To date, monochrome phosphorescent OLEDs, spanning the visible spectrum, have been demonstrated with external quantum efficiency exceeding 25% at 1000 cd/m². A table including various phosphorescent OLED characteristics reported by Universal Display Corporation, a leading phosphorescent OLED company (Table 6.1), as an example of the latest developments in highly efficient and long-lived phosphorescent OLEDs (Universal Display 2013).

References

Adachi, C., M. A. Baldo, S. R. Forrest, and M. E. Thompson. High-efficiency organic electrophosphorescent devices with tris(2-phenylpyridine)iridium doped into electron-transporting materials. *Appl. Phys. Lett.* 77, 904, 2000.

Adachi, C., M. A. Baldo, M. E. Thompson, and S. R. Forrest. Nearly 100% internal phosphorescence efficiency in an organic light-emitting device. *J. Appl. Phys.* 90, 5048, 2001.

Alvarado, S. F., P. F. Seidler, D. G. Lidzey, and D. D. C. Bradley. Direct determination of the exciton binding energy of conjugated polymers using a scanning tunneling microscope. *Phys. Rev. Lett.* 81, 1082, 1998.

Ashcroft, N. W. and N. D. Mermin. *Solid State Physics*. Philadelphia: Saunders College, 1976.

Atkins, P. W. *Quanta*. New York: Oxford University Press, 1991.

Atkins, P. W. *The 2nd Law: Energy, Chaos and Form*. NY: Scientific American Books, 1994.

Baldo, M. A., D. F. O'Brien, Y. You, A. Shoustikov, S. Sibley, M. E. Thompson, and S. R. Forrest. Highly efficient phosphorescent emission from organic electroluminescent devices. *Nature* 395, 151, 1998.

Baldo, M. A., S. Lamansky, P. E. Burrows, M. E. Thompson, and S. R. Forrest. Very high-efficiency green organic light-emitting devices based on electrophosphorescence. *Appl. Phys. Lett.* 75, 4, 1999.

Baldo, M. A., C. Adachi, and S. R. Forrest. Transient analysis of organic electrophosphorescence. II. Transient analysis of triplet–triplet annihilation. *Phys. Rev. B* **62**, 10967, 2000.

Barth, S. and H. Bassler. Intrinsic photoconduction in PPV-type conjugated polymers. *Phys. Rev. Lett.* **79**, 4445, 1997.

Bulovic, V., M. A. Baldo, and S. R. Forrest. Excitons and energy transfer in doped luminescent molecular organic materials. In R. Farchioni and G. Grosso (Eds.), *Organic Electronic Materials: Conjugated Polymers and Low Molecular Weight Organic Solids*. New York: Springer-Verlag, 2001.

Carlson, B., G. D. Phelan, W. Kaminsky, L. Dalton, X. Jiang, S. Liu, and A. K. Jen. Divalent osmium complexes: Synthesis, characterization, strong red phosphorescence, and electrophosphorescence. *J. Am. Chem. Soc.* **124**, 14162, 2002.

Chen, H. Y., C. H. Yang, Y. Chi, Y. M. Cheng, Y. S. Yeh, P. T. Chou, H. Y. Hsieh, C. S. Liu, S. M. Peng, and G. H. Lee. Room-temperature NIR phosphorescence of new iridium (III) complexes with ligands derived from benzoquinoxaline. *Can. J. Chem.* **84**, 309, 2006.

Cleave, V., G. Yahioglu, P. Le Barny, R. H. Friend, and N. Tessler. Harvesting singlet and triplet energy in Polymer LEDs. *Adv. Mater.* **11**, 285, 1999.

D'Andrade, B. W., M. Baldo, C. Adachi, J. Brooks, M. E. Thompson, and S. R. Forrest. High-efficiency yellow double-doped organic light-emitting devices based on phosphor-sensitized fluorescence. *Appl. Phys. Lett.* **79**, 1045, 2001.

Daudel, R., G. Leroy, D. Peeters, and M. Sana. *Quantum Chemistry*, 2nd edn. NY: Wiley, 1983.

Emin, D. and T. Holstein. Studies of small-polaron motion 4: Adiabatic theory of Hall Effect. *Ann. Phys.* **53**, 439 1969.

Forrest, S. R. Ultrathin organic films grown by organic molecular beam deposition and related techniques. *Chem. Rev.* **97**, 1793, 1997.

Gutmann, F., L. E. Lyons, and H. Keyzer. *Organic Semiconductors*. Malabar, Florida: R. E. Krieger, 1981.

Hill, I.G., A. Kahn, Z.G. Soos, and R.A. Pascal. Charge separation energy in films of π-conjugated organic molecules. *Chem. Phys. Lett.* **327**, 181, 2000.

Hoshino, S. and H. Suzuki. Electroluminescence from triplet excited states of benzophenone. *Appl. Phys. Lett.* **69**, 224, 1996.

Kalinowski, J., W. Stampor, J. Mężyk, M. Cocchi, D. Virgili, V. Fattori, and P. Di Marco. Quenching effects in organic electrophosphorescence. *Phys. Rev. B.* **66**, 235321, 2002.

Kepler, R. G. *Organic Semiconductors*. New York: Macmillan, 1962.

Kido, J., K. Nagai, and Y. Ohashi. Electroluminescence in a Terbium Complex. *Chem. Lett.* **19**, 657, 1990.

Knupfer, M., J. Fink, E. Zojer, G. Leising, U. Scherf, and K. Mullen. Localized and delocalized singlet excitons in ladder-type poly(paraphenylene). *Phys. Rev. B.* **57**, R4202, 1998.

Levine, I. *Quantum Chemistry*, 4th edn. Englewood Cliffs, NJ: Prentice-Hall, 1991.

Lowe, J. P. *Quantum Chemistry*, 2nd edn. NY: Academic Press, 1993.

Lunt, R. R., J. B. Benziger, and S. R. Forrest. Growth of an ordered organic heterojunction. *Adv. Mater.* **19**, 4229, 2007.

Ma, Y., H. Zhang, J. Shen, and C. Che. Electroluminescence from triplet metal-ligand charge-transfer excited state of transition metal complexes. *Synth. Met.* **94**, 245, 1998.

Marcus, R. A. Chemical and electrochemical electron-transfer theory. *Annu. Rev. Phys. Chem.* **15**, 155, 1964.

Peumans, P., V. Bulović, and S. R. Forrest. Efficient, high-bandwidth organic multilayer photodetectors. *Appl. Phys. Lett.* **76**, 3855, 2000.

Pfeiffera, M., K. Leoa, X. Zhou, J.S. Huang, M. Hofmann, A. Werner, and J. Blochwitz-Nimoth. Doped organic semiconductors: Physics and application in light emitting diodes. *Org. Electron.* **4**, 89, 2003.

Pope, M. and C. E. Swenberg. *Electronic Processes in Organic Crystals and Polymers*, 2nd edn. New York: Oxford University Press, 1999.

Rayabarapu, D. K., B. Paulose, J. P. Duan, and C. H. Cheng. New iridium complexes with cyclometalated alkenylquinoline ligands as highly efficient saturated red-light emitters for organic light-emitting diodes. *Adv. Mater.* **17**, 349, 2005.

Sajoto, T., P. I. Djurovich, A. Tamayo, M. Yousufuddin, R. Bau, M. E. Thompson, R. J. Holmes, and S. R. Forrest. Blue and near-UV phosphorescence from iridium complexes with cyclometalated pyrazolyl or N-heterocyclic carbine ligands. *Inorg. Chem.* **44**, 7992, 2005.

Silinsh, E.A. and V. Capek. *Organic Molecular Crystals: Interaction, Localization and Transport Phenomena*. New York: American Institute of Physics Press, 1994.

Sze, S. M. *Physics of Semiconductor Devices*, 2nd edn. New York: Wiley, 1981.

Tang, C. W. and S. A. Vanslyke. Organic electroluminescent diodes. *Appl. Phys. Lett.* **51**, 913, 1987.

Chapter 6

Tokito, S., T. Iijima, T. Tsuzuki, and F. Sato. High-efficiency white phosphorescent organic light-emitting devices with greenish-blue and red- emitting layers. *Appl. Phys. Lett.* **83**, 2459, 2003.

Tung, Y. L., S. W. Lee, Y. Chi, L. S. Chen, C. F. Shu, F. I, Wu, A. J. Carty, P. T. Chou, S. M. Peng, and G. H. Lee. Organic light-emitting diodes based on charge-neutral Ru^{II} phosphorescent emitters. *Adv. Mater.* **17**, 1059, 2005.

Turro, N. J. *Modern Molecular Photochemistry*. Sausalito, CA: University Science Books, 1991.

Universal Display Corporation. 2013. Available at: http://www.universaldisplay.com/default.asp?contentID=604.

Wright, J. D. *Molecular Crystals*, 2nd edn. Cambridge: Cambridge University Press, 1995.

7. Electron Transport Materials

Hisahiro Sasabe and Junji Kido

7.1 Introduction

Since the first report of a practical organic light-emitting device (OLED) by Tang and VanSlyke (1987), OLEDs have made significant progress toward becoming a viable technology for solid-state lighting and next-generation flat-panel displays. Tang and VanSlyke made a two-layer device that consisted of di-[4-(N,N-ditolyl-amino)-phenyl]cyclohexane (TAPC) as a hole transport material (HTM) and 8-hydroxyquinoline aluminum (Alq) as an electron transport material (ETM) shown in Figure 7.1. This device showed over 1000 cd/m^2, an external quantum efficiency (η_{ext}) of 1%, and power efficiency (η_p) of 1.5 lm/W at an applied voltage of only 10 V. The key points are (1) use of ultrathin organic layer of around 100 nm and (2) use of organic multilayers with different electrochemical properties.

Adachi et al. (1988) proposed a design concept using a three-layer structure device, so-called a double heterostructure OLED to improve the carrier recombination efficiency (Figure 7.2). The authors used N,N'-bis(3-methylphenyl)-N,N'-diphenylbenzidine (TPD) as HTM, perylene bis-benzimidazole (PV) as an electron transport layer (ETL), and anthracene, coronene, and perylene as an emissive layer (EML). These results demonstrated efficient carrier double injection into the EML by use of separate HTM and ETM. The HTM

Chapter 7

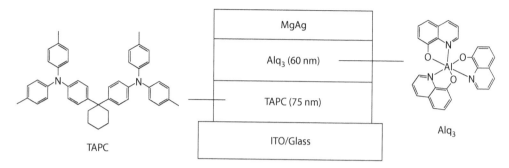

FIGURE 7.1   First practical OLED reported by Tang and VanSlyke (1987). (Adapted from Tang, C. W., VanSlyke, S. A. 1987. *Appl. Phys. Lett.* 51: 913–915.)

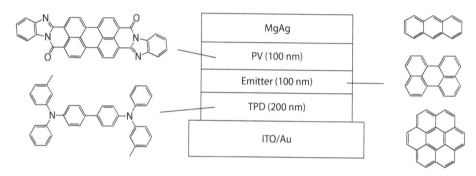

FIGURE 7.2   A double heterostructure OLED developed by Adachi. (Adapted from Adachi, C. et al. 1988. *Jpn. J. Appl. Phys.* 27: L269–L271; Adachi, C., Tsutsui, T., Saito, T. 1990. *Appl. Phys. Lett.* 57: 531–533.)

and ETM also block electrons and holes, respectively, to confine both carriers in the EML, creating improved carrier balance. This device is the prototype of present multi-layer OLEDs.

According to Equation 7.1 (Tsutsui 1997), a key measure of OLED performance, the external quantum efficiency (η_{ext}), can be described as a product of four factors: (1) carrier balance factor (γ), (2) exciton formation ratio (η_r), (3) photoluminescent quantum yield (η_{PL}), and (4) light outcoupling factor (η_{out}) shown in Figure 7.3. Therefore, it is critically important to maximize these four factors to develop a high-performance OLED. The rest of this chapter discusses in detail how the composition and molecular structure of the ETM and interface affect γ.

$$\eta_{ext} = \gamma \times \eta_\gamma \times \eta_{PL} \times \eta_{out} \tag{7.1}$$

In an NPD/Alq device, the mobility of holes in N,N'-di(1-naphthyl)-N,N'-diphenyl-(1,1'-biphenyl)-4,4'-diamine (NPD) film is >1000 times higher than that of electrons in Alq film. Thus, the carrier balance factor (γ) is insufficient causing a low efficiency. To improve the device efficiency, it is necessary to improve the electron mobility (μ_e) of the ETM. Use of an ETM with high μ_e can also improve the conductivity in an OLED leading to low driving voltages. Thus, development of novel high-performance ETMs is one of the key issues to reach the required power efficiency of OLEDs.

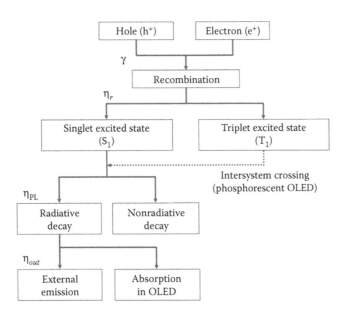

FIGURE 7.3 A schematic representation of the elementary process in OLED. (Adapted from Tsutsui, T. 1997. *MRS Bull.* 22: 39–45.)

The reduction of electrical power consumption is absolutely essential from a practical point of view. Since the power consumption is directly proportional to the driving voltage in an OLED, the driving voltage must be minimized for higher efficiency devices. One approach to reducing the driving voltage is to add an inorganic electron-injection layer (EIL) (see Chapter 5 for a more detailed discussion of charge injection and Chapter 9 for a detailed discussion on conductivity doping). One example is the use of an ultra-thin lithium fluoride (LiF) layer (0.5–1.0 nm) at the Alq/Al interface. In 1997, Mason and coworkers reported an NPD/Alq device with LiF as an EIL (Hung et al. 1997). Compared to the corresponding Mg/Ag-based device, the LiF/Al-based device showed a 7 V reduction in the driving voltage at 100 mA/cm² (Figure 7.4). Similarly, Wakimoto et al. (1997) have reported the use of inorganic salts, such as lithium oxide (Li_2O), sodium chloride (NaCl), and potassium chloride (KCl) as an EIL to reduce an operating voltage.

Another approach to reduce the driving voltage is use of a metal-doped organic layer as an EIL at organic/cathode interface. This methodology is called chemical doping, generating radical anions as intrinsic electron carriers, which results in a low barrier height for electron injection and high conductivity of the doped layer. In 1998, Kido and coworkers reported that a device with Li-doped Alq/Al layer showed high luminance of over 30,000 cd/m² at 10.5 V, while a device without the metal-doped Alq layer exhibited only 3400 cd/m² at 14 V (Kido and Matsumoto 1998). Similarly, in 2002, Huang and Leo reported an Alq device with Li-doped 4,7-diphenyl-1,10-phenanthroline (Bphen)/Al layer, which showed 1000 cd/m² at 2.9 V (Huang et al. 2002). Compared to Tang's initial Alq device (Tang and VanSlyke 1987), this Alq device showed 7 V reduction in driving voltage at 1000 cd/m². Furthermore, in 2002, Leo and coworkers reported an $Ir(ppy)_3$-based device with Cs-doped Bphen/Al layer, which showed 1000 cd/m² at 3.0 V (Pfeiffer et al. 2002). However, these highly active alkaline metals such as Li and Cs are not easy to handle and readily oxidize in the presence of ambient oxygen and water.

Chapter 7

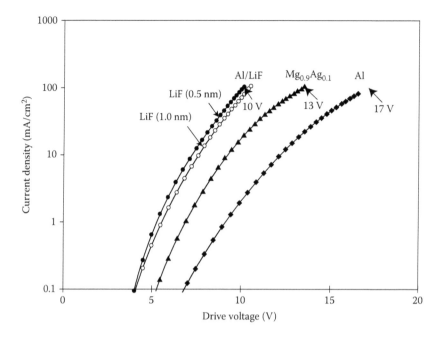

FIGURE 7.4 Current density–luminance characteristics of NPD/Alq device with LiF as an electron-injection layer. (From Hung, S., Tang, C. W., Mason, M. G. 1997. *Appl. Phys. Lett.* 70: 152–154. With permission.)

As an alternative strategy, a new type of cathode interface layer composed of metal complexes such as 8-quinolinolato lithium (Liq), 8-quinolinolato sodium (Naq), lithium acetylacetonate (Liacac), and lithium dipivaloylmethane (Lidpm) were reported by Kido and coworkers in 1998 (Figure 7.5) (Endo et al. 1998, 2002, Schmitz et al. 2000). These metal complexes can be evaporated at a relatively low temperature of 200°C–300°C compared with LiF (>700°C), and are easy to handle under ambient conditions. By using a codeposited layer of Alq and Liq as an EIL, the electron injection from the Al cathode to Alq layer can be facilitated effectively. Other than Liq and Naq, Qiao and Qiu have reported the use of 8-quinolinolato cesium (Csq) as a cathode interface layer in 2008 (Xie et al. 2008).

Subsequent reports by Kido describe the use of additional lithium compounds as an EIL (Figure 7.6). Fujihira et al. (1999) reported lithium carboxylates that showed superior performance to that of LiF (Ganzorig and Fujihira 1999). Kim et al. (1999) reported the use of 2-(2-hydroxyphenyl)benzoxazolate lithium (LiPBO) as an EIL, which can also be used as a blue emitter. Liang et al. (2003) reported a hydroxyloxadiazole lithium

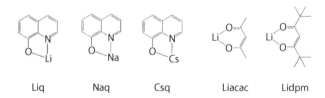

FIGURE 7.5 Li and Cs complexes used as an EIL.

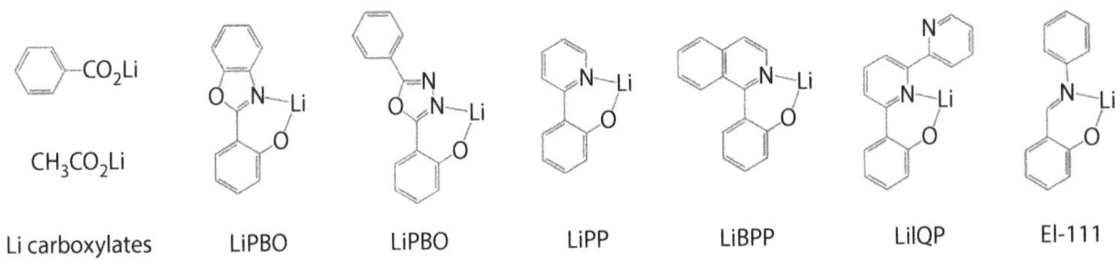

FIGURE 7.6   Li complexes used as an EIL.

complex as both an ETL/emitter material and a substitute for LiF interlayer between Alq and the aluminum cathode. Pu et al. (2009) reported lithium phenolate derivatives for an EIL. These materials show much lower sublimation temperature, around 300°C, than that of LiF (717°C), with comparable performance. Surprisingly, a 40-nm thick film of lithium 2-(2′,2″-bipyridine-6′-yl)phenolate (LiBPP) or lithium 2-(2-pyridyl)phenolate (LiPP) is effective as an EIL, providing low-driving voltages, while such a thick film of LiF serves as a complete insulator, resulting in high-driving voltages. Very recently, Kathirgamanathan et al. (2012) have reported novel lithium Schiff-base cluster complexes such as EI-111. These complexes also show superior performance compared to LiF in a N,N'-diphenylquinacridone (DPQA)-based green OLED.

7.2   Electron Transport Materials

7.2.1   Molecular Design for Electron Transport Materials

ETMs perform several important functions. The key requirements for multifunctional ETMs are: (1) good electron-injection properties, (2) high electron mobility, (3) high hole-blocking ability, and (4) sufficient triplet energy for exciton blocking. Generally, electron deficient aromatic moieties such as oxadiazoles, triazoles, imidazoles, pyridines, and pyrimidines are used as building blocks for an ETM to accept electrons from the cathode efficiently. In this section, the results of density functional theory (DFT) calculations are discussed in order to generate a better understanding of ETM molecular design (Sasabe, H., Kido, J. Unpublished results) (Figure 7.7).

In these calculations, the effect of incorporation of a nitrogen atom into benzene 1 on the frontier molecular orbitals (FMOs, highest occupied molecular orbital [HOMO]/ lowest unoccupied molecular orbital [LUMO]) energies was investigated. The FMOs of benzene 1 are calculated to be 7.07/0.49 eV, respectively. When a nitrogen atom is introduced into benzene 1 to give pyridine 2, the FMOs are changed to 7.25 and 1.15 eV, respectively. The HOMO of pyridine is 0.25 eV deeper than that of benzene, while the LUMO is 0.66 eV deeper. Therefore, the introduction of a nitrogen atom into a benzene ring improves its electron-accepting ability. The FMO energies shift further when two nitrogen atoms are introduced into benzene to give three isomers such as pyrazine 3, pyridazine 4, and pyrimidine 5. The LUMOs become much deeper, 1.65–1.91 eV, greatly enhancing the electron-accepting ability. Finally, triazine 6 gives the deepest LUMO (2.02 eV) among these compounds. From these results, it is clear that significant improvements in the electron-accepting nature can be made by introducing a nitrogen atom into the proper position in a molecule. However, the proper position of nitrogen

Chapter 7

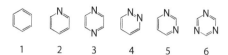

Compound	HOMO (eV)	LUMO (eV)
Benzene (1)	7.07	0.49
Pyridine (2)	7.25	1.15
Pyrazine (3)	7.20	1.91
Pyridazine (4)	6.75	1.89
Pyrimidine (5)	7.31	1.65
Triazine (6)	7.97	2.02

FIGURE 7.7   HOMO/LUMO energies of compound 1–6 obtained from DFT calculation at the B3LYP 6–311G + (d,p)//6–31G(d) level of theory.

in a molecule for a high-performance ETM depends upon a number of factors that are discussed later in this chapter.

7.2.2   Electron Transport Materials for Fluorescent and Red to Green Phosphorescent OLEDs

Compared to hole transporting materials, a much smaller number of ETMs has been reported for use in OLEDs so far (Hughes and Bryce 2005, Kulkarni et al. 2004, Sasabe and Kido 2011, Shirota 2005, Xiao et al. 2011), due to the difficulties in simultaneously satisfying all the necessary criteria for a stable high-performance ETL. In particular, electron mobility is typically lower in organic materials than hole mobility. In addition, chemical stability in an operating OLED is more difficult to achieve for an ETL material than for a hole transport layer material. This section will briefly describe a number of examples of molecules which have been used as ETMs, with an emphasis on applications in fluorescent (regardless of color) and red and green phosphorescent OLEDs. The unique challenges for blue phosphorescent OLEDs will be addressed below in a separate subsection.

For small molecule-based fluorescent and green and red phosphorescent OLEDs, imidazoles, siloles, boranes, and phosphine oxides as well as metal complexes were examined as potential ETL materials since the early stages of OLED research. Figure 7.8 shows some early representative examples.

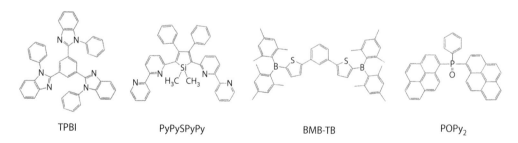

| TPBI | PyPySPyPy | BMB-TB | POPy₂ |

FIGURE 7.8   Representative examples of ETMs.

Early examples of ETMs possessed some limitations. A relatively high electron mobility (up to 10^{-4} cm^2/Vs), adequate electron-injection properties, and reasonable hole-blocking ability were achieved in individual materials rather than combining all these necessary elements in one ETM. This, in part, happened due to the lack of a better understanding of effective molecular design. As the field advanced, material chemists looked for structure–property relationships in order to design a next-generation ETM (Hughes and Bryce 2005, Kulkarni et al. 2004, Shirota 2005) to combine good charge transport, injection, triplet exciton energy, and chemical stability in a high-performance ETL material. While injection barriers at the layer interfaces are defined by the boundary orbital energies of the layer materials and their alignment, the charge transport through an individual layer, in this case through the ETL, is affected by a combination of factors related to both the electronic properties of individual molecules and intermolecular phenomena such as packing, orbital overlap, reorganization energies of electron-transfer reactions, and so on. This is typically more challenging to control compared to the HOMO/LUMO level positions. Nevertheless, there have been significant advances in addressing the electron mobility in ETMs. For instance, an oligoimidazole derivative developed by Shi in 1997, 2,2′,2″-(1,3, 5-benzinetriyl)-tris(1-phenyl-1-H-benzimidazole) (TPBI), showed a μ_e of 8×10^{-5} cm^2/Vs, and was commonly used for fluorescent and phosphorescent OLEDs. Compared to Alq, TPBI has a deeper I_p of 6.2 eV, indicating better hole-blocking ability (Shi et al. 1997). A silole derivative 2, 5-bis(6′-(2′, 2″-bipyridyl))-1, 1-dimethyl-3,4-diphenylsilole (PyPySPyPy) developed by Tamao and Yamaguchi had a μ_e on the order of 10^{-4} cm^2/Vs and gave superior performance in an NPD/Alq device (Tamao et al. 1996, Uchida et al. 2001). A dimesitylboryl-modified thiophene derivative, 1,2,3-tris[5-(dimesitylboryl)thiophen-2-yl]benzene (BMB-TB) developed by Shirota had a deep I_p of 6.3 eV ensuring good hole-blocking ability, and showed superior performance in blue fluorescent OLED (Kinoshita and Shirota 2001). A phosphine-oxide derivative, phenyldipyrenylphosphine oxide (POPy$_2$) developed by Adachi showed excellent electron injection and transport properties when doped with Cs metal (Oyamada et al. 2005). An NPD/Alq device with Cs-doped POPy$_2$ as an EIL showed a large reduction in driving voltage to give 100 mA/cm^2 at only 3.9 V.

Additional work by a number of groups of material chemists led to the development of several synthetic strategies to improve electron mobility (Sasabe and Kido 2011, Xiao et al. 2011). A number of recent examples of materials that illustrate progress in addressing the inferior electron mobility of ETMs are shown in Figure 7.9. An

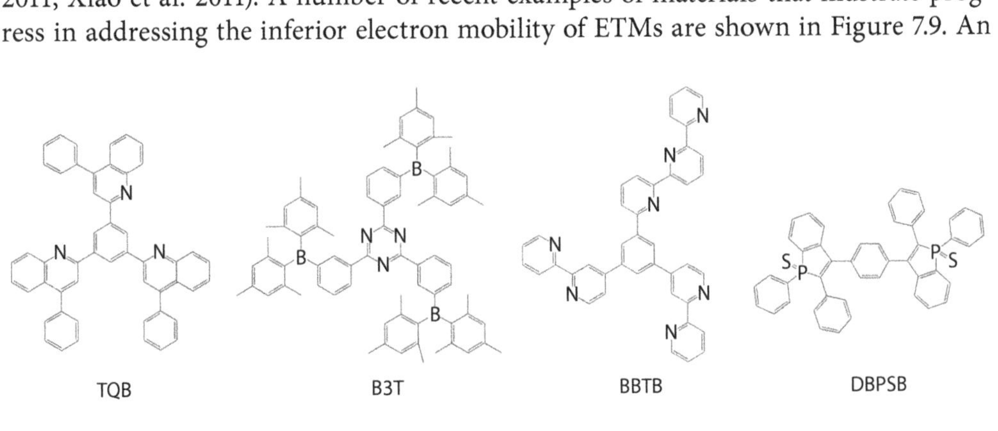

TQB   B3T   BBTB   DBPSB

FIGURE 7.9 Recent examples of ETMs.

Chapter 7

oligoquinoline derivative developed by Jenekhe, 1, 3, 5-tris(4-phenylquinolin-2-yl)benzene (TQB), has a μ_e of 3.6×10^{-4} cm²/Vs measured by the space charge limited current (SCLC) method. In addition to high mobility, this molecule has the additional advantage of being amenable to solution processing from a formic acid/H$_2$O mixture (Earmme et al. 2010). A polyboryl-functionalized triazine derivative, 2,4,6-tris(*m*-dimesitylborylphenyl)-1,3,5-triazine (B3T) developed by Wang has a low I_p of 6.7 eV improving on hole-blocking ability (Sun et al. 2011). An Ir(ppy)$_3$-based OLED employing B3T as the ETL showed a maximum current efficiency ($\eta_{c,max}$) of 68.9 cd/A (η_{ext} 19.8%). However, the device with B3T showed relatively high turn-on voltages probably due to inefficient electron injection from the LiF/Al cathode. Improving on the latter problem, a bipyridine derivative, 1,3-bisbipyridyl-5-terpyridylbenzene (BBTB) was developed by Ichikawa. In addition, BBTB achieved a superior μ_e of ~10^{-3} cm²/Vs measured by the time-of-flight (TOF) method (Ichikawa et al. 2012). Furthermore, a benzophosphole sulfide derivative, di(benzo[b]phosphole sulfide) (DBPSB) developed by Tsuji had a high μ_e of 2×10^{-3} cm²/Vs as measured by the TOF method (Tsuji et al. 2009). Compared to the corresponding oxide derivative, di(benzo[b]phosphole oxide) benzene (DBPOB), the sulfide derivative showed 1000 times higher μ_e. This is a result of the lower polarity of the $P = S$ moiety compared to the $P = O$ moiety, which can become a deep trap that reduces the μ_e.

7.2.3 Electron Transport Materials for Blue and White Phosphorescent OLEDs

Although the ETMs shown in Figures 7.8 and 7.9 are effective in fluorescent and red to green phosphorescent OLEDs, these materials, except TPBI and TQB, are not suitable for use in blue phosphorescent OLEDs because of their relatively low triplet exciton energies. In a phosphorescent OLED, the confinement of triplet excitons on the emitter is necessary to maximize device performance. Therefore, the host material, as well as the ETL and any other components of the EML should have a higher E_T than that of the phosphorescent emitter. Indeed, the above-mentioned ETMs have a lower E_T than that of a typical blue phosphorescent emitter, causing triplet-exciton quenching at EML/ETL interface. As an example, the well-known blue phosphorescent emitter, iridium(III)bis(4,6-(difluorophenyl)pyridinato-N,C′)picolinate (FIrpic), has an onset phosphorescence (E_T) of 2.77 eV (see Figure 7.10). ETL materials with triplet energies lower than 2.77 would result in significant efficiency loss in blue phosphorescent OLEDs.

3-(4-Biphenylyl)-4-phenyl-5-(4-tert-butylphenyl)-1,2,4-triazole (TAZ) and 2, 9-dimethyl-4,7-diphenyl-1,10-phenanthroline (BCP) are well-known wide-energy-gap materials with adequate electron mobility and chemical stability. They therefore have been used in research labs as ETMs for phosphorescent red and green and fluorescent devices. The solid-state phosphorescent spectra of TAZ and BCP at 5 K are shown in Figure 7.10. The arrows show the onset of phosphorescence (E_T) for these materials. These materials have >0.1 eV lower E_T than that of FIrpic, and are therefore not suitable ETMs for blue phosphorescent OLEDs. To maximize performance, an ETM with higher E_T is required.

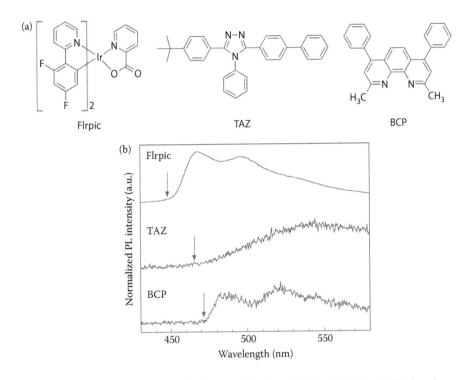

FIGURE 7.10   (a) Chemical structures of (left to right) FIrpic, TAZ and BCP, and (b) phosphorescence spectra of TAZ and BCP in the solid state at 5 K, compared to that of FIrpic.

However, a key challenge is to achieve both good electron transport and high E_T within a single molecule. The next section discusses methods to increase E_T, while increasing electron mobility is addressed below. One key principle can be demonstrated using benzene derivatives as a model system. The E_T's of oligophenylene derivatives are shown in Figure 7.11. Benzene has E_T of 3.65 eV, while biphenyl has 2.84 eV, which is 0.81 eV lower than that of benzene. For terphenyls, the *para*-derivative has E_T of 2.55 eV, while the *meta*-derivative E_T is 2.81 eV, which is comparable to that of biphenyl. Therefore, the *meta*-configuration provides a higher E_T in oligophenylene derivatives, as this configuration reduces the π-conjugation and increases the energy of the triplet state.

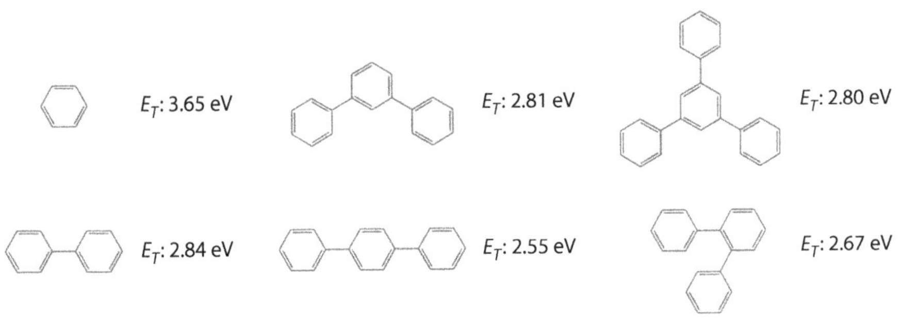

E_T: 3.65 eV E_T: 2.81 eV E_T: 2.80 eV

E_T: 2.84 eV E_T: 2.55 eV E_T: 2.67 eV

FIGURE 7.11   Relationships between chemical structure and E_T energy. (From Van Dijken, A. et al. 2008. *Highly Efficient OLEDs with Phosphorescent Materials.* Wiley-VCH, p. 321.)

Chapter 7

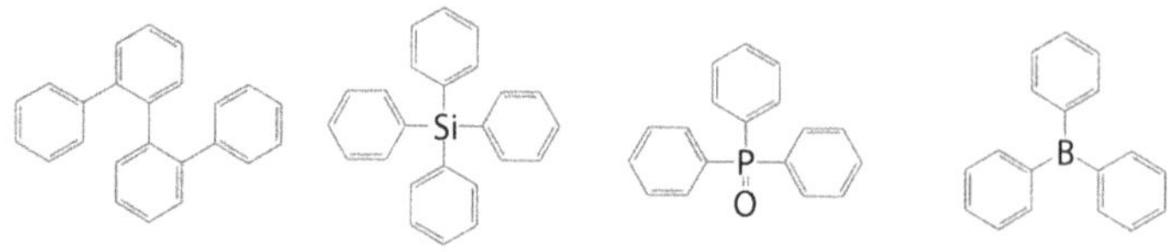

FIGURE 7.12　Typical building blocks used in wide-energy-gap materials.

Another strategy to reduce π-conjugation in the system is to use twisted and/or non-conjugated system(s), such as quaterphenylene (Agata et al. 2007), tetraphenylsilane (Ren et al. 2004), triphenylphosphine oxide (Padmaperuma et al. 2006), and triphenyl-borane (Tanaka et al. 2007a,c) moieties (Figure 7.12). These structures can also be used to generate molecules with a high E_T.

It usually takes a tremendous effort and time to synthesize and purify a material. Therefore, researchers have sought some indication of key optical and electronic properties of a target molecule before synthesis. A favored approach is the use of DFT calculations to roughly estimate properties such as the HOMO energy (i.e., hole-blocking ability), LUMO energy (i.e., electron-injection properties), and E_T. The following section describes a representative example of this approach.

Sasabe and Kido carried out DFT calculations for compounds 7–10 to predict their HOMO/LUMO energies (Sasabe, H., Kido, J. Unpublished results) (Figure 7.13). Compared to the oligophenylene derivative 7, the nitrogen-containing derivatives 8–10 showed deeper LUMO energies. In particular, compound 10 with peripheral pyridine rings had a ca. 0.7 eV deeper HOMO energy than that of compound 7, suggesting good hole-blocking ability. Thus, compound 10 is the most promising candidate among compounds 7–10. After DFT calculations, the synthesis of 3,5,3,5-tetra-4-pyridyl-[1,1;3,1]ter-phenyl (B4PyPB) (10) was carried out via a Suzuki coupling reaction. However, B4PyPB was subsequently found to be highly crystalline, which is not suitable for OLED applications due to resulting short circuits in the device. A candidate material which generates amorphous thin films was therefore sought. To this end, 3,5,3,5-tetra-3-pyridyl-[1,1;3,1] terphenyl (B3PyPB) was synthesized using a $Pd_2(dba)_3/PCy_3$ catalyzed Suzuki coupling

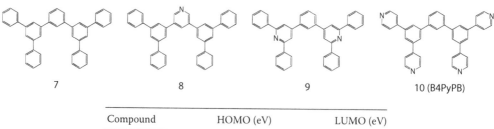

7	8	9	10 (B4PyPB)

Compound	HOMO (eV)	LUMO (eV)
7	6.23	1.42
8	6.39	1.55
9	6.34	1.90
10 (B4PyPB)	6.78	2.10

FIGURE 7.13　HOMO/LUMO energies of compound 7–10 obtained from DFT calculation at the B3LYP 6–311G + (d,p)//6–31G(d) level of theory.

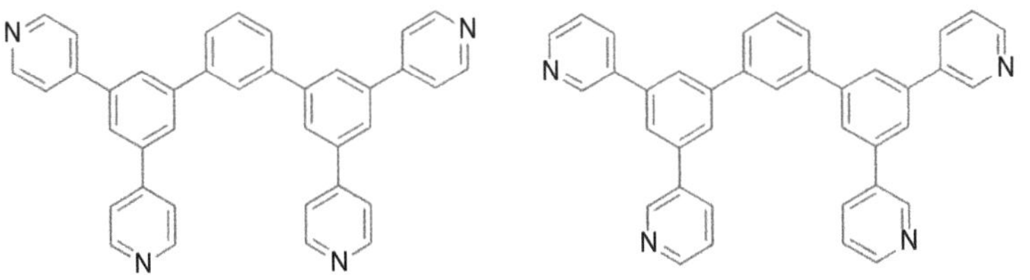

FIGURE 7.14 Chemical structures of BPyPB derivatives.

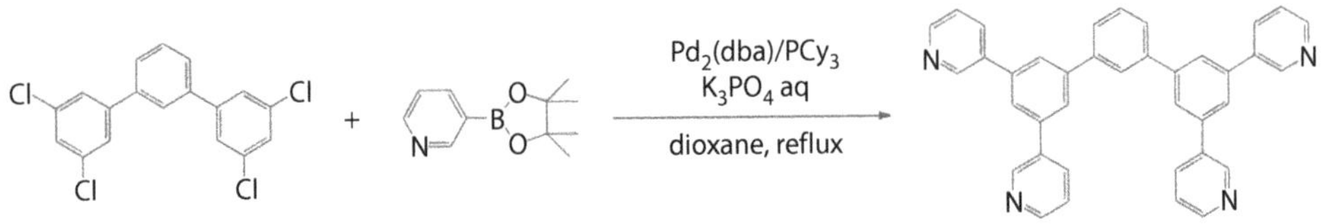

FIGURE 7.15 Synthetic route of B3PyPB.

reaction of the tetrachloride and 3-pyridine boronate ester (Sasabe et al. 2008b) (Figures 7.14 and 7.15). The resulting B3PyPB showed amorphous nature in a thin film.

The physical properties of B3PyPB are summarized in Table 7.1. The E_T, as determined by the onset of phosphorescence at 5 K in the solid state, was 2.77 eV (448 nm), equal to that of FIrpic ($E_T = 2.77$ eV). Furthermore, B3PyPB was found to have a high electron mobility, $\mu_e = 1.0 \times 10^{-4}$ cm²/Vs at an electric field of 3.3×10^{-5} V/cm¹, 100 times higher than that of ETMs such as Alq or TAZ. To investigate the electron transport properties of B3PyPB in a device, OLEDs with a structure of [ITO/NPD (50 nm)/Alq (40 nm)/ B3PyPB or Alq (30 nm)/LiF (0.5 nm)/Al (100 nm)] were fabricated. The current density–voltage (J–V) characteristics are shown in Figure 7.16. A current density of 100 mA/cm² was measured for B3PyPB at an applied voltage of 8.2 V, compared to 10.1 V for Alq. In other words, the device with NPD/Alq with B3PyPB shows ~2.0 V lower driving voltage than one fabricated using Alq.

A blue phosphorescent OLED incorporating a B3PyPB ETL using a 4,4,4-tris(N-carbazolyl)triphenylamine (TCTA)/FIrpic EML has been reported (Figure 7.17). This device uses two EML layers with a single host (TCTA), varying the FIrpic concentration

Table 7.1 Physical Properties of ETMs with 3,5-Dipyridylphenyl Moieties

Compound	$I_p/E_g/E_a$ (eV)[a]	E_T (eV)[b]	μ_e (cm²/Vs)[c]
B3PyPB	6.67/4.05/2.62	2.77	10^{-4}
B3PyMPM	6.97/3.53/3.44	2.75	10^{-5}
B3PyPPM	7.15/3.41/3.74	2.81	10^{-7}

[a] I_p was measured by an UPS or an AC-3. E_g was taken as the point of intersection of the normalized absorption spectra. E_a was calculated using I_p and E_g.

[b] Determined by onset of phosphorescence at 5 K.

[c] Mobility by TOF method.

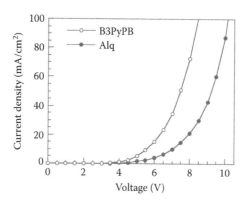

FIGURE 7.16   Current density–voltage (*J–V*) characteristics of NPD/Alq device using B3PyPB and Alq as an ETM. (Reprinted with permission from Sasabe, H. et al. 2008b. Wide-energy-gap electron-transporting materials containing 3,5-dipyridylphenyl moieties for an ultra high efficiency blue organic light-emitting device. *Chem. Mater.* 20: 5951–5953. Copyright 2008 American Chemical Society.)

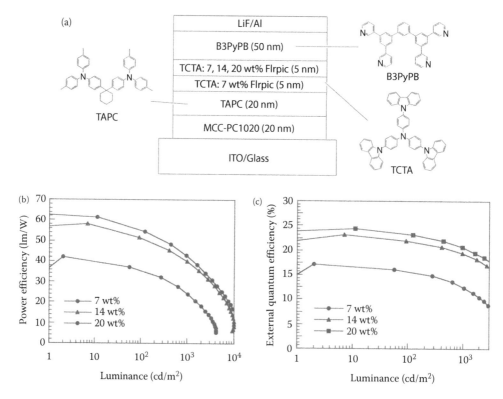

FIGURE 7.17   (a) Structure of high efficiency blue phosphorescent OLED utilizing B3PyPB ETL, with chemical structures of TCTA component of TCTA/FIrpic host, TAPC HTL, and B3PyPB ETL shown. Amount of FIrpic in top layer of TCTA/FIrpic host varied (7%, 14% or 20%). (b) Power efficiency–luminance (*PE–L*), and (c) external quantum efficiency–luminance (*EQE–L*) characteristics of OLEDs shown in (a). (Reprinted with permission from Sasabe, H., Kido, J. 2011. Multifunctional materials in high-performance OLEDs: Challenges for solid-state lighting. *Chem. Mater.* 23: 621–630. Copyright 2011 American Chemical Society.)

in the second EML. The device structure is [ITO/MCC-PC1020 (20 nm)/TAPC (20 nm)/ TCTA: 7 wt% FIrpic (5 nm)/TCTA: 7, 14, 20 wt% FIrpic (5 nm)/B3PyPB(50 nm)/LiF/ Al]. (The hole injection layer, MCC-PC1020, is a chemically doped polymer provided by Mitsubishi Chemical, Science and Technology Research Center, Inc.) The power efficiency–luminance (*PE–L*) and η_{ext}–luminance (*EQE–L*) characteristics are also shown in Figure 7.17. In this architecture, the layer with higher FIrpic doping concentration improves device performance, probably due to the improved carrier balance achieved by improved electron transport provided by the emitter. These results indicate direct carrier trapping and recombination on FIrpic molecules. The optimized device shows a maximum power efficiency ($\eta_{p,max}$) over 60 lm/W ($\eta_{ext,100}$ 24%), $\eta_{p,100}$ of 56 lm/W (2.96 V), and $\eta_{p,1000}$ of 42 lm/W (3.41 V).

A white OLED with high efficacy requires a highly efficient blue emissive system. Thus, there has been a strong emphasis on the development of efficient and robust blue phosphorescent devices. Improving the efficiency of exciton confinement and charge balance for such devices requires a high-performance ETL. Su et al. (2008a) demonstrated a high-efficiency white OLED based on the high-efficiency blue phosphorescent OLED shown in Figure 7.17 in combination with an orange phosphorescent emitter, iridium(III) bis-(2-phenylquinoly-N,C′)dipivaloylmethane (PQ$_2$Ir). 2,2-bis[3-(*N,N*-ditolylamino) phenyl]-biphenyl (3DTAPBP) and 2,6-bis[3′-(*N*-carbazole)phenyl]pyridine (DCzPPy) are used as HTM and host material, respectively, in this device (Figure 7.18).

The device structure is [ITO/MCC-PC1020 (20 nm)/3DTAPBP (20 nm)/TCTA:7 wt% FIrpic (4.75 nm)/TCTA:3 wt% PQ$_2$Ir (0.25 nm)/DCzPPy:3 wt% PQ$_2$Ir (0.25 nm)/ DCzPPy:20 wt% FIrpic(4.75 nm)/B3PyPB (50 nm)/LiF (0.5 nm)/Al]. The power efficiency–luminance–current efficiency (*PE–L–CE*) characteristics are shown in Figure 7.19. The optimized device shows $\eta_{p,100}$ of 52 lm/W ($\eta_{ext,100}$ = 26%) and $\eta_{p,1000}$ of 43 lm/W ($\eta_{ext,1000}$ = 25%). The color rendering index (CRI) and CIE$_{x,y}$ are 68 and (0.341, 0.396), respectively. B3PyPB was used as the ETM demonstrating its effectiveness in white devices as well.

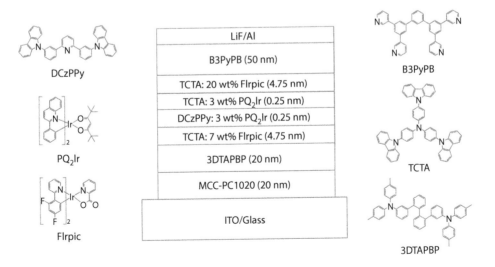

FIGURE 7.18 Device structure and materials used in white phosphorescent OLED.

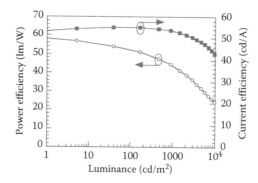

FIGURE 7.19   Power efficiency–luminance–current efficiency (*PE–L–CE*) characteristics of white phosphorescent OLEDs. (From Su, S.-J. et al. 2008b. *Adv. Mater.* 20: 4189. With permission.)

7.2.4   Multifunctional Electron Transport Materials for Extremely High-Efficiency OLEDs

As stated previously, ETMs must simultaneously fulfill several roles (electron injection, electron transport, hole blocking, exciton blocking). To simultaneously optimize performance in each of these roles, the relationship between the ETM chemical structure and key physical properties, and ultimately OLED performance, must be understood. To this end, several ETMs have been investigated to determine the effect of the core skeleton on I_p and E_a, physical properties, and OLED performance. For an OLED material, I_p and E_a are important because they correlate with the positions of the boundary orbital energy levels and thus affect charge injection and charge transport through the organic stack. For ETMs containing 3,5-dipyridylphenyl moieties, replacing the benzene skeleton with 2-methylpyrimidine or 2-phenylpyrimidine changes I_p from 6.7 through 7.2 eV and E_a from 2.6 through 3.7 eV (Table 7.1) (Tanaka et al. 2007b, Sasabe et al. 2008a,b), which corresponds to deeper HOMO and LUMO energies as a result of the substitution. Examples of the OLEDs that employ 3.5-dipyridylphenyl moieties are discussed below.

Ir(ppy)$_3$-based green phosphorescent OLEDs with different ETLs, that is, B3PyPB, 2-phenyl-4,6-bis(3,5-di-3-pyridylphenyl)primidine (B3PyPPM) and as a reference, BCP/ Alq, have been reported. The device structure is [ITO/TPDPES:10 wt% TBPAH (20 nm)/ TAPC (30 nm)/CBP: 8 wt% Ir(ppy)$_3$ (10 nm)/ETL (50 nm)/LiF/Al]. The *PE–L* characteristics are shown in Figure 7.20. B3PyPB- and B3PyPPM-based OLEDs show superior performance with $\eta_{p,max}$ of 118 lm/W ($\eta_{ext,max}$ = 29%) and 140 lm/W ($\eta_{ext,max}$ = 29%), respectively, which is approximately 1.5–1.7 times higher than the value for the reference OLED using BCP/Alq. The operating voltage is much lower for the B3PyPPM-based OLED than for the B3PyPB-based OLED (2.6 and 3.2 V, respectively, at 100 cd/m^2), even though μ_e for B3PyPPM is 1000 times lower than for B3PyPB. The reduced operating voltage is attributed to the deeper E_a level of B3PyPPM. Thus, introduction of a pyrimidine skeleton with two C=N double bonds greatly improves electron injection.

Similarly, Su et al. (2008a, 2009) investigated structure–property relationships for 1,3,5-tri(*m*-pyridyl-phenyl)benzene (TmPyPB) derivatives, and reported that the

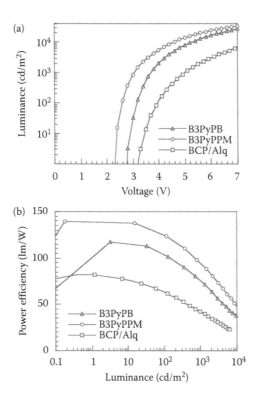

FIGURE 7.20 (a) Luminance-voltage (L–V), and (b) power efficiency-luminance (PE–L) characteristics of green phosphorescent OLEDs. (Reprinted with permission from Sasabe, H., Kido, J. 2011. Multifunctional materials in high-performance OLEDs: Challenges for solid-state lighting. *Chem. Mater.* 23: 621–630. Copyright 2011 American Chemical Society.)

orientation of nitrogen on the peripheral pyridines affects electron-injection properties and OLED performance (Figure 7.21).

For the FIrpic-based OLEDs shown in Figure 7.22, the device performance, as measured by the current density and luminance at the same voltage, decreases in the order Tm4PyPB > Tm3PyPB > Tm2PyPB. For TmPyPB derivatives, E_a decreases in the order Tm4PyPB (2.94 eV) > Tm3PyPB (2.90 eV) > Tm2PyPB (2.74 eV) as predicted by the DFT calculations. The measured values of μ_e are very similar for all the compounds and are of the order of 10^{-3} cm^2/Vs irrespective of nitrogen orientation. Thus, we deduce that differences in OLED performance are due to the differences in electron injection among these TmPyPB derivatives.

7.3 Use of Molecular Interaction for High-Performance Electron Transport Materials

7.3.1 Intramolecular Interactions

The performance of an amorphous solid film can be improved via sophisticated molecular design taking advantage of inter and/or intramolecular interactions. For example, the weak CH/n (n: lone pair) hydrogen bond, that is, CH/O and CH/N, is a representative

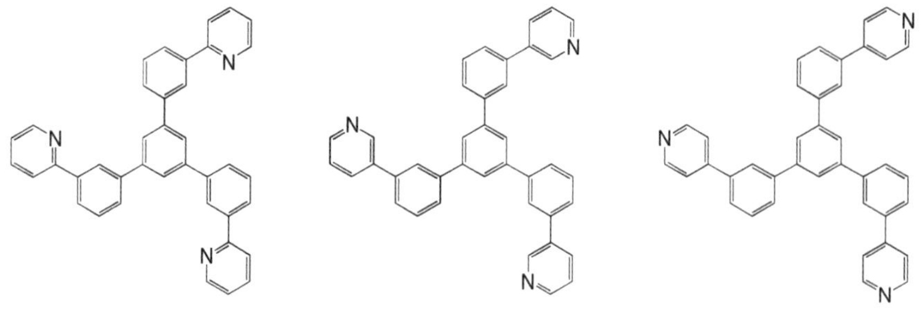

	Tm2PyPB		Tm3PyPB		Tm4PyPB

Compound	I_p (eV)[a]	E_a (eV)[b]	E_g (eV)[c]	T_g (°C)[d]	T_m (°C)[d]
Tm2PyPB	6.43	2.74	3.69	77	174
Tm3PyPB	6.68	2.90	3.78	79	181
Tm4PyPB	6.68	2.94	3.74	99	282

[a]Measured by an UPS.
[b]Taken as the point of intersection of the normalized absorption spectra.
[c]Calculated using I_p and E_g.
[d]Determined by a differential scanning calorimetry measurement.

FIGURE 7.21   Physical properties of TmPyPB derivatives.

example (Desiraju and Steiner 1999). The binding energy of this weak CH/N hydrogen bond is estimated to be 10–20 kJ/mol, which is about half the energy of a typical hydrogen bond (20–30 kJ/mol). This can be seen in the difference in the boiling points of benzene (80.1°C) and pyridine (115.2°C). Although the molecular weights of these molecules are almost the same ($MW_{benzene}$: 78.1 versus $MW_{pyridine}$: 79.1), the CH/N hydrogen bond interaction between pyridines leads to an increase in the energy required to evaporate the liquid. In fact, Hohenstein and Sherill (2009) have reported the binding energy of a pyridine dimer (shown in Figure 7.23) to be 13 kJ/mol using quantum chemical calculations.

An example of exploiting CH/N hydrogen bonding between pyridine units in OLED ETL materials has been given by Ichikawa et al. (2007, 2006) who reported a 2, 2′-bipyridine derivative, 1, 3-bis[2-(2, 2′-bipyridin-6-yl)-1, 3, 4-oxadiazo-5-yl]benzene (BpyOXD). BpyOXD has a high μ_e of 3.1×10^{-3} cm²/Vs. Compared to 1, 3-bis[2-(4-tert-butylphenyl)-1,3,4-oxadiazo-5-yl]benzene (OXD7) (structure shown in Figure 7. 24 with BpyOXD), BpyOXD showed superior performance in an NPD/Alq device with a much lower driving voltage (Figure 7.25).

Yokoyama et al. have investigated the molecular aggregation state of oxadiazole derivatives by variable angle spectroscopic ellipsometry (Yokoyama et al. 2009, Yokoyama 2011). They reported that intramolecular CH–N hydrogen bond interactions of pyridine rings cause the molecules to become planar and promote horizontal molecular orientation, leading to high mobility.

7.3.2   Use of Intermolecular Interaction

In the previous section, we discussed the use of "*intra*"-molecular interaction in a 2-pyridine derivative to improve the molecular packing and charge transport in ETLs

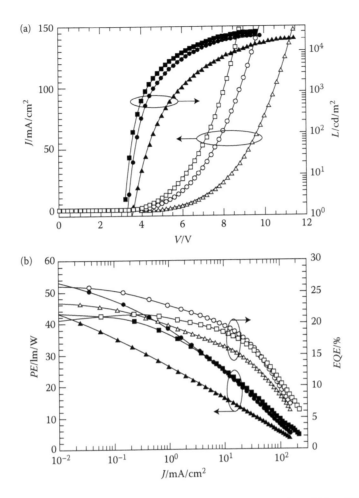

FIGURE 7.22 (a) *J–V–L* and (b) *PE–J–EQE* characteristics of TmPyPB-based blue OLEDs: Tm4PyPB (square), Tm3PyPB (circle), Tm2PyPB (triangle). (From Su, S.-J. et al. 2009. *Adv. Funct. Mater.* 19: 1260–1267. With permission.)

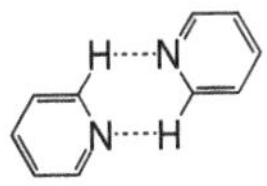

FIGURE 7.23 A weak CH/N hydrogen bond interaction between pyridines.

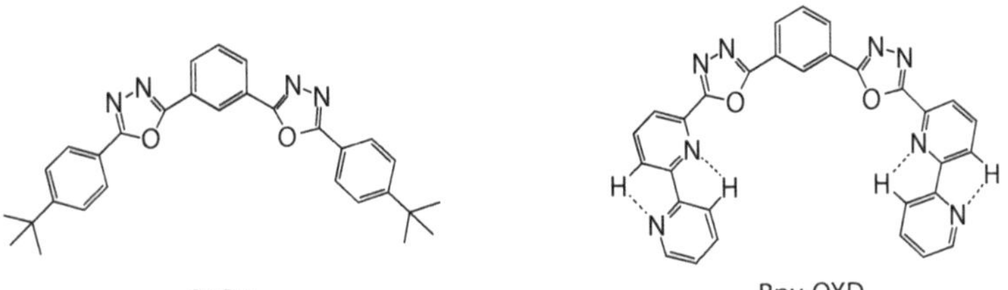

OXD7   Bpy-OXD

FIGURE 7.24 Chemical structure of OXD7 and BpyOXD.

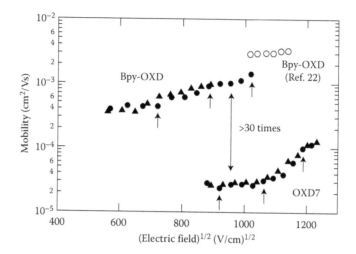

FIGURE 7.25 TOF mobility of OXD7 and BpyOXD. (From Yokoyama, D. et al. 2009. *Appl. Phys. Lett.* 95: 243303 1–3. With permission.)

(Figure 7.26). An alternative is to design a structure that, by using 3- and/or 4-pyridine derivative(s), takes advantage of an "*inter*"-molecular hydrogen bond network to engineer the structure of a solid film.

Recently, several bis-4, 6-(3, 5-dipyridylphenyl)-2-methylpyrimidine (BPyMPM) derivatives have been developed as ETL materials and used to investigate the effect of nitrogen orientation on physical properties (Figure 7.27). Each of these BPyMPM derivatives has a different nitrogen orientation in the peripheral pyridine rings (Sasabe et al. 2011). Below, we briefly discuss the role hydrogen bonding plays in their physical and chemical properties, and the implications this has for OLED performance.

Thin layer chromatography (TLC, silica gel) analysis using chloroform/methanol (30:1 v/v) as an eluent showed that the retention factor (R_f) was decreased in the order

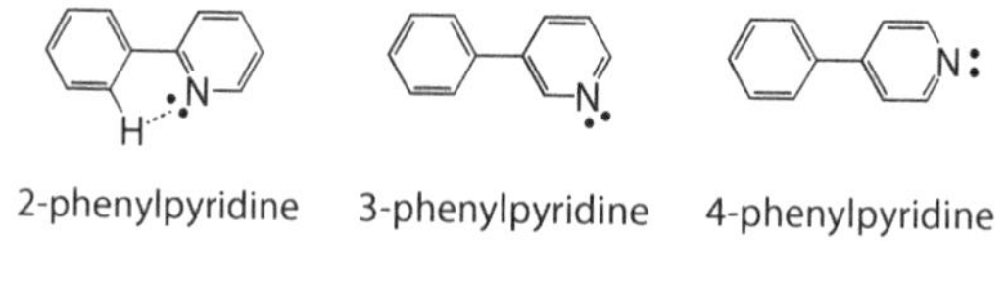

2-phenylpyridine   3-phenylpyridine   4-phenylpyridine

FIGURE 7.26 Chemical structure of phenylpyridine.

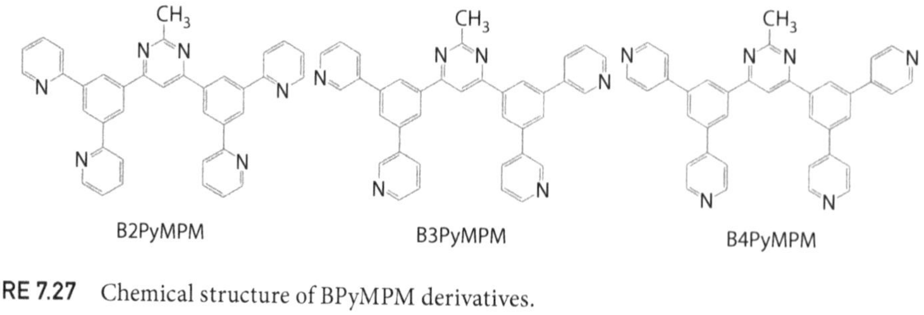

B2PyMPM   B3PyMPM   B4PyMPM

FIGURE 7.27 Chemical structure of BPyMPM derivatives.

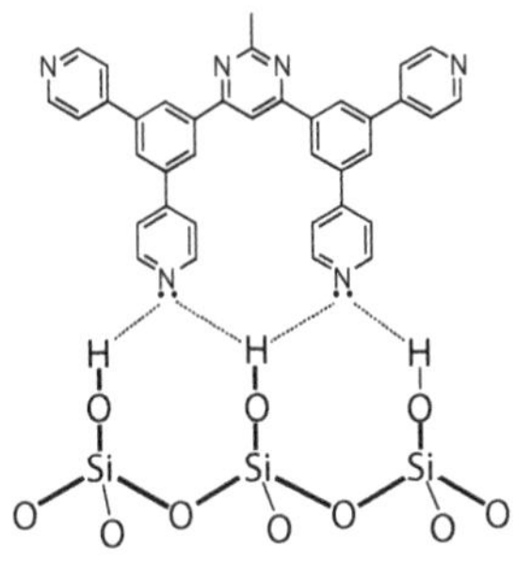

FIGURE 7.28 Hydrogen bond interaction between silanol and B4PyMPM on TLC.

B2PyMPM (0.80) ≫ B3PyMPM (0.25) > B4PyMPM (0.12). Therefore, the degree of interaction with silica gel, which shows the strength of the hydrogen bond interaction between a single substrate molecule and silanol, should increase in the order B2PyMPM ≪ B3PyMPM < B4PyMPM (Figure 7.28).

BPyMPM derivatives were only soluble in halogenated solvents, such as chloroform and dichloromethane, and B3PyMPM and B4PyMPM were found to be much less soluble in chloroform compared to B2PyMPM. On the other hand, all of the BPyMPM derivatives were highly soluble in a chloroform/methanol mixture (e.g., 30:1 v/v). Since the hydrogen bonds are cleaved by methanol, these observations strongly suggest the presence of intermolecular hydrogen bonds in the solid state and that the degree of the interactions increased in the order B2PyMPM ≪ B3PyMPM < B4PyMPM.

The glass transition temperature (T_g) of B2PyMPM was determined to be 107°C, indicating high thermal stability of the thin film. On the other hand, no T_g was detected for B3PyMPM and B4PyMPM thin films. Thermogravimetric analysis showed 5% weight loss (T_{d5}) of the BPyMPM derivatives occurring at temperatures >465°C, indicating high thermal stability. The melting point (T_m) of B4PyMPM was observed to be ca. 50°C higher than that of B3PyMPM, and ca. 120°C higher than that of B2PyMPM. Such thermal properties (summarized in Table 7.2) may come mainly from the above-mentioned hydrogen bonds in the solid state, because the difference between these BPyMPM derivatives is solely in the position of substituted pyridine rings. Thus, the degree of hydrogen bonding is expected to increase in the order B2PyMPM < B3PyMPM < B4PyMPM. This consideration is consistent with the TLC analysis.

The optical properties of B3PyMPM derivatives discussed above were determined by ultraviolet photoelectron spectroscopy (UPS) and UV–vis absorption spectroscopy. From the UPS data, the I_p's were determined to be 6.62 eV for B2PyMPM, 6.97 eV for B3PyMPM, and 7.30 eV for B4PyMPM. The E_a's were estimated by subtraction of the

Table 7.2 Thermal and Optical Properties of BPyMPM Derivatives

Compound	$I_p/E_g/E_a/E_T$ (eV)	$T_g/T_m/T_{d5}$ (°C)	μ_e (cm²/Vs)
B2PyMPM	6.62/3.55/3.07/2.75	107/257/465	1.6×10^{-6}
B3PyMPM	6.97/3.53/3.44/2.75	n.d./326/484	1.5×10^{-5}
B4PyMPM	7.30/3.59/3.71/2.80	n.d./374/472	1.0×10^{-4}

Note: n.d., not detected.

Chapter 7

optical energy gaps (E_g) from the I_p's, with values of 3.07 eV for B2PyMPM, 3.44 eV for B3PyMPM, and 3.71 eV for B4PyMPM, respectively. Although the structural differences among these compounds are very small, the differences in the I_p and E_a values are quite large. The position of substituted pyridine rings is the key factor determining the optical properties.

Time-of-fight measurements of vacuum-deposited films have also been carried out to determine electron mobilities. At room temperature, the μ_e of B4PyMPM is 10 times higher than that of B3PyMPM and 100 times higher than that of B2PyMPM (Figure 7.29).

To determine the charge-transport parameters for the BPyMPM derivatives, Kido and Sasabe investigated the temperature and field dependencies of μ_e. The energetic disorder (σ) and positional disorder (Σ) were also estimated using Bässler's disorder formalism. According to Monté Carlo simulations, the temperature and field dependencies of μ_e are given by Equation 7.2, where $\Sigma \geq 1.5$:

$$\mu_e = \mu_0 \exp\left[-\left(\frac{2\sigma}{3k_B T}\right)^2\right]\exp\left\{C\left[\left(\frac{\sigma}{k_B T}\right)^2 - \Sigma^2\right]\sqrt{E}\right\} \tag{7.2}$$

If $\Sigma < 1.5$, the temperature and field dependencies of μ_e are given by

$$\mu_e = \mu_0 \exp\left[-\left(\frac{2\sigma}{3k_B T}\right)^2\right]\exp\left\{C\left[\left(\frac{\sigma}{k_B T}\right)^2 - 2.25\right]\sqrt{E}\right\} \tag{7.3}$$

where μ_0 is a hypothetical mobility in the disorder-free system, k_B is Boltzmann's constant, E is the electric field, T is the temperature, and C is an empirical constant.

The values of the electron-transport parameters are listed in Table 7.3. The degree of energetic disorder σ, which describes fluctuations of the hopping site energy, is estimated to decrease in the order B2PyMPM (91 meV) > B3PyMPM (88 meV) > B4PyMPM

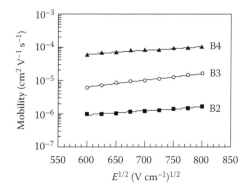

FIGURE 7.29 TOF mobility of BPyMPM derivatives. (From Sasabe, H. et al. 2011. *Adv. Funct. Mater.* 21: 336–342. With permission.)

Table 7.3   Charge Transport Parameters of BPyMPM Derivatives

Compound	μ_0 (cm²/Vs)[a]	σ (meV)[b]	Σ[c]	C (cm V)$^{1/2}$[d]
B2PyMPM	5.1×10^{-5}	91	2.7	4.4×10^{-4}
B3PyMPM	6.9×10^{-5}	88	1.2	4.4×10^{-4}
B4PyMPM	4.5×10^{-4}	76	0.6	3.3×10^{-4}

[a] Hypothetical mobility in the disorder-free system.
[b] Energetic disorder.
[c] Positional disorder.
[d] Empirical constant.

(76 meV). In BPyMPM derivatives, because the number of conformers increases in the order B4PyMPM < B3PyMPM = B2PyMPM, the differences in σ can be attributed mainly to the number of conformers. The value of positional disorder Σ, which describes fluctuations of the intermolecular distance, is calculated as 2.7 for B2PyMPM, and <1.5 for B3PyMPM and B4PyMPM. Thus, because μ_0 for B4PyMPM is relatively large, partially well-ordered molecular orientations should be obtainable in a vacuum-deposited B4PyMPM film.

Winkler and Houk (2007) predicted using computational methods that nitrogen-rich oligoacenes can form self-complementary systems to make highly ordered 2D sheets by means of intermolecular CH–N hydrogen bond interactions. Although CH–N hydrogen bonds are relatively weak, the large number of bonds is expected to facilitate formation of 2D sheets. In addition, Ziener and coworkers determined using scanning tunneling microscopy that an oligopyridine derivative can be assembled into 2D sheets at the liquid/solid interface on highly oriented pyrolytic graphite (HOPG) (Meier et al. 2005). This oligopyridine derivative has a structure very similar to that of BPyMPM derivatives. The authors hypothesize that the results of electron-transport parameters of BPyMPM derivatives indicate the vacuum-deposited B3PyMPM and B4PyMPM films have partially well-ordered molecular orientations as well as dense packing derived from the CH–N hydrogen bond interactions leading to high μ_e (Figure 7.30). To confirm this hypothesis, it is necessary to further investigate molecular aggregation states.

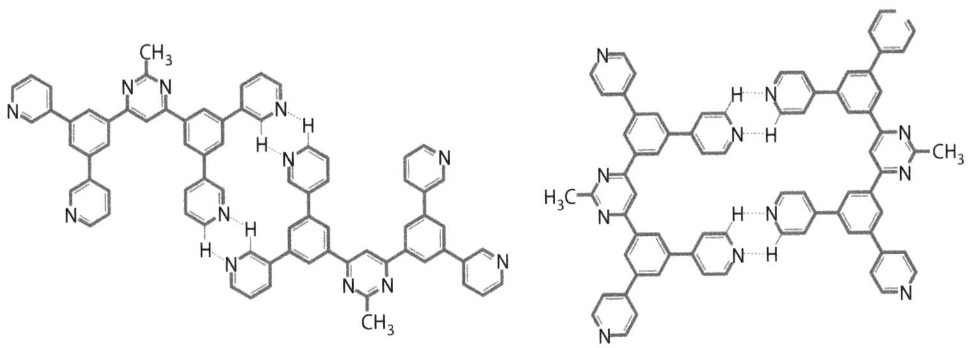

FIGURE 7.30   Examples of proposed network formation of B3PyMPM and B4PyMPM based on weak CH–N hydrogen-bonding interaction.

One example of such measurements is the work by Yokoyama et al., wherein they have reported the molecular orientation of BPyMPM derivatives using variable angle spectroscopic ellipsometry (Yokoyama et al. 2011). They determined that B3PyMPM and B4PyMPM gave highly oriented films that are likely driven by CH–N hydrogen bonding.

7.4　Conclusion and Outlook

This chapter describes the range of materials that have been explored for use as ETMs, along with design concepts and recent advances in multifunctional ETMs for high-performance OLEDs. To realize power efficiencies which would make OLEDs competitive with fluorescent lighting, development of novel materials is required, especially ETMs with high μ_e. Furthermore, the unique problems for the development of an ETM with multiple functionalities, such as high E_T, high μ_e, and high stability, have not been fully resolved, and have impeded the development of stable and efficient blue and white OLEDs. These unique problems also explain the limited number of ETLs reported, compared to other types of OLED materials. A joint computational and experimental study is needed to facilitate such development, and a study that includes contributions from quantum chemistry, material synthesis, and device physics would be ideal (Kim et al. 2011, Marsal et al. 2004, Salman et al. 2011). Organic materials offer flexibility in material design, and an unlimited number of organic materials can be synthesized. Several pyridine-containing ETMs show not only outstanding performance but also a unique self-assembled nature in the solid state which contributes to their performance. Thus, a key conclusion is that it is important to control the 2D and 3D molecular aggregation states for the development of next-generation multifunctional ETMs, for example, by means of weak intra- and/or intermolecular hydrogen bond interactions (Bartels 2010). Researchers in the field of organic electronics must consider supramolecular design principles to create a novel multifunctional ETM.

Acknowledgments

We thank all the researchers who participated in our work discussed in the present chapter and whose names appear in references. We also greatly acknowledge the financial support in part by New Energy and Industrial Technology Development Organization (NEDO) through the "Advanced Organic Device Project," by the Japan Regional Innovation Strategy Program by the Excellence (J-RISE) (creating an international research hub for advanced organic electronics) of the Japan Science and Technology Agency (JST), and by KAKENHI (23750204).

References

Adachi, C., Tokito, S., Tsutsui, T., Saito, S. 1988. Electroluminescence in organic films with three-layer structure. *Jpn. J. Appl. Phys.* 27: L269–L271.
Adachi, C., Tsutsui, T., Saito, T. 1990. Confinement of charge carriers and molecular excitons within 5-nm-thick emitter layer in organic electroluminescent devices with a double heterostructure. *Appl. Phys. Lett.* 57: 531–533.

Agata, Y., Shimizu, H., Kido, J. 2007. Syntheses and properties of novel quarterphenylene-based materials for blue organic light-emitting devices. *Chem. Lett.* 36: 316–317.

Bartels, L. 2010. Tailoring molecular layers at metal surfaces. *Nat. Chem.* 2: 87–95.

Desiraju, G. R., Steiner, T. 1999. *The Weak Hydrogen Bond–In Structural Chemistry and Biology.* IUCr: Oxford University Press, Chichester, UK.

Endo, J., Kido, J., Matsumoto, T. 1998. *Ext. Abst.* (59th Autum Meet.). *Jpn. Soc. Appl. Phys.* 16a-YH-10, p. 1086.

Endo, J., Matsumoto, T., Kido, J. 2002. Organic electroluminescent devices having metal complexes as cathode interface layer. *Jpn. J. Appl. Phys.* 41: L800–L803.

Earmme, T., Ahmed, E., Jenekhe, S. A. 2010. Solution-processed highly efficient blue phosphorescent polymer light-emitting diodes enabled by a new electron transport material. *Adv. Mater.* 22: 4744–4748.

Ganzorig, C., Fujihira, M. 1999. A lithium carboxylate ultrathin film on an aluminum cathode for enhanced electron injection in organic electroluminescent devices. *Jpn. J. Appl. Phys.* 38: L1348–L1350.

Hohenstein, E. G., Sherrill, C. D. 2009. Effects of heteroatoms on aromatic π–π interactions: Benzene–pyridine and pyridine dimer. *J. Phys. Chem. A* 113: 878–886.

Huang, J., Pfeiffer, M., Werner, A., Blochwitz, J., Leo, K., Liu, S. 2002. Low-voltage organic electroluminescent devices using *pin* structures. *Appl. Phys. Lett.* 80: 139–141.

Hughes, L. G., Bryce, M. R. 2005. Electron-transporting materials for organic electroluminescent and electrophosphorescent devices. *J. Mater. Chem.* 15: 94–107.

Hung, S., Tang, C. W., Mason, M. G. 1997. Enhanced electron injection in organic electroluminescence devices using an Al/LiF electrode. *Appl. Phys. Lett.* 70: 152–154.

Ichikawa, M., Kawaguchi, T., Kobayashi, K. et al. 2006. Bipyridyl oxadiazoles as efficient and durable electron-transporting and hole-blocking molecular materials. *J. Mater. Chem.* 16: 221–225.

Ichikawa, M., Hiramatsu, N., Yokoyama, N. et al. 2007. Electron transport with mobility above 10^{-3} cm²/ Vs in amorphous film of co-planar bipyridyl-substituted oxadiazole. *Phys. Status Solidi (RRL)* 1: R37–R39.

Ichikawa, M., Yamamoto, T., Jeon, H.-G. et al. 2012. Benzene substituted with bipyridine and terpyridine as electron-transporting materials for organic light-emitting devices. *J. Mater. Chem.* 22: 6765–6773.

Kathirgamanathan, P., Surendrakumar, S., Antipan-Lara, et al. 2012. Novel lithium Schiff-base cluster complexes as electron injectors: Synthesis, crystal structure, thin film characterisation and their performance in OLEDs. *J. Mater. Chem.* 22: 6104–6116.

Kido, J., Matsumoto, T. 1998. Bright organic electroluminescent devices having a metal-doped electron-injecting layer. *Appl. Phys. Lett.* 73: 2866–2868.

Kim, D., Coropceanu, V., Brédas, J. L. 2011. Design of efficient ambipolar host materials for organic blue electrophosphorescence: Theoretical characterization of hosts based on carbazole derivatives. *J. Am. Chem. Soc.* 133: 17895–17900.

Kim, Y., Lee, J.-G., Kim, S. 1999. Blue light-emitting device based on a unidentate organometallic complex containing lithium as an emission layer. *Adv. Mater.* 11: 1463–1466.

Kinoshita, M., Shirota, Y. 2001. 1,3-Bis[5-(dimesitylboryl)thiophen-2-yl]benzene and 1,3,5-Tris[5-(dimesitylboryl)thiophen-2-yl]benzene as a novel family of electron-transporting hole blockers for organic electroluminescent devices. *Chem. Lett.* 11: 614–615.

Kulkarni, A. P., Tonzola, C. J., Babel, A., Jenekhe, S. A. 2004. Electron transport materials for organic light-emitting diodes. *Chem. Mater.* 16: 4556–4573.

Liang, F., Chen, J., Wang, L., Ma, D., Jing, X., Wang, F. 2003. A hydroxyphenyloxadiazole lithium complex as a highly efficient blue emitter and interface material in organic light-emitting diodes. *J. Mater. Chem.* 13: 2922–2926.

Marsal, P., Avilov, I., da Silva Filho, D. A., Brédas, J. L., Beljonne, D. 2004. Molecular hosts for triplet emission in light emitting diodes: A quantum-chemical study. *Chem. Phys. Lett.* 392: 521–528.

Meier, C., Ziener, U., Landfester, K., Weihrich, P. 2005. Weak hydrogen bonds as a structural motif for two-dimensional assemblies of oligopyridines on highly oriented pyrolytic graphite: An STM investigation. *J. Phys. Chem. B* 109: 21015–21027.

Oyamada, T., Sasabe, H., Adachi, C., Murase, S., Tominaga, T., Maeda, C. 2005. Extremely low-voltage driving of organic light-emitting diodes with a Cs-doped phenyldipyrenylphosphine oxide layer as an electron-injection layer. *Appl. Phys. Lett.* 86: 033503 1–3.

Padmaperuma, A. B., Sapochak, L. S., Burrows, P. E. 2006. New charge transporting host material for short wavelength organic electrophosphorescence: 2,7-Bis(diphenylphosphine oxide)-9,9-dimethylfluorene. *Chem. Mater.* 18: 2389–2396.

Chapter 7

Pfeiffer, M., Forrest, S. R., Leo, K., Thompson, M. E. 2002. Electrophosphorescent p–i–n organic light-emitting devices for very-high-efficiency flat-panel displays. *Adv. Mater.* 14: 1633–1636.

Pu, Y.-J., Miyamoto, M., Nakayama, K., Oyama, T., Yokoyama, M., Kido, J. 2009. Lithium phenolate complexes for an electron injection layer in organic light-emitting diodes. *Org. Electr.* 10: 228–232.

Ren, X., Li, J., Holmes, R. J., Djurovich, P. I., Forrest, S. R., Thompson, M. E. 2004. Ultrahigh energy gap hosts in deep blue organic electrophosphorescent devices. *Chem. Mater.* 16: 4743–4747.

Salman, S., Kim, D., Coropceanu, V., Brédas, J. L. 2011. Theoretical investigation of triscarbazole derivatives as host materials for blue electrophosphorescence: Effects of topology. *Chem. Mater.* 23:5223–5230.

Sasabe, H., Chiba, T., Su, S.-J., Pu, Y.-J., Nakayama, K., Kido, J. 2008a. 2-Phenylpyrimidine skeleton-based electron-transport materials for extremely efficient green organic light-emitting devices. *Chem. Commun.* 44: 5821–5823.

Sasabe, H., Gonmori, E., Chiba, T. et al. 2008b. Wide-energy-gap electron-transporting materials containing 3, 5-dipyridylphenyl moieties for an ultra high efficiency blue organic light-emitting device. *Chem. Mater.* 20: 5951–5953.

Sasabe, H., Tanaka, D., Yokoyama, D. et al. 2011. Influence of substituted pyridine rings on physical properties and electron mobilities of 2-methylpyrimidine skeleton-based electron transporters. *Adv. Funct. Mater.* 21: 336–342.

Sasabe, H., Kido, J. 2011. Multifunctional materials in high-performance OLEDs: Challenges for solid-state lighting. *Chem. Mater.* 23: 621–630.

Schmitz, C., Schmidt, H.-W., Thelakkat, M. 2000. Lithium–quinolate complexes as emitter and interface materials in organic light-emitting diodes. *Chem. Mater.* 12: 3012–3019.

Shi, J., Tang, C. W., Chen, C. H. 1997. U.S. Patent. 5646948.

Shirota, Y. 2005. Photo- and electroactive amorphous molecular materials—Molecular design, syntheses, reactions, properties, and applications. *J. Mater. Chem.* 15: 75–93.

Su, S.-J., Chiba, T., Takeda, T., Kido, J. 2008a. Pyridine-containing triphenylbenzene derivatives with high electron mobility for highly efficient phosphorescent OLEDs. *Adv. Mater.* 20: 2125–2130.

Su, S.-J., Gonmori, E., Sasabe, H., Kido, J. 2008b. Highly efficient organic blue-and white-light-emitting devices having a carrier- and exciton-confining structure for reduced efficiency roll-off. *Adv. Mater.* 20: 4189.

Su, S.-J., Takahashi, Y., Chiba, T., Takeda, T., Kido, J. 2009. Structure–property relationship of pyridine-containing triphenyl benzene electron-transport materials for highly efficient blue phosphorescent OLEDs. *Adv. Funct. Mater.* 19: 1260–1267.

Sun, C., Hudson, Z. M., Helander, M. G., Lu, Z.-H., Wang, S. 2011. A polyboryl-functionalized triazine as an electron transport material for OLEDs. *Organometallics* 30: 5552–5555.

Tamao, K., Uchida, M., Izumizawa, T., Furukawa, K., Yamaguchi, S. 1996. Silole derivatives as efficient electron transporting materials. *J. Am. Chem. Soc.* 118: 11974–11975.

Tanaka, D., Takeda, T., Chiba, T., Watanabe, S., Kido, J. 2007a. Novel electron-transport material containing boron atom with a high triplet excited energy level. *Chem. Lett.* 36: 262–263.

Tanaka, D., Sasabe, H., Li, Y.-J., Su, S.-J., Takeda, T., Kido, J. 2007b. Ultra high efficiency green organic light-emitting devices. *Jpn. J. Appl. Phys.* 46: L10–L12.

Tanaka, D., Agata, Y., Takeda, T., Watanabe, S., Kido, J. 2007c. High luminous efficiency blue organic light-emitting devices using high triplet excited energy materials. *Jpn. J. Appl. Phys.* 46: L117–L119.

Tang, C. W., VanSlyke, S. A. 1987. Organic electroluminescent diodes. *Appl. Phys. Lett.* 51: 913–915.

Tsuji, H., Sato, K., Sato, Y., Nakamura, E. 2009. Benzo[*b*]phosphole sulfides. Highly electron-transporting and thermally stable molecular materials for organic semiconductor devices. *J. Mater. Chem.* 19: 3364–3366.

Tsutsui, T. 1997. Progress in electroluminescent devices using molecular thin films. *MRS Bull.* 22:39–45.

Uchida, M., Izumizawa, T., Nakano, T., Yamaguchi, S., Tamao, K., Furukawa, K. 2001. Structural optimization of 2, 5-diarylsiloles as excellent electron-transporting materials for organic electroluminescent devices. *Chem. Mater.* 13: 2680–2683.

Van Dijken, A., Brunner, K., Börner, H., Langeveld, B. M. W. 2008. High-efficiency phosphorescent polymer LEDs. In: H. Yersin (Ed.), *Highly Efficient OLEDs with Phosphorescent Materials*. Wiley-VCH, p. 321.

Wakimoto, T., Fukuda, Y., Nagayama, K., Yokoi, A., Nakada, H., Tsuchida, M. 1997. Organic EL cells using alkaline metal compounds as electron injection materials. *IEEE Trans. Electron Device* 44:1245–1248.

Winkler, M., Houk, K. N. 2007. Nitrogen-rich oligoacenes: Candidates for n-channel organic semiconductors. *J. Am. Chem. Soc.* 129: 1805–1815.

Xiao, L., Chen, Z., Qu, B., Luo, J., Kong, S., Gong, Q., Kido, J. 2011. Recent progresses on materials for elec-
    trophosphorescent organic light-emitting devices. *Adv. Mater.* 23: 926–952.
Xie, K., Qiao, J., Duan, L., Li, Y., Zhang, D., Dong, G., Wang, L., Qiu, Y. 2008. Organic cesium salt as an
    efficient electron injection material for organic light-emitting diodes. *Appl. Phys. Lett.* 93: 183302 1–3.
Yokoyama, D., Sakaguchi, A., Suzuki, M., Adachi, C. 2009. Enhancement of electron transport by horizontal
    molecular orientation of oxadiazole planar molecules in organic amorphous films. *Appl. Phys. Lett.* 95:
    243303 1–3.
Yokoyama, D., Sasabe, H., Furukawa, H., Adachi, C., Kido, J. 2011. Molecular stacking induced by intermo-
    lecular C–H⋯N hydrogen bonds leading to high carrier mobility in vacuum-deposited organic films.
    Adv. Funct. Mater. 21: 1375–1382.
Yokoyama, D. 2011. Molecular orientation in small-molecule organic light-emitting diodes. *J. Mater. Chem.*
    21: 19187–19202.

Chapter 7

8. Hole Transport Materials

Evgueni Polikarpov

8.1   Introduction

The success of the first organic light-emitting diodes (OLEDs) in overcoming the low efficiency observed in earlier attempts at harnessing organic electroluminescence came from switching to thin film heterostructures. The key advantage of a heterostructure in this application is in confining charges to facilitate efficient recombination and emissive exciton formation. Originating from the respective electrodes, the charges of opposite signs travel through the dedicated single carrier charge transporting layers to meet and recombine in the emission zone located in the middle of the device stack. In order for the charges to recombine and for the entire device stack to be reasonably conductive, the materials comprising the heterostructure must satisfy two criteria: high charge mobility of the bulk material and frontier orbital alignment at the heterostructure interface (frontier orbitals, also called boundary orbitals, are highest occupied molecular orbital [HOMO] and lowest unoccupied molecular orbital [LUMO] in a molecular material). In other words, a charge transport material must be able to accept charge from the injecting electrode and inject a carrier of one sign into the emissive zone. Furthermore, a charge transporting material should be able to block the carriers of the opposite sign from passing through the organics and recombining at the electrodes. In the case of OLED materials assigned specifically to transport holes, the need to block charges translates into requirements for the frontier orbitals: a shallow LUMO level to prevent electrons entering the hole transport layer (HTL) from the emissive layer, and a HOMO level matching that of the emissive layer material and the work function of the anode. Typically, hole-injecting electrodes such as indium tin oxide (ITO), Ga-doped ZnO, and others discussed in Chapter 4 of this book have work functions between 4 and 5 eV. The compounds that have HOMO levels matching these values are air stable weak electron donors. A typical example of such compounds would be an aromatic heterocycle containing nitrogen heteroatoms, such as triarylamine or carbazole derivatives. Indeed, the first heterostructure OLED employed 1,1′-bis[(di-4-tolylamino)phenyl]cyclohexane (TAPC), in which the hole transport was provided by two tolylamine functionalities. TAPC has fairly high hole mobility for an organic compound, 2.9×10^{-4} cm^2/Vs (Muellen and Scherf 2006; Aonuma et al. 2007). In addition to sufficient charge mobility, TAPC has high enough triplet exciton energy (2.87 eV, [Goushi et al. 2004]) to provide exciton confinement in the emissive layer even for phosphorescent blue-emitting devices. Finally, since the lowest singlet excited state energy of an organic molecule is higher than the lowest triplet. TAPC possesses a shallow LUMO as well, making it a good electron blocker for most device structures. All these desirable characteristics of TAPC mean that, in fact, this first hole transport material (HTM) ever used in a modern OLED has been the material to be included in some of the most efficient blue devices (Chopra et al. 2010) to date. However, as an OLED component, TAPC has one crucial issue that keeps the search for new HTLs an ongoing problem: its stability in an operating device is insufficient due to chemical degradation via bond breaking at the saturated carbon site (Kondakov 2008). Stability continues to be an issue for many materials; chemical degradation and crystallization are both known to lead to loss of performance. Some of the materials reviewed further in this chapter achieved outstanding charge transport characteristics, others—excellent exciton and charge confining properties, others still showed very long operational life times. But a material that combines all these desired properties has not yet been found at the time of writing. First, this chapter will focus on methods to characterize HTL materials, followed by a discussion of the technologically relevant issue of achieving high charge conductivity in thick layers of organic films. Finally, we will conclude with a review of the current state of material development and remaining challenges in HTMs to enable wider OLED commercialization.

8.2 Characterization of Hole Transport Properties

8.2.1 Frontier Orbital Energies: Direct and Indirect Measurements

Since an OLED consists of several layers of various stacked organic films, offsets in the energies between layers act as potential energy barriers to the flow of charge. The HOMO and LUMO energies are used to describe isolated molecules. Energy levels in solid-state organic films, bonded by weak intermolecular interactions, can be derived from these orbital energies. For an HTL material, E_{HOMO} determines the charge injection barrier from the transparent conducting oxide (TCO) into the HTL and from the HTL material into the emissive layer, whereas sufficiently high E_{LUMO} provides an HTL material with electron-blocking ability needed for the charge confinement in the emissive zone of the device.

The conventional methods to ascertain frontier orbital energies are cyclic voltammetry (CV) (D'Andrade et al. 2005) and photoelectron spectroscopy (ultraviolet photoelectron [UPS] for E_{HOMO} and inverse photoelectron spectroscopy [IPES] for E_{LUMO}). UPS is the direct method to obtain the HOMO energy values (E_{HOMO}) (Rajagopal et al.

1998). The UPS experiments determine the ionization energy (E_i) of a molecule on the surface of a thin film, where $E_i = -E_{HOMO}$ (Tsiper et al. 2002). IPES is a direct probe for the energy of the LUMO (E_{LUMO}) (Cahen and Kahn 2003; Seki and Kanai 2006). Solution-based CV experiments determine the relative molecular oxidation and reduction potentials (V_{CV}), which are indirectly related to the E_{HOMO} and E_{LUMO}, respectively. Whereas the UPS and IPES are direct methods of probing the frontier orbital energies, their large error bars, complicated data interpretation and expensive hardware favor the use of solution electrochemistry for estimation of E_{HOMO} and E_{LUMO}. The correlation between solution electrochemistry results and the direct spectroscopic methods of the orbital energy measurements are described in Djurovich et al. (2009) and D'Andrade et al. (2005) for LUMO and HOMO, respectively. In D'Andrade et al. (2005), the E_{HOMO}'s of a number of organic semiconductors commonly employed in thin film electronic devices were determined by UPS, and compared to the relative oxidation potential V_{OX} measured using CV, leading to the relationship

$$E_{HOMO} = -(1.4 \pm 0.1) \times (qV_{OX}) - (4.6 \pm 0.08) \, eV \qquad (8.1)$$

Equation 8.1 shows that there is a linear relationship between the CV and UPS values. The difference between the measured UPS and CV E_{HOMO}'s are due to solvation and image charge effects present under the CV experimental conditions.

The energy of the LUMO level can be calculated from the measured solution reduction potential V_{RED} using the following relationship derived in Djurovich et al. (2009), provided here as

$$E_{LUMO} = (1.19 \pm 0.08) \times (qV_{RED}) - (4.78 \pm 0.17) \, eV \qquad (8.2)$$

It is not uncommon to see the E_{LUMO} derived from the combination of a known E_{HOMO} value (measured by either UPS or CV) and the optical gap obtained from absorption spectra. It should be pointed out however that this type of estimation does not take into account the exciton binding energy that can be quite significant for many organic materials. The correlation between the IPES LUMO and the values obtained from the optical gap (equal to |Eopt + HOMO|) is poor, according to Djurovich et al. (2009), as a result of ambiguities in assigning the energy of the HOMO–LUMO gap from optical data.

8.2.2   Bulk Hole Transport through the Film

Besides the frontier orbital energies of the materials, the key parameter is the charge mobility of materials the device is composed of. Measurement of the charge mobility of OLED materials has its specific challenges. The large mobility in inorganic semiconductors is determined by the Hall Effect and conductivity measurements (Hamaguchi 2001). The drift mobility is then related to the Hall mobility by a scattering factor that depends upon the scattering mechanisms and distribution function of carriers. This technique is not suitable for relatively low-mobility organic materials. Therefore, other approaches have been developed. The charge carrier mobility in organic films can be determined by several methods. Among them, time of flight (TOF) (Aonuma et al. 2007), analysis of space-charge-limited current (SCLC) (An et al. 2005; Diouf et al. 2011), and field-effect mobility derived from the characteristics of organic field-effect transistors (OFETs) (Swensen et al. 2012) are the most used. We will briefly review their principles of application. Excellent reviews of these techniques as applied to organic electronic materials can be found in Shirota and Kageyama (2007) and Tiwari and Greenham (2009).

8.2.3 TOF Mobility Measurements

In the TOF method, a sample of an organic material is sandwiched between two electrodes, one of which is transparent. The charge carriers that are generated near one of the electrodes by pulsed laser radiation drift across the sample to the other electrode under an applied electric field. The key observable in this method is the carrier transit time, τ, which is determined by the velocity of charge carriers v and the sample thickness v according to

$$\tau = d/v \tag{8.3}$$

Given that $v = \mu F$ and $F = V/d$, where F is the strength of the electric field and V is the potential, the carrier drift mobility μ can be determined from the TOF measurements via

$$\mu = d^2/V\tau \tag{8.4}$$

The samples for this method can be prepared using vacuum evaporation, solution processing, or by any other methods that are used in OLED thin film deposition. The thickness of samples is usually in the range from 5 through 20 μm. This is significantly greater than a typical thickness of even the thickest layers in the OLED stack. Therefore, since charge mobility is field dependent, the values extracted from the TOF experiments need to be used with caution in application to OLEDs. The TOF method typically provides the upper boundary for a mobility value for a given material. It is free from complications arising from the interface and contact resistance that other methods such as field-effect transistor measurements or single-carrier devices are known for. Therefore, the TOF technique is arguably the most OLED-relevant mobility measurement, providing useful comparisons of the charge mobilities between different materials. As an example, a detailed study of hole transporting arylamine-based materials reports TOF mobilities for a series of oligomer compounds and compared to common OLED HTL materials α-NPD and TAPC (Aonuma et al. 2007), the structures of which are shown in Figure 8.1.

This study helps explain the implications of using thicker films of HTL materials in OLEDs, and the findings will be discussed further in this chapter.

8.2.4 Space–Charge–Limited Current Analysis

Charge carrier mobility can be extracted from the J–V characteristics measured in the dark. At low voltages, the J–V characteristics typically show ohmic behavior. At high

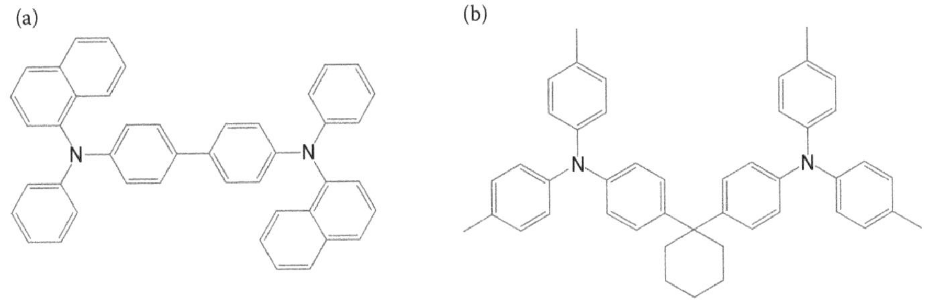

(a)   (b)

FIGURE 8.1   Chemical structures of (a) α-NPD and (b) TAPC.

applied voltages, the current becomes space-charge-limited. When the contact between the electrode and the organic layer is ohmic and the current is transport limited instead of injection limited, the SCLC J is given by Mott–Gurney equation (Mott and Grunew 1940)

$$J = \frac{9}{8}\epsilon\mu\frac{V^2}{d^3}\theta = \frac{9}{8}\epsilon\mu\frac{1}{d}F^2\theta \qquad (8.5)$$

where ϵ and d are the permittivity and thickness of the sample, and θ is a factor that considers the presence of charge carrier traps, that is, the ratio of the number of free carriers to the total number of carriers. When the current flow is in agreement with SCLC, J should be proportional to the square of the electric field (F^2), which is dependent upon the sample thickness. When θ is equal to 1, the current becomes trap-free SCLC. The charge carrier mobility can be evaluated from Equation 8.5 based on the assumption that the contact between the electrode and the organic layer is ohmic without any energy barrier for charge injection. Equation 8.5 is valid if the mobility is independent of the electric field. Since the charge carrier mobility of organic amorphous solids is usually electric field dependent, Equation 8.6 is modified to include the field dependence multiplier (Murgatroyd 1970)

$$J = \frac{9}{8}\epsilon\mu_0\exp(\beta F^{\frac{1}{2}})\frac{1}{d}F^2\theta \qquad (8.6)$$

where μ_0 is the mobility when $F = 0$. If the mobility is independent of the electric field, $\beta = 0$.

The SCLC method enables mobility measurements in a complete OLED device as opposed to a single (relatively thick) film of a given material. The SCLC data presented in Diouf et al. (2011) suggest that the hole mobility of an HTL can be influenced by the device architecture and electronic environment of the HTL. Thus, an unusually high hole mobility of 4.7×10^{-1} cm²/Vs is reported for the device containing α-NPD and $N^4,N^{4'}$-Bis[4-[bis(3-methylphenyl)amino]phenyl]-$N^4,N^{4'}$-diphenyl-[1,1′-biphenyl]-4,4′-diamine (DNTPD) hole transport layers in combinations with a deep-LUMO electron transport layer (ETL) HAT-CN. The structures of DNTPD and 1,4,5,8,9,11-hexaazatri-phenylene-hexanitrile (HAT-CN) are shown in Figure 8.2.

The high hole mobility is attributed to the extremely small gap between the high HOMO of the HTL and the deep LUMO of the ETL that leads to enhanced hole

Chapter 8

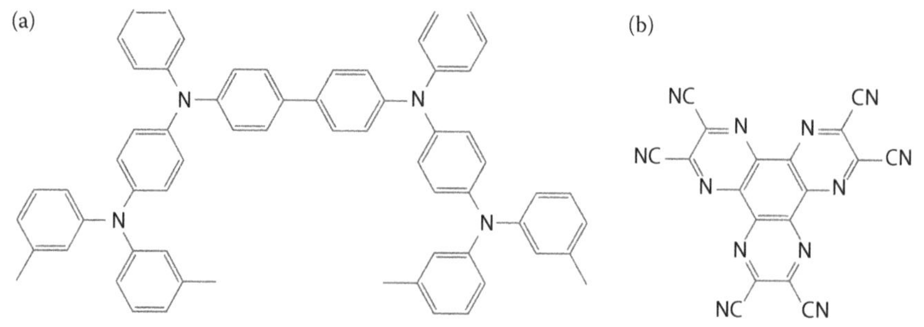

(a)   (b)

FIGURE 8.2 Chemical structures of the hole transporting material (a) DNTPD and electron-transporting material (b) HAT-CN.

conduction through charge recombination at the interface. At the same time, the hole mobility in neat α-NPD films measured by SCLC is much higher—measured as 1.6×10^{-5} and 7.6×10^{-7} for 50 and 1000 nm films, respectively (Chu and Song 2007). In this work, the SCLC and TOF mobilites for α-NPD are in excellent agreement for the thicker films, where the bulk properties should dominate the properties. The decrease in mobility in the thinner films is attributed to the interfacial trap states.

8.2.5   Organic Field Effect Transistor Hole Mobility

OFETs use organic materials as the channel. The gate dielectric materials in OFETs can be either inorganic such as SiO_2, or organic materials such as polymethylmethacrylate (PMMA). When negative voltage is applied to the gate electrode, holes in the organic semiconductor layer accumulate at the interface with the gate dielectric, and holes are transported from the source to the drain electrode. This type of device is called a *p*-channel device. Likewise, the application of positive voltage to the gate electrode causes electron transport in the case of *n*-channel devices. The current flow between source and drain (I_{SD}) can be modulated by both the gate voltage (V_G) and the source/drain voltage (V_{SD}). I_{SD} under a given V_G increases almost linearly with the increasing V_{SD} and gradually becomes saturated as shown in Figure 8.3.

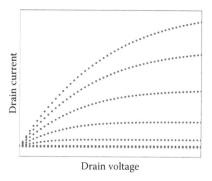

FIGURE 8.3   A typical I_{SD}–V_{SD} curve for an OFET.

The current (I_{SD}) is given by

$$I_{SD} = \frac{C_i W \mu_{FE}}{L}\left[\left(V_G - V_T\right)V_{SD} - \frac{V_{SD}^2}{2}\right] \tag{8.7}$$

where μ_{FE} is the field-effect mobility of the charge carrier, L is the channel length, W is the channel width, C_i is the capacitance per unit area of the gate dielectric, and V_T is the threshold voltage. The I_{SD} versus V_D plots are characterized by linear and saturation regimes. The field-effect mobility (μ_{FET}) can be determined from the slope of the linear plots of $(I_{SD,sat})^{1/2}$ versus V_G, according to the following equation:

$$I_{SD,sat} = \frac{C_i W \mu_{FET}}{2L}\left(V_G - V_T\right)^2 \tag{8.8}$$

Unlike the TOF method, the use of transistors to obtain the mobility values is complicated by the existence of contributions from contact resistance at various interfaces of the device. The contributions from the bulk and interface charge transport can be difficult to decouple. The correlation between the mobility data obtained through OFET and TOF methods is discussed in detail elsewhere (Weis et al. 2009). Relating the mobility values obtained from these methods to the materials in the OLED stack is further complicated by the fact that mobility can be dependent on the thickness of the conducting layer (Borsenberger et al. 1992; Okumoto et al. 2000; Poplavskyy et al. 2003). Comparative studies of charge transport using different techniques have been carried out on, for example, 4,4′,4″-tris(3-methyphenylphenylamino)-triphenylamine (*m*-MTDATA) (Giebeler et al. 1998; Shirota et al. 1998), 2,2′,7,7′-tetrakis(N,N-di-p-methoxyphenylamine)9,9′-spirobifluorene (OMeTAD) (Kimura et al. 2000), α-NPD (Kimura et al. 2000; Shirota et al. 2005; Saragi et al. 2006), and CuPc (Kitamura et al. 2004). The hole mobilities of *m*-MTDATA and α-NPD determined by the dark-injection space-charge-limited current (DI-SCLC) method have been shown to be in agreement with those determined by the TOF method (Tse et al. 2006). Room-temperature mobilities of spiro-OMeTAD measured by three independent methods—TOF, DI-SCLC, and steady-state trap-free space charge limited current methods—have been shown to agree over a range of sample thicknesses from 4 through 135 nm (Kimura et al. 2000). The hole mobility of CuPc measured by the TOF and field-effect transistor (FET) methods produces comparable results: FET—$(0.94-1.3) \times 10^{-3}$ cm²/Vs and TOF—$(1.5-2.0) \times 10^{-3}$ cm²/Vs (Kitamura et al. 2004). Likewise, mobility data for *m*-MTDATA determined by the TOF and FET methods were almost the same (Shirota et al. 2005). On the other hand, FET values for the mobility of N,N′-bis(3-methylphenyl)-N,N′-diphenylbenzidine (TPD) and α-NPD were approximately 2 orders of magnitude smaller than those determined by the TOF method (Kimura et al. 2000; Shirota et al. 2005; Saragi et al. 2006).

In addition to the TOF, FET and SCLC methods reviewed in the previous section, charge mobility in organic solids has been studied using delayed exciton generation (Reineke et al. 2008) (this method enables mobility measurements for mixed films), pulse-radiolysis time-resolved microwave conductivity (Hoofman et al. 1998; Prins et al. 2005), photo-induced transient Stark spectroscopy (PTSS) (Cabanillas-Gonzalez

Chapter 8

et al. 2006), and charge extraction by linearly increasing voltage (CELIV) (Pivrikas et al. 2005), among other techniques. In all cases, the results of mobility measurements may be strongly dependent on the sample geometry, especially for highly crystalline materials. For example, in TOF and SCLC, the sample is sandwiched between two electrodes and the conduction of the charges is perpendicular to the substrate plane, whereas in an FET the charge mobility is characterized within the plane of the substrate. These considerations are mitigated in the measurement of amorphous films, the preferred situation for OLED materials.

8.3   Hole Transport in Thick HTLs

Thicker organic layers require a higher voltage during operation due to their higher resistivity and take longer to deposit (increasing manufacturing time). However, thicker organic layers, especially on the hole side of a heterojunction device, can smooth out the surface of the underlying electrode and cover asperities, thereby reducing the number of short circuit paths and improving the overall device yield (Parker 1994; Aonuma et al. 2007). One approach to reducing operating voltage is conductivity doping, either bulk or interface (discussed in detail in Chapter 9). Conductivity doping may be effective when used properly, although this approach may also have side effects that include luminescence quenching and absorption in the visible region by the conductivity dopants, as well as potentially negative impact on operational stability (D'Andrade et al. 2003). Another approach to reducing operating voltage is through careful materials selection to minimize the charge injection barrier. According to Swensen et al. (2013), minimal current losses in a thick HTL can be achieved when the HTL material has high enough hole mobility (such as α-NPD) in combination with a low charge injection barrier from the TCO into the HTL. In this case, the voltage increase is small with increasing HTL thickness. The presence of a significant charge injection barrier leads to a concomitant voltage increase with HTL thickness even when the HTL mobility is high.

An interesting correlation between the drive voltage in devices with a thick HTL and materials properties is described in Aonuma et al. (2007). In this work, the authors studied a series of triphenylamine-based HTL materials where the members in the series differ from each other by degree of oligomerization. The structure of the monomers and dimers are shown in Figure 8.4, while the trimers and tetramers are shown in Figure 8.5. It was observed that the glass transition temperatures (T_g) and thermal stability increased with increased oligomerization in the series. This work showed a good correlation between T_g and the amount of increase in the device drive voltage when the HTL thickness was increased. The use of materials with higher T_g consistently resulted in weaker dependence of the device drive voltage on the HTL thickness.

These results suggested that the high T_g resulted not only from high molecular weight, but also from increasing molecular rigidity that improved intermolecular interactions to reduce the drive voltage increase as the HTL thickness increased.

8.4   HTL Materials in OLEDs

The first heterostructure OLED demonstrated by Tang and VanSlyke used a layer of TAPC as an HTL (see Figure 8.4, top right for the chemical structure). Since then,

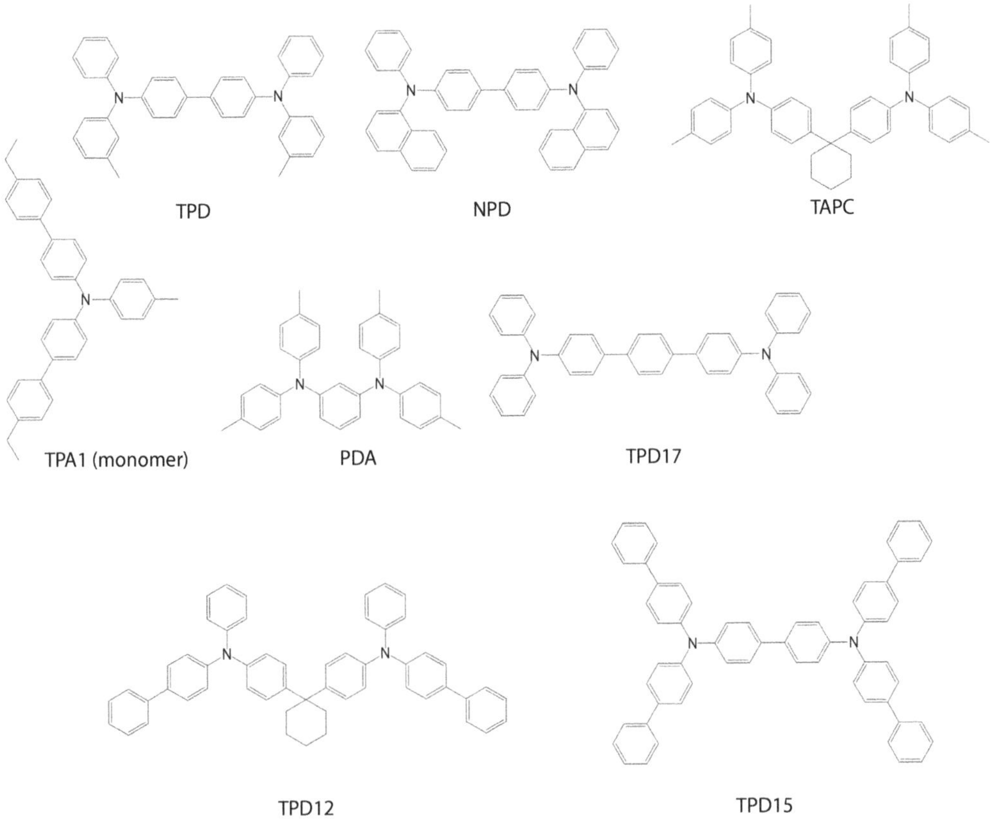

FIGURE 8.4 Monomeric and dimeric HTL materials based on the TPA unit.

aromatic amine-based HTL materials have been an almost indispensable building block for hole transporting materials. The reason for this is the π-conjugation through the lone-pair electrons on the nitrogen in the HOMO level leads to low ionization potentials and a small energy barrier for hole creation. This is coupled with a typically low reorganization energy for the electron-transfer reactions between adjacent molecules, further reducing the energy barrier for charge hopping.

In small-molecule triphenylamine-based materials, low thermal stability and propensity to crystallization can be a problem. Due to its relatively high mobility and chemical stability, TPD (see Figure 8.4, top left) used to be one of the most common choices for HTL materials for OLEDs. However, the T_g of TPD is only 65°C, leading to issues with crystallization and loss of performance in OLEDs. To address this issue, Koene et al. (1998) hypothesized that highly asymmetric triarylamine-based materials would exhibit increased T_g and reduced propensity for crystallization. To this end, they synthesized a library of materials with the structure shown in Figure 8.6, where $n = 1$ or 2 and Ar_i are all different. The T_g values of these materials ranged up to 155°C, depending on the degree of asymmetry. Some of these compounds did not crystallize, even while going through their glass transition. Other examples of demonstrating the ability to design HTL materials from arylamine fragments to achieve desired thermal and transport properties can be found in Figure 8.4 and the accompanying discussion.

Chapter 8

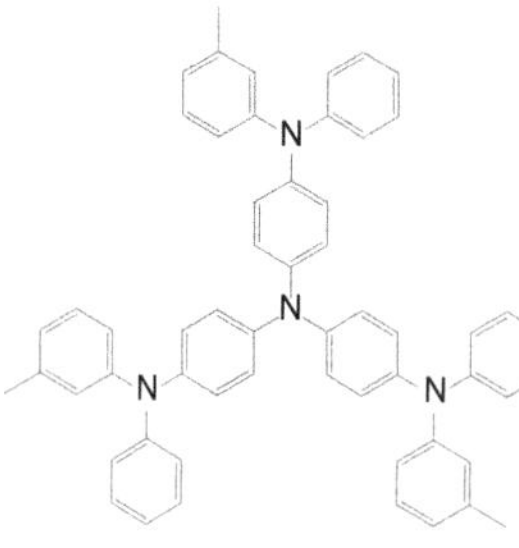

m-MTDATA (trimer)

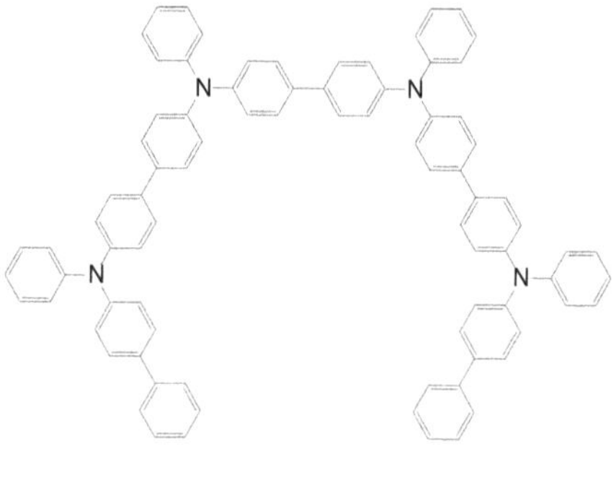

TPT9

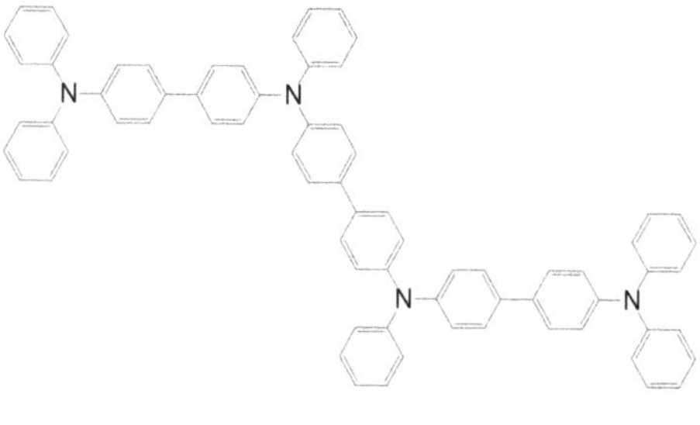

TPT1

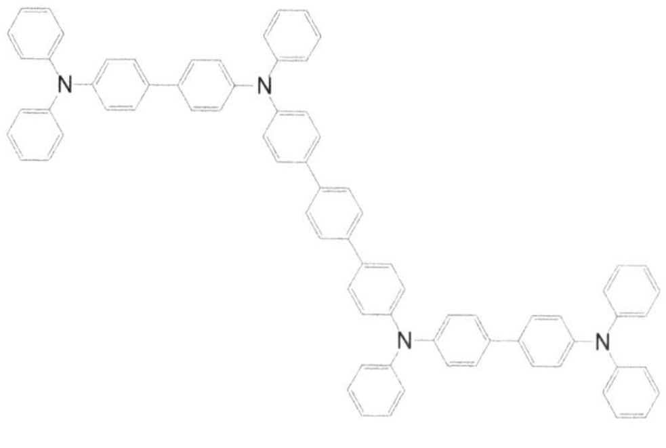

TPT2

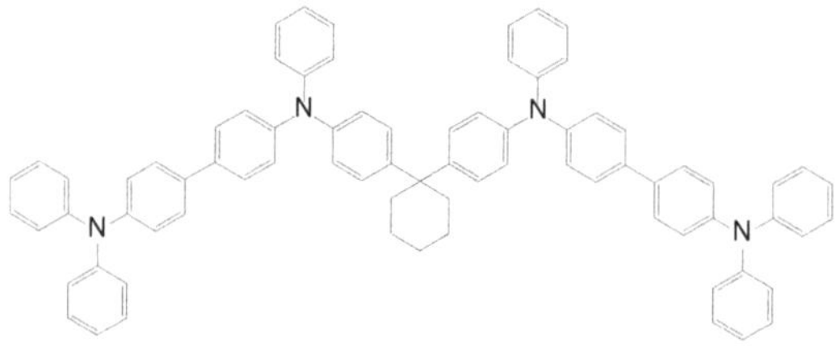

TPT7

FIGURE 8.5 The trimeric (*m*-MTDATA) and tetrameric HTL materials based on the TPA unit.

The carbazole moiety can perform in an analogous fashion to the triarylamine moiety in HTL materials. Substitution of a carbazole for triarylamine generally leads to an increase in T_g and triplet exciton energy E_T (which can improve exciton confinement and thus efficiency, particularly for higher energy excitons). This substitution is illustrated in Figure 8.7, below, where two so-called "starburst" HTL materials, TPA-based *m*-MTDATA and its carbazole-based analog 4,4′,4″-tris(carbazol-9-yl)triphenylamine (TCTA).

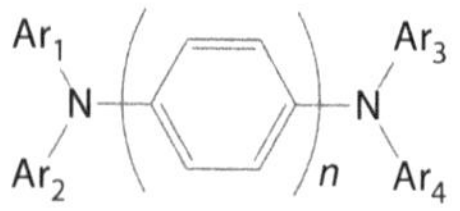

FIGURE 8.6 Triarylamine framework used to generate highly asymmetric HTL candidates.

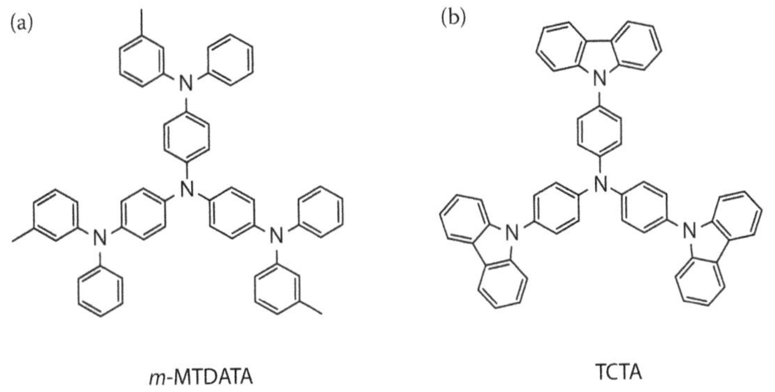

FIGURE 8.7 (a) *m*-MTDATA and (b) TCTA.

Compared to the canonical HTL material α-NPD, which has a glass transition temperature of 95°C, the carbazole-based TCTA exhibits a higher T_g of 151°C likely due to the higher rigidity, whereas the triarylamine-based *m*-MTDATA is only 75°C (Kuwabara et al. 1994). Both of these starburst HTMs show lower hole mobilities than that of α-NPD. As measured by the TOF method, the hole mobilities are 3×10^{-5} cm²/Vs for *m*-MTDATA (Aonuma et al. 2007), 3×10^{-4} cm²/Vs for TCTA (Kang et al. 2007), and 2.2×10^{-3} cm²/Vs for α-NPD (Aonuma et al. 2007). Finally, the E_T of TCTA is 2.8 eV (Reineke et al. 2007) versus 2.3 eV for *m*-MTDATA (Adachi et al. 2001).

Starburst structures and conjugated π-systems capped with aromatic amines (Noda et al. 1999; Shirota et al. 2000) represent widely used structural motifs. Other promising classes of HTL materials in which hole transport is provided by aromatic amine units include spiro-compounds (Salbeck et al. 1997; Spreitzer et al. 1999), fluorine-containing aromatic amines (Okumoto and Shirota 2001), indolocarbazoles (Hu et al. 1999), and phenylazomethines and their metal complexes (Yamamoto et al. 2002; Cho et al. 2005). A variety of modifications to these core structures have been implemented to fine-tune the specific characteristics of these compounds, such as frontier orbital energy levels, sublimation temperature, triplet exciton energy, and hole mobility. To date, though, no molecule meets all performance requirements.

In addition to the aromatic amine and carbazole functionalities, other functional groups and heteroatoms have been used. These include the silane moiety in DTASi (bis[4-(p,p′-ditolylamino)phenyl]diphenylsilane), for example (Ide et al. 2006; Tanaka et al. 2007), shown in Figure 8.8.

Here the silane group serves the purpose of breaking conjugation within the structure to increase E_T, rather than providing any benefits to hole transport *per se*. DTASi was designed to be such high E_T material and therefore contains a saturated silane fragment that breaks conjugation, which improves suitability for use in blue OLEDs. The triarylamines still provide the hole transport functionality in DTASi.

An example of an HTL material that does not rely on aromatic amine functionality for charge transport is a triphenylmethane derivative abbreviated MPMP, bis (4-N, N-diethylamino-2-methylphenyl)-4-methylphenylmethane (Borsenberger 1992), shown in Figure 8.9.

Chapter 8

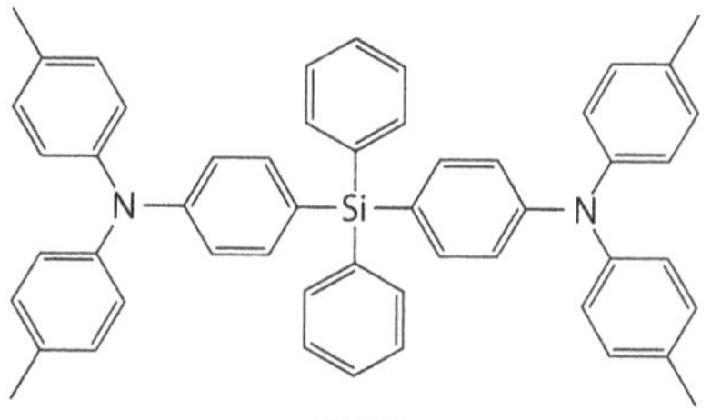

FIGURE 8.8   DTASi.

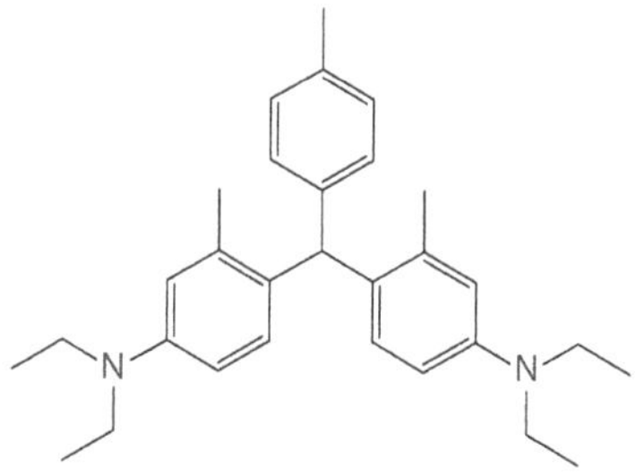

FIGURE 8.9   MPMP.

Borsenberger et al. fabricated blue–green OLEDs using MPMP as the HTL and Ir(ftmf-ppy)₃ as the emitter. These OLEDs demonstrated low operating voltages and high emission efficiencies, due in part to the comparatively high hole mobility of MPMP, within the range 10^{-4}–10^{-3} cm²/Vs (Borsenberger et al. 1996) and high triplet energy, >3 eV (Wang 2004). No information was provided regarding the stability of MPMP.

Other classes of small molecular compounds based on heteroatoms have yet to become widely used in either research or manufacturing. A discussion of the incorporation of hole transport functionality into other OLED materials such as ambipolar hosts is discussed in other chapters of this monograph.

8.5   Conclusions

The properties of some representative HTMs are given in Table 8.1. The requirements imposed on HTL materials are chemical stability under device operation, resistance to phase change (crystallization), high charge mobility and carrier concentration to enable sufficient conductivity through the film, and, finally, proper orbital energy level alignment to ensure hole injection from an anode and into the EML, and electron blocking. This also includes appropriate triplet exciton energy level positioning to confine emissive excitons within the EML and avoid their quenching at the HTL/EML interface. Various methods to characterize these properties have been discussed in this chapter. The ultimate test of success in the design of an HTL material is performance in an actual device. The development of an HTL material that would simultaneously satisfy all the criteria stated above still continues to be the grand challenge. To date, no single HTL material meets these requirements. A common and well-studied material α-NPD

Table 8.1 Properties of Representative HTL Materials

	Hole Mobility (cm²/Vs⁻¹)	E_T (eV)	T_g (°C)	HOMO (eV)	LUMO (eV)	References
TAPC	2.9×10^{-4}	2.9	79	5.8	2.4	a
α-NPD	2.2×10^{-3}	2.3	96	5.5	2.5	b
TCTA	2.0×10^{-5}	2.7	151	5.2	2.7	c
TPD	1.5×10^{-3}	2.3	65	5.4	2.2	d
CuPc	$(1.5–2.0) \times 10^{-3}$	1.6	>200	5.2	3.5	e
m-MTDATA	3.0×10^{-5}	2.4	75	5.1	1.9	f
DITASi	1.0×10^{-3}	2.9	106	5.6	2.2	g

[a] Aonuma et al. (2007).
[b] Aonuma et al. (2007), Lee et al. (2008).
[c] Kuwabara et al. (1994), Muellen and Scherf (2006), Park et al. (2009), Ramchandra et al. (2010).
[d] Aonuma et al. (2007), Nayak et al. (2010).
[e] McVie et al. (1978), Mueller (1999), Kitamura et al. (2004), Chu and Song (2006).
[f] Aonuma et al. (2007), Kondakova et al. (2008).
[g] Tanaka et al. (2007), Lee et al. (2009).

satisfies two important criteria—it has high hole mobility and long-term stability under device operation. Thus, α-NPD is a viable choice for green and red OLEDs. However, its low triplet energy of 2.3 eV results in emission quenching in blue phosphorescent devices. Since an efficient blue OLED is necessary to produce white light or full color displays, the low triplet energy precludes α-NPD from being used as the sole HTL material, requiring at least one extra layer of high E_T material between the HTL and the emissive zone to provide exciton blocking. TAPC, the hole transporting material used in the first efficient OLED in 1987, has adequate hole mobility and, unlike α-NPD, a high triplet energy (Chopra et al. 2010). Unfortunately, TAPC is not chemically stable under device operating conditions (Sivasubramaniam et al. 2009). The weak link in the TAPC structure is the quaternary carbon atom in the cyclohexyl ring. This carbon forms a relatively stable carbocation, leading to chemical degradation under operating conditions. Ideally, this structure could be modified to eliminate the weak chemical bonds while retaining the hole mobility and high triplet exciton energy by replacing the cyclohexyl with a stable but saturated link that prevents π-conjugation in the modified molecule. To date, this has not been achieved.

The alternative approach is to use an exciton blocking layer (EBL) between the HTL and EML. However, the EBL material must satisfy the same set of criteria as were established for the HTL material itself: high triplet exciton energy, efficient hole transport, and stability. Nonetheless, in this approach is pursued, the hole mobility does not have to be quite as good because the EBL is typically kept very thin. TCTA is commonly chosen for exciton blocking at the HTL/EML interface in phosphorescent blue devices. Though its hole mobility compared to α-NPD is low, it is adequate if the TCTA layer is <10 nm thick, which is sufficient to confine the blue-emitting excitons in the emission zone. However, the stability of TCTA in an operating device raises concerns (Sivasubramaniam et al. 2009). Thus, the search for an HTL material that can meet all requirements simultaneously continues.

Chapter 8

References

Adachi, C., R. Kwong, and S. R. Forrest. 2001. Efficient electrophosphorescence using a doped ambipolar conductive molecular organic thin film. *Organic Electronics* 2(1): 37–43.

An, Z., J. Yu, S. C. Jones, S. Barlow, S. Yoo, B. Domercq et al. 2005. High electron mobility in room-temperature discotic liquid-crystalline perylene diimides. *Advanced Materials* 17(21): 2580–2583.

Aonuma, M., T. Oyamada, H. Sasabe, T. Miki, and C. Adachi. 2007. Material design of hole transport materials capable of thick-film formation in organic light emitting diodes. *Applied Physics Letters* 90(18): 183503.

Borsenberger, P., W. Gruenbaum, and E. Magin. 1996. Hole transport in vapor-deposited triphenylmethane glasses. *Japanese Journal of Applied Physics* 35(part 1): 2698–2703.

Borsenberger, P. M. 1992. Hole Transport in bis (4-N, N-diethylamino-2-methylphenyl)-4-methylphenylmethane doped polymers. *Physica Status Solidi (B)* 173(2): 671–680.

Borsenberger, P. M., L. T. Pautmeier, and H. Bässler. 1992. Nondispersive-to-dispersive charge-transport transition in disordered molecular solids. *Physical Review B* 46(19): 12145–12153.

Cabanillas-Gonzalez, J., T. Virgili, A. Gambetta, G. Lanzani, T. D. Anthopoulos, and D. M. de Leeuw. 2006. Photoinduced transient stark spectroscopy in organic semiconductors: A method for charge mobility determination in the picosecond regime. *Physical Review Letters* 96(10): 106601.

Cahen, D. and A. Kahn. 2003. Electron energetics at surfaces and interfaces: Concepts and experiments. *Advanced Materials* 15(4): 271–277.

Cho, J.-S., A. Kimoto, M. Higuchi, and K. Yamamoto. 2005. Synthesis of diphenylamine-substituted phenyl-azomethine dendrimers and the performance of organic light-emitting diodes. *Macromolecular Chemistry and Physics* 206(6): 635–641.

Chopra, N., J. S. Swensen, E. Polikarpov, L. Cosimbescu, F. So, and A. B. Padmaperuma. 2010. High efficiency and low roll-off blue phosphorescent organic light-emitting devices using mixed host architecture. *Applied Physics Letters* 97(3): 033304 (1–3).

Chu, C.-W., V. Shrotriya, G. Li and Y. Yang. 2006. Tuning acceptor energy level for efficient charge collection in copper-phthalocyanine-based organic solar cells. *Applied Physics Letters* 88(15): 153504 (1–3).

Chu, T.-Y. and O.-K. Song. 2007. Hole mobility of N,N[sup [prime]]-bis(naphthalen-1-yl)-N,N[sup [prime]]-bis(phenyl) benzidine investigated by using space-charge-limited currents. *Applied Physics Letters* 90(20): 203512–203513.

D'Andrade, B. W., S. R. Forrest, and A. B. Chwang. 2003. Operational stability of electrophosphorescent devices containing p and n doped transport layers. *Applied Physics Letters* 83(19): 3858–3860.

D'Andrade, B. W., S. Datta, S. R. Forrest, P. Djurovich, E. Polikarpov, and M. E. Thompson. 2005. Relationship between the ionization and oxidation potentials of molecular organic semiconductors. *Organic Electronics* 6(1): 11–20.

Diouf, B. B., W. S. Jeon, J. S. Park, J. W. Choi, Y. H. Son, D. C. Lim et al. 2011. High hole mobility through charge recombination interface in organic light-emitting diodes. *Synthetic Metals* 161(19–20): 2087–2091.

Djurovich, P. I., E. I. Mayo, S. R. Forrest, and M. E. Thompson. 2009. Measurement of the lowest unoccupied molecular orbital energies of molecular organic semiconductors. *Organic Electronics* 10(3): 515–520.

Giebeler, C., H. Antoniadis, D. D. C. Bradley, and Y. Shirota. 1998. Space-charge-limited charge injection from indium tin oxide into a starburst amine and its implications for organic light-emitting diodes. *Applied Physics Letters* 72(19): 2448–2450.

Goushi, K., R. Kwong, J. J. Brown, H. Sasabe, and C. Adachi. 2004. Triplet exciton confinement and unconfinement by adjacent hole-transport layers. *Journal of Applied Physics* 95(12): 7798–7802.

Hamaguchi, C. 2001. *Basic Semiconductor Physics*. New Dehli: Springer.

Hoofman, R. J. O. M., M. P. de Haas, L. D. A. Siebbeles, and J. M. Warman. 1998. Highly mobile electrons and holes on isolated chains of the semiconducting polymer poly(phenylene vinylene). *Nature* 392(6671): 54–56.

Hu, N.-X., S. Xie, Z. Popovic, B. Ong, A.-M. Hor, and S. Wang. 1999. 5,11-Dihydro-5,11-di-1-naphthylindolo[3,2-b]carbazole: Atropisomerism in a novel hole-transport molecule for organic light-emitting diodes. *Journal of the American Chemical Society* 121(21): 5097–5098.

Ide, N., T. Komoda, and J. Kido. 2006. Organic light-emitting diode (OLED) and its application to lighting devices. *SPIE Proceedings* 6333: 63330M–1.

Kang, J.-W., S.-H. Lee, H.-D. Park, W.-I. Jeong, K.-M. Yoo, Y.-S. Park et al. 2007. Low roll-off of efficiency at high current density in phosphorescent organic light emitting diodes. *Applied Physics Letters* 90(22): 223508 (1–3).

Kimura, M., S.-I. Inoue, K. Shimada, S. Tokito, K. Noda, Y. Taga et al. 2000. Spirocycle-incorporated tri-phenylamine derivatives as an advanced organic electroluminescent material. *Chemistry Letters* (2): 192–193.

Kitamura, M., T. Imada, S. Kako, and Y. Arakawa. 2004. Time-of-flight measurement of lateral carrier mobility in organic thin films. *Japanese Journal of Applied Physics* 43: 2326.

Koene, B. E., D. E. Loy, and M. E. Thompson. 1998. Asymmetric triaryldiamines as thermally stable hole transporting layers for organic light-emitting devices. *Chemistry of Materials* 10(8): 2235–2250.

Kondakov, D. Y. 2008. Role of chemical reactions of arylamine hole transport materials in operational degradation of organic light-emitting diodes. *Journal of Applied Physics* 104(8): 084520.

Kondakova, M. E., T. D. Pawlik, R. H. Young, D. J. Giesen, D. Y. Kondakov, C. T. Brown et al. 2008. High-efficiency, low-voltage phosphorescent organic light-emitting diode devices with mixed host. *Journal of Applied Physics* 104(9): 094501.

Kuwabara, Y., H. Ogawa, H. Inada, N. Noma, and Y. Shirota. 1994. Thermally stable multilared organic electroluminescent devices using novel starburst molecules, 4,4′,4″-tri (*N*-carbazolyl) triphenylamine (TCTA) and 4,4′,4″-tris (3-methylphenylphenylamino) triphenylamine (m-MTDATA), as hole-transport materials. *Advanced Materials* 6(9): 677–679.

Lee, D., N. Chopra, S.-H. Eom, Y. Zheng, J. Xue, and F. So. 2008. Effects of triplet energy confinement by charge transporting layers on blue phosphorescent organic light emitting diodes. *Proceedings of SPIE* 7051: 7051T.

Lee, J., J.-I. Lee, J. Y. Lee, and H. Y. Chu. 2009. High efficiency deep blue phosphorescent organic light-emitting diodes using a double emissive layer structure. *Synthetic Metals* 159(19–20): 1956–1959.

McVie, J., R. S. Sinclair and T. G. Truscott. 1978. Triplet states of copper and metal-free phthalocyanines. *Journal of the Chemical Society, Faraday Transactions 2: Molecular and Chemical Physics* 74: 1870–1879.

Mott, N. F. and R. W. Grunew. 1940. *Electronic Processes in Ionic Crystals*. London: Oxford University Press.

Muellen, K. and U. Scherf. 2006. *Organic Light-Emitting Devices, Synthesis, Properties, and Applications*. Weinheim: Wiley-VCH.

Mueller, G. (Ed.) 1999. *Electroluminescence. Semiconductors and Semimetals*. San Diego: Academic Press.

Murgatroyd, P. N. 1970. Theory of space-charge-limited current enhanced by Frenkel effect. *Journal of Physics D: Applied Physics* 3(2): 151.

Nayak, P. K., N. Agarwal, F. Ali, M. P. Patankar, K. L. Narasimhan, and P. N. 2010. Blue and white light electroluminescence in a multilayer OLED using a new aluminum complex. *Journal of Chemical Sciences* 122(6): 847–855.

Noda, T., H. Ogawa, N. Noma, and Y. Shirota. 1999. Organic light-emitting diodes using a novel family of amorphous molecular materials containing an oligothiophene moiety as colour-tunable emitting materials. *Journal of Materials Chemistry* 9(9): 2177–2181.

Okumoto, K. and Y. Shirota. 2001. Development of new hole-transporting amorphous molecular materials for organic electroluminescent devices and their charge-transport properties. *Materials Science and Engineering: B* 85(2–3): 135–139.

Okumoto, K., K. Wayaku, T. Noda, H. Kageyama, and Y. Shirota. 2000. Amorphous molecular materials: charge transport in the glassy state of N,N′-di(biphenylyl)-N,N′-diphenyl-[1,1′-biphenyl]-4,4′-diamines. *Synthetic Metals* 111–112: 473–476.

Park, J. J., T. J. Park, W. S. Jeon, R. Pode, J. Jang, J. H. Kwon et al. 2009. Small molecule interlayer for solution processed phosphorescent organic light emitting device. *Organic Electronics* 10(1): 189–193.

Parker, I. D. 1994. Carrier tunneling and device characteristics in polymer light-emitting diodes. *Journal of Applied Physics* 75(3): 1656–1666.

Pivrikas, A., G. Juška, K. Arlauskas, M. Scharber, A. Mozer, N. Sariciftci et al. 2005. Charge carrier transport and recombination in bulk-heterojunction solar-cells. *SPIE Proceedings* 5938: 59380N–59381.

Poplavskyy, D. and J. Nelson. 2003. Nondispersive hole transport in amorphous films of methoxy-spirofluorene-arylamine organic compound. *Journal of Applied Physics* 93(1): 341–346.

Prins, P., L. P. Candeias, A. J. J. M. van Breemen, J. Sweelssen, P. T. Herwig, H. F. M. Schoo et al. 2005. Electron and hole dynamics on isolated chains of a solution-processable poly(thienylenevinylene) derivative in dilute solution. *Advanced Materials* 17(6): 718–723.

Rajagopal, A., C. Wu, and A. Kahn. 1998. Energy level offset at organic semiconductor heterojunctions. *Journal of Applied Physics* 83(5): 2649–2655.

Ramchandra, P., L. Seung-Joon, L. Sung-Ho, K. Suhkmann, and K. Jang Hyuk. 2010. Solution processed efficient orange phosphorescent organic light-emitting device with small molecule host. *Journal of Physics D: Applied Physics* 43(2): 025101.

Reineke, S., F. Lindner, Q. Huang, G. Schwartz, K. Walzer, and K. Leo. 2008. Measuring carrier mobility in conventional multilayer organic light emitting devices by delayed exciton generation. *Physica Status Solidi (B)* 245(5): 804–809.

Reineke, S., G. Schwartz, K. Walzer, and K. Leo. 2007. Reduced efficiency roll-off in phosphorescent organic light emitting diodes by suppression of triplet–triplet annihilation. *Applied Physics Letters* 91(12): 123508.

Salbeck, J., N. Yu, J. Bauer, F. Weissörtel, and H. Bestgen. 1997. Low molecular organic glasses for blue electroluminescence. *Synthetic Metals* 91(1–3): 209–215.

Saragi, T. P., T. Fuhrmann-Lieker, and J. Salbeck. 2006. Comparison of charge-carrier transport in thin films of spiro-linked compounds and their corresponding parent compounds. *Advanced Functional Materials* 16(7): 966–974.

Seki, K. and K. Kanai. 2006. Development of experimental methods for determining the electronic structure of organic materials. *Molecular Crystals and Liquid Crystals* 455(1): 145–181.

Shirota, Y. and H. Kageyama. 2007. Charge carrier transporting molecular materials and their applications in devices. *Chemical Reviews* 107(4): 953–1010.

Shirota, Y., S. Nomura, and H. Kageyama. 1998. Charge transport in amorphous molecular materials. *SPIE's International Symposium on Optical Science, Engineering, and Instrumentation*, San Diego, CA, USA, pp. 132–141.

Shirota, Y., K. Okumoto, and H. Inada. 2000. Thermally stable organic light-emitting diodes using new families of hole-transporting amorphous molecular materials. *Synthetic Metals* 111–112: 387–391.

Shirota, Y., K. Okumoto, H. Ohishi, M. Tanaka, M. Nakao, K. Wayaku et al. 2005. Charge transport in amorphous molecular materials. *Proceedings of SPIE-International Society for Optical Engineering* 5937: 593717.

Sivasubramaniam, V., F. Brodkorb, S. Hanning, O. Buttler, H. P. Loebl, V. Van Elsbergen et al. 2009. Degradation of HTL layers during device operation in PhOLEDs. *Solid State Sciences* 11(11): 1933–1940.

Spreitzer, H., H. W. Schenk, J. Salbeck, F. Weissoertel, H. Reil, and W. Riess. 1999. Temperature stability of OLEDs using amorphous compounds with spiro-bifluorene core. *SPIE Proceedings* 3797: 316–324.

Swensen, J. S., L. Wang, E. Polikarpov, J. E. Rainbolt, P. K. Koech, L. Cosimbescu et al. 2013. Near independence of OLED operating voltage on transport layer thickness. *Synthetic Metals* 163: 29–32.

Swensen, J. S., L. Wang, J. E. Rainbolt, P. K. Koech, E. Polikarpov, D. J. Gaspar et al. 2012. Characterization of solution processed, p-doped films using hole-only devices and organic field-effect transistors. *Organic Electronics* 13(12): 3085–3090.

Tanaka, D., Y. Agata, T. Takeda, S. Watanabe, and J. Kido. 2007. High luminous efficiency blue organic light-emitting devices using high triplet excited energy materials. *Japanese Journal of Applied Physics* 46(5): L117–L119.

Tiwari, S. and N. C. Greenham. 2009. Charge mobility measurement techniques in organic semiconductors. *Optical and Quantum Electronics* 41(2): 69–89.

Tse, S., S. Tsang, and S. So. 2006. Polymeric conducting anode for small organic transporting molecules in dark injection experiments. *Journal of Applied Physics* 100(6): 063708 (1–5).

Tsiper, E., Z. Soos, W. Gao, and A. Kahn. 2002. Electronic polarization at surfaces and thin films of organic molecular crystals: PTCDA. *Chemical Physics Letters* 360(1): 47–52.

Wang, Y. 2004. Dramatic effects of hole transport layer on the efficiency of iridium-based organic light-emitting diodes. *Applied Physics Letters* 85(21): 4848–4850.

Weis, M., J. Lin, D. Taguchi, T. Manaka, and M. Iwamoto. 2009. Analysis of transient currents in organic field effect transistor: The time-of-flight method. *The Journal of Physical Chemistry C* 113(43): 18459–18461.

Yamamoto, K., M. Higuchi, S. Shiki, M. Tsuruta, and H. Chiba. 2002. Stepwise radial complexation of imine groups in phenylazomethine dendrimers. *Nature* 415: 509–511.

9. Conductivity Doping

Falk Loeser, Max Tietze, Björn Lüssem, and Jan Blochwitz-Nimoth

9.1   Introduction

9.1.1   Role of Conductivity Dopants

As described in Chapter 1, an OLED is an optoelectronic device, which is based on

- Injection of holes from the anode and injection of electrons from the cathode
- Transport of holes through the hole-transporting layer(s) and transport of electrons through the electron-transporting layer(s) toward the emission zone located in the middle of the OLED device

Chapter 9

- Recombination of holes and electrons in the emission zone and subsequent emission of photons

One key approach to enhance injection and increase transport behavior is the use of conductivity dopants.

Conductivity dopants are additives (organic or inorganic materials) that controllably and stably increase the charge carrier conductivity of the respective transport layers by enhancing the amount of available mobile charge carriers. The conductivity σ of a semiconducting layer is to a first approximation defined by

$$\sigma = e \cdot \mu \cdot n \tag{9.1}$$

where e is the elementary charge, μ the charge carrier mobility, and n the number of mobile charge carriers. If the density of charges is enhanced without altering the mobility of this charge carrier, the conductivity gets linear proportional to the density of charge carriers. Conductivity doping has two main effects in an OLED:

- It enhances the charge carrier injection from the contacts (mostly metal or transparent degenerate semiconductors such as indium tin oxide, ITO) into the neighboring organic transport material (for details, see Section 9.3.3).
- It enhances the transport through the transport layers owing to higher charge carrier conductivity (for details, see Section 9.3.1).

Consequently, conductivity doping reduces the operating voltage of an OLED, thereby enhancing significantly the power efficacy of the OLED (provided that the other power efficacy determining factors, as internal quantum efficiency or charge carrier balance are not negatively affected). This effect can be seen by a variety of measurements and observations, which will be described in Section 9.3. For example, from ultraviolet photoelectron spectroscopy (UPS) measurements, it can be concluded that upon doping, the Fermi level of the organic semiconducting layer is moved toward the transport level (which is a consequence of the higher density of mobile charge carriers).

Beyond these main effects, several second-order effects happen by the use of conductivity doping:

- Charge carrier balance: the use of doping may change the distribution of holes and electrons inside the emission zone, since doping may have a larger effect on the carrier injection and transport mainly for one charge carrier. Thereby, a new dimension in the OLED device design and optimization is available, which can be used to enhance the overall OLED performance.
- The energetic properties of the contacts become less important, which in turn means that the effects of surface potential fluctuations over a large area become less pronounced. This leads to a better microscopic homogeneity of the OLED area.
- Thick layers can be easily incorporated into the OLED: the higher conductivity of doped transport layers allows increasing the thickness of the layers without increasing the operating voltage. This can be used to optimize the optical cavity of the

OLED (optical cavity here refers to the distance between the reflective and semireflective contacts and the distance of the emission zone to the contacts), which helps optical outcoupling of photons out of the OLED (see Section 9.5). Additionally, thicker transport layers and hence a larger distance between the OLED electrodes may help to improve production yields (see Section 9.5, and Chapters 8 and 16).

- When conductivity doping is used in a transport layer, no additional charge injection layer between the electrode and the transport layer is needed, making the device setup easier.

The following example (taken from Pfeiffer 1999, Pfeiffer et al. 1999) illustrates how conductivity doping using a molecular dopant influences the conductivity of a doped layer. This seminal measurement investigated the example of zinc-phthalocyanine (ZnPc) and vanadyl-phthalocyanine (VOPc), two prototypical organic semiconductors, doped with tetra-fluoro-tetra-cyano-chinodimethane (F_4-TCNQ), a prototypical organic p-dopant (see Figure 9.1). When a small amount of the dopant is added, the conductivity increases several orders of magnitude over the undoped layer, which for a properly purified material is below 10^{-10} S/cm for ZnPc and below 10^{-12} S/cm for VOPc). The increase in conductivity is not linear but tends to be superlinear for many host-dopant systems. Doping concentrations are either given by molecular (as in Figure 9.1) or by weight ratios and are in the low percent range. The conductivities are low compared with crystalline (inorganic) semiconductors because of the lower mobilities of organic semiconductors and the higher density of traps in organic semiconductors.

The conductivity doping approach has been applied to OLEDs since the late 1990s. The earlier-described VOPc:F_4-TCNQ example system was first used in OLEDs by Blochwitz et al. in 1998; these measurements demonstrated that the operating voltage in OLEDs can indeed be reduced drastically upon p-type doping.

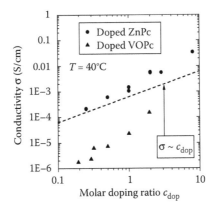

FIGURE 9.1 Conductivity of ZnPc and VOPc thin films doped with F_4-TCNQ measured in the coplanar contact geometry. The dashed line represents linear dependence. Even a low concentration of only 0.2 mol% increases the conductivity of ZnPc to 10^{-4} S/cm (intrinsic: <10^{-10} S/cm). The differences between ZnPc and VOPc can be attributed to differences in the mobility of the different phthalocyanines. (Taken from M. Pfeiffer, Controlled doping of organic vacuum deposited dye layers: Basics and applications, PhD thesis, TU Dresden, 1999.)

Chapter 9

9.1.2   Key Concepts and Definitions

Many terms for conductivity doping have been and are being used in the OLED literature. In this section, we will define four key terms and introduce several important concepts that will be used throughout the rest of this chapter.

1. We distinguish conductivity doping from the use of emissive molecules incorporated into a nonemissive host layer to enhance the internal quantum efficiency of light creation—often the word "doping" is used here as well. This doping can be referred to as "emitter doping."

2. The following terms are also in use for conductivity doping:
   - *Doping:* used as a short form of conductivity doping, not to be confused with emitter doping. In many cases, the context is clear, in which case this term can be used as well.
   - *Electrical doping:* refers to the fact that the electrical conductivity is enhanced by conductivity doping. Nonexperts in the field might believe this implies the application of an electrical field during the doping process; this is not the case. Otherwise, this term can be used as well (and is widely used in scientific literature).
   - *Redox doping:* this term comes from the fact that a reduction/oxidation reaction takes place between the host molecules and the dopant molecules during the doping process. This is indeed the case as it is later described.

     In this chapter, we will use the terms "conductivity doping" and "doping" both to refer to conductivity doping.

3. Some papers refer to a doping process in the case where a second host material with higher mobility is mixed into the first host material to enhance the conductivity of a layer. This effect cannot be called "doping" because no charge carrier transfer and consequently no increase of mobile charge carrier density takes place. Instead, the effect of this addition is to increase the overall layer mobility because the added second host material is having a higher mobility and thereby the total mobility of the layer is enhanced (see, e.g., Tardy et al. 2006: they use TBu-PBD as "n-dopant" for the conductivity enhancement of BCP as an electron transport material).

     To differentiate this "mixing" from conductivity doping, an analysis of the energy levels of the host and the additive molecules can be conducted. Doping is likely if

     For p-doping: the lowest unoccupied molecular orbital (LUMO) of the p-dopant is close (or below) the highest occupied molecular orbital (HOMO) of the HTL host material.

     For n-doping: the HOMO of the (active) n-dopant is close or above the LUMO of the ETL host material.

     "Close" means less than 0.5 eV above (p-doping) or below (n-doping) the relevant energy level. Furthermore, one should confirm that the mobile charge carrier density is indeed enhanced (see Section 9.3.1). The definition of 0.5 eV as "close" is to a certain extent empirical, especially since the HOMO and LUMO states are not single energy levels but rather a distribution of states (DOS) in a solid thin film, leading to variations of the LUMO (p-dopants) or HOMO (n-dopants) level of a single dopant molecule depending on the actual host molecule environment. Another indicator of "real" doping is the movement of the Fermi level toward the transport level as will be discussed in Section 9.3.3 (Figure 9.2).

From a chemical point of view, the p-type doping of an electron-donating matrix molecule M with an electron-attracting (acceptor like) dopant molecule A can be described with the mass action law:

$$M + M'A \rightleftharpoons M + [M'^+ A^-] \rightleftharpoons M^+ + M'A^- \tag{9.2}$$

where M is a random matrix molecule and M' is a matrix molecule in proximity to a dopant A. The first step of the reaction 9.2 is the initial charge transfer describing the electron transfer from a matrix molecule to the nearest acceptor molecule. This intermediate local charge transfer state $[M'^+ A^-]$ may dissociate into a quasi-unbound state, which means that the matrix molecule carrying the positive charge is too far away from the ionized dopant molecule to feel a Coulomb attraction. Thus, the positive charge can move through the organic layer by hopping and the density of M^+ determines the density of free equilibrium holes $p_{f,0}$ in the layer. Doping of M with A is efficient if the balance of the above reaction is on the right side, that is, the density of free holes $p_{f,0}$ approaches the doping density N_A. Reaction 9.2 describes the case of p-type doping, and n-type doping may be described similarly. However, as we will learn later (see Section 9.4) in most practical material systems and at typical doping concentrations of a few percent, only a minority of the dopants contribute at a given time to the mobile charge carrier density. A main difference between p- and n-type doping of OLED charge-transporting materials is that the HOMO of a potentially strong n-type dopant must be in the range of 3 eV. This means that these materials are prone to be air-sensitive (i.e., reacting with oxygen) and generally unstable (easy to chemically destroy) materials because their LUMO level will be close to the vacuum level. Figure 9.3 shows the challenge: ETL host materials that are used for other electronic devices or for organic photovoltaic (OPV) devices have a much lower lying LUMO level and are easier to dope than OLED-suited ETL materials (Walzer et al. 2007).

4. Sometimes, the term doping is used in conjunction with chemically altering the majority of host molecules by the use of a high amount of additives to create a new host molecule with higher intrinsic charge carrier mobility. This is the case when, for example, doping poly-3,4-ethylenedioxythiophene (PEDOT) with polystyrene

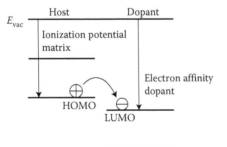

FIGURE 9.2 Schematic energy level alignment of a p-doped matrix:dopant thin-film system. Commonly, p-doping is assumed by electron transfer from the HOMO of a matrix molecule to the LUMO of a dopant molecule.

Chapter 9

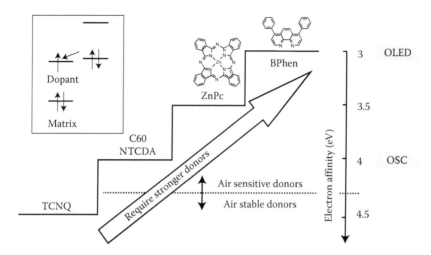

FIGURE 9.3 A schematic view of the energetics of n-doping of organic semiconductor layers for different applications. It is more difficult to n-dope OLED ETL hosts since the HOMO (or electron affinity) of the n-dopant must lie very much above the oxidation potential (or electron affinity) of oxygen and will hence easily be oxidized. (Taken from K. Walzer et al., *Chem. Rev.* 107, 2007: 1233–1271.)

sulfonate (PSS) to form PEDOT:PSS. This type of doping can be referred to as "chemical doping" (see, e.g., Watanabe et al. 2007). We address this process more specifically in Section 9.2.4 of this chapter.

The doping process is often carried out by coevaporation in vacuum. In this process, host and dopant materials are mixed in the organic layer during the deposition process, with the ratio controlled by their respective rates of evaporation (Figure 9.4). There, a (partial) reaction and charge carrier transfer can take place.

Alternatively, for solution-processed layers (see Chapter 15), doping can be accomplished by mixing the dopant molecules into the ink containing the charge transport materials to be doped. However, unintended chemical reaction effects may take place before the layer is applied, hampering the doping effect. Although the rest of this chapter primarily refers to examples based on vacuum deposition, the teaching in this chapter can be similarly applied to a large extent for solution-processed doping.

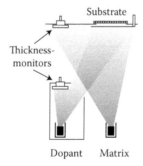

FIGURE 9.4 A schematic view of the coevaporation method to make doped organic layers. (Taken from J. Blochwitz, Organic light-emitting diodes with doped charge transport layers, PhD thesis, TU Dresden, 2001.)

9.2   Historical Development and Examples of Conductivity Dopants for OLEDs and Organic Photovoltaics

9.2.1   Early Work with Gases, Metals, and Ions

The first publication describing doping of charge carrier transport layers in OLEDs was actually a patent application from Egusa (1992). Here, in a simple translation from inorganic semiconductor physics, it was proposed that strong acceptor atoms or molecules could be used to enable charge carrier transfer and subsequent mobile charge carrier density (in this case, holes) enhancement. Examples included numerous metals (as Au, Pt, W, Ir), vapors, and gases (Cl, Br, I), inorganic ions (POCl3, AsF6), and smaller organic molecules such as TCNQ, F_4-TCNQ, dicyano-dichloro-quinone (DDQ), chloranyl, bromanyl, and others. For n-type doping, strong donor atoms and molecules were proposed. Examples include again (less noble) metals (Al, Ag, Au, In), alkali metals (as the later becoming important lithium), alkali earth metals, rare earth elements, and small molecules such as NH_3, aniline, phenylenediamine, and tetrathiafulvalene (TTF). Later work demonstrated that many of these materials are too weak acceptors or donors, or are strong enough donors or acceptors, but are physically too small to yield stable p–n junctions.

In the scientific literature, early doping work was focused on the doping of organic layers by mixing with elemental materials (analogous to the mixing of silicon single crystals with bromine or other atoms) or gases. The basic problem encountered here was that this doping with very small additives does not lead to stable doped layers because the dopants can easily diffuse and hence layered p–n or p–i–n junctions cannot be prepared (e.g., Yamamoto et al. 1979).

Most of this early work was directed toward p-type doping, meaning addition of an acceptor molecule. One of the first model systems used weakly donor-like organic semiconductors such as phthalocyanines (Pc's). Exposing them to strongly oxidizing gases such as iodine or bromine leads to very high conductivities (Yamamoto et al. 1979). The main problem with this approach is that stopping the exposure of the organic semiconducting layer to these gases leads to a relatively fast decrease of the conductivity owing to dedoping (removal of the dopants). This dedoping becomes even faster at elevated temperatures.

The conductivity of organic dyes can also be influenced by doping with donor or acceptor molecules having an extended pi-electron system. Phthalocyanines, for example, have been p-type-doped with acceptor molecules such as ortho-chloranil as early as 1960 (Kearns et al. 1960); later molecules such as tetracyano-quinodimethane (TCNQ) (Boudjema et al. 1988) and DDQ (Marks et al. 1985, Maitrot et al. 1986, El-Khatib et al. 1988) have been used. A good overview of the early work on phthalocyanine doping can also be found in Simon and Andre (1985). The dopants described in that work are relatively weak (meaning their LUMO levels lie well above the HOMO level of the Pc's) and hence the doping effect was rather weak. For this reason, only very high doping (or rather "mixing") ratios between 1:20 and 1:1 were studied. A maximum conductivity of 6×10^{-6} S/cm was achieved at a doping ratio of 1:5. In addition, phthalocyanines are not suitable for use in high-efficiency OLEDs owing to their small band gap (hence substantial absorption in the visible range) and HOMOs that are still too high. Doping of other classes of matrix materials have been examined, such as oligothiophenes with DDQ (Lous et al. 1995) and polyacetylenes with AsF_5 (Park et al. 1980).

In addition, doping with Lewis acids has been investigated, for example, by Kido et al., who used $FeCl_3$ to dope hole transport materials (Kido et al. 1998, Endo et al. 2002b). Lewis acids are molecules that are able to accept an electron lone pair. The doping effect comes from a disproportionation reaction of the Lewis acid leading to the formation of a cationic and a neutral molecule; details can be found in Werner (2003).

Early work applying doped layers in organic semiconductor devices include Lous et al. (using DDQ as dopant) and Hiller et al. (working with F16-ZnPc as host and Pc derivates as dopants) in 1995 and 1998. These authors investigated rectification diodes (Lous et al. 1995, Hiller et al. 1998), addressing metal/doped semiconductor junctions (not yet organic–organic junctions). Generally, one problem with these doping approaches was the partial instability of the doped layers. Other problems had to do with the low volatility of some of the materials, particularly the dopants. Thus, the doping effect could not be totally controlled, especially at p–n junctions (interfaces between p-type-doped and n-type-doped layers) or pi-junctions (interfaces between p-type-doped and intrinsic, or nondoped, layers).

Another early approach to p-type doping was the use of metal oxides, such as molybdenum oxide, MoO_3. These materials have been used as a hole injection layer (see Chapter 5) adjacent to the anode (tuning the anode work function and improving charge injection; see Tokito et al. 1996). These materials have also been added to hole transport materials as conductivity dopants (Ikeda et al. 2006). Alternative materials to MoO_3 include WO_3 (Chang et al. 2006, Hsieh et al. 2006), ReO_3 (Leem et al. 2007, Krause et al. 2011), and others. These metal oxides typically have very low LUMO levels making charge transfer from the host to the dopant possible. Typical doping concentrations are in the range of close to 10% to several 10% (Kröger et al. 2009). One weakness of metal oxides is their high evaporation temperature—above 800°C. This large difference in the host and dopant molecule evaporation temperature can cause problems during long-term manufacturing (see Chapter 14). As explained above, finding suitable n-type dopants is an even more difficult task.

Early approaches to enhance the electron conductivity of OLED suitable electron transport layers focused on the use of strongly reducing gases, atoms, or ionic compounds, based on these atoms. Prime examples are first row main group elements (or alkali metals) such as Li and Cs. As early as the 1970s, alkali metals were proposed for n-type doping of organic materials (Ivory et al. 1979), owing to their strong donor properties. The first reports using alkali metals (Li in most cases) in OLEDs date back to the early 1990s (Itoh et al. 1990, Kido et al. 1993, Hung et al. 1997, Endo et al. 2002a). In these works, alkali metals, alkali metal complexes (e.g., LiF), or oxides of alkali metals (e.g., Li_2O) were used as a thin layer between the ETL and the remaining cathode material. Later work demonstrated that Li or LiF deposited along with the ETL matrix molecules yielded a doped layer. Together, these studies revealed the working principle of alkali metal compounds such as LiF: upon subsequent exposure to hot metal cathode atoms (e.g., Al), LiF dissociates, and Li migrates into the underlying ETL materials, leading to a doping effect (e.g., Mason et al. 2001). A more comprehensive review article covering this topic can be found here in Walzer et al. (2007). Efficient OLEDs made with Li-doped ETL layers have been described by Kido and Matsumoto (1998), Parthasarathy et al. (2001), and Huang et al. (2002). The usability of this approach at large-scale production is still under debate: Li migration (which apparently happens as described earlier) may lead to unwanted diffusion into or close to the emitter layer, consequently reducing the OLED efficiency by quenching excitons (Kido and Matsumoto 1998, D'Andrade et al. 2003).

Another often-used alkali metal is cesium (Cs) (Meerheim et al. 2006, Oyamada et al. 2005). The doping process is very similar to Li. Owing to the difference in atomic mass (133 g/mol vs. 7 g/mol) and ionic radius (181 pm vs. 90 pm for the respective cations (Shannon 1976), it can be expected that Cs may diffuse less in organic matrix materials. On the other hand, Cs is more volatile than Li under vacuum conditions and may diffuse more easily throughout the vacuum coating equipment. Similar to LiF, CsF has been used in OLEDs (Piromreun et al. 2000, Watanabe et al. 2007). As a variant of CsF, Cs_2CO_3 was used as an n-dopant for OLED ETLs (Hasegawa et al. 2004, Wu et al. 2006, Krause et al. 2011). The working principle is not quite clear yet, as Cs_2CO_3 is chemically too inactive material to be an efficient donor of electrons to OLED ETLs. One speculation is that pure Cs vapor is released when Cs_2CO_3 is heated, while others claim that decomposition of Cs_2CO_3 into Cs_2O and CO_2 is responsible for the doping effect (Chen et al. 2006). There is some evidence that Cs_2CO_3 is more compatible with a wider range of cathode materials than LiF (which works best with Al as top contact), which makes it more suitable for top emission OLEDs (Cho et al. 2011). Other compounds under study have been CsN_3 (Yook et al. 2010) or cesium phosphate (Cs_3PO_4) (Wemken et al. 2012). A common weakness of all these alkali metal and alkali metal compound n-type dopants is that the active dopant Cs or Li easily reacts with oxygen. Thus, large amounts of wasted Cs or Li in a big vacuum coating tool pose a potential safety issue.

9.2.2   Organic Molecules as Conductivity Dopants

Although organic layers can be successfully doped by gases, metals, and ions, these doped layers tend to be unstable, which leads to a reduced device lifetime (Kido and Matsumoto 1998, D'Andrade et al. 2003). This shortcoming triggered research on molecular dopants, which have a higher mass and are less mobile in the organic film.

The strong electron acceptor F_4-TCNQ (electron affinity (EA) ~5.4 eV and ionization potential 8.35 eV) (Blochwitz 2001) was first proposed as a molecular p-dopant by Pfeiffer et al. (1998). It has been extensively used to improve the efficiency of OLEDs and organic solar cells (Blochwitz et al. 1998, Walzer et al. 2007). Various methods have been used to elucidate the doping mechanism, including conductivity measurements, UPS (Olthof et al. 2009a,b, Gao and Kahn 2001), measurement of the Seebeck coefficient (Pfeiffer et al. 1998), impedance spectroscopy (Drechsel et al. 2002), and UV/Vis spectroscopy (Gao et al. 2008, Yim et al. 2008) (cf. Section 9.3). Although F_4-TCNQ has mainly been used to dope small molecular films (Walzer et al. 2007 and references

Chapter 9

therein), it has also been used to increase the conductivity of polymeric films (Yim et al. 2008, Zhang et al. 2009).

However, even F_4-TCNQ is too volatile for many applications, since evaporation of the molecule can lead to contamination of the deposition chamber and unstable doping profiles in devices. This shortcoming triggered research on numerous variations of the F_4-TCNQ molecule with increased vapor pressure (Cosimbescu et al. 2011). Some of the successful variants are F_2-HCNQ (Gao et al. 2008), which has a lower LUMO (5.59 eV) and better thermal stability, and F_6-TNAP (or more chemically correctly referred to as F_6-TCNNQ), which has a significantly increased molecular weight compared to F_4-TCNQ (Koech et al. 2010). Alternatively, a fluorinated fullerene—$C_{60}F_{36}$—has been shown to be a strong p-dopant as well (Solomeshch et al. 2009). It was used to realize highly efficient OLEDs and organic solar cells, which are on par with F_4-TCNQ doped devices, but with lower volatility and no deposition chamber contamination (Meerheim et al. 2011) (Figure 9.5).

Dopants reported in the literature are bis(ethylenedithio)-tetrathiafulvalene (BEDT-TTF) (Nollau et al. 2000), tetrathianaphthacene (TTN) (Tanaka et al. 2005), bis(cyclopentadienyl)-cobalt(II) cobaltocene $CoCp_2$ (Chan et al. 2006b), or a dinuclear complexes of chromium or tungsten with the anion of 1,3,4,6,7,8-hexahydro-2H-pyrimido[1,2-a]pyrimidine (hpp) ($Cr_2(hpp)_4$ or $W_2(hpp)_4$), the latter even being able to efficiently dope OLED electron transport materials (Cotton et al. 2002, Menke et al. 2012). However, as these dopants need to have a very high lying HOMO level or ionization potential to efficiently transfer an electron to a suitable ETL host material (for OLEDs, LUMOs are around 2 eV; see Section 9.1.2, Figure 9.3), they are intrinsically reactive with oxygen and must hence be handled under inert atmosphere. This holds especially true for the strongest dopant in this list, $W_2(hpp)_4$.

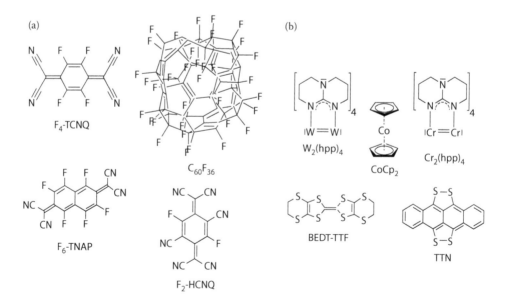

FIGURE 9.5   Commonly used organic (a) p- and (b) n-dopants.

9.2.3 Precursor Molecules as Conductivity Dopants

Historically, n-type doping of electron transport layers in OLEDs was realized for the first time by using alkali metals such as Li and Cs (see Section 9.2.1). However, for reaching the desired conductivities, high doping ratios of 1:1 were necessary. Thus, the high amount of ionized counterions within the mixed film disturbs the transport of free electrons by the formation of trap states as a result of Coulomb attraction. Furthermore, the formation of well-defined doping profiles is prevented by the high diffusivity of the small metal atoms through the matrix material negatively impacting the device stability.

Using the so-called precursor molecules as conductivity dopants is a sophisticated way to circumvent the disadvantages of alkali metals and nonair-stable molecular n-type dopants. This approach to conductivity doping is based on the *in situ* activation of the donor from a stable precursor compound by light or thermal energy during or after the coevaporation of the dopant and the matrix materials. In the mixed film, electron transfer to the matrix occurs accompanied by the formation of donor cations. Figure 9.6 summarizes various precursor compounds, which will be presented in this section.

Conductivity doping by precursor molecules was introduced by A. Werner et al. in 2003. They suggested cationic dyes as precursor compounds because these air-stable molecules typically already contain an organic cationic part. The intention was to separate the cationic structure from the precursor compound and to use it as a donor. n-Type doping of 1,4,5,8-naphthalenetetracarboxylic dianhydride (NTCDA) by coevaporation with Pyronine B (PyB) chloride (Figure 9.6a) was shown through an increase of conductivity of NTCDA films by four orders of magnitude from 3.3e-8 S/cm (undoped) to 2.0e-4 S/cm (3% PyB) (Werner et al. 2003). From Fourier transform infrared spectroscopy (FTIR), UV/Vis, and mass spectroscopy measurements, Werner et al. (2004)

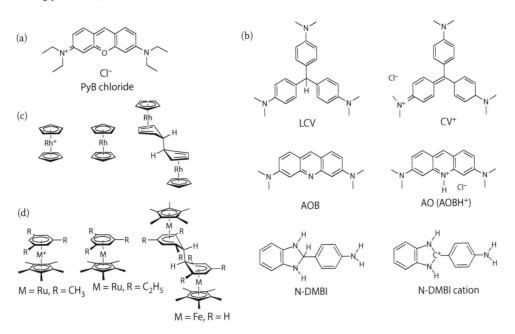

FIGURE 9.6 Summary of various "precursor" compounds, which are used as air-stable n-dopants (for further discussion see text).

conclude leuco pyronine B formed during evaporation. However, the doping effect vanishes when the sample is exposed to air.

An ultraviolet photoelectron spectroscopy/inverse photoelectron spectroscopy (UPS/IPES; see Section 9.3.3) study on this material system reveals a clear shift of the Fermi level toward the LUMO of NTCDA for both surface and bulk doping, indicating n-type doping although the ionization potential (IP) of pure PyB (6.06 eV) significantly exceeds the EA of NTCDA (4.02 eV) (Chan et al. 2006a,b). However, as supported by density functional theory (DFT) calculations, DOS features arising ~4.3 eV below the vacuum level within the "pure" PyB spectrum could be attributed to the presence of both the leuco and neutral radical forms of PyB in the condensed film. These features are likely responsible for n-type doping because of much better level matching to the LUMO of NTCDA. The resulting doping mechanism is thus given by the generation of reducing radicals from thermal deposition of salts of stable cations (Figure 9.7a). Similar n-type doping of NTCDA with rhodamine B (RhB) chloride was shown (Werner 2003).

A prominent electron-transporting material widely used in solar cells is the buckminsterfullerene C_{60} (EA = 4.0 eV (Zhao and Kahn 2009), IP = 6.4 eV). Conductivity doping of fullerenes by photo-induced electron transfer followed by hydrogen atom transfer is a second approach to applying precursor molecules as conductivity dopants. Examples of these compounds are crystal violet and its leuco base (CV/LCV) (Li et al. 2004), acridine orange base (AOB) (Li et al. 2006), and a newer material class based on 1,3-dimethyl-2-phenyl-2,3-dihydro-1H-benzoimidazole (DMBI) derivatives, for example, N-DMBI (Wei et al. 2010) (see Figure 9.6b).

Figure 9.8 shows the principle of doping C_{60} by LCV. Since the IP of LCV (5.1 eV, Nelson 1961) exceeds the EA of C_{60} by 1 eV, a direct charge transfer is very unlikely. However, by illuminating or heating the doped film, electrons in LCV are promoted to an excited state, thus enabling charge transfer to the LUMO of the fullerene (Figure 9.8a). This process results in the formation of two radical molecules LCV$^{\cdot+}$ and $C_{60}^{\cdot-}$. The energetically preferred electron back-transfer to LCV is prevented by a second reaction channel, namely hydrogen abstraction from the LCV$^+$ radical to a neutral C_{60} molecule. Thus the LCV is transformed into the cationic CV with a completely filled HOMO making the n-doping of C_{60} irreversible and permanent (Figure 9.8b and c). The generalized reaction is given in Figure 9.7b. By absorption and FTIR measurements of pure LCV and CV thin films (Li et al. 2004), it was shown that the leuco base of crystal violet is

(a) $\quad M^+Cl^- \xrightarrow{\Delta} M^{\cdot} \xrightarrow{A} M^+ + A^{\cdot-}$
$\quad\quad\quad\quad\quad$ and side products

(b) $\quad MH + C_{60} \xrightarrow{h\nu} MH^{\cdot+} + C_{60}^{\cdot-} \xrightarrow{\frac{1}{x}C_{60}} M^+ + C_{60}^{\cdot-} + \frac{1}{x}C_{60}H_x$

(c) $\quad \frac{1}{2}M_2 \rightleftharpoons M^{\cdot} \xrightarrow{A} M^+ + A^{\cdot-}$

(d) $\quad M_2 + A \rightleftharpoons M_2^{\cdot+} + A^{\cdot-} \longrightarrow M^{\cdot} + M^+ + A^{\cdot-} \xrightarrow{A} 2M^+ + 2A^{\cdot-}$

FIGURE 9.7 Possible reaction channels of precursor compounds applied as n-dopants M to an acceptor molecule matrix A. The details of each approach (a)–(d) are described in the text.

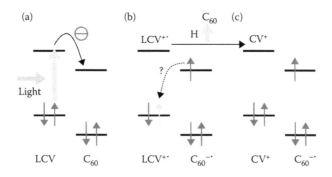

FIGURE 9.8 Principle of n-type doping of the fullerene C_{60} by leuco crystal violet (LCV).

formed during the vacuum deposition of CV, thus being the driving force of the doping process for both precursor compounds. Doping of C_{60} by CV with a molar ratio of 27:1 increases the conductivity by nearly six orders of magnitude up to 8e-3 S/cm. Doping of 5.6 mol% AOB into the buckminsterfullerene reaches an even higher conductivity of up to 3e-2 S/cm (Li et al. 2006), but still two orders of magnitude lower than using an air-unstable molecular n-type dopant like $W_2(hpp)_4$ (see Section 9.2.2).

According to Wei et al. the previously introduced precursor compounds need to be processed by vacuum thermal evaporation to function effectively, and are therefore inherently incompatible with solution processing. The class of DMBI compounds overcomes this drawback. Air-stable solution-processed n-channel OFETs based on [6,6]-phenyl C_{61} butyric acid methyl ester (PCBM) doped by 2 wt% N-DMBI were shown by Wei et al. (2010). The conductivity of PCBM films is considerably increased by more than four orders of magnitude up to 1.9e-3 S/cm owing to doping with N-DMBI. However, the proposed doping mechanism is slightly different than with CV/LCV. In the first step, N-DMBI radicals are created by hydrogen abstraction. These radicals are supposed to provide the electron transfer to PCBM (HOMO = −3.8 eV, Liu et al. 2012) from their single occupied molecular orbital (SOMO), which is −2.4 eV, in the second step. Electron transfer from neutral N-DMBI can be excluded because of its low HOMO of −4.7 eV (Wei et al. 2010). However, the doping still has to be activated by heating (80°C). Furthermore, unintentional side products are consequently present in the thin films owing to chemical reactions taking place during the doping process. This is also the case for the other previously described precursor compounds.

Most recently, a third approach of n-type doping based on "relatively" nonreactive dimers formed by highly reducing monomers was suggested by Guo et al. (2012). The generalized reaction equations of these doping mechanisms are given in Figure 9.7c and d. In the former case, the reversible unfavorable cleavage of a dimer M_2 (indicated by unequal double arrows) leads to radical monomers M, readily donating an electron to a neighboring acceptor molecule. In the latter case, an unfavorable reversible electron transfer from a dimer to an acceptor is followed by the rapid cleavage of the remaining dimer radical and thus creating two ionized acceptors and monomers. Dimeric forms of rhodocene (Figure 9.6c) and (pentamethylcyclopentadienyl)(arene)ruthenium and iron compounds (Figure 9.6d) are shown to be precursors to generate n-type dopants for various electron transport materials (ETMs), in particular for materials with low electron affinities such as pentacene (EA = 2.8 eV) or copper-phthalocyanine (CuPc, EA = 3.1 eV), processed either by vacuum evaporation or from solution. The n-doping effect is confirmed by UPS and j-V measurements on vertical Au/n-ETL/Au structures. For instance, doping of 3.5 wt% rhodocene dimers into CuPc yields a Fermi level shift of 0.7 eV away from the occupied states and an increase in the current density by six orders of magnitude. Furthermore, Guo et al. report on p–i–n-diodes based on CuPc, that is, p-doped with $Mo(tfd)_3$ (molybdenum tris[1,2-bis(trifluoromethyl)ethane-1,2-dithiolene]) and n-doped with the rhodocene dimers (each 1 wt%), achieving a rectification of 10^6 at 4 V. The dimeric molecules are supposed to be stable in air over several weeks and are also moderately stable in solution (benzene-d_6: decomposition within a few days) (Guo et al. 2012). An earlier publication on using this approach for OLEDs stems from the patent literature (Limmert et al. 2007).

In this section, a short overview of air-stable precursor compounds as n-type conductivity dopants was presented. In direct comparison to low IP molecular n-dopants these materials tend to be inferior in terms of achievable conductivities of typical ETMs. Nevertheless, precursor molecules are important for future development owing to their handling properties, in particular for applications in solution processed devices.

9.2.4 Doping of Polymers

Many of the early attempts to use at least one doped or conductivity-tuned layer in an OLED device were focused on polymeric p-type-doped organic semiconductors. The remaining layers in these early works were polymeric or vacuum deposited. A number of material systems were tested.

In the 1980s, Hayashi et al. inserted a doped poly(3-methylthiophene), P3MT, layer between the anode and a perylene emitter layer. The dopant was the counterion $ClO4^-$. Owing to the perylene layer, the device efficiency was rather poor. The doped polymeric layer in this work was used as a degenerate semiconductor for use as a metal contact replacement, which indeed is a better description of the effects taking place in such polymeric layers (Hayashi et al. 1986).

In the mid-1990s, Yang and Heeger (1994) and Antoniadis et al. (1995) introduced doped polyaniline for polymer LEDs. Again, the polymer layers were used as a polymeric anode in these devices. The higher conductivity of this polymeric anode was achieved by doping with camphor sulfonic acid (CSA) (Yang and Heeger 1994).

Around the same time, a number of other systems doped with inorganic salts were demonstrated. $FeCl_3$-doped polythiophene was used by Romero et al. (1995) and Ganzorig et al. doped a spin-coatable form of TPD with $SbCl_5$ (Ganzorig 2000).

Iodine-doped MEH-PPV (poly[2-methoxy,5′(2′-ethyl-hexyloxyl)-1,4-phenylene vinylene]) was used by Huang et al. (1997), and Yamamori et al. showed thick hole transport layers of polycarbonate doped with a chloroantimonate (TBAHA) (Yamamori et al. 1998, 1999).

However, over the past decades, the most common and widely used hole injection and transport layer in bi-layer polymer LEDs was a mixture of poly(3,4-ethylenedioxythiophene) and poly(4-styrenesulfonate) (Kugler et al. 1999) (PEDOT:PSS; commercially available from Bayer at that time, now acquired by Hereaus).

Not all these examples are conductivity-doped as described in the beginning of this chapter. Some change the nature of the host molecules by chemically altering them instead; this we call chemical doping. A prime example for chemical doping is PEDOT:PSS.

The majority of the work in polymers has been directed toward chemical doping. Chemical doping describes the chemical change of a host by an added compound (the "dopant") with the target to make it more conductive. Examples include the prototypical PEDOT:PSS, doped polyaniline, doped methylthiophene and eventually polycarbonate. For chemical doping, higher doping concentrations are needed (high in this case means approaching or exceeding one dopant molecule per polymer conjugated unit) because the host molecules/polymers must be almost completely modified. One of the explanations given (e.g., for PEDOT:PSS) is that PEDOT:PSS forms a degenerately doped semiconductor, which means that metallic PEDOT:PSS complexes—indicated by occupied density of states up to the Fermi level (Crispin et al. 2006, Hwang et al. 2006)—are surrounded by isolating PSS and electrical transport is either via percolation or hopping from metallic to metallic polymer chains. This hopping-like transport can be seen by a different temperature dependence of the PEDOT:PSS layer than expected from nondegenerately doped semiconductors: A log(conductivity) versus one over the square-root of the temperature plot can be fitted by a straight line (see Figure 9.9, Aleshin et al. 1998), whereas for small molecule and conductivity doping the Arrhenius plot (log(conductivity) versus one over temperature) yields a straight line (see Figure 9.9a, Pfeiffer et al. 1998). Consequently, the PEDOT:PSS hole injection layer is often referred to as being part of the anode (Brown et al. 1999). Interestingly, when a colored polymer (absorption bands of the neutral species is

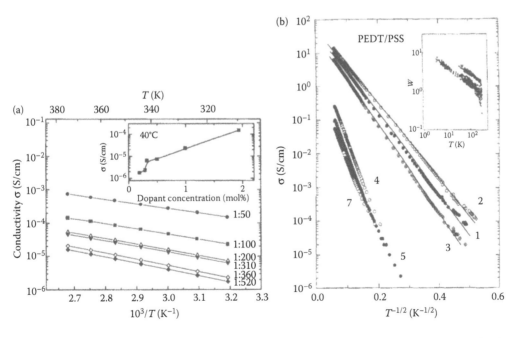

FIGURE 9.9 (a) Behavior of conductivity versus temperature for a conductivity-doped small-molecule layer (Pfeiffer et al. 1998) and (b) a chemically doped polymer, here PEDOT:PSS (Aleshin et al. 1998). The small molecule system is VOPc doped with F_4-TCNQ; the polymer system is PEDOT:PSS.

in the visible range) is chemically doped, it will typically become transparent. The reason is that upon doping with for example, acids, the neutral polymer chain becomes charged as radicals are formed. The radical absorption is blue shifted toward the UV and the next absorption band still lies in the IR.

More recent work demonstrates real conductivity doping of polymers (Aziz et al. 2007, Yim et al. 2008). As an aside, the dopant concentration should be determined by (and compared to) the number of dopants per conjugation length of the polymer under study as that is the smallest unit in the polymeric systems over which free charge carriers are delocalized. However, this is not always known in detail. In any event, Yim et al. reported on doping of different polymers such as poly(3-hexylthiophen-2,5-diyl) (P3HT), poly(9,9-di-n-octylfluorene-alt-bis-N,N-(4-butylphenyl)-bis-N,N-phenyl-1,4-phenyl-enediamine) (PFB), poly(9,9-di-n-octylfluorene-alt-(1,4-phenylene-((4-sec-butylphe-nyl)imino)-1,4-phenylene) (TFB), and poly(9,9-dioctylfluorene-alt-benzothiadiazole), poly[(9,9-di-n-octylfluorenyl-2,7-diyl)-alt-(benzo[2,1,3]thiadiazol-4,8-diyl)] (F8BT) by a typical small molecular p-type dopant, namely F_4-TCNQ. This was done by adding dopant solutions into host solutions using the same solvent. Their findings (conductivity vs. doping concentration, FTIR, absorption spectra) showed that behavior of these doped layers are very similar to that seen in studies addressing doping of small molecule layers by codeposition. The material that yielded the highest conductivities and the largest shift of the FTIR absorption of the carbon–nitrogen triple bond, indicating a complete charge transfer from the host to the dopant F_4-TCNQ, was P3HT. The overall absorption of these doped layers did not change much owing to the low doping level.

9.3 Characterization of Conductivity Doping Effects

9.3.1 Conductivity Measurements

The conductivity σ is given by the Equation 9.1, $\sigma = e\,n\,\mu$, were e is the electronic charge, n is the carrier density (of the respective transported charges), and μa the mobility of these charges inside the material. The mobile charge carrier density can for example be obtained from thermoelectric (Seebeck) measurements as described in Section 9.3.2. The conductivity of the organic layers can be directly measured as described earlier, and both results can be used to calculate the mobility; alternatively, the mobility can be measured via field effect measurements. Typical conductivities for undoped and properly purified organic semiconductor materials are in the range of 10^{-10} S/cm while doping increases the conductivity to at least 10^{-5} S/cm.

To measure the lateral conductivity, usually thin films of (doped) organic material are coated onto two separated contacts. A simple determination of the layer thickness and the measured resistance can be used for calculation of conductivity. These types of measurements can be easily performed *in situ* in a vacuum chamber.

Figure 9.10 describes an easy setup to measure conductivity. Conductive electrodes are arranged on a substrate as shown in Figure 9.10a. The electrode structure can be produced by lithography with very high accuracy. The conductive material can be ITO or any metal. The organic material is evaporated on top of the structure. When voltage is applied, the current flows through the gaps (50–100 μm) between the specially arranged electrodes. The finger-like arrangement as shown in Figure 9.10b provides a

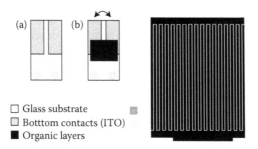

(a) (b)

☐ Glass substrate
☐ Botttom contacts (ITO)
■ Organic layers

FIGURE 9.10 Conductivity measurement with a coplanar setup. A thin organic layer is evaporated on a conductive structure and current between bottom contacts is measured (a). A more advanced layout can be used to improve the signal-to-noise ratio of conductivity measurements. A narrow gap between the two electrodes combined with a long electrode length. Using this setup, conductivities down to 10^{-10} S/cm can be measured (b).

long electrode length in a compact design. This long electrode enables the transport of high currents and a good signal-to-noise ratio. A simple four-point measurement setup finally allows the determination of conductivity with high accuracy. The lateral method to measure the conductivity is important because doped organic layers are too conductive to create a significant voltage drop in thin films (~100 nm) necessary for a voltage measurement. A useful layout consists of several resistors with a channel length L between the contact pads of 20–80 µm and a width of the stripes of 100 mm. The thickness d of the organic material can be determined *in situ* by a quartz-crystal monitor or with a profilometer. The resistance R of the setup is measured *in situ* and the conductivity σ is calculated using Equation 9.3:

$$\sigma = 1/R * L/Wd \tag{9.3}$$

9.3.2 Thermoelectric Effects (Seebeck Effect)

When a temperature difference is applied to the contact interfaces of a metal/semiconductor/metal structure, a voltage difference can be observed between the two contacts. This so-called Seebeck effect was discovered in 1821 by Thomas Johann Seebeck. The thermoelectric voltage has its origin in the different densities of charge carriers at the edges of the semiconductor, causing an electron diffusion current from the "hot" (2) to the "cold" (1) side. Especially in the case of an electrically doped semiconductor, the remaining ionized dopants lead to an electric field, causing a backward drift current. In steady state, this can be observed as the Seebeck voltage $V_{See} = V_1(T) - V_2(T)$. The so-called Seebeck coefficient S at the temperature T is defined by Equation 9.4:

$$S(T) = \lim_{\Delta T \to 0} \frac{V_1(T,\Delta T) - V_2(T,\Delta T)}{\Delta T} = \lim_{\Delta T \to 0} \frac{V_{Sec}(T)}{\Delta T} \tag{9.4}$$

Here, $\Delta T = T_2 - T_1$ denotes the temperature difference between the two electrodes, with $T_2 > T_1$, $T_2 = T + \Delta T/2$ and $T_2 > T_1$, $T_1 = T - \Delta T/2$. The sign of the Seebeck coefficient indicates the conduction type of the semiconductor: $S < 0$ for electron and $S > 0$ for hole transport. Since the charge carrier transport in disordered organic semiconductors takes place by hopping via localized states commonly assumed to have a Gaussian distribution of energies (Bässler 1993), the electrical conductivity is a complicated function of the charge carrier density $n(T)$ and the mobility $\mu(n,T)$, shown in Equation 9.5:

$$\sigma(T)\,dE = E \times n(T) \times \mu(n,T)\,dE \tag{9.5}$$

or more generally, in Equation 9.6:

$$\sigma(T) = \int \delta\sigma(E,T)\,dE \tag{9.6}$$

Here, $\delta\sigma(E,T)$ is the differential conductivity of the states in the interval $[E, E + \Delta E]$ within the DOS of the material. Fritzsche (1971) derived a general expression associating the Seebeck coefficient S with the differential conductivity by Equation 9.7:

$$S(T) = -\frac{1}{eT}\int (E - E_F(T))\frac{\delta\sigma(E,T)}{\sigma(T)}\,dE = -\frac{E_\sigma(T) - E_F(T)}{eT} \tag{9.7}$$

with

$$E_\sigma(T) = \int E \frac{\delta\sigma(E,T)}{\sigma(T)}\,dE \tag{9.8}$$

denoting an average transport level. The integration extends over the entire energy range. Equation 9.7 is valid regardless of the prevailing transport mechanism (e.g., hopping or band transport) and assumes only the application of Fermi–Dirac statistics to the system. Thus, the final expression for $S(T)$ depends on the actual DOS in which the transport takes place.

Assuming, for example, electron transport with a sharp energy distribution around E_μ several $k_B T$ above the Fermi level, the Seebeck coefficient is given by Equation 9.9:

$$S(T) = -\frac{k_B}{e}\left(\frac{E_\mu - E_F(T)}{k_B T} + A\right) \tag{9.9}$$

Here, A is the so-called scattering constant and depends on the transport mechanism. For instance, considering transport via a discrete level, the scattering constant becomes zero.

Usually, charge carrier transport in organic semiconductors is described by hopping via a Gaussian distribution of localized states (Bässler 1993). The actual states contributing to transport are further distributed around an average transport level. Effectively, this can be expressed by a differential mobility function $\delta\mu(E,T)$ depending on the molecular properties and morphology of the material under investigation. Usually, $\mu(n,E,T)$ is unknown and the derivation of an exact expression for the Seebeck coefficient $S(T)$ by Equation 9.6 is not possible. However, for characterizing conductivity doping of organic semiconductors, commonly, Equation 9.8 is applied ($A = 0$) treating E_μ as the effective transport level. Thus, the Fermi level shift due to conductivity doping can be directly deduced by measuring the Seebeck coefficient. Nevertheless, conclusions about the energetic distribution of the actual transporting states within the DOS are hard to draw.

Sometimes, the charge carrier density is estimated in an analogous way to the classical semiconductor theory assuming the Maxwell–Boltzmann approximation (Equation 9.10):

$$n(T) = N_\mu \, \exp\!\left(\frac{e}{k_B} S(T) \right) \tag{9.10}$$

Here, N_μ is the effective density of states at the transport level E_μ, usually assumed to be equal to the molecular density of the matrix material. However, the real value of N_μ is unknown and for a comprehensive understanding one has to go beyond the simple assumption of only one transport energy.

In early work, temperature-dependent Seebeck measurements were used for characterizing the charge carrier transport in organic semiconducting thin films, that is, in order to determine the conduction type (electrons or holes) of the organic material (Schlettwein et al. 1994, Meyer et al. 1995) or to check whether the conductivity is intrinsic or dominated by impurities (Böhm et al. 1997).

In 1991, intentional doping of polyacetylene films with iodine was investigated by Seebeck and conductivity measurements by Zuzok et al. (1991). The polymer films were doped by exposing them to vapor of iodine up to saturation, resulting in 0.8%–14% iodine-doped films. The Seebeck coefficient decreased at room temperature from 120 to only 20 μV/K (p-type behavior) upon doping and decreases strongly with decreasing temperature toward zero at $T = 0$.

In 1998, Pfeiffer et al. presented a detailed conductivity and Seebeck study on vanadyl-phthalocyanine (VOPc) intentionally doped with the strong acceptor molecule 2,3,5,6-tetrafluoro-7,7,8,8-tetracyanoquinodimethane F_4-TCNQ by vacuum cosublimation (Pfeiffer et al. 1998). The measured Seebeck coefficients at different doping ratios are shown in Figure 9.11a versus the film temperature. Increasing the doping ratio from 1:520 to 1:50 provides a decrease of S by more than 200 μV/K. The Seebeck coefficients are positive, indicating p-type conduction that confirms the transport of holes via matrix molecules rather than electron hopping via dopants. A similar study concerning n-type doping of NTCDA with BEDT-TTF was published by Nollau et al. (2000) shortly thereafter. In this work, the Seebeck coefficients were negative and thus unambiguously indicated electron transport (see Figure 9.11b). The Seebeck coefficient decreased by more than 600 μV/K upon conductivity doping, but was still two times higher compared with the p-doped system VOPc:F_4-TCNQ even at a high doping ratio of 1:45. The doping effects can directly be seen in the enhancement of the conductivity. Doping of 1:45 BEDT-TTF into NTDCA at room temperature improves the conductivity by one order of magnitude (to 1e-6S/cm), whereas doping of F_4-TCNQ into VOPc (1:50) yields an improvement by two orders of magnitude (to 2e-4 S/cm), each compared to the lowest doped sample. On

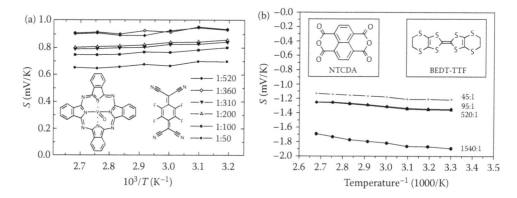

FIGURE 9.11 Seebeck coefficients versus temperature for varied doping concentrations of F_4-TCNQ doped into VOPc (a) and BEDT-TTF into NTCDA (b). (Taken from M. Pfeiffer et al., *Appl. Phys. Lett.* 73, 1998: 3202–3204; A. Nollau et al., *J. Appl. Phys.* 87, 2000: 4340–4343.)

the p-side, similar results were reported by Männig et al. in 2001 for p-type doping of zinc-phthalocyanine (ZnPc) with F_4-TCNQ (Maennig et al. 2001). In this case, an increase in the molar doping ratio from 0.002 to 0.02 decreases S from about 800 μV/K to only 400 μV/K accompanied by an increase of the conductivity by two orders of magnitude. Further, Maennig et al. demonstrated similarly strong doping effects for both amorphous (obtained by substrate cooling to −100°C during the deposition) and polycrystalline ZnPc (deposited at room temperature), differing only in the absolute values of σ and S.

As described in Section 9.2.3, precursor compounds are a promising material class providing air-stable n-dopants. For instance, conductivity doping of the fullerene C_{60} by AOB was shown by a comprehensive conductivity and Seebeck study in 2006 by Li et al. (2006). Light-activated AOB yields a decrease of S from approximately −1.2 mV/K (at 0.5 mol%) to −0.8 mV/K (at 2.5 mol%) in C_{60}. The conductivity reaches 3e-2 S/cm (5.6 mol%).

Most recently, Menke et al. published conductivity and Seebeck data of C_{60} n-type doped with the air-sensitive molecules tetrakis(1,3,4,6,7,8-hexahydro-2H-pyrimido[1,2-a]pyrimidinato)ditungsten(II) ($W_2(hpp)_4$) and tetrakis(1,3,4,6,7,8-hexahydro-2H-pyrimido[1,2-a]pyrimidinato)dichromium(II) ($Cr_2(hpp)_4$) (Menke et al. 2012). As seen in Figure 9.12b, for these dopants, the Seebeck coefficient decreases from about −550 μV/K to only just above −100 μV/K, and thus 300 μV/K lower than for the system C_{60}:AOB. Hence, a much stronger doping effect compared to AOB is expected, which is directly confirmed by conductivity data. Very high values of up to 4 S/cm are achieved at high molar ratios (Figure 9.12a), exceeding the maximum conductivity of C_{60}:AOB films previously published by Li et al. by two orders of magnitude.

Seebeck measurements on the prototypical doped system pentacene:F_4-TCNQ were published by Harada et al. in 2010 and included variations of the doping concentration from 0.5 to 10 mol%. Here, the positive Seebeck coefficient (hole conductive) decreased from just above 400 μV/K to around 150 μV/K accompanied by an increase in conductivity up to 4.1e-2 S/cm at 2 mol%. For higher doping ratios, the conductivity again decreases, in this case by one order of magnitude owing to a large mobility decrease.

The last example shows the general problem of unambiguous interpretation of the obtained σ and S data. For a comprehensive theoretical understanding, an independent

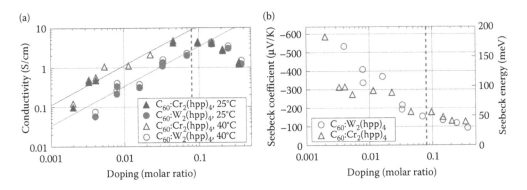

FIGURE 9.12 Conductivity (a) and Seebeck coefficients (b) of the fullerene C_{60} n-type doped by $W_2(hpp)_4$ of $Cr_2(hpp)_4$. (Taken from T. Menke et al., *Appl. Phys. Lett.* 100, 2012: 093304.)

measurement of the actual charge carrier density and mobility is necessary. Commonly, to overcome these issues, the mobility is determined by field-effect measurements on OFETs, whereas the charge carrier density is estimated by Equation 9.9 and temperature-dependent measurements of the Seebeck coefficient (Pfeiffer et al. 1998, Nollau et al. 2000, Maennig et al. 2001, Menke et al. 2012). However, the data are often not consistent, in particular, in terms of a description based on deep or shallow dopant levels (Pfeiffer et al. 1998, Nollau et al. 2000). Further effects, such as additional unintended doping caused by impurities, intrinsically present trap states and/or morphology changes due to incorporation of dopant molecules into a matrix crystal (Pfeiffer et al. 1998, Nollau et al. 2000, Harada et al. 2010, Kleemann et al. 2012a,b), are supposed to cause the inconsistencies. Up to now, only a few attempts were published aiming a clarification and more comprehensive theoretical description (Maennig et al. 2001). The scientific discussion is still going on, but mainly hampered by the different properties (e.g., morphology changes) of the various material systems as well as by a missing common theory on the activation of molecular doping.

The main conclusion we have to draw is that for an unambiguous characterization of conductivity doping of organic semiconductors, additional experimental techniques such as UPS (Section 9.3.3), impedance spectroscopy (Section 9.3.4), or infrared spectroscopy (Section 9.3.5) are needed. A further interesting approach, which will not be covered here in detail, is the combination of Seebeck and OFET measurements within the same device. This is because by measuring S, it is feasible to resolve a shift of the Fermi level with a change in the gate voltage (von Muehlen et al. 2007, Pernstich et al. 2008). This concept could also be adapted to doped layers.

9.3.3 Energetic Alignment at Interfaces (UPS)

In this section, we want to briefly introduce UPS as a widely used technique aimed at the determination of the energetic alignment at metal/organic (m/o) and/or organic/organic (o/o) interfaces. Owing to the extreme surface sensitivity of the photoemission process, UPS allows for resolution of the electronic properties of m/o and o/o interfaces on a subnanometer scale.

Chapter 9

Photoelectron spectroscopy is based on the external photoeffect, first discovered by H. Hertz in 1887 and explained by *A.* Einstein in 1905 invoking the quantum nature of light. Illuminating a material with monochromatic light leads to the emission of photoelectrons if the photon energy $h\nu$ exceeds the work function *Wf* of the material. The maximum kinetic energy of the photoelectrons is given by Equation 9.11:

$$E_{\text{kin,max}} = h\nu - Wf \tag{9.11}$$

In practice, a whole spectrum of photoelectrons is emitted from the surface with a kinetic energy distribution characteristic of the material under investigation. The number of electrons with a certain kinetic energy E_{kin} directly correlates to the material-specific density of (occupied) states (DOS) at the binding energy E_B (Equation 9.12):

$$E_{\text{kin}} = h\nu - Wf - E_B \tag{9.12}$$

Schematically, this is shown for a metal and an organic semiconductor in Figure 9.13a. For a metal, the fastest electrons originate from states close to the Fermi level and in the case of a molecular semiconductor, from the HOMO. The actual distribution of kinetic energy, the measurement of which is the purpose of UPS, reflects directly the DOS of the material. For lower kinetic energies, the intensity increases strongly as a result of inelastic scattering effects that the electrons experience during their transit

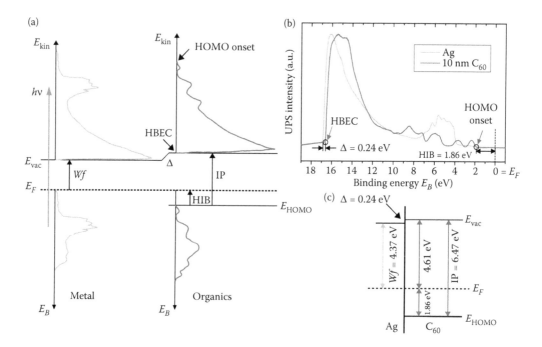

FIGURE 9.13 (a) Principles of ultraviolet photoelectron spectroscopy (UPS). (b) The UPS spectra of a 10 nm C_{60} thin film and the underlying Ag substrate and (c) the corresponding energy diagram are also shown. By stepwise deposition, the energy alignment directly at the interface can be determined.

through the bulk to the sample surface and thus overlap the signal originating from the DOS. At zero kinetic energy, the spectrum breaks abruptly at the so-called high binding energy cutoff (HBEC). Thus, the work function of a material can be derived from the width of the corresponding UPS spectrum (Equation 9.13):

$$Wf = h\nu - E_{HBEC} \qquad (9.13)$$

For a molecular semiconductor, the fastest electrons originate from the HOMO, seen in a solid thin-film UPS spectrum as the so-called HOMO onset, and defining the hole injection barrier (HIB) from the underlying metal substrate (HIB $= E_F - E_{HOMO}$). Consequently, the ionization potential of the organic material is determined by Equation 9.14:

$$IP = h\nu - (E_{HBEC} - HIB) \qquad (9.14)$$

By stepwise deposition of two materials, the evolution of the HOMO onset and the HBEC position allows the determination of the energy level alignment at this interface, that is, the appearance of an interface dipole Δ by a shift of the HBEC (vacuum level) or level bending effects by a shift of the HOMO onset. In Figure 9.13b, the UPS spectra of 10 nm C_{60} as well as of the underlying Ag substrate are shown as exemplary examples (dipole 0.24 eV).

The mean free path λ of electrons within a solid film depends strongly on the kinetic energy of the electrons and the density of the solid. For UPS, λ typically is in the range of 2–3 monolayers (1–2 nm) (Hüfner 2003). Owing to this low photoelectron escape depth, UPS is an extremely surface-sensitive method and a clean sample surface is mandatory. Therefore, UPS measurements are carried out under UHV conditions with typical base pressures $<10^{-10}$ mbar preventing a premature adsorption of contaminants. The photoemission spectra are commonly measured by combination of a hemispherical deflection analyzer with an electron detector (e.g., channel electron multiplier), reaching a typical UPS energy resolution of about 100 meV at room temperature. He gas-discharge tubes (HeI = 21.22 eV) or synchrotron radiation sources are usually used as excitation sources.

Before concentrating on molecular doping, the main issues regarding energy level alignment at intrinsic m/o and o/o interfaces will be briefly touched. It has been shown by numerous studies that interface dipoles arise at m/o and even at o/o contacts rather than a strict vacuum level alignment, crucially determining the hole or electron injection barriers at these contacts. Reviews describing the electronic properties of various m/o and o/o interfaces causing the formation of dipoles and thus influencing device performances are given in Ishii et al. (1999), Hill et al. (2000), Kahn et al. (2003), Koch (2007), and Braun et al. (2009). Possible processes described in these references include

i. Actual electron transfer from the organics to metal or vice versa
ii. Chemical reaction of molecules with the metal surface changing the electronic structure directly at the interface
iii. Formation of interface (gap) states
iv. Interface rearrangement due to reduction of the electronic tailing of the metal into the vacuum by presence of organic molecules (push-back effect)

Chapter 9

In the simplest picture, a dipole induced by charge transfer occurs if either Wf_{met} < EA (pinning of E_F near to the LUMO) or Wf_{met} > IP (pinning of E_F near to the HOMO). For IP < Wf_{met} < EA, vacuum level alignment is possible. However, in this case, processes (ii) through (iv) often control the level alignment. In addition to adjustment of the Fermi level by chemical reactions inducing gap states concentrated at the m/o interface, for example, Alq3 on Mg or Al (Shen 2001), direct metal-induced gap states (MIGS) were suggested to explain the level alignment at such interfaces with physisorbed molecules (Vazquez et al. 2004).

Owing to the low density of intrinsic charge carriers in undoped organics and weak intermolecular interactions, intrinsic o/o contacts are prone to align without the formation of interface dipoles. A compilation of various o/o interfaces studied by UPS proving evidence for this assumption was published by Hill et al. in 2000. However, exceptions were observed for materials with (partial) charge transfer due to nonoverlapping gaps, for example, an interface of CuPc (IP = 4.8 eV) and $F_{16}CuPc$ (EA = 5.2 eV) (Tang 2007).

Molecular doping was first investigated using UPS by Blochwitz et al. in 2001 for p-type-doped ZnPc thin films evaporated on Au and ITO substrates. A clear shift of the Fermi level toward the HOMO onset of ZnPc thin films doped by the strong acceptor molecule F_4-TCNQ (molar ratio of 1:30) was shown (Figure 9.14) for both Au and ITO. Level bending as well as a very narrow depletion region (<5 nm) at the contacts as known from inorganic semiconductor theory (Schottky contacts) were observed by means of thickness-resolved UPS measurements. Molecular doping strongly reduced the final HIB to only about 0.21 eV, independent of the substrate; this indicates a strong increase of the charge carrier density in the doped thin film compared with the intrinsic layer. Shortly thereafter, similar results were independently published by Gao and Kahn (2001, 2002). These findings validate the previously published observation of p-doping in this system shown by Pfeiffer et al. (1998) by means of conductivity and Seebeck measurements.

Reducing and adjusting the final HIB independently of the contact material is one key effect of molecular doping. It avoids the above-described interface effects at intrinsic m/o and o/o interfaces and thus allowing for a controllable level alignment at such interfaces within complete devices. For instance, the substrate independence of the alignment was shown by Olthof et al. (2009a,b) for MeO-TPD films p-type-doped by

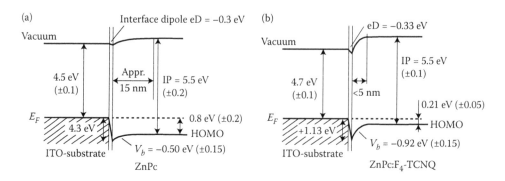

FIGURE 9.14 Energy level alignment at the contact between undoped (a) and doped (b) ZnPc and ITO measured by UPS.

0.04 molar ratio of F_4-TCNQ evaporated on sputter cleaned silver foil, ITO, or spin-coated PEDOT:PSS, respectively. In each case, a space charge region of approximately 5 nm with a final HIB of 0.48 eV was observed. As a consequence, it can be expected that hole injection from the contact into the doped layer is largely independent of the work function of the contact material. Moreover, by systematically varying the doping concentration, the position of the Fermi level in the gap can be tuned precisely. For illustration, the final HIB of MeO-TPD films p-doped by F_4-TCNQ on silver is plotted versus the doping concentration in Figure 9.15a. Additionally, recent experimental UPS data published by Tietze et al. in 2012 on MeO-TPD thin films doped by the p-dopant molecules F_6-TCNNQ and $C_{60}F_{36}$ are given for comparison. Although the latter dopants cause similar Fermi level shifts in MeO-TPD, the observed HIBs achieved by doping with F_4-TCNQ are significantly larger for medium and low doping concentrations (i.e., a lower doping effect by F_4-TCNQ). However, a stronger shift than expected from classical semiconductor physics (i.e., slope of k_BT in a log plot of the Fermi level shift vs. the doping concentration) as well as a kind of Fermi level pinning at high doping concentrations are characteristic features of molecular doping. This effect was also reported for other matrix:dopant systems, for example, for α-NPD:F_4-TCNQ (Gao and Kahn 2003), CBP:MoO_x and α-NPD:MoO_x (Kröger et al. 2009). The underlying physics is still a matter of scientific discussion and prevailing attempts at explanation will be briefly presented in Section 9.4.

For n-type doping, similar interface studies applying UPS were published, for example, for Cs doped into CuPc (Yan et al. 2001, Gao and Yan 2003b), NTCDA n-doped by PyB (Chan et al. 2006a,b), or more recently solution-based n-type doping of a polymer by air-stable dimers of rhodocene (Qi et al. 2012) (see Section 9.2.3). However, the

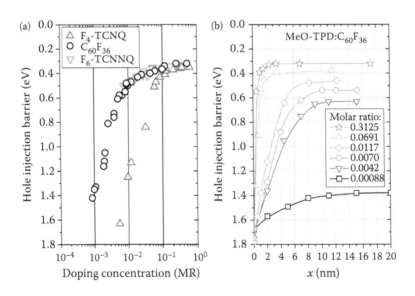

FIGURE 9.15 (a) Hole injection barriers of MeO-TPD thin films p-doped by three different dopants on silver. SCL formation at an Ag/MeO-TPD/MeO-TPD:$C_{60}F_{36}$(x/nm) interface varying the doping concentration over more than two orders of magnitude. x indicates the thickness of the p-doped MeO-TPD film. (b) Hole injection barrier level as a function of molar ratio. (Data are taken from S. Olthof et al., *J. Appl. Phys.* 106, 2009a: 103711; M. Tietze et al., *Phys. Rev. B* 86, 2012: 035320.)

LUMO is not accessible because UPS measures the density of occupied states. Hence, for investigations of molecular n-type doping the IPES is more suitable. IPES measurements are initiated by irradiation with an electron beam of defined kinetic energy E_i. If an electron falls down into an unoccupied state E_f during its transit through the material, a photon with the energy $E_i - E_f$ is emitted. Hence, the distribution of all unoccupied states of a material can be probed by the measurement of the corresponding whole photon spectrum. In comparison with UPS, the cross-section of the IPES process as well as its energy resolution is significantly lower (>300 meV).

In addition to the increase of the charge carrier density (and therefore conductivity) accompanied by a defined shift of the Fermi level toward the HOMO/LUMO (p-/n-doping) with rising doping concentration, another important feature of conductivity doping is the reduction of the space charge layer (SCL) width inside a doped organic layer close to a metal contact, as illustrated in Figure 9.15b. In this figure, the evolution of the Fermi level at an Ag/intrinsic(5 nm)/p-MeO-TPD(x/nm) contact is shown (rather than the direct Ag/p-MeO-TPD contact level bending since interface doping effects are thus avoided). The decrease in the SCL thickness is a crucial point, since the actual HIB experienced by the charge carriers directly at the m/o contact is still independent of the doping concentration, for example, 1.7 eV for Ag/p-MeO-TPD. However, for very high doping concentrations (molar ratio > 0.1) thin SCL thicknesses of less than 3 nm are obtained (Olthof et al. 2009a,b, Tietze et al. 2012). Consequently, the probability of hole injection from metal to organics via tunneling through this thin barrier is strongly enhanced, providing a quasi-ohmic contact as proven by current–voltage characteristics of diode devices (Walzer et al. 2007) or vertical single layer structures (Gao and Kahn 2003c). For low doping ratios (molar ratio < 0.001) much thicker space charge zones (tunnel barrier widths) are present (Figure 9.15b) resulting in a reduced enhancement of carrier injection compared to the intrinsic m/o contact (Gao and Kahn 2003c). Owing to the significant injection improvement, charge carrier transport within the state-of-the-art p–i–n OLEDs and solar cells is mainly limited by the intrinsic interlayer transport rather than by the contact of the p- and n-doped transport layers to the adjacent electrodes (Walzer et al. 2007, see also Section 9.5.1).

A complete picture of the energy level alignment within a p–i–n OLED based on UPS measurements was presented by Olthof et al. (2009a,b) (Figure 9.16). By analyzing all m/o and o/o interfaces within a red phosphorescent OLED stack using UPS at each layer deposition step, it was shown that doping the transport layers defines the built-in field of the device (defined as the vacuum level offset between the doped layers). This work showed that the built-in voltage arises mainly from the voltage drop across the intrinsic emission layer. Furthermore, this study confirmed that there is no common vacuum level over the whole device as it was generally believed in the past.

Finally, by thickness-resolved UPS measurements at m/o or o/o contacts, the doping efficiency, given by the ratio of ionized N_A^- to all acceptor molecules N_A, can be estimated by determination of the SCL thickness w and applying the Poisson equation under the abrupt approximation (Equation 9.15):

$$N_A^- = \frac{2\varepsilon \cdot \Delta V}{ew^2} \tag{9.15}$$

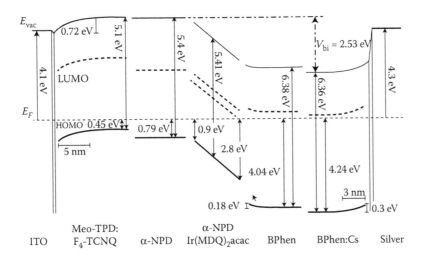

FIGURE 9.16 Schematic energy level diagram of a complete red phosphorescent OLED stack measured by UPS analyzing all m/o and o/o contacts contained in the device. (Taken from S. Olthof et al., *Phys. Rev. B* 79, 2009b: 245308.)

Here, ΔV is the built-in potential caused by level bending via the SCL, e the elementary charge, and ε the permittivity of the organic thin film. Typically, for molecular doping at higher molar ratios, low doping efficiencies of only a few percent have been reported, for example, 4% for MeO-TPD:F_4-TCNQ (Olthof et al. 2009a,b), <10% MeO-TPD:F_6-TCNNQ and MeO-TPD:$C_{60}F_{36}$ (Tietze et al. 2012) or <2% for CBP:MoO_3 (Kröger et al. 2009, Hamwi et al. 2009). The reasons for such low values are still under debate and will be briefly discussed in Section 9.4.

In summary, UPS is a powerful tool to analyze the energy level alignment at m/o and o/o interfaces using either intrinsic or conductivity-doped organic thin films. Molecular doping processes can be studied aiming at the determination of the Fermi level position and the doping efficiency. The latter can be further analyzed by impedance spectroscopy, and the amount of actual charge transfer from/to dopant molecules can be deduced by other optical techniques (UV/Vis, FTIR). These characterization methods will be presented in the following section(s).

9.3.4 Impedance Spectroscopy

Impedance spectroscopy is a standardized and well-established method to characterize doping in inorganic semiconductors. It relies on the principle that the width of a charge depletion zone can be evaluated from its capacitive response according to the Mott–Schottky rule. Adopting this method to doped organic semiconductors, however, turns out to be difficult owing to the large doping ratios typically used for organic semiconductors. In particular, considering a doping ratio of, for example, 4 mol%, a typical width of a charge depletion zone is expected to be <10 nm (Kleemann et al. 2010). Hence, the usage of MIS (metal–insulator–semiconductor) structures, as commonly used for such analysis, is not suited since the capacitance of the dielectric material has to be significantly larger than the capacitance of the depletion zone. Moreover, MIS structures suffer from the fact that interface states and deep trap states can drastically influence the depletion width making a quantitative analysis impossible (Meijer et al. 2001, Stallinga et al. 2001). A solution to this problem are p–i–n diodes with low to moderate dopant concentrations and extremely thin intrinsic interlayers. These devices can be used to study the formation of charge depletion zones (Kleemann et al. 2010, 2012b). Moreover, such diode structures avoid the presence of the organic/dielectric (or m/o) interface, an advantage over MIS devices. As shown by Kleemann et al. (2012b), the number of ionized donor/acceptor states can be determined with a high accuracy. This analysis is based on the Mott–Schottky rule, which predicts that the derivative of the squared inverse capacitance (d/dV ($1/C^2$), C depletion capacitance, V applied DC voltage) is inversely proportional to the number of ionized dopant states. Moreover, the shape of the Mott–Schottky curve ($1/C^2$ vs. V, see Figure 9.17) as well as the frequency dependence of the capacitance response can provide important information on the presence of deep trap states. In particular, only in the case where a linear relation between $1/C^2$ and V is observed in the Mott–Schottky plot, and the capacitance is frequency independent, can the contribution of deep trap states be excluded. An example evaluation is shown in Figure 9.17.

Under these conditions, Kleemann et al. were able to demonstrate that the doping efficiency in the host:dopant system Ir(piq)$_3$ (Tris(1-phenylisoquinoline)iridium(III)):NDP2 (a proprietary dopant of Novaled AG) can be larger than 60%, demonstrating doping activation energies smaller than k_BT. This means that more than every second hole generated on the matrix is free and contributes to the charge transport. These extremely large doping efficiencies are in line with FTIR measurements (Pfeiffer et al. 1999). However, they are contradicted by UPS investigations showing doping efficiencies of below 10% for the commonly used MeO-TPD:F$_4$-TCNQ hole transport layer (Olthof et al. 2009a,b, Tietze et al. 2012). Hence, the large doping efficiencies obtained by Kleemann et al. cannot be generalized and are presumably a consequence of the particular host dopant:system. Furthermore, it should be emphasized that rather large electric fields of ~1 MV/cm are

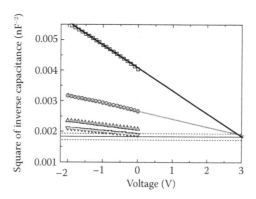

FIGURE 9.17 Mott–Schottky plot of a pin$^+$ junction consisting of Al/RE68:NDP2 (50 nm)/RE68 (7 nm)/RE68:NDN1 (16 wt%, 50 nm)/Al. The p-doping concentration is varied from 4 wt% (down triangle), 2 wt% (up triangle), 1 wt% (circle) to 0.5 wt% (square) p-doping concentration (n-doping concentration is 16 wt%). The slope of the plot is inversely proportional to the number of free charge carriers (ionized acceptors), which can be used to calculate the doping efficiency. (Adapted from H. Kleemann, B. Luessem, K. Leo, *J. Appl. Phys.* 111, 2012b: 123722.)

present in the p–i–n diodes during the capacitance measurements. This might support the dissociation of CT complexes into free charge carriers and ionized dopant states.

9.3.5 Optical Characterization Methods

Doping effects in organic layers can be studied by a variety of optical methods to obtain important information regarding new energy levels and charge transfer. For example, the absorption of doped organic films measured by UV/Vis spectroscopy often shows additional subgap features at longer wavelengths (Yim et al. 2008, Solomeshch et al. 2009, Koech et al. 2010, Burschka et al. 2011). These additional peaks can be explained by the absorption of charged dopant or host molecules indicating that a charge transfer from the dopant to the host has taken place. Thus, UV/Vis spectroscopy is a fast method to test for successful doping in a wide range of matrix:dopant material combinations. Optical absorption features, coming from new hybrid states, may also show up in the IR (i.e., at lower energy). This will be discussed in Section 9.4 in more detail.

Doping and formation of ionized dopant states can be measured by FTIR (Yim et al. 2008). FTIR has also been used to quantify the strength of the charge transfer in doped organic layers. For example, it has been shown that the position of the absorption peak of the CN-stretching mode of F_4-TCNQ depends on the charge state Z of the dopant (Pfeiffer 1999). For this measurement, the IR absorption band change of the vibration mode under investigation from the neutral molecule to a single negatively charged molecule (in case of p-dopants) must be known, which can be achieved by comparison to other measurements (as, e.g., spectro-electrochemistry on intentionally mixed acceptor–donor systems or x-ray diffraction) and quantum chemical calculations of the molecular structures in the neutral and charged states. For example, this procedure for F_4-TCNQ is described by Chapell et al. (1981), Meneghetti and Pecile (1986), and Ploennigs (1999). In particular, it has been shown that the degree of charge transfer strongly depends on the ionization energies of the dopant and host molecule (Walzer et al. 2007). One example is the full charge transfer achieved for the combination ZnPc:F_4-TCNQ, where the electron affinity of the dopant exceeds the ionization potential of the matrix (cf. Table 9.1).

However, a complete charge transfer between host and dopant does not necessarily imply that one free electron is generated per dopant molecule. Owing to the strong Coulomb interactions and strong binding energies of charge transfer complexes in organic materials, the hole can still be bound by Coulomb forces to the ionized dopant molecule.

Table 9.1 Ionization Potential of Different Matrix Materials and the Degree of Charge Transfer (Z) for Different Dopant:Matrix Combinations

Matrix	ZnPc	ZnPc	m-MTDATA	TPD	MeO-TPD
Dopant	F_4-TCNQ	TCNQ	F_4-TCNQ	F_4-TCNQ	F_4-TCNQ
IP (eV)	5.1	5.1	5.1	5.4	5.1
Z	1	0.2	1	0.64	0.74

Source: Adapted from K. Walzer et al. *Chem. Rev.* 107, 2007: 1233–1271.

Note: The electron affinity of F_4-TCNQ is 5.4 eV and of TCNQ 4.2 eV. If the electron affinity of the dopant exceeds the ionization potential of the host, a charge transfer is efficient.

Chapter 9

9.4 Molecular Doping Mechanisms: Current Views

The theory describing electrical doping of inorganic, covalently bound semiconductors is well established. Doping is described by shallow (or deep) impurity levels in the band gap arising from the perturbation of the single crystal structure. Additional charge carriers are thermally activated to the respective transport band edges E_C (n-doping) or E_V (p-doping). However, for amorphous molecular semiconductors, the situation becomes more complicated owing to their low permittivities and thus strong Coulomb coupling effects (formation of Frenkel excitons). Hence, charge carriers are transported by hopping via localized states (sites) rather than via extended bands (Bässler 1993). Furthermore, for organic thin films, the transport properties are strongly influenced by the morphology (e.g., polycrystalline vs. amorphous), resulting in effective charge carrier mobilities μ varying over many orders of magnitude for different materials, for example, $\sim 10^{-1}$ cm^2/Vs (C_{60}) versus $\sim 10^{-5}$ cm^2/Vs (MeO-TPD). Furthermore, μ is prone to be charge carrier concentration dependent, which complicates the description of a doped organic semiconductor even more. However, despite the material-specific uncertainties, the conductivity usually increases superlinearly with the doping concentration (Pfeiffer et al. 1998, Maennig et al. 2001).

At this point, the actual charge transfer (CT) process between a matrix and a dopant molecule as well as the release of a free carrier (polaron) from a created CT state are not completely understood. Hence, in this section, we will briefly summarize the partly contradictory prevailing attempts at explaining the molecular doping process.

Molecular doping may be considered a two-step process: (i) formation of a CT state and (ii) dissociation of the CT state into an ionized dopant molecule and a free charge carrier, for example, for p-doping of a matrix M by acceptor molecules A as shown in Equations 9.16 and 9.17:

$$MMAM \Leftrightarrow M[M^+A^-]M \tag{9.16}$$

$$M[M^+A^-]M \Leftrightarrow MM^+A^-M \Rightarrow M^+MA^-M \tag{9.17}$$

In the simple and commonly accepted picture, step (i), as shown in Equation 9.16, is possible if the LUMO of the acceptor lies below the HOMO of an adjacent matrix molecule (p-doping, see Figure 9.2), which can lead to high CT ratios, for example, 1 for ZnPc doped by F_4-TCNQ (Section 9.3.5). Alternatively, the formation of a new dopant molecule by hybridization of one matrix molecule with one original acceptor molecule has been proposed by Salzmann et al. in 2012 with the example of pentacene p-doped by F_4-TCNQ (Salzmann et al. 2012). This will be discussed shortly.

Assuming the simple charge transfer picture between a matrix and an original dopant molecule, Mityashin et al. (2012) calculated the charge carrier generation efficiency for the prototypical system pentacene:F_4-TCNQ by DFT, considering the probabilities of generation, recombination, and dissociation of the CT states taking various electrostatic interactions between host and dopant molecules into account. As a consequence, a minimal dopant concentration (1%) is necessary for CT dissociation, that is, for overcoming the Coulomb barrier. Above this boundary, the (still low) generation efficiency increases with increasing doping concentration and thus enhances the conductivity superlinearly. For lower doping ratios, the generation efficiency and thus the doping efficiency is zero. However, incontrovertible experimental evidence is still missing, partially because of the complexity of reliably preparing layers with such low doping levels, and partially because of the complex polycrystalline layer growth of pentacene.

An extended photoemission and theoretical study on MeO-TPD doped by two different acceptor molecules was published by Tietze et al. in 2012. In this work, the strong Fermi level shift at a certain concentration, as measured by UPS (see Section 9.3.3 and Figure 9.18a), is explained by intrinsically present gap (trap) states with a density of approximately 3.5×10^{18} cm^{-3} (note that it is commonly assumed that trap densities in organic semiconductors are much higher than in inorganic semiconductors). At lower doping ratios, doping fills up these traps. Free charge carriers are generated only when the doping concentration exceeds the trap density. Tietze et al. calculated the Fermi level shift as well as the doping efficiency p/N_A by numerically solving the neutrality equation using a semiclassical approach (i.e., assuming thermal activation of dopants effectively described by an acceptor level E_A in the gap, a Gaussian trap distribution, and a Gaussian or exponential DOS). The results are plotted in Figure 9.18 (continuous lines) and show a very good agreement with the UPS measurements. In addition, the UPS results of MeO-TPD films doped by F$_4$-TCNQ published by Olthof et al. (2009a,b) are also shown in Figure 9.18. Apparently, the doping efficiencies are lower for F$_4$-TCNQ doped films compared with those for C$_{60}$F$_{36}$ and F$_6$-TCNNQ. This can be understood by the presence of additional trap states introduced by the F$_4$-TCNQ dopant molecules, which is supported by a similar calculation (dashed lines). Hence, in order to shift the Fermi level significantly toward the HOMO onset, a higher doping concentration of F$_4$-TCNQ molecules is required. However, for each p-dopant, the doping efficiency

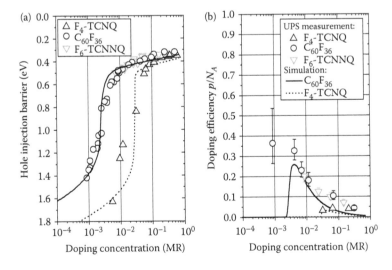

FIGURE 9.18 Fermi level shift (a) and doping efficiency (b) versus the doping concentration (molar ratio) of p-doped MeO-TPD thin films measured either by UPS (symbols indicate the used dopant) or calculated (lines) by a semiclassical approach. (Data are taken from M. Tietze et al., *Phys. Rev. B* 86, 2012: 035320; S. Olthof et al., *J. Appl. Phys.* 106, 2009a: 103711.)

Chapter 9

decreases drastically to only a few percent with increasing doping concentration and the Fermi level saturates for molar ratios >0.1. Tietze et al. attributed this phenomenon to the formation of a deep acceptor state, describing effectively the dissociation probability (step (ii) described in Equation 9.16) of a matrix:dopant CT state within MeO-TPD, that is, the equilibrium between the release of charge carrier from a CT state and the again capture by an isolated ionized dopant molecule. Owing to the low permittivity of organic materials, the overall carrier generation probability is low, depending on the specific material and decreasing with increasing doping concentration (Tietze et al. 2012). This hypothesis is in clear contradiction to the proposals from Mityashin et al., but supported by UPS measurements of Olthof et al., since doping the same dopant (F_4-TCNQ) into different matrix materials at a (high) doping ratio shows different HIBs for each matrix (Olthof et al. 2009a,b). Thus, we can conclude that once an efficient charge transfer (step (i)) is achieved by an appropriate molecular level adjustment, the whole doping process is mainly limited by the dissociation step (ii).

In 2004, Gregg et al. investigated this issue by defining a concentration-dependent acceptor level E_A and calculating the conductivity with classical semiconductor equations. They argue that owing to the formation of local CT states and the presence of highly polarizable dopants, the low dielectric constant of the intrinsic system increases with increasing doping concentration, which is effectively described by a decreasing E_A resulting in a superlinear increase of the conductivity. However, a systematic experimental study (UPS, Seebeck) validating this proposal is still missing.

Beside the number of (free or mobile) charge carriers created by doping, the prevailing transport mechanism (hopping) crucially determines the actual conductivity of a doped thin film, that is, the actual mobility μ of created free charge carriers. In organic semiconductors, μ is a complicated function of the carrier density, temperature, and electric field strength. This issue is not included in the studies presented earlier. However, in 2003, R. Schmechel published a comprehensive calculation applying a mobility function $\mu(E)$ proportional to the hopping rate (neE) to determine the actual conductive states (around an effective transport level) in the DOS depending on the doping concentration (Schmechel 2003). The hopping rates were based on the Miller–Abraham model (Miller and Abrahams 1960). The calculation was compared to Seebeck and conductivity measurements on ZnPc:F_4-TCNQ previously published by Maennig et al. (2001), where the transport was explained by a percolation model. It turns out that deep trap states with a density of just below 10^{18} cm^{-3} had to be considered for good agreement between theory and experiment. Qualitatively, this was already suggested by Maennig et al. because the conductivity of the ZnPc:F_4-TCNQ films decreases relatively quickly going from medium to very low doping ratios. Tietze et al. explained the strong Fermi level shift of doped MeO-TPD films measured by UPS (Olthof et al. 2009a,b, Tietze et al. 2012) in the same manner. Therefore, trap (gap) states introduced intrinsically or by doping seem to strongly affect the properties of molecular conductivity doping for various material systems. However, for clarification, more experimental evidence is necessary.

So far, we have assumed that organic dopant molecules and host molecules undergo a more or less complete charge carrier transfer (step (i)), creating charged host molecules. This surely happens for many host:dopant combinations, especially for those with matching energy levels between (in the p-doping case) the HOMO of the host and the LUMO of the dopant. However, for some material systems, an alternative explanation of the first step of doping may be more appropriate. Such examples are given in the work of the Koch group in Berlin (see, e.g., Salzmann et al. 2012). The authors state that a complete charge carrier transfer should yield singly occupied states in the gap of the intrinsic semiconductor host (polaron states close to the Fermi energy), which is not the case. This argument is supported by UPS measurements on p- and n-doped organic layers.

Thus, Salzmann et al. developed a new interpretation of the initial charge transfer process, called the hybridization model. In this model, complete charge transfer does not take place, but rather an intermolecular orbital hybridization between the host HOMO and the dopant LUMO occurs, leading to the formation of a doubly occupied bonding state and an empty antibonding state. This is substantiated in the paper by experimental studies on pentacene doped with F_4-TCNQ and two other polymeric hosts, as well as by DFT calculations of the new hybrid states. In fact, the bonding state ends up being below the HOMO of the host material (and hence not visible via UPS measurements), while the antibonding state is above the HOMO of the host and may be considered to be the energy level to which the Fermi level moves (and is pinned at) at higher doping concentrations. Salzmann et al. also found new absorption features in the IR range (i.e., at lower energies) supporting the creation of a new optical transition between the bonding and antibonding hybrid states. Previous FTIR measurements on the same system, pentacene:F_4-TCNQ, suggested an incomplete charge transfer (Blochwitz 2001). So, it seems the hybridization model better describes dopant:host systems where no complete charge transfer is possible. However, this model only describes the first step of the doping process, but does not explain how mobile charges on the host are subsequently created.

Future comprehensive studies with the use of experimental data from absorption measurements in the visible and the IR range, UPS studies, and others (and interpreted together) for a variety of host:dopant systems will be needed to reveal the real process or processes behind successful conductivity doping. Important information to correctly interpret measurements on doped organic semiconductor systems is still missing or not easily accessible. In particular, the exact field and doping concentration-dependent mobilities and the exact energy levels of the hybrid material M^+D^- (positively charged matrix and negatively charged dopant) surrounded by many more matrix molecules are not known. Thus, at this point, no comprehensive picture valid for a large variety of material combinations can be determined.

9.5 Applications of Conductivity Dopants

9.5.1 Application in OLEDs

Doping of organic layers allows the realization of high-efficiency OLED devices with low operating voltage and long lifetime. For example, in Figure 9.19, the operating voltage is reduced from 6 to 3 V at 1.000 cd/m², and at low luminance close to the thermodynamic limit of about 2.4 eV for this green OLED (Pfeiffer et al. 2002).

The high conductivity of the doped transport layers (typically 10^{-5}–10^{-4} S/cm) prevents virtually any voltage loss by carrier transport. Even for thicker layers of some 100 nm (see Figure 9.20), there is no voltage penalty. Thicker transport layers are needed for different reasons. One reason is the optical design of the organic stack, which is crucial to reach high efficiencies (see Chapter 12). This optical design allows

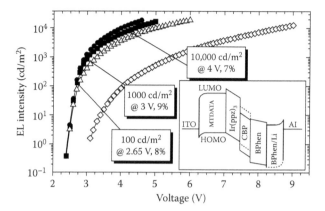

FIGURE 9.19   Electroluminescence intensity versus voltage for three different p–i–n OLEDs with doped transport layers compared to an undoped reference device (diamonds). (Taken from S.R. Pfeiffer et al., *Adv. Mater.* 14, 2002: 1633–1636.)

efficient outcoupling of the internally generated light (sometimes referred to as micro-cavity tuning).

Another benefit of using thicker transport layers is yield enhancement. Especially on larger-area devices as needed for lighting applications, a certain thickness of the first layer on the substrate is needed to smooth substrate defects and avoid electrical shorts. This typically requires hole transport layers be within the second optical maximum (for the emission zone) of around 200 nm. It was shown that the device yield, particularly for larger-area devices, is enhanced with thicker HTL or ETL (Blochwitz-Nimoth et al. 2010). Conductivity doping of the transport layers allows such high layer thickness without operating voltage penalties.

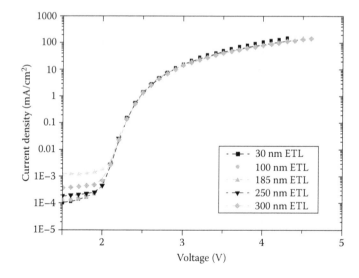

FIGURE 9.20   *I–V* curves of p–i–n OLEDs with different ETL thicknesses. All other layers are kept constant. The *I–V* curves are identical (except for some deviation at very low currents). (Adapted from J. Birnstock et al., *SID 08 Digest* 40(1), 2008: 822–825.)

Optimal stack alignment is even more critical when OLEDs consist of several units stacked on top of each other (Figure 9.21). See Chapter 16 for a detailed discussion of tandem or stacked OLEDs. One particular advantage of p–i–n OLEDs is that, since each p–i–n OLED starts from a p-type conductivity-doped HTL and ends with an n-type-doped ETL, one can just simply stack one p–i–n OLED on top of another p–i–n OLED, with possible addition of a thin stabilization layer to enhance OLED lifetime (these layers are not shown in Figure 9.21). The created n–p junction inside the device serves as a charge generation layer (Hatwar et al. 2009, Kleemann et al. 2010).

Furthermore, conductivity-doped layers are compatible with internal outcoupling materials (e.g., the Novaled material NET61) without any performance or lifetime loss. The outcoupling layer is sandwiched inside an n-doped ETL and does not hinder charge carrier transport as might the incorporation of noncharge-carrying nanoparticles, for instance (Figure 9.21). The rough layer formed by the regularly evaporated organic material NET61 also enhances the optical output of the device by scattering and preventing plasmon losses at the adjacent cathode. Combining internal and external outcoupling (e.g., microlens array) allows the production of a large-area OLED device with 60 lm/W at 1000 cd/m² with and extrapolated lifetime (50%) of 90,000 h (Loeser et al. 2012). The good transport properties even allow high efficiencies at very high luminances. So, the power efficiency is only reduced by about 20% at a brightness of 9000 cd/m² when compared with that at 1000 cd/m² (Figure 9.22). Recently, Reinecke et al. have shown that using several flat outcoupling methods, a carefully chosen emitter layer arrangement as well as conductivity doping of transport layers allows for a white OLED with 90 lm/W

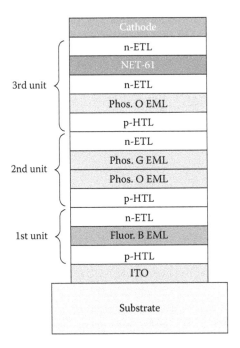

FIGURE 9.21 Stacked p–i–n doped device with an outcoupling layer. This rough layer allows the enhancement of trapped light either from surface plasmons as well as wave guided in the OLED. The doped layers around the outcoupling layer allow an undisturbed carrier transport. (Adapted from F. Loeser et al., *J. Photon. Energy* 2, 2012: 021207.)

Chapter 9

(Reineke et al. 2009), this efficacy being higher than the power efficacy of fluorescent tubes (see Chapter 12 for a detailed discussion of optical outcoupling).

Another important advantage of transport layer doping is the improvement of carrier injection from the electrodes. Without doping, the work functions of the different electrodes must be tuned to the organic material. In contrast to tuning the electrodes (see Chapter 4 for a discussion of transparent electrodes and Chapter 16 for a discussion of opaque electrode design), conductivity doping allows a wide range of electrode materials on the p and the n side to be suitable and is therefore compatible with virtually all process requirements (backplane types, electrode materials). Figure 9.23 shows the IVL data for OLEDs on standard ITO and on ZAO (aluminum-doped zinc oxide) anodes. There is no difference in carrier injection into the OLED between the two anode materials, as demonstrated by similar p–i–n OLED performance.

The doping concept allows the optimization of OLED devices by tuning the charge carrier balance with respect to the desired application, including, for example, blocking layers. One example of such an optimization process based on the p–i–n approach is shown by Huang et al., where a careful selection of materials for ETL and n-dopants allowed the simultaneous improvement of lifetime and efficiency (Huang et al. 2012).

In today's OLEDs, blocking or interlayers are usually used between the charge carrier transport layers and the emission layer zone (see the discussions in Chapters 5, 6, 7, and 8). This is for reasons of minority charge carrier blocking (to avoid them leaving the emission zone), exciton confinement (to avoid excitons leaving the emission zone), or to avoid triplet exciton quenching at the lower triplet levels of the transport host molecules.

If conductivity doping is used in the transport layers in principle, no additional blocking layers need to be introduced unless the following is believed to take place: (i) speculated migration of dopant into the first few nanometers of emitter layer, or (ii)

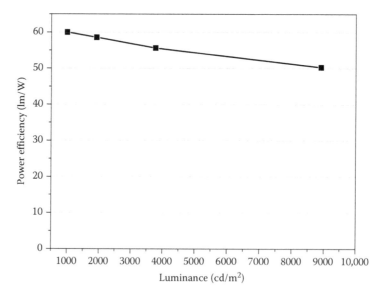

FIGURE 9.22 Power efficiency of a stacked PIN OLED (three units) with internal and external outcoupling. The OLED can be operated at very high brightness of 9000 cd/m² with an efficiency loss of only 20% compared to 1000 cd/m². (Adapted from F. Loeser et al., *J. Photon. Energy* 2, 2012: 021207.)

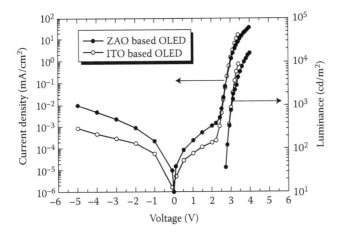

FIGURE 9.23 *I–V–L* curves of PIN OLEDs on ITO and ZAO substrates. The electrical and optical properties are comparable for both anode materials. (Adapted from C. May et al., *Thin Solid Films* 516, 2008: 4609–4612, doi: 10.1016/j.tsf.2007.06.014.)

a high charge carrier density at the interface between the emission layer and the respective doped layer. In any case, it is difficult to separate the effects of a blocking/interlayer inside an OLED. In many cases, such blocking layers are needed anyway to avoid charge carrier or exciton leakage or to avoid exciton quenching. Consequently, these considerations will usually lead to the introduction of a blocking/interlayer at at least one side of the emission zone. Such interlayers have usually very low/high energy levels, making it hard to efficiently dope them. So, these layers end up being thin undoped interlayers (usually thin enough for charges to tunnel through the low conductivity layer).

9.5.2 Applications of Conductivity Dopants in Other Electronic Devices

Doping can be used to realize or improve other devices in addition to OLEDs and organic solar cells. In particular, the control of the Fermi level and of the extension of the depletion zones in doped layers can be beneficial in realizing novel devices.

Organic Zener diodes are a good example of the use of dopants in electronic devices (Kleemann et al. 2010). Zener diodes are p–i–n diodes that show a reversible and defined breakdown at reverse bias. Zener diodes can be used for overvoltage protection, as voltage references, or as selection elements in passive matrix memory structures.

The current/voltage characteristic of an organic Zener diode is shown in Figure 9.24. The device consists of Al(100 nm)/RE68 (tris(1-phenylisoquinoline)iridium(III)) doped with the Novaled dopant NDP2 (varying doping concentration, 50 nm), 7 nm of intrinsic RE68, 50 nm of RE68 doped with the Novaled n-dopant NDN1 (16 wt%), and finally 100 nm of Aluminum. In the forward direction, the device shows normal diode characteristics with a turn-on voltage of approximately 2 V. In the reverse direction, a reversible breakdown can be observed. Most interestingly, the breakdown voltage depends on the doping concentration, that is, a higher p-doping concentration leads to a lower breakdown voltage.

The operation of the device in the reverse direction can be explained by tunneling of electrons from the HOMO of the p-doped region to the LUMO of the n-doped region,

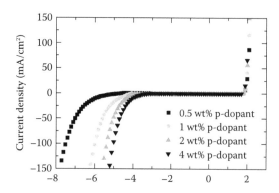

FIGURE 9.24 Operation of an organic Zener diode. The breakdown voltage can be tuned by the doping concentration. (Adapted from H. Kleemann et al., *Nano Lett.* 10, 2010: 4929.)

which generates free electrons and holes and leads to a drastic increase in the backward current (Kleemann et al. 2010). By increasing the doping concentration in the p-doped layers, the width of the depletion zones at the interface between the doped and the intrinsic layer is reduced, which shortens the tunneling distance and increases the electrical field drop across the depletion zones and the intrinsic layer and thus reduces the breakdown voltage.

Doping has also been used to improve the characteristics of organic field effect transistors, but, considering the importance of doping in inorganic metal–oxide–semiconductor field-effect transistors (MOSFETs), doping is surprisingly much less common in organic devices. Mostly, doping has been used to reduce the contact resistance at the source and drain contacts (Ante et al. 2011, Tiwari et al. 2011). A doped layer at the contacts to the organic semiconductor introduces a strong level bending and a thin depletion zone at the interface, which can be easily tunneled through by the injected charge carrier. Doping of the channel of the transistor has also been reported (Ma et al. 2008, Lee et al. 2011). For example, doping of a C60 transistor with an air-stable n-dopant led to an increase in the transistor stability (Oh et al. 2010, Wei et al. 2010). However, doped transistor channels usually lead to an increased off-current and lower transistor performance, which limits the use of doping for organic transistors.

9.6 Conclusion: Conductivity Dopants

OLEDs (and other organic optoelectronic devices) rely on charge carrier injection and transport. In order to achieve best performance (e.g., power efficacy in case of OLEDs or conversion efficiency in case of OPV), both these properties must be optimized to avoid electrical losses. Conductivity doping is one key method that can help achieve this purpose.

Conductivity doping is a process that introduces additives to the charge transport material with the goal to bring the Fermi level closer to the transport level of a semiconductor transport layer and hence increase the density of mobile charge carriers in such devices.

It has been shown in this chapter that conductivity doping has two main effects in an OLED:

- Enhancement of the charge carrier injection from the (mostly metal or degenerate oxides) contacts into the neighboring organic transport materials
- Enhancement of the transport through the transport layers due to increased charge carrier conductivity

In addition, many other benefits may arise from conductivity doping and can be used to improve the overall performance of the doped devices, such as charge carrier balance tuning, more uniform properties of the contact materials over large areas, increased production yield via the use of thicker charge carrier transport layers, optical cavity tuning to enhance light extraction or capture of OLEDs and OPV devices, respectively, and more facile use of stacked OLED and OPV architectures.

In this chapter, we presented today's understanding of the doping process. In a first step (at least partial), charge transfer between a host and a dopant molecule takes place. In this respect, doping efficiency seems linked to the energy difference between host HOMO and dopant LUMO (in case of p-type doping, vice versa for n-type doping). This process is followed by a partial release of holes (or electrons) from the host molecules, leading to an increase of mobile charge carrier density. This is only possible when the Coulomb attraction between ionized host molecules and dopant molecules is overcome. The detailed mechanism for free carrier generation is not quite clear yet. Several studies have been performed revealing single aspects of the doping process for certain material systems. However, a comprehensive and congruent picture of the doping process in organic molecules is still missing and will require additional research effort.

Today, doping is used in a wide variety of OLED display, OLED lighting, and OPV devices, including devices currently in mass production. Conductivity doping is established as one key technology in the domain of organic electronics. Today's work focuses on a deeper understanding of the doping process and the extension of conductivity doping to more OLED and OPV device structures and toward electronic devices using organic semiconductor materials.

References

A.N. Aleshin, S.R. Williams, A.J. Heeger, Transport properties of poly(3,4-ethylenedioxythiophene)/poly(styrenesulfonate), *Synthetic Metals* 94, 1998: 173–177.

F. Ante, D. Kälblein, U. Zschieschang, T.W. Canzler, A. Werner, K. Takimiya, M. Ikeda, T. Sekitani, T. Someya, H. Klauk, Contact doping and ultrathin gate dielectrics for nanoscale organic thin-film transistors, *Small* 7, 2011: 1186.

H. Antoniadis, D.B. Roitman, J.N. Miller, Organic Electroluminescent Device, Patent Application US 5719467, 1995.

E.F. Aziz, A. Vollmer, S. Eisebitt, W. Eberhardt, P. Pingel, D. Neher, N. Koch, Localized charge transfer in a molecularly doped conducting polymer, *Adv. Mater.* 19, 2007: 3257–3260.

H. Bässler, Charge transport in disordered organic photoinductors—A Monte Carlo simulation study, *Phys. Stat. Sol. B* 175, 1993: 15–56.

J. Birnstock, T. Canzler, M. Hofmann, A. Lux, S. Murano, P. Wellmann, A. Werner, Improved structures and materials to enhance device lifetime, *SID 08 Digest* 40 (1), 2008: 822–825.

J. Blochwitz, M. Pfeiffer, T. Fritz, K. Leo, Low voltage organic light emitting diodes featuring doped phthalo-cyanine as hole transport material, *Appl. Phys. Lett.* 73, 1998: 729–731.

J. Blochwitz, T. Fritz, M. Pfeiffer, K. Leo, D.M. Alloway, P.A. Lee, N.R. Armstrong, Interface electronic structure of organic semiconductors with controlled doping levels, *Org. Electron.* 2, 2001: 97–104.

J. Blochwitz, Organic light-emitting diodes with doped charge transport layers, PhD thesis, TU Dresden, 2001.

J. Blochwitz-Nimoth, O. Langguth, S. Murano, G. He, T. Romainczyk, J. Birnstock, PIN-OLEDs for active-matrix-display use, *J. SID* 18/8, 2010: 596.

B. Boudjema, N. El-Khatib, M. Gamoudi, G. Guillaud, M. Maitrot, Thermally stimulated currents in undoped and doped zinc phthalocyanine films, *Rev. Phys. Appl.* 23, 1988: 1127–1134.

W. Böhm, T. Fritz, K. Leo, Charge transport in thin organic semiconducting films: A Seebeck and conductivity study, *Phys. Stat. Sol. (a)* 160, 1997: 81–87.

T.M. Brown, J.S. Kim, R.H. Friend, F. Caciallia, R. Daik, W.J. Feast, Built-in field electroabsorption spectroscopy of polymer light-emitting diodes incorporating a doped poly(3,4-ethylene dioxythiophene) hole injection layer, *Appl. Phys. Lett.* 75 (12), 1999: 1679ff.

S. Braun, W.R. Salaneck, M. Fahlmann, Energy level alignment at organic/metal and organic/organic interfaces, *Adv. Mater.* 21, 2009: 1450–1472.

J. Burschka, A. Dualeh, F. Kessler, E. Baranoff, N.-L. Cevey-Ha, C. Yi, M.K. Nazeeruddin, M. Grätzel, Tris(2-(1*H*-pyrazol-1-yl)pyridine)cobalt(III) as p-type dopant for organic semiconductors and its application in highly efficient solid-state dye-sensitized solar cells, *J. Am. Chem. Soc.* 133, 2011: 18042.

C.-C. Chang, M.-T. Hsieh, J.-F. Chen, S.-W. Hwang, C.H. Chen, Highly power efficient organic light-emitting diodes with a p-doping layer, *Appl. Phys. Lett.* 89, 2006: 253504.

C.K. Chan, E.-G. Kim, J.-L. Bredas, A. Kahn, Molecular n-type doping of 1,4,5,8-naphthalene tetracarboxylic dianhydride by pyronin B studied using direct and inverse photoelectron spectroscopies, *Adv. Funct. Mater.* 16, 2006a: 831–837.

C.-K. Chan, F. Amy, Q. Zhang, S. Barlow, S. Marder, A. Kahn, N-type doping of an electron-transport material by controlled gas-phase incorporation of cobaltocene, *Chem. Phys. Lett.* 431, 2006b: 67.

J.S. Chappell, A.N. Bloch, W.A. Bryden, M. Maxfield, T.O. Poehler, D.O. Cowan, Degree of charge transfer in organic conductors by infrared absorption spectroscopy, *Am. Chem. Soc.* 103, 1981: 2442.

S.Y. Chen, T.Y. Chu, J.F. Chen, C.Y. Su, C.H. Chen, Stable inverted bottom-emitting organic electroluminescent devices with molecular doping and morphology improvement, *Appl. Phys. Lett.* 89, 2006: 053518.

H. Cho, J.-M. Choi, S. Yoo, Highly transparent organic light-emitting diodes with a metallic top electrode: the dual role of a Cs2CO3 layer, *Opt. Express* 19 (2), 2011: 1113.

L. Cosimbescu, A.B. Padmaperuma, D.J. Gaspar, 7,7,8,8-Tetracyanoquinodimethane-based molecular dopants for p-type doping of OLEDs: A theoretical investigation, *J. Phys. Chem. A* 115, 2011: 13498.

F.-A. Cotton, N.E. Gruhn, J. Gu, P. Huang, D.L. Lichtenberger, C.A. Murillo, L.O. Van Dorn, C.C. Wilkinson, Closed-shell molecules that ionize more readily than cesium, *Science* 298, 2002: 1971.

X. Crispin, F.L.E. Jakobsson, A. Crispin, P.C.M. Grim, P. Andersson, A. Volodin, C. van Haesendonck, M. Van der Auweraer, W.R. Salaneck, M. Berggren, The origin of the high conductivity of Poly(3,4-ethylenedioxythiophene)-Poly(styrenesulfonate) (PEDOT-PSS) plastic electrodes, *Chem. Mater.* 18, 2006: 4354–4360.

B.W. D'Andrade, S.R. Forrest, A.B. Chwang, Operational stability of electrophosphorescent devices containing p and n doped transport layers, *Appl. Phys. Lett.* 83 (19), 2003: 3858.

J. Drechsel, M. Pfeiffer, X. Zhou, A. Nollau, K. Leo, Organic Mip-diodes by p-doping of amorphous wide-gap semiconductors: *CV* and impedance spectroscopy, *Synthetic Metals* 127, 2002: 201.

S. Egusa, Organic Electroluminescent Device, European Patent Application EP0498979A1, published 19.8.1992.

N. El-Khatib, B. Boudjema, G. Guillaud, M. Maitrot, Theoretical and experimental doping of molecular materials: p and n doping of zinc phthalocyanine, *J. Less Comm. Metals* 143, 1988: 101–112.

J. Endo, T. Matsumoto, J. Kido, Organic electroluminescent devices having metal complexes as cathode interface layer, *Jpn. J. Appl. Phys.* 41, 2002a: 800.

J. Endo, T. Matsumoto, J. Kido, Organic electroluminescent devices with a vacuum-deposited Lewis-acid-doped hole-injecting layer, *Jpn. J. Appl. Phys.* 41, 2002b: L358–L360.

H. Fritzsche, A general expression for the thermoelectric power, *Solid State Commun.* 9, 1971: 1813–1815.

C. Ganzorig, M. Fujihara, Improved drive voltages of organic electroluminescent devices with an efficient p-type aromatic diamine hole-injection layer, *Appl. Phys. Lett.* 77 (25), 2000: 4211–4213.

W. Gao, A. Kahn, Controlled p-doping of zinc phthalocyanine by coevaporation with tetrafluorotetracyanoquinodimethane: A direct and inverse photoemission study, *Appl. Phys. Lett.* 79, 2001: 4040–4042.

W. Gao, A. Kahn, Electronic structure and current injection in zinc phthalocyanine doped with tetrafluoro-tetracyanoquinodimethane: Interface versus bulk effects, *Org. Electron.* 3, 2002: 53–63.

W. Gao, A. Kahn, Controlled p doping of the hole-transport molecular material N,N8-diphenyl-N,N8-bis(1-naphthyl)-1,18-biphenyl-4,48-diamine with tetrafluorotetracyanoquinodimethane, *J. Appl. Phys.* 94, 2003: 359–366.

W. Gao, A. Kahn, Electrical doping: The impact on interfaces of Pi-conjugated molecular films, *J. Phys. Condens. Matter* 15, 2003: S2757–S2770.

Y. Gao, L. Yan, Cs doping and energy level shift in CuPc, *Chem. Phys. Lett.* 380, 2003: 451–455.

Z.Q. Gao, B.X. Mi, G.Z. Xu, Y.Q. Wan, M.L. Gong, K.W. Cheah, C.H. Chen, An organic *p*-type dopant with high thermal stability for an organic semiconductor, *Chem. Commun.* 117, 2008: 451–455.

B.A. Gregg, S.-G. Chen, R.A. Cormier, Coulomb forces and doping in organic semiconductors, *Chem. Mater.* 16, 2004: 4586–4599.

S. Guo, S.B. Kim, S.K. Mohapatra, Y. Qi, T. Sajoto, A. Kahn, S.R. Marder, S. Barlow, n-Doping of organic electronic materials using air-stable organometallics, *Adv. Mater.* 24, 2012: 699–703.

S. Hamwi, J. Meyer, T. Winkler, T. Riedl, W. Kowalsky, p-Type doping efficiency of MoO3 in organic hole transport materials, *Appl. Phys. Lett.* 94, 2009: 253307.

K. Harada, M. Sumino, C. Adachi, S. Tanaka, K. Miyazaki, Improved thermoelectric performance of organic thin-film elements utilizing a bilayer structure of pentacene and 2,3,5,6-tetrafluoro-7,7,8,8-tetracyano-quinodimethane (F4-TCNQ), *Appl. Phys. Lett.* 96, 2010: 253304.

T. Hatwar, J.P. Spindler, W.J. Begley, D.J. Giesen, D.Y. Kondakov, S. van Slyke, S. Murano, E. Kucur, G. He, J. Blochwitz-Nimoth, High-performance tandem white OLEDs using a Li-free "P-N" connector, *SID Symp. Dig. Tech. Papers*, 40(1), 2009: 499–502.

T. Hasegawa, S. Miura, T. Moriyama, T. Kimura, I. Takaya, Y. Osato, H. Mizutani, Novel electron-injection layer for top emission OLEDs, *SID Int. Symp. Digest Tech. Papers* 35 (1), 2004: 154–157.

S. Hayashi, H. Etoh, S. Saito, Electroluminescence of perylene films with a conducting polymer as an anode, *Jpn. J. Appl. Phys.* 25 (9), 1986: L773–L775.

I.G. Hill, D. Milliron, J. Schwartz, A. Kahn, Organic semiconductor interfaces: Electronic structure and transport properties, *Appl. Surf. Sci.* 166, 2000: 354–362.

S. Hiller, D. Schlettwein, N.R. Armstrong, D. Wöhrle, Influence of surface reactions and ionization gradients on junction properties of F16-PcZn, *J. Mater. Chem.* 8 (4), 1998: 945–954.

M.-T. Hsieh, C.-C. Chang, J.-F. Chen, C.H. Chen, Study of hole concentration of 1,4-bis[N-(1-naphthyl)-N′-phenylamino]-4,4′diamine doped with tungsten oxide by admittance spectroscopy, *Appl. Phys. Lett.* 89, 2006: 103510.

L.S. Hung, C.W. Tang, M.G. Mason, Enhanced electron injection in organic electroluminescence devices using an Al/LiF electrode, *Appl. Phys. Lett.* 70 (2), 1997: 152–154.

F. Huang, A.G. MacDiarmid, B.R. Hsieh, An iodine-doped polymer light-emitting diode, *Appl. Phys. Lett.* 71 (17), 1997: 2415–2417.

J. Huang, M. Pfeiffer, A. Werner, J. Blochwitz, K. Leo, S. Liu, Low-voltage organic electroluminescent devices using pin structures, *Appl. Phys. Lett.* 80 (1), 2002: 139.

Q. Huang, T. Rosenow, T.W. Canzler, M. Furno, C. Rothe, S. Dorok, U. Denker, O. Fadhel, J. Birnstock, Electron-transport layers with air-stable dopants for display applications, SID *Symp. Dig. Tech. Papers* 43 (1), 2012: 275–278.

S. Hüfner, *Photoelectron Spectroscopy—Principles and Applications*, 3rd edition, Springer-Verlag, Berlin, Heidelberg, New York, 2003.

J. Hwang, F. Amy, A. Kahn, Spectroscopic study on sputtered PEDOT:PSS: Role of surface PSS layer, *Org. Electron.* 7, 2006: 387–396.

H. Ikeda, J. Sakata, M. Hayakawa, T. Aoyama, T. Kawakami, K. Kamata, Y. Iwaki, S. Seo, Y. Noda, R. Nomura, S. Yamazaki, P-185: Low-Drive-Voltage OLEDs with a Buffer Layer Having Molybdenum Oxide, *SID Symp. Dig. Tech. Papers* 2006: 923.

H. Ishii, K. Sugiyama, E. Ito, K. Seki, Energy level alignment and interfacial electronic structures at organic metal and organic organic interfaces, *Adv. Mater.* 11, 1999: 605.

Y. Itoh, N. Tomikawa, S. Kobayashi, T. Minato, Extended Abstracts. The 51th Autumn Meeting, *The Japan Society of Applied Physics, Tokyo* 1990: 1040.

D.M. Ivory, G.G. Miller, J.M. Sowa, L.W. Shacklette, R.R. Chance, R.H. Baughman, Highly conducting charge-transfer complexes of poly(*p*-phenylene), *J. Chem. Phys.* 71, 1979: 1506.

A. Kahn, N. Koch, W.Y. Gao, Electronic structure and electrical properties of interfaces between metals and pi-conjugated molecular films, *J. Polym. Sci. B Polym. Phys.* 41, 2003: 2529.

D.R. Kearns, G. Tollin, M. Calvin, Electrical properties of organic solids. II. Effects of added electron acceptor on metall-free phthalocyanine, *J. Chem. Phys.* 32 (4), 1960: 1020–1025.

H. Kleemann, R. Gutierrez, F. Lindner, S. Avdoshenko, P.D. Manrique, Björn Luessem, G. Cuniberti, Karl Leo, Organic Zener diodes: Tunneling across the gap in organic semiconductor materials, *Nano Lett.* 10, 2010: 4929.

H. Kleemann, C. Schuenemann, A.A. Zakhidov, M. Riede, B. Luessem, K. Leo, Structural phase transition in pentacene caused by molecular doping and its effect on the charge carrier mobility, *Org. Electron.* 13, 2012a: 85–65.

H. Kleemann, B. Luessem, K. Leo, Controlled formation of charge depletion zones by molecular doping in organic pin-diodes and its description by the Mott-Schottky relation, *J. Appl. Phys.* 111, 2012b: 123722.

J. Kido, K. Nagai, Y. Okamoto, Bright organic electroluminescence devices with double layer cathode, *IEEE Trans. Electron Devices* 40 (7), 1993: 1342ff.

J. Kido, T. Matsumoto, Bright organic electroluminescent devices having a metal-doped electron-injecting layer, *Appl. Phys. Lett.* 73 (20), 1998: 2866–2868.

J. Kido, J. Endo, T. Matsumoto, *Polym. Prepr. (Am. Chem. Soc. Div. Polym. Chem.)* 47, 1998: 1940ff.

N. Koch, Organic electronic devices and their functional interfaces, *ChemPhysChem* 8, 2007: 1438–1455.

P.K. Koech, A.B. Padmaperuma, L. Wang, J.S. Swensen, E. Polikarpov, J.T. Darsell, J.E. Rainbolt, D.J. Gaspar, Synthesis and application of 1,3,4,5,7,8-hexafluorotetracyanonaphthoquinodimethane (F6-TNAP): A conductivity dopant for organic light-emitting devices, *Chem. Mater.* 22, 2010: 3926.

R. Krause, F. Steinbacher, G. Schmid, J.H. Wemken, A. Hunze, Cheap p- and n-doping for highly efficient organic devices, *J. Photon. Energy* 1, 2011: 011022-1.

M. Kröger, S. Hamwib, J. Meyer, T. Riedl, W. Kowalsky, A. Kahn, P-type doping of organic wide band gap materials by transition metal oxides: A case-study on molybdenum trioxide, *Org. Electron.* 10, 2009: 932–938.

T. Kugler, W.R. Salaneck, H. Rost, A.B. Holmes, Polymer band alignment at the interface with indium tin oxide: Consequences for light emitting devices, *Chem. Phys. Lett.* 310 (5–6), 1999: 391–396.

C.-T. Lee, H.-C. Chen, Performance improvement mechanisms of organic thin-film transistors using MoO_x-doped pentacene as channel layer, *Org. Electron.* 12, 2011: 1852.

D.-S. Leem, H.-D. Park, J.-W. Kang, J.-H. Lee, J.W. Kim, J.-J. Kim, Low driving voltage and high stability organic light-emitting diodes with rhenium oxide-doped hole transporting layer, *Appl. Phys. Lett.* 91, 2007: 011113.

F. Li, A. Werner, M. Pfeiffer, X. Liu, K. Leo, Leuco crystal violet as a dopant for n-doping of organic thin films of fullerene C60, *J. Phys. Chem. B* 108, 2004: 17076–17082.

F. Li, A. Werner, M. Pfeiffer et al., Acridine orange base as a dopant for n doping of C60 thin films, *J. Appl. Phys.* 100, 2006: 023716.

M. Limmert, H. Hartmann, O. Zeika, A. Werner, M. Ammann, EP1837926, Publ. 26.9.2007.

J. Liu, X. Guo, Y. Qin, S. Liang, Z.-X. Guo, Y. Li, Dumb-belled PCBM derivative with better photovoltaic performance, *J. Mater. Chem.* 22, 2012: 1758–1761.

F. Loeser, T. Romainczyk, C. Rothe, D. Pavicic, A. Haldi, M. Hofmann, S. Murano, T. Canzler, J. Birnstock, Improvement of device efficiency in PIN-OLEDs by controlling the charge carrier balance and intrinsic outcoupling methods, *J. Photon. Energy* 2, 2012, 021207.

E.J. Lous, P.W.M. Blom, L.W. Molenkamp, D.M. deLeeuw, Schottky contacts on a highly doped organic semiconductor, *Phys. Rev. B Condens. Matter* 51 (23), 1995: 17251–17254.

L. Ma, W.H. Lee, Y.D. Park, J.S. Kim, H.S. Lee, K. Cho, High performance polythiophene thin-film transistors doped with very small amounts of an electron acceptor, *Appl. Phys. Lett.* 92, 2008: 063310.

B. Maennig, M. Pfeiffer, A. Nollau, X. Zhou, P. Simon, K. Leo, Controlled p-type doping of polycrystalline and amorphous organic layers: Self-consistent description of conductivity and field-effect mobility by a microscopic percolation model, *Phys. Rev. B* 64, 2001: 195208.

M. Maitrot, G. Guillaud, B. Boudjema, J.J. Andre, J. Simon, Molecular material based junctions: Formation of a Schottky contact with metallophthalocyanine thin films doped by the cosublimation method, *J. Appl. Phys.* 60 (7), 1986: 2396–2400.

T.J. Marks et al., Electrically conductive metallomacrocyclic assemblies, *Science* 227, 1985: 881–889.

M.G. Mason, C.W. Tang, L.-S. Hung, P. Raychaudhuri, J. Madathil, D.J. Giesen, L. Yan, Q.T. Le, Y. Gao, S.-T. Lee, L.S. Liao, L.F. Cheng, W.R. Salaneck, D.A. dos Santos, J.L. Bredas, Interfacial chemistry of alq3 and LiF with reactive metals, *J. Appl. Phys.* 89 (5), 2001: 2756–2765.

C. May, Y. Tomita, M. Toerker, M. Eritt, F. Loeffler, J. Amelung, K. Leo, In-line deposition of organic light-emitting devices for large area applications, *Thin Solid Films* 516, 2008: 4609–4612, doi: 10.1016/j.tsf.2007.06.014.

R. Meerheim, K. Walzer, M. Pfeiffer, K. Leo, Ultrastable and efficient red organic light emitting diodes with doped transport layers, *Appl. Phys. Lett.* 89, 2006: 061111.

R. Meerheim, S. Olthof, M. Hermenau, S. Scholz, A. Petrich, N. Tessler, O. Solomeshch, B. Lüssem, M. Riede, K. Leo, Investigation of $C_{60}F_{36}$ as low-volatility p-dopant in organic optoelectronic devices, *Appl. Phys.* 109, 2011: 103102.

E.-J. Meijer, A.V.G. Mangnus, C.M. Hart, D.M. de Leeuw, T.M. Klapwijk, Frequency behavior and the Mott–Schottky analysis in poly(3-hexyl thiophene) metal–insulator–semiconductor diodes, *Appl. Phys. Lett.* 78, 2001: 3902–3904.

M. Meneghetti, C. Pecile, Charge–transfer organic crystals: Molecular vibrations and spectroscopic effects of electron–molecular vibration coupling of the strong electron acceptor $TCNQF_4$, *J. Chem. Phys.* 84, 1986: 4149.

T. Menke, D. Ray, J. Meiss, K. Leo, M. Riede, In-situ conductivity and Seebeck measurements of highly efficient n-dopants in fullerene C60, *Appl. Phys. Lett.* 100, 2012: 093304.

J.-P. Meyer, D. Schlettwein, D. Wöhrle, N.I. Jaeger, Charge transport in thin films of molecular semiconductors as investigated by measurements of thermoelectric power and electrical conductivity, *Thin Solid Films* 258, 1995: 317–324.

A. Miller, E Abrahams, Impurity conduction at low concentrations, *Phys. Rev.* 120, 1960: 745–755.

A. Mityashin, Y. Olivier et al., Unrevealing the mechanism of molecular doping, *Adv. Mater.* 24, 2012: 1535–1539.

R.C. Nelson, Energy transfers between sensitizer and substrate. IV. Energy levels in solid dyes, *J. Opt. Soc. Am.* 51, 1961: 1186–1191.

A. Nollau, M. Pfeiffer, T. Fritz, K. Leo, Controlled n-type doping of a molecular organic semiconductor: Naphthalenetetracarboxylic dianhydride (NTCDA) doped with bis(ethylenedithio)-tetrathiafulvalene (BEDT-TTF), *J. Appl. Phys.* 87, 2000: 4340–4343.

J.H. Oh, P. Wei, Z. Bao, Molecular n-type doping for air-stable electron transport in vacuum-processed n-channel organic transistors, *Appl. Phys. Lett.* 97, 2010: 243305.

S. Olthof, W. Tress, R. Meerheim, B. Luessem, K. Leo, Photoelectron spectroscopy study of systematically varied doping concentrations in an organic semiconductor layer using a molecular p-dopant, *J. Appl. Phys.* 106, 2009a: 103711.

S. Olthof, R. Meerheim, M. Schober, K. Leo, Energy level alignment at the interfaces in a multilayer organic light-emitting diode structure, *Phys. Rev. B* 79, 2009b: 245308.

T. Oyamada, H. Sasabe, Ch. Adachi, S. Murase, T. Tominaga, Ch. Maeda, Extremely low-voltage driving of organic light-emitting diodes with a Cs-doped phenyldipyrenylphosphine oxide layer as an electron-injection layer, *Appl. Phys. Lett.* 86, 2005: 033503.

Y.-W. Park, A.J. Heeger, M.A. Druy, A.G. MacDiarmid, Electrical transport in doped polyacetylene, *J. Chem. Phys.* 73, 1980: 946–957.

G. Parthasarathy, C. Shen, A. Kahn, S.R. Forrest, Lithium doping of organic semiconducting organic charge transport materials, *J. Appl. Phys.* 89 (9), 2001: 4986–4992.

K.P. Pernstich, B. Rössner, B. Batlogg, Field-effect-modulated Seebeck coefficient in organic semiconductors, *Nat. Mater.* 7, 2008: 321–325.

M. Pfeiffer, A. Beyer, T. Fritz, K. Leo, Controlled doping of phthalocyanine layers by cosublimation with acceptor molecules: A systematic Seebeck and conductivity study, *Appl. Phys. Lett.* 73, 1998: 3202–3204.

M. Pfeiffer, Controlled doping of organic vacuum deposited dye layers: Basics and applications, PhD thesis, TU Dresden, 1999.

M. Pfeiffer, T. Fritz, J. Blochwitz, A. Nollau, B. Plönnigs, A. Beyer, K. Leo, *Adv. Solid State Phys.* 39, 1999: 77–90.

M. Pfeiffer, S.R. Forrest, K. Loe, M.E. Thompson, Electrophosphorescent p-i-n organic light-emitting devices for very-high-efficiency-flat-panel displays, *Adv. Mater.* 14, 2002: 1633–1636.

P. Piromreun, H. Oh, Y. Shen, G.G. Malliaras, J.C. Scott, P.J. Brock, Role of CsF on electron injection into a conjugated polymer, *Appl. Phys. Lett.* 77, 2000: 2403–2405.

B. Ploennigs, Elektrische und strukurelle Charakterisierung dotierter Farbstoffaufdampfschichten, Diplomarbeit, TU Dresden, 1999.

Y. Qi, S.K. Mohapatra, S.B. Kim, S. Barlow, S.R. Marder, A. Kahn, Solution doping of organic semiconductors using air-stable n-dopants, *Appl. Phys. Lett.* 100, 2012: 083305.

S. Reineke, F. Lindner, G. Schwartz, N. Seidler, K. Walzer, B. Luessem1, K. Leo, White organic light-emitting diodes with fluorescent tube efficiency, *Nature*, 459, 2009: 234–238.

D.B. Romero, M. Schaer, L. Zuppiroli, B. Cesar, B. Francois, *Appl. Phys. Lett.* 67, 1995: 1659–1661.

I. Salzmann, G. Heimel, S. Duhm, M. Oehzelt, P. Pingel, B.M. George, A. Schnegg et al. Intermolecular hybridization governs molecular electrical doping, *Phys. Rev. Lett.* 108, 2012: 035502.

D. Schlettwein, D. Wöhrle, E. Karmann, U. Melville, Conduction type of substituted tetraazaporphyrins and perylene tetracarboxylic acid diimides as detected by thermoelectric power measurements, *Chem. Mater.* 6, 1994: 3–6.

R. Schmechel, Hopping transport in doped organic semiconductors: A theoretical approach and its application to p-doped zinc-phatocyanine, *J. Appl. Phys.* 93, 2003: 4653–4660

R.D. Shannon, Revised effective ionic radii and systematic studies of interatomic distances in halides and chalcogenides, *Acta Cryst.* A32, 1976: 751–767. doi: 10.1107/S0567739476001551.

Shen, 2001, Chemical and electrical properties of interfaces between magnesium and aluminum and tris-(8-hydroxy quinoline) aluminum, *J. Appl. Phys.* 89, 2001: 449–459.

J. Simon, J.-J. Andre (Ed.), *Molecular Semiconductors: Photoelectrical Properties and Solar Cells*, Springer-Verlag, Berlin, Heidelberg, New York, Tokyo, 1985.

O. Solomeshch, Y.J. Yu, A.A. Goryunkov, L.N. Sidorov, R.F. Tuktarov, D.H. Choi, J.-I. Jin, N. Tessler, Ground-state interaction and electrical doping of fluorinated C60 in conjugated polymers, *Adv. Mater.* 21, 2009: 4456.

P. Stallinga, H.L. Gomes, H. Rost, A.B. Holmes, M.G. Harrison, R.H. Friend, Minority-carrier effects in polyphenylenevinylene as studied by electrical characterization, *J. Appl. Phys.* 89, 2001: 1713–1724.

J.X. Tang, C.S. Lee, S.T. Lee, Electronic structures of organic/organic heterojunctions: From vacuum level alignment to Fermi level pinning, *J. Appl. Phys.* 101, 2007: 064504.

S. Tanaka, K. Kanai, E. Kawabe, T. Iwahashi, T. Nishi, Y. Ouchi, K. Seki, Doping effect of tetrathianaphthacene molecule in organic semiconductors on their interfacial electronic structures studied by UV photoemission spectroscopy, *Jpn. J. Appl. Phys.* 44, 2005: 3760.

J. Tardy, M. Ben Khalifa, D. Vaufrey, Organic light emitting devices with doped electron transport and hole blocking layers, *Mater. Sci. Eng. C* 26, 2006: 196–201.

M. Tietze, L. Burtone, M. Riede, B. Luessem, K. Leo, Fermi-level shift and doping efficiency in p-doped small molecule organic semiconductors: A photoelectron spectroscopy and theoretical study, *Phys. Rev. B* 86, 2012: 035320.

S.P. Tiwari, W.J. Potscavage Jr., T. Sajoto, S. Barlow, S.R. Marder, B. Kippelen, Pentacene organic field-effect transistors with doped electrode-semiconductor contacts, *Org. Electron.* 11, 2010: 860.

S. Tokito, K. Noda, Y. Taga, Metal oxides as a hole injecting layer for organic electroluminescent devices, *J. Phys. D Appl. Phys.* 29 (11), 1996: 2750.

H. Vazquez, F. Florez, R. Oswaldowski, J. Ortega, R. Perez, A. Kahn, Barrier formation at metal–organic interfaces: Dipole formation and the charge neutrality level, *Appl. Surf. Sci.* 234, 2004: 107–112.

A. von Muehlen, N. Errien, M. Schaer, M.-N. Bussac, L. Zuppiroli, Thermopower measurements on pentacene transistors, *Phys. Rev. B* 75, 2007: 115338.

K. Walzer, B. Maennig, M. Pfeiffer, K. Leo, Highly efficient organic devices based on electrically doped transport layers, *Chem. Rev.* 107, 2007: 1233–1271.

S. Watanabe, N. Ide, J. Kido, High-efficiency green phosphorescent organic light-emitting devices with chemically doped layers, *Jpn. J. Appl. Phys.* 46 (3A), 2007: 1186–1188.

P. Wei, J.H. Oh, G. Dong, Z. Bao, Use of a 1H-benzoimidazole derivative as an n-type dopant and to enable air-stable solution-processed n-channel organic thin-film transistors, *J. Am. Chem. Soc.* 132, 2010: 8852–8853.

J.H. Wemken, R. Krause, T. Mikolajick, G. Schmid, Low-cost caesium phosphate as n-dopant for organic light-emitting diodes, *J. Appl. Phys.* 111, 2012: 074502.

A. Werner, N-type doping of organic thin films using a novel class of dopants, PhD thesis, TU Dresden, 2003.

A.G. Werner, F. Li, K. Harada, M. Pfeiffer, T. Fritz, K. Leo, Pyronine B as a donor for n-type doping of organic thin films, *Appl. Phys. Lett.* 82, 2003: 4495–4497.

A.G. Werner, F. Li, K. Harada et al., N-type doping of organic thin films using cationic dyes, *Adv. Funct. Mater.* 24, 2004: 255–260.

C.-I. Wu, C.T. Lin, Y.H. Chen, M.-H. Chen, Y.J. Lu, C.C. Wu, Electronic structures and electron-injection mechanisms of cesium-carbonate-incorporated cathode structures for organic light-emitting-devices, *Appl. Phys. Lett.* 88 (15), 2006: 152104.

A. Yamamori, C. Adachi, T. Koyama, Y. Taniguchi, Doped organic light emitting diodes having a 650-nm-thick hole transport layer, *Appl. Phys. Lett.* 72 (17), 1998: 2147–2149.

A. Yamamori, C. Adachi, T. Koyama, Y. Taniguchi, Electroluminescence of organic light emitting diodes with a thick hole transport layer composed of a triphenylamine based polymer doped with an antimonium compound, *J. Appl. Phys.* 86 (8), 1999: 4369–4376.

Y. Yamamoto, K. Yoshino, Y. Inuishi, Electrical properties of phthalocyanine-halogen complexes, *J. Phys. Soc. Jpn.* 47 (6), 1979: 1887–1891.

L. Yan, N.J. Watkins, S. Zorba, Y. Gao, C.W. Tang, Direct observation of Fermi-level pinning in Cs-doped CuPc film, *Appl. Phys. Lett.* 79, 2001: 4148–4150.

Y. Yang, A.J. Heeger, Polyaniline as a transparent electrode for polymer light-emitting diodes: Lower operating voltage and higher efficiency, *Appl. Phys. Lett.* 64, 1994: 1245–1247.

K.-H. Yim, G.L. Whiting, C.E. Murphy, J.J.M. Halls, J.H. Burroughes, R.H. Friend, Ji-S. Kim, Controlling electrical properties of conjugated polymers via a solution-based p-type doping, *Adv. Mater.* 20, 2008: 3319–3324.

K.S. Yook, S.O. Jeon, S.-Y. Min, J.Y. Lee, H.-J. Yang, T. Noh, S.-K. Kang, T.-W. Lee, Highly efficient p-i-n and tandem organic light-emitting devices using an air-stable and low-temperature-evaporable metal azide as n-dopant, *Adv. Funct. Mater.* 20, 2010: 1797–1802.

Y. Zhang, B. de Boer, P.W.M. Blom, Controllable molecular doping and charge transport in solution-processed polymer semiconducting layers, *Adv. Funct. Mat.* 19, 2009: 1901.

W. Zhao, A. Kahn, Charge transfer at n-doped organic-organic heterojunctions, *J. Appl. Phys.* 105, 2009: 123711.

R. Zuzok, A.B. Kaiser, W. Pukacki, S. Roth, Thermoelectric power and conductivity of iodine-doped new polyacetylene, *J. Chem. Phys.* 95, 1991: 1270–1275.

10. Development of Host Materials for High-Efficiency Organic Light-Emitting Devices

Asanga B. Padmaperuma

10.1 Introduction

Organic electroluminescence was discovered in the 1960s (Pope et al. 1963; Short and Hercules 1965) but it was not until 1987 that the Kodak team pioneered the use of a two-layer thin-film device, called a heterojunction device, substituting the crystalline materials that had been previously used. One of the most important subsequent innovations was the use of phosphorescent emitters doped into a matrix, often called the *host material*, which serves several functions. In a host-dopant device, the excitons are formed on the host and then rapidly transferred by a nonradiative process to the phosphor. This chapter focuses on the performance requirements and development of host materials, including charge transport, stability, and energy levels. This chapter also covers computational approaches to molecular property prediction, measurement, and characterization of host materials.

Several challenges to the development of high-efficiency organic light-emitting diodes (OLEDs) are addressed by the use of a host material. These include prevention of concentration quenching (a form of self-quenching), providing efficient charge injection, charge

Chapter 10

transport, and exciton transfer, and ensuring good stability. In practice, the host material is optimized for charge injection from the hole transport layer (HTL) and electron transport layer (ETL) and charge transport to facilitate charge recombination and exciton formation. This suggests that the host highest occupied molecular orbital (HOMO) and lowest unoccupied molecular orbital (LUMO) should be similar to the ETL LUMO and HTL HOMO to facilitate efficient electron and hole transfer, respectively. In practice, some energy level mismatch is typically present, but the energy level mismatch should be small to ensure low device operating voltage.

Development of efficient blue OLEDs based on phosphorescence has been particularly challenging compared to other colors, from both the host and dopant design perspectives. To efficiently transfer energy to a phosphorescent dopant, both the singlet- and triplet-excited states of the host material must be higher in energy than the triplet-excited state of the dopant (Adachi et al. 2001). In practice, this means a triplet exciton energy ≥2.75 eV to achieve efficient (exothermic) energy transfer, and to prevent exciton quenching due to back transfer. In the first devices using phosphorescent emitters, the quantum efficiency of the device increased dramatically, but the power efficiency of these systems was still limited by a high operating voltage relative to the energy of the emitted photons. A particular drawback of early wide bandgap host materials was that they were not able to transport both electrons and holes. As a result, the emitter is required to transport one sign of charge. In order to improve charge transport of both electrons and holes, the concentration of the dopant must be increased. However, this leads to self-quenching, or concentration quenching. The next section discusses some of the history of the interplay between charge transport and concentration quenching in blue OLEDs.

One of the earliest reports on wide bandgap host materials were the ultra-wide bandgap tetra-aryl silanes (e.g., *p*-bis(triphenylsilylyl) [UGH2]) demonstrated by the Thompson group as high-triplet-energy host materials for a deep-blue phosphor, iridium(III) bis(4′,6′- difluorophenylpyridinato)tetrakis(1-pyrazolyl)borate (FIr6) (Holmes et al. 2003). These materials have deep HOMO levels that prevent hole injection from commonly used HTLs. The HOMO and LUMO energies of the UGH2 hosts were −7.2 and −2.8 eV, respectively. Comparison of these values with those of the phosphorescent dopant (−6.1 and −3.1 eV, respectively) suggested that charge transport in the EML relied mainly on the dopant, which was confirmed by measuring the dependence of the device operating voltages on the dopant concentration. At lower dopant levels (<5%), the operating voltage increased with increasing FIr6 concentration, due to charge trapping at the HTL/EML interface. However, at dopant concentrations above 5%, the voltage dropped and light emission efficiency increased, producing a device with an external quantum efficiency (EQE) above 5%. Overall, the device operating voltages were relatively high (>10 V) (Ren et al. 2004).

In the UGH2 devices, charge trapping on the dopant was beneficial because the host material did not have good charge transport properties. Increasing the power efficiency in such a device has proven to be difficult because of a trade-off between operating voltage and light emission efficiency, which declines with the increasing dopant concentration due to concentration quenching. Later reports demonstrated the feasibility of improving the power efficiency of a blue phosphorescent device in a predominantly electron-transporting host. Using the sky blue organometallic complex, iridium(III) bis[(4,6-difluorophenyl)-pyridinato-*N,C2'*] picolinate (FIrpic) as the dopant, a similar EQE compared to UGH2 and related host materials was achieved, but at a lower operating voltage. In some cases, the improvement led to twice as much the luminous power efficiency of the device (Burrows et al. 2006; Padmaperuma et al. 2006; Sapochak et al. 2006; Vecchi et al. 2006). However, similar to UGH2, the operating voltage and EQE of these devices made with electron-transporting phosphine oxide (PO) host materials were dependent on dopant concentration (see Figure 10.1).

Indeed, when FIrpic was doped into 3,6-bis(diphenylphosphoryl)-9-ethylcarbazole (PO10) at concentrations below 5%, the operating voltage increased compared to a device containing no dopant, and exhibited low EQE. At this low dopant concentration, the charge balance in the EML was poor and the voltage higher because the PO hosts have deeper HOMO energies (E_{HOMO} (PO) < −6.0 eV) than the emitter (E_{HOMO}(FIrpic) = −5.8 eV) preventing efficient hole injection. Therefore, the holes are directly injected into the dopant molecule in the EML and trapped at the HTL/EML interface. As the dopant concentration is increased, the EQE also increases and operating voltage decreases slightly. However, the concentration of dopant (20%) that yields the maximum EQE is almost twice the concentration of dopant that gives maximum φ_{PL} in a solid-state film (2%–12%) (Sapochak et al. 2008). This means that at dopant concentrations >12%, the EQE is likely limited by self-quenching and that at dopant concentrations <12%, the EQE is limited by poor charge balance. Although current state-of-the-art PO compounds have excellent hole-blocking and electron injection properties (Oyamada et al. 2005; Matsushima and Adachi 2006), they are not optimized as host materials because of their inability to transport holes.

OLEDs using PO hosts that show poor hole transport have similar (though less severe) limitations compared to UGH2 devices in that the charge trapping on the emitter is necessary to achieve maximum EQE. Host materials that can transport both electrons

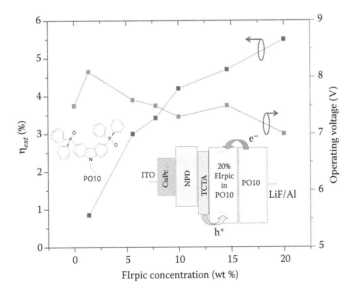

FIGURE 10.1 Voltage and quantum efficiency as a function of dopant concentration. The device used FIrpic doped into the host PO10 and was operated at a current density of 13 mA/cm².

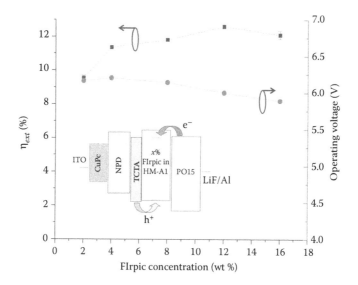

FIGURE 10.2 Voltage and quantum efficiency as a function of dopant concentration for a device using FIrpic doped into the host HM-A1. All devices were measured at a current density of 13 mA/cm².

and holes are called bipolar or ambipolar hosts. We will use the term ambipolar for the remaining part of this chapter for consistency. It was shown in the reports that followed the original PO work that ambipolar host materials based on POs can be developed by chemically integrating hole-transporting capabilities into a PO molecule (Cai et al. 2008; Polikarpov et al. 2009). In contrast to all preceding reports on the PO hosts, the ambipolar hosts from showed much weaker dependence of EQE and operational voltage on dopant concentration (see Figure 10.2).

Designing materials that have small singlet–triplet gaps can help achieve sufficiently high triplet energies for the host while retaining good charge transport properties. Since charge transport does not take place on the emitter, devices can maintain low driving voltages at low emitter concentrations, which helps avoid concentration quenching, reduce the use of costly emitter materials, and improves the device power efficiency. The current state-of-the-art host materials almost always accept and transport both electrons and holes.

10.2 Design Criteria for Host Materials

This section describes the properties of a good host for a high-efficiency OLED. As pointed out earlier, the challenges are most pressing and difficult to address for the blue-emitting devices due to the unique requirements for charge transport, triplet energy (E_T), and bandgap that are difficult to fulfill simultaneously. Ideally, a host material for a blue OLED system will transport both holes and electrons effectively, that is, have good electron (σ_e) and hole (σ_h) mobility. In practice, this means $\sigma > 10^{-4}$ cm²/Vs, although this number may be controversial and the method of measurement will affect the absolute number. Furthermore, the host should be energy level-matched to the adjacent transport layers to ensure good charge injection. Additionally, the host needs to have higher triplet energy than the emitter, which must be accomplished without sacrificing the charge-transporting properties of the host–guest system. The wide bandgap requirement for the host is dictated by the high energy of the blue exciton—the bandgap of the host should be larger than the emitter. The wide bandgap can negatively impact charge injection and transport, and therefore makes it difficult to simultaneously optimize the charge injection and transport properties of the host and the light emission properties (i.e., maximum luminescent efficiency) of the dopant. Thus, in blue OLEDs, using hosts with unipolar transport, at least one sign of charge is trapped by the dopant due to the energy level mismatch of the charge transport layers and the host. As a result, electrons or holes trapped by the dopant cannot be transferred to the host and must await transport of the other charge to form an exciton which then leads to emission. In addition to the device characteristics described in the previous section (dopant concentration-dependent voltage), this also leads to reduced device efficiency at higher current densities due to exciton–exciton and exciton–polaron quenching (Giebink et al. 2008). For example, PO molecular linkages can be used to build ambipolar host materials that can be vacuum deposited, among other examples. These ambipolar hosts have been used in devices demonstrating high efficiency and low operating voltages. Finally, the host material must exhibit chemical stability under operating conditions, including avoiding negative interactions with adjacent layers and dopants (see Chapter 13 for more details on chemical degradation of OLED materials).

Based on our experience, the host design must adhere to the following rules:

$E_{HOMO}(\text{host}) \sim E_{HOMO}(\text{HTL})$
$E_{LUMO}(\text{host}) \sim E_{LUMO}(\text{ETL})$
High E_T (~the triplet energy of the emitter)
Inherent chemical stability
Ambipolar transport with $\sigma_e \sim \sigma_h$ assuming injection rate is similar

10.3 A Chemist's Best Friend: Computational Prediction

Recent advances in computational algorithms and high-performance computing hardware coupled with user-friendly computational software have made theoretical predictions a valuable tool in molecular design for organic electronic applications. Computational results can be particularly valuable when used in predicting trends of molecular properties within a group of molecules, rather than being the source of absolute energy values or other stand-alone parameters for individual molecules. The important point to stress here is that no matter what level of theory one uses, the predictions are most relevant when comparing computed molecular properties in a library of molecules. As with any system, the accuracy of the predictions increases with the size of the library. Also, it is essential to validate some fraction of the library of computed values with available experimental values. Once an accurate method is developed and validated against known baseline materials, one can continue to predict novel unknown systems. The rest of this section describes one set of methods that has been shown to be useful in predicting properties of interest to the OLED community.

The most important quantities to compute for host materials are related to the electronic and molecular structure of the ground and excited states. The ability of a molecule to accept a given charge carrier from a transport layer largely depends on the relative positions of the frontier orbital energy levels of the donor and acceptor. These values can be easily predicted carrying out ground-state optimization of a molecular structure. Methods are also available to estimate the excited singlet and the triplet state energies. Estimations of the charge mobility for a given organic molecular material can be estimated by calculating the reorganization energy. The electronic structure of atoms, molecules, and even condensed phases can be investigated using density functional theory (DFT) (Parr and Yang 1995). DFT methods have found applications in biological, physical, and chemical systems to investigate and predict the electronic structure in the ground state of systems. In DFT, properties of multielectron systems are determined by using functions of another function (or functionals), which also gives rise to the name of the method. Although DFT has been used since the 1970s, it was not considered to be accurate enough for calculations in quantum chemistry until the 1990s. DFT has its conceptual roots in the Thomas–Fermi model but it was the two Hohenberg-Kohn (H–K) theorems (Hohenberg and Kohn 1964) that has unlocked the potential of this method. The refinement of approximations to model the exchange and correlation interactions have led to an explosion of application of DFT methods in computational chemistry and physics.

10.3.1   Electronic Density Distributions and Molecular Orbital Energies

From a computational standpoint, the energy and the location of the molecular orbitals (MOs) are one of the easiest tasks for modern quantum chemical methods. Irrespective of whether the molecule exists in nature or whether it can be synthesized, if it can be constructed in the user interface of a computational chemistry program, its properties can be computed. The energies of the MOs provide an indication as to where the HOMO or the LUMO energy might lie, which can help determine whether the "hypothetical host compound" can be used in a device with known transport layers. For instance, let us assume we want to design a host molecule to work with a hole transport material with a predicted E_{HOMO} of −4.5 eV. In this case, we will eliminate from contention any host material with a predicted E_{HOMO} deeper than ~−4.5 eV. An example of the use of the predictive capabilities of quantum chemical calculations in developing a series of host materials is shown in Figure 10.3. The HOMO and LUMO energies of hypothetical molecules were computed from optimized geometries using the NWChem computational package developed at Pacific Northwest National Laboratory (Valiev et al. 2010).

These computed values are compared to 4,4′-cyclohexylidenebis[N,N-bis(4-methylphenyl)benzenamine] (TAPC) and dibenzo[b,d]thiophene-2,8-diylbis(diphenylphosphine oxide) (PO15), an HTL and ETL, respectively, used in a baseline blue OLED. The HTL material TAPC was first reported by Kido, and has a triplet energy higher than that of FIrpic and high hole mobility. The ETL material PO15 was first synthesized in the laboratories of PNNL; it has a good combination of high E_T and electron mobility. Computational studies of these molecular systems suggest that all proposed targets have similar E_{HOMO}, within acceptable levels for efficient hole transport (about −5 eV) and comparable to that of TAPC. The E_{LUMO} varies depending on the nature of the electron deficient group used in the design (i.e., pyridine, pyrazine, or an aryl group such as

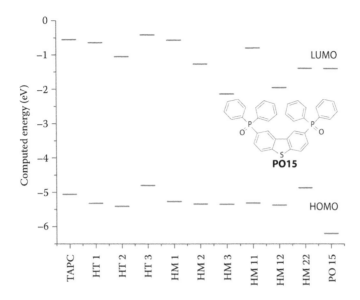

FIGURE 10.3 Calculated HOMO and LUMO energies of a series of potential compounds compared to representative hole transport (TAPC) and electron transport (PO15) materials.

phenyl, biphenyl, naphthyl, and pyrenyl). The HOMO levels of the unknown molecules are well suited to accept holes from the hole transport material.

Each of the host materials studied here has a HOMO energy shallower than that of PO15. This ensures the holes are confined to the emissive layer. As such, any of these molecules have potentially favorable hole injection and transport. However, the LUMO energies of some of the molecules are not suitable for facile electron injection from PO15. Specifically, HT 3, HM 1, and HM 11 have E_{LUMO} higher than that of PO15: for molecules such as these it will be hard to inject electrons from the ETL. Consequently, the driving voltage will increase. Another factor to consider is the relationship of the LUMO energy of the host to that of TAPC. If E_{LUMO}(host) is deeper than that of TAPC, electrons are blocked from entering the HTL. As a consequence, only HM 2, HM 4, HM 12, and HM 22 among the materials in this series, are suitable host materials for the system where TAPC and PO15 are used as the hole and electron transport material.

The location of the electron density of the HOMO and LUMO provides valuable information in the design of new host molecules. The extent of localization of the ground state provides an indication as to the singlet and triplet energies of the molecule. In particular, a molecule that has a long conjugation length (extended conjugation) of the electron density will have a lower singlet and triplet energy than a molecule that has a shorter localization length. Additional useful insight can be gained from the position of the nodes. To demonstrate how ground-state electron density maps can be used to design molecules, triphenylamine (TPA) was substituted with (a) phenyl, (b) diphenylborane, and (c) diphenylphosphine oxide moieties (Figure 10.4). In each case, the substitution on the TPA moiety affected the E_{HOMO} and E_{LUMO}. Nonetheless, all materials retained a HOMO level shallow enough for hole injection. However, coupling of TPA with a phenyl ring extended the conjugation of the ground state, with the direct result that the triplet energy was decreased. Coupling the diphenylboryl moiety

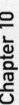

Chapter 10

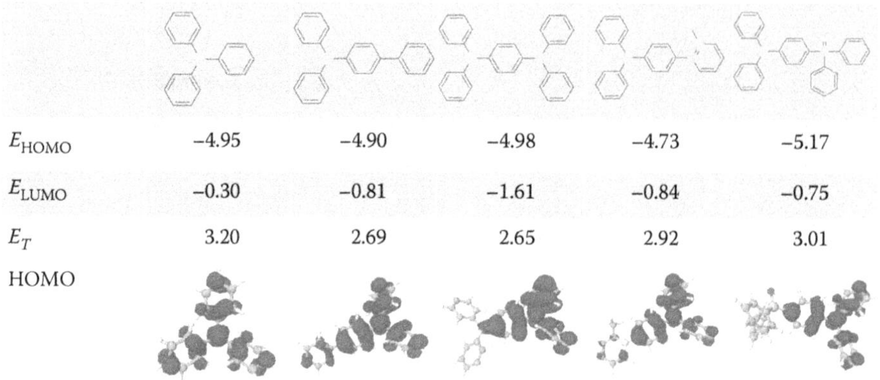

E_{HOMO}	-4.95	-4.90	-4.98	-4.73	-5.17
E_{LUMO}	-0.30	-0.81	-1.61	-0.84	-0.75
E_T	3.20	2.69	2.65	2.92	3.01
HOMO					

FIGURE 10.4 Chemical structure, energy of the triplet state, energies of HOMO and LUMO levels, and the electron density maps of substituted triphenylamine moieties.

with TPA results in intramolecular electron transfer from the amine to the boron atom through the bridging phenyl group. This intramolecular electron transfer also extends conjugation (as in the phenyl), with the same result—lower triplet exciton energy. This phenomenon has been seen experimentally for a molecule with a dimesitylphenylboryl moiety coupled with a tertiary aromatic amine (Tamoto et al. 1997). In the case of the diphenylphosphine oxide moiety, the saturated P=O linkage acts to inductively withdraw electron density from each of the attached aryl groups to generate a polar system with weak intramolecular charge transfer, but no extension of the molecular system's conjugation length. Thus, the inert PO moiety provides the ability to control the exciton energy, leading to an experimental E_T of 2.9 eV.

10.3.2 Excited-State Energy Levels

To investigate the properties and dynamics of molecular systems in the excited state (or in the presence of time-dependent potentials), time-dependent density functional theory (TD-DFT) can be used. Individual molecules or solids can be studied with TD-DFT to predict features such as excitation energies. As evident from the name, TD-DFT is an extension of DFT (Burke et al. 2005). The conceptual and computational foundations of TD-DFT are analogous to DFT, but the formal foundation is the Runge–Gross (RG) theorem proposed in 1984 (Runge and Gross 1984), also considered the time-dependent analog of the H–K theorem (Hohenberg and Kohn 1964). The RG theorem shows that, for a given initial wavefunction, there is a unique mapping between the time-dependent external potential of a system and its time-dependent density. The most popular application of TD-DFT is in the prediction of the energies of excited states of isolated molecules, that is, single molecule in gas phase. However, it is important to note that these predictions must be done for a series of similar molecules and when taken within the context of the series predictions are fairly accurate. In one example of the application of this method, researchers at PNNL have designed, synthesized, and studied several PO containing ambipolar host materials. Here, two of those molecules are used to demonstrate the correlation between TD-DFT studies and experimental measurements.

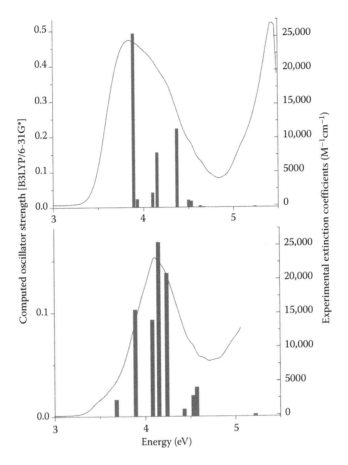

FIGURE 10.5   TD-DFT analysis of two ambipolar hosts, HM-A1 and HM-A2 (structures given in the figure), based on triphenylamine and phosphine oxide moieties.

The TD-DFT analysis of HM-A1 and its structural isomer HM-A2 is shown in Figure 10.5. The experimentally measured solution absorbance spectra of these molecules are also plotted for comparison. The lowest-energy singlet transitions for the two molecules are different; this is reflected in both the theoretically predicted absorbance and the experimentally determined absorption spectra. The lowest-energy transitions of both HM-A1 and HM-A2 originate from the HOMO state, which is localized on the TPA moiety. All singlet-excited states are due to π to π* transitions. However, the *meta* configuration of the bridging phenyl ring of HM-A2 changes both the energy and the oscillator strength (f) of the lowest energy transition compared to the para-isomer HM-A1. For HM-A2, the LUMO ← HOMO transition is red-shifted compared to HM-A1, and has lower oscillator strength. For HM-A1, the two lowest singlet states are close (within 0.04 eV) while the gap between these two states in HM-A2 widens to 0.21 eV. For HM-A1, the LUMO ← HOMO transition is the dominant transition, whereas for HM-A2, it is the LUMO + 1 ← HOMO transition. The close correlation between the predicted and experimental data indicates that we can use TD-DFT studies to estimate the energies of the singlet- and triplet-excited states.

Chapter 10

10.3.3 Mobility

A complete theoretical characterization of transport properties of organic materials was carried out by Brédas using DFT. The reliability of DFT to reproduce experimental findings was demonstrated (Coropceanu et al. 2007). In this work, it was shown that charge carrier mobility (both hole and electrons) occurs via a band hopping transition occurring at about room temperature. This hopping mechanism can be described as electron transfer from one molecule (charged) to the adjacent relaxed molecule. The electron transport in the ETL can be described as a self-exchange reaction (Equation 10.1), where the M^- refers to the charged molecule in the anionic state, and M refers to the adjacent relaxed molecule in the neutral state.

$$M^- + M \rightarrow M^- + M \tag{10.1}$$

The hopping rates for the self-exchange reaction can be explained by semiclassical Marcus theory, given below (Marcus 1993):

$$K_{ET} = \frac{4\pi^2}{h} \frac{1}{\sqrt{4\pi k_B T}} t^2 \exp\left(\frac{\lambda}{4\pi k_B T}\right) \tag{10.2}$$

The rate constant is dependent on only two factors: the electronic coupling term (t) between two adjacent molecules and the reorganization energy (λ). The electronic coupling term is also called the transfer integral, as it largely depends upon the orbital overlap between the molecules involved, and needs to be large for high mobility (Brédas et al. 2002; Park et al. 2007). The reorganization energy (λ) is a measure of the change in nuclear coordinates that occurs when an electron is added or removed from the molecular system, as well as from the changes in the surrounding environment due to polarization effects. Sakanoue demonstrated that the reorganization energy is considered a good measure for hole mobility and that DFT methods are reliable for investigation of hole mobility (Sakanoue et al. 1999). The intramolecular reorganization energy (λ), which is an intrinsic property of a material, can be estimated with an adiabatic process using adiabatic potential energy surface (shown in Figure 10.6). The total adiabatic reorganization energy is the sum of the reorganization energies given in Equations 10.3 and 10.4 (Marcus 1956a,b; Marcus and Sutin 1985; Silinsh et al. 1995; Reimers 2001; Bixon and Jortner 2007):

$$\lambda_N = E_N(\text{rel}) - E_N \tag{10.3}$$

$$\lambda_C = E_C(\text{rel}) - E_C \tag{10.4}$$

The energy terms $E_N(\text{rel})$ and $E_C(\text{rel})$ are the energies of the neutral state in the optimized (relaxed) geometry of a charged (cationic or anionic) molecule and the energies of a charged state in the optimized geometry of a neutral molecule, respectively. The terms E_N and E_C refer to the energies of the neutral state in the optimized geometry of a neutral molecule and the energies of the charged states in the optimized geometry of a charged molecule, respectively.

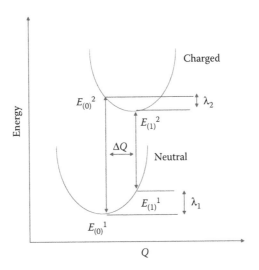

FIGURE 10.6   Schematic diagram of general adiabatic energy surfaces corresponding to the ionization process. Q is a normal-mode displacement. λ_C and λ_N indicate the reorganization energy of the neutral and charged states.

The measured reorganization energies can be used to predict the mobility of a given molecule. However, a complete mobility study of a series of related molecules is needed to check the fidelity of the method. Work by Aonuma, where hole mobility of hole-transporting molecules were measured and reported using time of flight (TOF) methods, provides a useful data set for comparison of calculated and measured values (Aonuma et al. 2007). For comparison, Padmaperuma et al. calculated the reorganization energy using the method described by Brédas et al. (2002). The results are shown in Figure 10.7. From Figure 10.7, it is clear that hole transport materials with a smaller

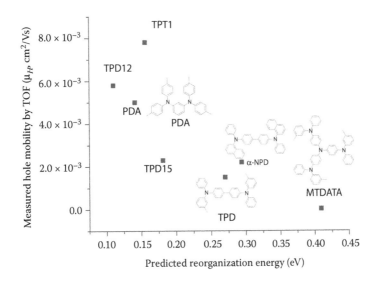

FIGURE 10.7   Time of flight mobilities (Aonuma et al. 2007) versus computed reorganization energies for common OLED hole-transporting materials.

reorganization energy have higher mobility. This work indicates that the reorganization energy is a good measure for σ_h, and that DFT methods can be used for estimation of σ_h. In general, therefore, computing the reorganization energies of a library of molecules gives a relative measure of σ_h. In another example, a similar approach was used to screen novel derivatives of the dinaphtho[2,3-*b*:2′,3′-*f*]thieno[3,2-*b*]thiophene semiconductor with high hole mobility and air stability, leading to the discovery of a new high-performance semiconductor (Sokolov et al. 2011).

10.3.4 Computational Design of Host Molecules

Given the results of the previous section, the logical first step in considering a new class of molecules in the design of new host materials is to evaluate the electronic properties of a library of molecules derived from a common core. Once a promising series of molecules of a particular class is identified, a thorough computational analysis can be carried out. For instance, let us assume we intend to develop a new set of ambipolar compounds with electron- and hole-transporting properties tuned to a range of values to match adjacent hole- and electron-transporting layers. We can combine an electron-transporting moiety (ETm) and the hole-transporting moiety (HTm) in one molecule to give rise to an ambipolar molecule. However, these molecules must be connected—and the chemical moiety used for connection should be a point of saturation (i.e., incapable of participating in a molecular order with extended conjugation). In other words, the electrons in the HTm should not be coupled to those in the ETm. This can be done by using a variety of species, though most commonly, a silane, carbon, or a PO moiety is used. In the rest of this section, the design of ambipolar host molecules using theory is described using a series of molecules as a convenient example.

A number of software packages exists to carry out DFT calculations. The work described in this chapter was carried out employing the *NWChem* (www.emsl.pnl.gov/NWChem) computational package developed at PNNL. Geometry optimization was carried out at B3LYP/6-31G*, which is a typical choice given its reasonable size and adequate description of electronic effects of interest. From the optimized geometries, the energies and electron densities of occupied and virtual orbital can estimated. For this purpose, TPA or *N*-phenylcarbazole (NPC) can be used as primary hole-transporting moieties (Molecule A and B, respectively, in Figure 10.8).

The HOMO level clearly resides on the hole-transporting moiety of the molecule. As a result, the energy of the HOMO level will also change as the structure is modified. The TPA moiety has a shallower E_{HOMO} than that of NPC; therefore, the use of TPA as the HTm results in a shallower HOMO energy compared to the carbazole derivatives. If a substituted TPA or NPC was used, the energy will change accordingly. This provides the ability to tune the HOMO energy to match the application. For the next set of molecules, TPA was used as HTm, and the ETm was changed (molecules C, D, and E in Figure 10.8). Although not shown, the HOMO level of each of these molecules was localized on the TPA moiety. As expected, the HOMO energy did not change due to modifications of the ETm. However, the LUMO level was localized on the lowest-energy ETm (phenyl vs. biphenyl). Therefore, the energy of the LUMO level changed according to the molecule used.

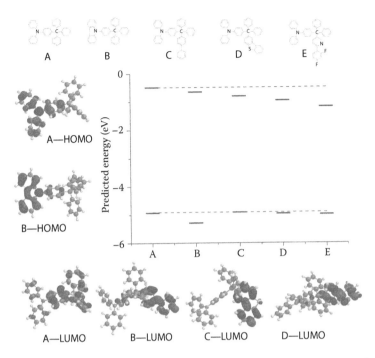

FIGURE 10.8 Computed energy of the HOMO and LUMO states and corresponding electron density maps of a series of ambipolar host molecules.

We can draw the following conclusions from the initial computational analysis: (a) the HOMO of the molecule is localized on the HTm, (b) the LUMO is localized on the ETm, (c) E_{HOMO} is independent of the design approach, (d) E_{HOMO} of the HM is similar to E_{HOMO} of HTm, (e) and the E_{LUMO} varies with the design approach. The take-home message here is that if we can design a molecule by drawing it on a cocktail napkin, we can use computational chemistry to evaluate the electronic properties of that molecule, provided that we have a library of data for similar molecules.

10.4 Physical and Chemical Characterization of Host Materials

Computational prediction of molecular properties, materials synthesis, and device evaluation comprise an iterative process of materials development, with built-in feedback loops from theory to experiment. This chapter will not address materials synthesis or purification—a variety of methods is used for both, but are drawn from standard synthetic chemistry procedures. However, once new molecules are synthesized and purified, they must be characterized, some methods of which are fairly specialized for charge transport materials. The information obtained from characterization tools guides further tuning of new materials and helps explain observed device behavior.

The electronic energy level structure of a host material plays a critical role in determining charge transport, recombination efficiency, and drive voltage of an OLED. The boundary orbital energy positions determine the value of energy barriers for charge injection and the ability of the organic stack as a whole to provide the necessary charge confinement in the emissive zone, while the triplet exciton energy of the host affects the host–guest energy transfer and emission efficiency and quenching. The energy of the excited states (i.e., singlet and triplet) can be experimentally estimated from absorbance, and photoluminescence spectra, and the methods for doing so have been well documented in the literature (Turro 1991). The direct method to determine the frontier orbital energy values is photoelectron spectroscopy. However, it has been shown that there exists a correlation between the frontier orbital energies derived from direct methods such as ultraviolet photoelectron spectroscopy (UPS) and inverse photoemission spectroscopy (IPES), and much simpler solution-based electrochemical methods (D'Andrade et al. 2005; Djurovich et al. 2009). Therefore, solution electrochemical values can be used to estimate the HOMO and LUMO energies of a molecule.

The unique requirement for the host as opposed to all other constituents of the device is that it needs to provide charge transport of both electrons and holes to maintain charge balance in the emissive zone. Direct measurement methods to determine charge transport in molecules are not trivial (see Chapter 7 for another discussion of measurement methods). The most commonly used method is TOF measurement. TOF mobility is often higher than other methods, which may be more sensitive to disorder, grain boundaries, and so on (Coropceanu et al. 2007). One of the shortcomings of the TOF method is that measuring TOF mobilities requires very thick films of the material, which is not an ideal tool to screen potential materials that are intended to be used in very thin films. It is also very difficult to measure the mobility of both electrons and holes in the same material under relevant conditions. Most host materials, even ambipolar hosts, preferentially transport carriers of one sign over the other. As an alternative, the ability of a new material to transport a given carrier can be probed by studying the location of the emission zone. The rest of this chapter will briefly describe electrochemical methods, unipolar devices, and the use of emission zone location to determine relative charge transport rates.

10.4.1 Electrochemical Methods

Solution cyclic voltammetry measurements have become routinely used for experimental estimation of HOMO and LUMO energies. Typically, a reduction (indicative of the LUMO level) or oxidation (related to the HOMO level) potential is measured in solution in relation to a known reference with a stable redox potential. Further, the solution electrochemical value is converted to a quantity corresponding to a gas-phase molecular orbital energy using a well-established reference compound such as ferrocene (D'Andrade et al. 2005; Djurovich et al. 2009). The E_{HOMO} and E_{LUMO} values of a new molecule can be derived from the cyclic voltammetry data using Equations 10.5 and 10.6:

$$E_{LUMO} = -1.19E_{red} - 4.78$$

(10.5)

$$E_{HOMO} = -1.4E_{ox} - 4.6$$

(10.6)

These methods are useful in estimating trends. As an example, the energies of the HOMO and LUMO of some PO-based host molecules were predicted using theoretical methods at the B3LYP/6-31G* level. Subsequent characterization indicated that the calculations routinely overestimate unoccupied orbital energies, but accurately predict trends within a series of molecules (Marsal et al. 2004).

10.4.2 Use of Unipolar Devices to Characterize Charge Transport

The ability of a material to transport charge of either or both carriers can be evaluated using single-carrier devices. The comparison of J–V characteristics of hole-only and electron-only devices can provide an indication of ambipolar charge transport in a host, provided that the proper electrodes are selected for the respective devices. To evaluate the electron and hole currents of two different narrow bandgap ambipolar hosts, devices with the structure ITO/100 nm organics/20 nm Au/100 nm Al (hole-only) and ITO/20 nm Al/100 nm organics/1 nm LiF/100 nm Al (electron-only) were studied. Figure 10.9 shows single-carrier device results—both electron-only and hole-only—for two host materials. Host 1 is electron-dominant as seen by the large discrepancy of the two current densities, while Host 2 is closer to truly ambipolar. The measurements were made with similar current densities for both carriers. These measurements provide useful information regarding the relative transport rates for electrons and holes. This is especially important if the host has poor mobility for a given carrier, in which case it can trap the minority carrier in the emissive layer. Although this method cannot quantitatively determine the electron and hole mobility of a given material, it provides a qualitative estimate of the material's charge transport ability compared to a known reference.

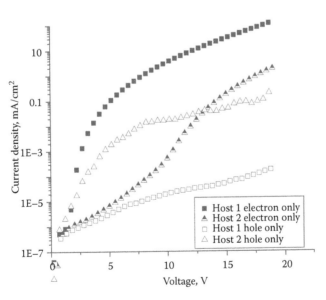

FIGURE 10.9 Results of J–V measurements of single-carrier devices for an electron-dominant host and a charge-balanced host.

10.4.3 Relationship between Charge Transport in the Host and the Emission Zone Location

The chemical structure of a host molecule defines its ability to transport both charge carriers. It has been shown that even materials with seemingly similar structures can perform differently when it comes to charge transport. An example of such differences within a group of interrelated molecules is discussed in Polikarpov et al. (2010), where a study of OLEDs with FIrpic as an emitter doped into a series of PO-based host materials is discussed. The ability of the host to transport charge carriers controls the charge balance in the light-emitting region of the OLED under operation, which strongly influences the quantum efficiency of the emitting device (Giebink and Forrest 2008). One can use computational tools and knowledge of chemistry to design functionalized host materials to control the charge balance in the emissive zone of an OLED. It is natural to assume that the location of the emission zone during device operation correlates with the charge transport character of the host materials (Polikarpov et al. 2010). For a host with ambipolar character, the emission zone is expected to be broad with the maximum emission coming from the bulk of the emissive layer (EML). However, if the host preferentially transports one sign of carriers, then the emission zone is narrowed and moved to the interface of the EML with the opposite sign charge-transporting layer. A schematic of the position of the emission zone for a host with hole-dominated, balanced, and electron-dominated charge transport is shown in Figure 10.10. Based on the location of the region with maximum emission, conclusions can be drawn regarding the balance of charge transport in a host material.

To determine how well balanced the charge transport is on a given host molecule, OLEDs with partially doped emissive layers can be used (Polikarpov et al. 2010). One way to represent the relationship between the emission zone distribution and the host charge transport is to plot the EQE of the devices with partially doped emissive layers as a function of the doping zone location, as shown in Figure 10.11. These results were generated from OLEDs using a series of phosphine-oxide host with different chemical structure

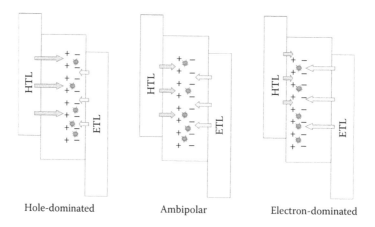

Hole-dominated Ambipolar Electron-dominated

FIGURE 10.10 Schematic representation of the emission region of an OLED with hole-dominated charge transport, balanced charge transport, and electron-dominated charge transport (left to right). The position and width of the emission zone changes according to the charge transport properties of the host.

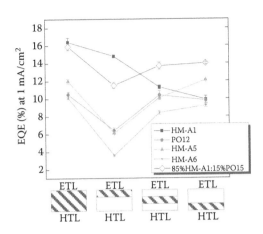

FIGURE 10.11   External quantum efficiency of a series of OLEDs using FIrpic doped into a PO host, where the position of the FIrpic has been systematically varied in the host layer to elucidate the charge transport properties of the PO host.

designed to alter the charge transport properties. The pattern-filled areas within the EML in the schematic representation of the OLEDs in Figure 10.11 correspond to the FIrpic-doped regions of the host. These EQE results were all generated using a current density of 1 mA/cm² for each doping location to ensure comparability. It can be seen from Figure 10.11 that most of the emission in the OLEDs using HM-A1 as the host comes from the EML region adjacent to the ETL interface. This can be seen by noting that the device EQE decreases as the FIrpic-doped zone moves away from the ETL interface. In a host material that preferentially transports holes, the holes injected from the HTL travel to the EML/ETL interface where they recombine with the electrons injected from the ETL to produce excitons and emission. This is exactly what is seen for HM-A1.

In contrast to HM-A1, HM-A5 predominantly conducts electrons (Polikarpov et al. 2010). Accordingly, the location of the emission zone shifts to the HTL interface. The maximum luminance comes from the devices with the FIrpic-doped portion of HM-A5 located next to the HTL. Devices fabricated using PO12 as the host material represent an intermediate case for a host, which is closer to charge-balanced. Based on single-carrier device data, PO12 has the most ambipolar character among the hosts shown in Figure 10.11. The single-carrier device data correlates well with the wide recombination zone centered within the EML. Finally, the emission zone for HM-A6 devices lies between the location of the emission zone for HM-A5 and PO12 devices. HM-A6 is not as ambipolar as PO12, but also not as electron-dominant as HM-A5. As a result, the EQE maxima puts the emission zone of HM-A6 next to the HTL because of the higher electron transport, but because of the higher hole transport in HM-A6 (in comparison to HM-A5 and similar to PO-12) there is also a significant amount of emission from the center doped region.

Figure 10.11 also shows results for OLEDs with a three-component emissive layer—a so-called cohost or mixed host device. In this case, the host consists of 5% FIrpic doped into an equal mix of HM-A1 and PO15. PO15 preferentially transports electrons and is generally used as an ETL material in the OLEDs; this provides good balance with a minimal injection barrier. Figure 10.11 demonstrates how mixing physically mixing

Chapter 10

components with complementary charge transport properties can improve the charge transport of the host matrix. The results shown in Figure 10.11 demonstrate that adding 15% PO15 into HM-A1 can, in fact, shift the emission zone from the EML/ETL interface (where it originally resided in hole-rich HM-A1 OLEDs) to the EML/HTL interface. This cohost emissive zone device shows that introducing electroactive additives that affect the location of the recombination zone and help tune charge balance in the emissive layer can also modify the charge transport properties of the host matrix. This is an alternative to chemically modifying the host structure, although processing and stability may be affected by adding in the cohost.

10.5 Summary

The host material plays a critical role in charge transport and confinement, in ensuring efficient exciton generation and limiting quenching. Together, these roles are essential in ensuring high efficiency and long lifetime in OLEDs. Charge mobility, electronic structure, processibility, and chemical stability are the most important attributes of a host material. The HOMO and LUMO energy levels must be compatible with adjacent layers and the emitter, while the triplet energy should be about the same as that of the emitter to prevent quenching by back transfer. The mobility should be high for both electrons and holes—and they should be similar in magnitude as well. The host should be stable, and not contribute to degradation of any of the other components of the device. Finally, the host must be amenable to whatever processing methods are to be used for deposition and device fabrication.

We have shown in this chapter methods for designing and characterizing host materials, including ways to elucidate charge transport properties and their effect in device efficiency under relevant OLED operating conditions. We alluded to the need to optimize device performance in the context of the whole system (device components, architecture, and fabrication), as opposed to meeting an ideal ambipolarity, which may not be necessary, and in fact could be detrimental. It may be advantageous from a stability standpoint to locate the emission zone near an interface in order to minimize the amount of electron current (which may be implicated in the degradation of certain materials; see Chapter 13 for a detailed discussion of chemical stability in OLEDs). The introduction of a charge-transporting host in a guest–host architecture opened up the possibility of very high quantum efficiency devices. This has led to a very bright future for a field that was first discovered in the 1960s, and was nearly dormant until the 1980s—a future that shines as bright as a modern OLED.

References

Adachi, C., R. C. Kwong, P. Djurovich, V. Adamovich, M. A. Baldo, M. E. Thompson et al. 2001. Endothermic energy transfer: A mechanism for generating very efficient high-energy phosphorescent emission in organic materials. *Applied Physics Letters* 79(13): 2082–2084.

Aonuma, M., T. Oyamada, H. Sasabe, T. Miki and C. Adachi. 2007. Material design of hole transport materials capable of thick-film formation in organic light emitting diodes. *Applied Physics Letters* 90(18): 183503.

Bixon, M. and J. Jortner. 2007. Electron transfer—from isolated molecules to biomolecules. *Advances in Chemical Physics*. I. Prigogine and S. A. Rice. Hoboken, NJ, USA, John Wiley & Sons, Inc. 106: 35–202.

Brédas, J. L., J. P. Calbert, D. A. da Silva and J. Cornil. 2002. Organic semiconductors: A theoretical characterization of the basic parameters governing charge transport. *Proceedings of the National Academy of Sciences of the United States of America* 99(9): 5804–5809.

Burke, K., J. Werschnik and E. K. U. Gross. 2005. Time-dependent density functional theory: Past, present, and future. *Journal of Chemical Physics* 123(6): 062206.

Burrows, P. E., A. B. Padmaperuma, L. S. Sapochak, P. Djurovich and M. E. Thompson. 2006. Ultraviolet electroluminescence and blue-green phosphorescence using an organic diphosphine oxide charge transporting layer. *Applied Physics Letters* 88(18): 183503.

Cai, X., A. B. Padmaperuma, L. S. Sapochak, P. A. Vecchi and P. E. Burrows. 2008. Electron and hole transport in a wide bandgap organic phosphine oxide for blue electrophosphorescence. *Applied Physics Letters* 92(8): 083308.

Coropceanu, V., J. Cornil, D. A. da Silva Filho, Y. Olivier, R. Silbey and J.-L. Bredas. 2007. Charge transport in organic semiconductors. *Chemical Reviews* 107(4): 926–952.

D'Andrade, B. W., S. Datta, S. R. Forrest, P. Djurovich, E. Polikarpov and M. E. Thompson. 2005. Relationship between the ionization and oxidation potentials of molecular organic semiconductors. *Organic Electronics* 6(1): 11–20.

Djurovich, P. I., E. I. Mayo, S. R. Forrest and M. E. Thompson. 2009. Measurement of the lowest unoccupied molecular orbital energies of molecular organic semiconductors. *Organic Electronics* 10(3): 515–520.

Giebink, N. C., B. W. D'Andrade, M. S. Weaver, P. B. Mackenzie, J. J. Brown, M. E. Thompson et al. 2008. Intrinsic luminance loss in phosphorescent small-molecule organic light emitting devices due to bimolecular annihilation reactions. *Journal of Applied Physics* 103(4): 044509.

Giebink, N. C. and S. R. Forrest. 2008. Quantum efficiency roll-off at high brightness in fluorescent and phosphorescent organic light emitting diodes. *Physical Review B* 77(23): 235215.

Hohenberg, P. and W. Kohn. 1964. Inhomogeneous electron gas. *Physical Review B* 136(3B): B864.

Holmes, R. J., B. W. D'Andrade, S. R. Forrest, X. Ren, J. Li and M. E. Thompson. 2003. Efficient, deep-blue organic electrophosphorescence by guest charge trapping. *Applied Physics Letters* 83(18): 3818.

Marcus, R. A. 1956a. Electrostatic free energy and other properties of states having nonequilibrium polarization. 1. *Journal of Chemical Physics* 24(5): 979–989.

Marcus, R. A. 1956b. On the theory of oxidation-reduction reactions involving electron transfer. 1. *Journal of Chemical Physics* 24(5): 966–978.

Marcus, R. A. 1993. Electron-transfer reactions in chemistry—Theory and experiment. *Reviews of Modern Physics* 65(3): 599–610.

Marcus, R. A. and N. Sutin. 1985. Electron transfers in chemistry and biology. *Biochimica et Biophysica Acta* 811(3): 265–322.

Marsal, P., I. Avilov, D. A. da Silva, J. L. Bredas and D. Beljonne. 2004. Molecular hosts for triplet emission in light emitting diodes: A quantum-chemical study. *Chemical Physics Letters* 392(4–6): 521–528.

Matsushima, T. and C. Adachi. 2006. Extremely low voltage organic light-emitting diodes with p-doped alpha-sexithiophene hole transport and n-doped phenyldipyrenylphosphine oxide electron transport layers. *Applied Physics Letters* 89(25): 253506.

Oyamada, T., H. Sasabe, C. Adachi, S. Murase, T. Tominaga and C. Maeda. 2005. Extremely low-voltage driving of organic light-emitting diodes with a Cs-doped phenyldipyrenylphosphine oxide layer as an electron-injection layer. *Applied Physics Letters* 86(3): 033503.

Padmaperuma, A. B., L. S. Sapochak and P. E. Burrows. 2006. New charge transporting host material for short wavelength organic electrophosphorescence: 2,7-bis(diphenylphosphine oxide)-9,9-dimethylfluorene. *Chemistry of Materials* 18(9): 2389–2396.

Park, Y. H., K. Yang, Y. H. Kim and S. K. Kwon. 2007. Ab initio studies on acene tetramers: Herringbone structure. *Bulletin of the Korean Chemical Society* 28(8): 1358–1362.

Parr, R. G. and W. T. Yang. 1995. Density-functional theory of the electronic-structure of molecules. *Annual Review of Physical Chemistry* 46: 701–728.

Polikarpov, E., J. S. Swensen, N. Chopra, F. So and A. B. Padmaperuma. 2009. An ambipolar phosphine oxide-based host for high power efficiency blue phosphorescent organic light emitting devices. *Applied Physics Letters* 94(22): 223304.

Polikarpov, E., J. S. Swensen, L. Cosimbescu, P. K. Koech, J. E. Rainbolt and A. B. Padmaperuma. 2010. Emission zone control in blue organic electrophosphorescent devices through chemical modification of host materials. *Applied Physics Letters* 96(5): 053306–053309.

Pope, M., H. P. Kallmann and P. Magnante. 1963. Electroluminescence in organic crystals. *The Journal of Chemical Physics* 38(8): 2042–2043.

Chapter 10

Reimers, J. R. 2001. A practical method for the use of curvilinear coordinates in calculations of normal-mode-projected displacements and Duschinsky rotation matrices for large molecules. *Journal of Chemical Physics* 115(20): 9103–9109.

Ren, X. F., J. Li, R. J. Holmes, P. I. Djurovich, S. R. Forrest and M. E. Thompson. 2004. Ultrahigh energy gap hosts in deep blue organic electrophosphorescent devices. *Chemistry of Materials* 16(23): 4743–4747.

Runge, E. and E. K. U. Gross. 1984. Density-functional theory for time-dependent systems. *Physical Review Letters* 52(12): 997–1000.

Sakanoue, K., M. Motoda, M. Sugimoto and S. Sakaki. 1999. A molecular orbital study on the hole transport property of organic amine compounds. *Journal of Physical Chemistry A* 103(28): 5551–5556.

Sapochak, L. S., A. B. Padmaperuma, X. Cai, J. L. Male and P. E. Burrows. 2008. Inductive effects of diphenylphosphoryl moieties on carbazole host materials: Design rules for blue electrophosphorescent organic light-emitting devices. *Journal of Physical Chemistry C* 112(21): 7989–7996.

Sapochak, L. S., A. B. Padmaperuma, P. A. Vecchi, H. Qiao and P. E. Burrows. 2006. Design strategies for achieving high triplet energy electron transporting host materials for blue electrophosphorescence—Article no. 63330F. *Organic Light Emitting Materials and Devices X.* Z. H. Kafafi. SPIE—International Society for Optical Engineering, Bellingham, Washington. 6333: F3330–F3330.

Short, G. D. and D. M. Hercules. 1965. Electroluminescence of organic compounds: The role of gaseous discharge in excitation process. *Journal of the American Chemical Society* 87(7): 1439–1442.

Silinsh, E. A., A. Klimkans, S. Larsson and V. Capek. 1995. Molecular polaron states in polyacene crystals—formation and transfer processes. *Chemical Physics* 198(3): 311–331.

Sokolov, A. N., S. Atahan-Evrenk, R. Mondal, H. B. Akkerman, R. S. Sanchez-Carrera, S. Granados-Focil et al. 2011. From computational discovery to experimental characterization of a high hole mobility organic crystal. *Nature Communications* 2: 437.

Tamoto, N., C. Adachi and K. Nagai. 1997. Electroluminescence of 1,3,4-oxadiazole and triphenylamine-containing molecules as an emitter in organic multilayer light emitting diodes. *Chemistry of Materials* 9(5): 1077–1085.

Turro, N. J. 1991. *Modern Molecular Photochemistry*. Sausalito, CA, University Science Books.

Valiev, M., E. J. Bylaska, N. Govind, K. Kowalski, T. P. Straatsma, H. J. J. Van Dam et al. 2010. NWChem: A comprehensive and scalable open-source solution for large scale molecular simulations. *Computer Physics Communications* 181(9): 1477–1489.

Vecchi, P. A., A. B. Padmaperuma, H. Qiao, L. S. Sapochak and P. E. Burrows. 2006. A dibenzofuran-based host material for blue electrophosphorescence. *Organic Letters* 8(19): 4211–4214.

11. Phosphorescent Emitters

Valentina A. Krylova and Mark E. Thompson

11.1   Emitters for High Efficiency OLEDs

The emissive electroluminescent layer is the fundamental element of an organic light-emitting device (OLED). The choice of emitter for an OLED is essential to achieve high efficiency and desired emission color. Early OLEDs relied on fluorescent materials (singlet emitters) and had rather low efficiencies (Burroughes et al., 1990; Tang and Vanslyke, 1987). In the late 1990s it was shown that four times greater efficiency can be achieved by using phosphorescent materials (triplet emitters) in OLEDs (Baldo et al., 1998; O'Brien et al., 1999). Since this breakthrough phosphorescent materials have been considered as the foremost emitters for high efficiency OLEDs as they enable attainment of the maximum theoretical efficiency.

The basic mechanism of electroluminescence (EL) involves carrier recombination, leading to formation of an electron–hole pair (exciton), which radiatively relaxes to give the observed EL. The hole and electron in an OLED carry a spin of +1/2 or −1/2. When a hole and an electron recombine to generate an exciton, four different spin combinations are possible (one singlet and three triplets) (Figure 11.1). Simple spin statistics predicts that

Chapter 11

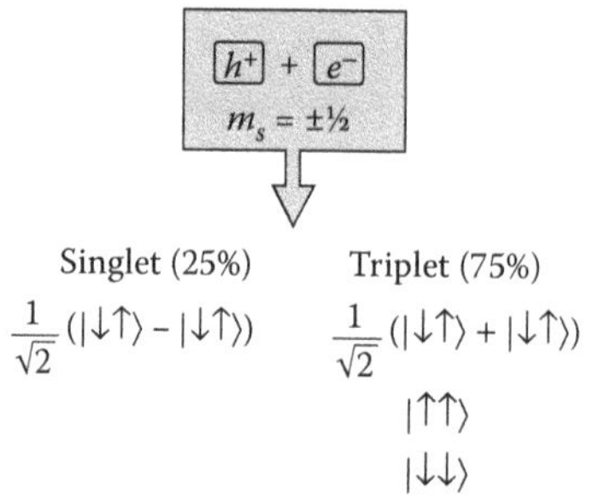

$$\boxed{h^+} + \boxed{e^-}$$
$$m_s = \pm\tfrac{1}{2}$$

Singlet (25%) Triplet (75%)

$$\frac{1}{\sqrt{2}}(|\downarrow\uparrow\rangle - |\downarrow\uparrow\rangle)$$ $$\frac{1}{\sqrt{2}}(|\downarrow\uparrow\rangle + |\downarrow\uparrow\rangle)$$

$$|\uparrow\uparrow\rangle$$
$$|\downarrow\downarrow\rangle$$

FIGURE 11.1 Electron–hole recombination gives a statistical mixture of 25% singlet and 75% triplet excitons.

25% of formed excitons are singlet (total spin 0) and 75% are triplet (total spin 1) excitons (Baldo et al., 1999b; Segal et al., 2003). Fluorescent dopants emit from singlet states, so only singlet excitons are utilized and triplet excitons are lost to nonradiative decay processes. As a result, fluorescent OLEDs are limited to 25% internal quantum efficiency, which is only $\approx$5% external quantum efficiency (EQE; see Chapter 12 for a discussion of optical losses which reduce the external quantum efficiency). Phosphorescent dopants can efficiently harvest both singlet and triplet excitons, thus 100% internal quantum efficiency can be achieved (20%–25% external efficiency). Therefore, phosphorescent emitters are crucial for high efficiency OLEDs.

Not all phosphorescent emitters are well suited for OLED applications. The key requirement for a phosphorescent dopant is that it should have a high radiative rate (k_r), together with a low nonradiative rate. This means that it should have a high phosphorescent quantum yield (Φ) (ideally close to 100%) and short radiative lifetime (τ) (1–10 μs):

$$k_r = \frac{\Phi}{\tau} \tag{11.1}$$

This condition is determined by the resistance–capacitance (RC) time constant of the OLED, which is typically 200–500 ns. If the rate of repopulation of excited states is markedly slower than the rate of exciton formation within the device, then a dopant will not efficiently trap excitons and this will limit device efficiency (Blumstengel and Dorsinville, 1999; Hoshino and Suzuki, 1996). According to first-order perturbation theory, the triplet radiative rate is directly proportional to the spin-orbit coupling (SOC) between singlet and triplet states and inversely proportional to the energy separation between these states (McGlynn et al., 1969):

$$k_r \propto \frac{\langle \psi_{S_1} | H_{SO} | \psi_{T_1} \rangle^2}{(\Delta E_{S_1-T_1})^2} \tag{11.2}$$

Therefore, efficient SOC is very important to achieve fast triplet radiative rates. Purely organic phosphorescent emitters generally have very long radiative lifetimes due to weak SOC between singlet and triplet (π–π^*) or (n–π^*) states (Turro et al., 2009) and consequently are not good phosphorescent dopants for OLEDs. Heavy-metal complexes (especially those containing third row transition metals such as Ir, Pt, Os, Ru) usually have strong SOC due to the "heavy-atom" effect. They can possess metal-to-ligand charge transfer states (MLCT) (d–π^*), which involve metal d-orbitals. A singlet MLCT state can couple strongly with a triplet MLCT state to promote fast intersystem crossing (ISC) to the triplet state. In addition, the energy separation $\Delta E(^1\text{MLCT-}^3\text{MLCT})$ is much smaller than $\Delta E(^1(\pi–\pi^*)–^3(\pi–\pi^*))$ and this, according to Equation 11.2, also leads to a faster radiative rate. It is not uncommon that the lowest excited state of a phosphorescent metal complex is a triplet ligand centered (^{3}LC) $^3(\pi–\pi^*)$ state. If a close lying ^{1}MLCT state is available it can indirectly couple with the lowest ^{3}LC state by configuration interaction resulting in high phosphorescence efficiencies and short radiative lifetimes (Rausch et al., 2010). Thus, organometallic transition metal complexes are considered the best phosphorescent dopants for OLEDs.

11.2 Heavy Metal Complexes as Emitters for OLEDs

11.2.1 Ir(III)-Based Emitters

Ir(III) complexes are the leading phosphorescent OLED emitter materials. This is due to the advantageous combination of photophysical and electrochemical properties, including photo- and thermal stabilities. Cyclometallated Ir(III) complexes have high photoluminescent (PL) quantum yields, up to 100%, and relatively short phosphorescent lifetimes of several microseconds, thus their triplet radiative rates can reach 10^5 s^{-1}. Emission colors can be tuned throughout the visible spectrum. Owing to their huge potential for applications in electroluminescent devices, Ir(III) complexes have been well-developed over the past 15 years. To date, their properties can be controlled and adjusted in a predictable way to satisfy specific application needs (Baranoff et al., 2009; Flamigni et al., 2007). In fact, Ir(III)-based OLEDs that perform close to the theoretical limit have been realized (Tanaka et al., 2007). From a large variety of phosphorescent emitters developed for OLED applications, it is Ir(III)-based emitters that have reached the phosphorescent OLED display market. The first Ir(III) complex used in OLEDs was *facial*-tris(2-phenylpyridinato-N,C^2)iridium(III), *fac*-Ir(ppy)$_3$ (Baldo et al., 1999a). It was synthesized by Watts and coworkers in 1985 (King et al., 1985) and exhibits green phosphorescence (PL λ_{max} = 515 nm, triplet energy E_T = 2.48 eV), a short emission lifetime and high quantum yields, both in degassed dichloromethane solution (1.6 µs; 90%) and in a doped film (1.4 µs; 96%) (Hofbeck and Yersin, 2010; Kawamura et al., 2005). The first Ir(ppy)$_3$-based OLED, in which Ir(ppy)$_3$ was doped into a CBP host (CBP = 4,4′-di(N-carbazolyl)biphenyl), had a structure of ITO/NPD/CBP:Ir(ppy)$_3$ (*x* wt%)/BCP/Alq$_3$/Mg-Al (NPD = N,N′-diphenyl-N,N′-bis(1-naphthyl)-1,1′-biphenyl-4,4′-diamine; BCP = bathocuproine; Alq$_3$ = tris(8-hydroxyquinolinato)aluminum) and showed an external quantum efficiency of 9% at optimal doping concentration of 6 wt% (Baldo et al., 1999a). After optimization, in particular when the CBP host material was replaced with an electron transporting triazole-containing host with higher triplet energy, TAZ (3-(biphenyl-4-yl)-4-phenyl-5-(4-tert-butylphenyl)-1,2,4-triazole), the device external efficiency increased up to 15% (Adachi et al., 2000). In 2007, Kido and coworkers reported an Ir(ppy)$_3$-based OLED with an external quantum efficiency >25% which corresponds to an internal efficiency of $\approx$100% (Tanaka et al., 2007).

Chapter 11

11.2.1.1   Cyclometallated Ir(III) Complexes

Two major types of cyclometallated Ir(III) complexes have been employed as emitting components of OLEDs, that is, homoleptic Ir(C^N)$_3$ and heteroleptic (C^N)$_2$Ir(L^X). The cyclometallating ligand is represented as C^N, where the "C" atom corresponds to an anionic moiety, such as a phenyl, that binds to a metal via a covalent metal–carbon bond; the "N" atom represents a neutral group, such as pyridyl, that forms a coordinative bond with a metal; L^X is a bidentate ancillary ligand. The general synthetic routes to prepare most of these iridium complexes are shown in Scheme 11.1. The first step involves the preparation of a μ-dichloro-bridged dimer complex from IrCl$_3$ · nH$_2$O and a ligand precursor (path a) (Lamansky et al., 2001a; Nonoyama, 1974). In the next step, substitution of the chlorides with a L^X chelate (path b) gives a heteroleptic (C^N)$_2$Ir(L^X) complex with a *trans*-N,N configuration of the C^N ligands. Alternatively, substitution with a third C^N ligand yields a homoleptic tris-cyclometallated complex Ir(C^N)$_3$ (path c), which has two geometric isomers: facial (*fac-*) or meridional (*mer-*). By careful control of reaction conditions (relatively low temperature (<140°C) and choice of the C^N ligand), it is possible to improve selectivity and obtain almost solely the *mer*-isomer. The *fac*-isomer is usually obtained as a major product at high temperatures (>200°C) or by thermal or photochemical isomerization of the *mer*-isomer in solution (path d) (Tamayo et al., 2003). Thus, the *fac*-isomer is a thermodynamic product and the *mer*-isomer is a kinetic product. The facial isomer may also be prepared by refluxing Ir(acac)$_3$ (acac = acetylacetonate) with the appropriate C^N ligand in glycerol (path e) (Dedeian et al., 1991). It is important to point out that these two isomers exhibit different photophysical behavior (Karatsu et al., 2003, 2006; Tamayo et al., 2003). As a result of two strongly donating phenyl groups being trans to each other, meridional isomers are

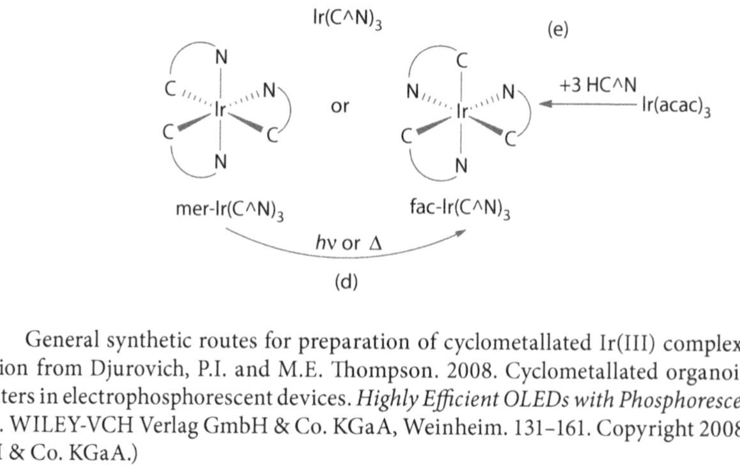

SCHEME 11.1   General synthetic routes for preparation of cyclometallated Ir(III) complexes. (Adapted with permission from Djurovich, P.I. and M.E. Thompson. 2008. Cyclometallated organoiridium complexes as emitters in electrophosphorescent devices. *Highly Efficient OLEDs with Phosphorescent Materials*, Yersin, H., ed. WILEY-VCH Verlag GmbH & Co. KGaA, Weinheim. 131–161. Copyright 2008 Wiley-VCH Verlag GmbH & Co. KGaA.)

easier to oxidize and therefore have lower emission energies than facial ones. They also have lower luminescent quantum yields due to higher nonradiative rates caused by bond dissociation and photoisomerization in the excited state. However, this process is suppressed in rigid media so *mer*-isomers of Ir(C^N)$_3$ can perform in devices almost as efficiently as *fac*-isomers (Yang et al., 2004a).

In order to tune emission properties and to obtain a dopant with desirable photoluminescent properties it is important to understand the nature of its emission. The excited states of cyclometallated iridium(III) complexes have been well-investigated. It was determined early on that luminescence from these complexes originates from an excited state that is an admixture of a triplet LC state and singlet and triplet MLCT states (Scheme 11.2) (Komada et al., 1986; Vanhelmont et al., 1997; Yersin and Humbs, 1999). Application of first-order perturbation theory to describe the mixing between ^{1}MLCT and ^{3}LC states gives Equation 11.3, that describes the lowest triplet state (Lever, 1984):

$$\Psi_{T_1} = \sqrt{1-\alpha^2} \mid {}^3\mathrm{LC}\rangle + \alpha \mid {}^1\mathrm{MLCT}\rangle \tag{11.3}$$

where Ψ_{T_1} is the wave function of the lowest triplet excited state, and α is the coefficient that describes the degree of singlet MLCT character mixed in the triplet LC state. α is directly proportional to the spin-orbit coupling matrix element and inversely proportional to the energy separation between the ^{3}LC and ^{1}MLCT states (Vanhelmont et al., 1997):

$$\alpha = \frac{\langle {}^3\mathrm{LC}|H_{SO}|{}^1\mathrm{MLCT}\rangle}{\Delta E} \tag{11.4}$$

The extent of ^{1}MLCT character (described by the mixing coefficient α) in the lowest excited state T_1 determines its radiative rate (k_r) (compare Equation 11.2). The radiative rate for emission can be estimated using Equation 11.5 (Komada et al., 1986; Li et al., 2005b):

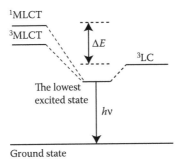

SCHEME 11.2 Schematic representation of the energy level mixing in cyclometallated Ir(III) complexes. (Reprinted with permission from Li, J. et al. 2005b. Synthetic control of excited-state properties in cyclometalated Ir(III) complexes using ancillary ligands. *Inorganic Chemistry* 44(6):1713–1727. Copyright 2005 American Chemical Society.)

$$k_r = k_r\left(^1\text{MLCT}\right)\left(\frac{\langle ^3\text{LC}|H_{SO}|^1\text{MLCT}\rangle}{\Delta E}\right)^2\left(\frac{\nu_{T_1}}{\nu^1_{\text{MLCT}}}\right)^3 \qquad (11.5)$$

where $k_r(^1\text{MLCT})$ is the radiative rate of the perturbing state and ν_{T_1} and ν^1_{MLCT} are the respective absorption transition energies. Large MLCT character leads to significantly increased radiative rate (higher quantum yield, shorter lifetime) of the $T_1 \rightarrow S_0$ transition. The heavy atom effect of iridium promotes efficient SOC and strong-field cyclometallating ligands stabilize the ^{1}MLCT state relative to the ^{3}LC state. These two effects contribute to a higher α value. Consequently, cyclometallated Ir(III) complexes frequently exhibit high phosphorescent quantum yields and short lifetimes.

Theoretical calculations using density functional theory (DFT) methods support this assignment (Hay, 2002; Vlcek and Zalis, 2007). Molecular orbital analysis of Ir(ppy)$_3$ and (ppy)$_2$Ir(acac) (Figure 11.2) shows that the HOMO is composed of approximately equal contributions from the metal d orbitals and π orbitals of pyridine. The LUMO is mainly localized on pyridine, with little metal character. Time-dependent DFT calculations revealed that the lowest triplet transition corresponds to excitation from a metal d orbital mixed with a ligand π orbital, to ligand π^* orbital $((5d, \pi) \rightarrow \pi^*)$ (Hay, 2002; Nozaki, 2006). The result of MO calculations of (ppy)$_2$Ir(acac) shows that acac is an ancillary ligand, as it does not significantly contribute to frontier orbitals of the complex and the excited states are governed by the "(C^N)$_2$Ir" fragment. Indeed, it was observed that the photophysical properties of (ppy)$_2$Ir(acac) and *fac*-Ir(ppy)$_3$ are almost identical (Figure 11.3, entries 1,2).

11.2.1.2   Tuning Emission Properties of Cyclometallated Ir(III) Complexes

The major advantage of cyclometallated iridium(III) complexes as phosphorescent dopants for OLED applications is the tunability of their PL properties. Several strategies have proved to be successful. Complexes shown in Figures 11.3 through 11.5 illustrate the effects of C^N ligand modifications on the emission properties of Ir(C^N)$_3$ and (C^N)$_2$Ir(acac) complexes. As noted above, the photophysical properties of these complexes are governed by the cyclometallating ligand.

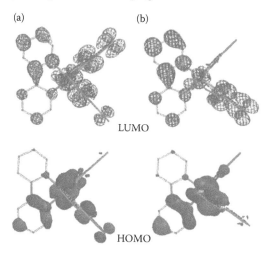

(a)   (b)

LUMO

HOMO

FIGURE 11.2   The calculated HOMO (bottom) and LUMO (top) orbitals for Ir(ppy)$_3$ (a) and (ppy)$_2$Ir(acac) (b).

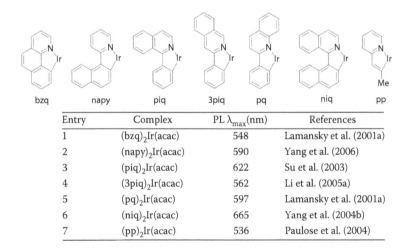

Entry	Complex	PL λ_{max}(nm)	References
1	fac-Ir(ppy)$_3$	519	Hofbeck and Yersin (2010)
2	(ppy)$_2$Ir(acac)	516	Lamansky et al. (2001a)
3	fac-Ir(4′,6′-dfppy)$_3$	467	Endo et al. (2008)
4	(4′,6′-dfppy)$_2$Ir (acac)	482	Li et al. (2005b)
5	fac-Ir(F$_4$ppy)$_3$	468	Ragni et al. (2006)
6	(F$_4$ppy)$_2$Ir(acac)	479	Ragni et al. (2006)
7	fac-Ir(4′-CF$_3$-ppy)$_3$	494 (77 K)	Dedeian et al. (1991)
8	fac-Ir(4′-MeO-ppy)$_3$	481 (77 K)	Dedeian et al. (1991)
9	fac-Ir(5′-MeO-ppy)$_3$	539 (77 K)	Dedeian et al. (1991)
10	fac-Ir(atpy)$_3$	581	Hisamatsu and Aoki (2011)

FIGURE 11.3 The effect of ligand substitution on emission energies of selected Ir(C^N)$_3$ and (C^N)$_2$Ir(acac) complexes in solution.

One of the tools for tuning emission color involves the introduction of electron donating and accepting groups in the C^N ligand (Dedeian et al., 1991; Grushin et al., 2001; Hisamatsu and Aoki, 2011). In general, addition of electron-withdrawing substituents (–F, –CF$_3$) to the phenyl ring reduces electron density on the metal and stabilizes the HOMO, therefore a blue shift is observed. Electron-donating groups (–OCH$_3$, –NH$_2$) on the phenyl ring lead to red shifted emission, due to destabilization of the HOMO. The position of substitution as well as mesomeric and inductive effects of substituents should

Entry	Complex	PL λ_{max}(nm)	References
1	(bzq)$_2$Ir(acac)	548	Lamansky et al. (2001a)
2	(napy)$_2$Ir(acac)	590	Yang et al. (2006)
3	(piq)$_2$Ir(acac)	622	Su et al. (2003)
4	(3piq)$_2$Ir(acac)	562	Li et al. (2005a)
5	(pq)$_2$Ir(acac)	597	Lamansky et al. (2001a)
6	(niq)$_2$Ir(acac)	665	Yang et al. (2004b)
7	(pp)$_2$Ir(acac)	536	Paulose et al. (2004)

FIGURE 11.4 The effect of the size of the C^N π-system on emission energies of (C^N)$_2$Ir(acac) complexes in solution.

Chapter 11

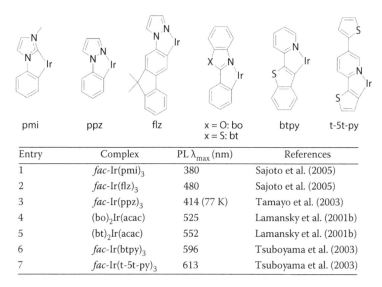

pmi	ppz	flz	x = O: bo x = S: bt	btpy	t-5t-py	

Entry	Complex	PL λ_{max} (nm)	References
1	*fac*-Ir(pmi)$_3$	380	Sajoto et al. (2005)
2	*fac*-Ir(flz)$_3$	480	Sajoto et al. (2005)
3	*fac*-Ir(ppz)$_3$	414 (77 K)	Tamayo et al. (2003)
4	(bo)$_2$Ir(acac)	525	Lamansky et al. (2001b)
5	(bt)$_2$Ir(acac)	552	Lamansky et al. (2001b)
6	*fac*-Ir(btpy)$_3$	596	Tsuboyama et al. (2003)
7	*fac*-Ir(t-5t-py)$_3$	613	Tsuboyama et al. (2003)

FIGURE 11.5 The effect of C^N ligand modification on emission energies of selected Ir(C^N)$_3$ and (C^N)$_2$Ir(acac) complexes in solution.

be taken into account. For example, a methoxy group at the 5-position of the ppy ligand (*para* to the site of coordination) is electron-donating through resonance, thus leads to red-shifted emission. At the 4-position of ppy ligand (*meta* to the site of coordination), a methoxy group has an electron-withdrawing inductive effect due to the electronegativity of oxygen, leading to a blue shift (Figure 11.3, entries 8, 9).

Another effective method to alter emission properties is to change the degree of π conjugation of the C^N ligand (Figure 11.4). Expanding the size of the π-system leads to a bathochromic shift as a result of a decrease in the HOMO-LUMO gap, due to stabilization of LUMO or destabilization of HOMO energy levels. For instance, the (bzq)$_2$Ir(acac) complex, where the π-system of the ppy ligand is expanded by the introduction of a bridging vinyl group, shows a 33 nm red-shift compared to parent (ppy)$_2$Ir(acac). Benzannulation of phenyl or pyridine moieties leads to an even larger red-shift, for example, (napy)$_2$Ir(acac) (PL λ_{max} = 590 nm) and (piq)$_2$Ir(acac) (PL λ_{max} = 622 nm). The site of π-extension is important. This can be illustrated by three different isomers of phenyl-quinoline ligand, that is, 3piq, pq and piq, that give Ir (III) complexes with rather different emission wavelengths of 562, 597, and 622 nm, respectively. Interestingly, attempts to blue shift emission color by reducing the size of π-system of ppy ligand also led to a bathochromic shift in (pp)$_2$Ir(acac) (Figure 11.4, entry 7). This is attributed to increased electron–electron repulsion in the triplet state due to reduction of the π-system, which leads to increased singlet-triplet splitting and therefore lower triplet energy (Turro et al., 2009).

Another approach to tune emission properties is by employing different cyclometallating ligands (Figure 11.5). A C^N ligand with a ^{3}LC state of low energy will red-shift emission relative to a ^{3}LC state with a higher energy. For example, incorporation of sulfur-containing heterocycles into cyclometallating ligands leads to Ir(III) complexes that emit in the yellow–red part of the spectrum. To obtain blue emitters, a pyridine moiety in ppy ligand was replaced with triazole, N-pyrazole or N-heterocyclic carbene-based group that increased the LUMO of the cyclometallating ligand.

Heteroleptic complexes (C^N)$_2$Ir(L^X) provide more room for altering the properties of complexes, as each ligand can be modified independently to achieve desired properties, that is, emission color, stability, solubility. If the singlet and triplet energies of L^X ligand are higher than that of the cyclometallated moiety, then the excited state properties are governed by the "(C^N)$_2$Ir" fragment. The L^X in this case is a nonchromophoric ancillary ligand. For example, the photophysical properties of (C^N)$_2$Ir(acac) and *fac*-Ir(C^N)$_3$ are almost identical. Nonetheless, excited state properties of (C^N)$_2$Ir(L^X) complexes can be fine-tuned by modification of the ancillary ligand. The role of the ancillary ligand was thoroughly investigated for a series of heteroleptic complexes (C^N)$_2$Ir(L^X) having the same cyclometallating ligand 2-(4′-tolylpyridine) (tpy) and different nonchromophoric L^X ligands (Figure 11.6) (Li et al., 2005b). It was shown that variation of the electron withdrawing/donating ability of L^X alters the oxidation potentials of complexes (i.e., varies the HOMO energies), while reduction potentials are effectively unchanged (i.e., the LUMO is unperturbed). The absorption and emission spectra shift to higher energy as the oxidation potentials increase. Stabilization of the HOMO level leads to an increase of the MLCT energy, while the energy of the LC state remains constant (Scheme 11.2). Therefore, the energy of the mixed lowest excited ^{3}LC/MLCT state increases, but the amount of MLCT character is reduced, so consequently the radiative rates decrease. This decrease is compensated somewhat by a decrease in the nonradiative rate for the complexes with high triplet energy.

If the triplet energy of the L^X ligand is lower than that of the C^N ligand, then it is the L^X ligand that principally controls the emission properties of the (C^N)$_2$Ir(L^X) complex. In this case, the L^X ligand is no longer nonchromophoric, and the term "ancillary" loses its meaning. Examples of cyclometallated iridium complexes where emission is dominated by the ancillary ligand include derivatives of β-diketonate (Lamansky

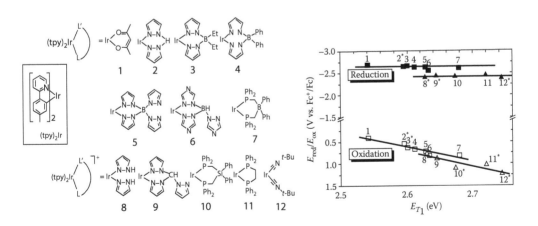

FIGURE 11.6 Molecular structures of the complexes (tpy)$_2$Ir(L^X) and plot of red/ox potentials vs. emission energy for a range of (tpy)$_2$Ir(L^X) complexes. Neutral (tpy)$_2$Ir(L^X) complexes (squares) and cationic (tpy)$_2$Ir(L^X) complexes (triangles). The symbol * indicates an irreversible oxidation or reduction process; otherwise the electrochemical process is reversible or quasi-reversible. (Reproduced with permission from Li, J. et al. 2005b. Synthetic control of excited-state properties in cyclometalated Ir(III) complexes using ancillary ligands. *Inorganic Chemistry* 44(6):1713–1727. Copyright 2005 American Chemical Society.)

et al., 2001b), N^O chelate (N^O = picolinate, quinolinate) (Kwon et al., 2005; You and Park, 2005), complexes possessing high triplet cyclometallating ligands, that is, phenylpyrazole derivatives (Chang et al., 2007), and N-heterocyclic carbenes (NHC) (Lu et al., 2011). Furthermore, the excited state properties can be tuned by modification of L^X ligand. This approach was utilized by Lu et al. to vary the color of the bis-carbene complex (mpmi)$_2$Ir(L^X) (mpmi = 1-(4-methyl-phenyl)-3-methylimidazolin-2-ylidene) by over 130 nm (Figure 11.7) (Lu et al., 2011). The triplet level of the NHC ligand (mpmi) lies above the energies of the chosen L^X ligands. Therefore, as the triplet energy of the chromophoric "ancillary" ligand decreases, the emission color of the complex in solution (CH$_2$Cl$_2$) changed from blue (PL λ_{max} = 466 nm) in the complex (mpmi)$_2$Ir(dmpypz) (dmpypz = 3,5-dimethyl-2-(1H-pyrazol-5-yl)pyridine) to green (PL λ_{max} = 530 nm) in the complex (mpmi)$_2$Ir(pybi) (pybi = 2-(pyridin-2-yl)-1H-benzo[d]imidazole) and to red (PL λ_{max} = 599 nm) in the complex (mpmi)$_2$Ir(priq) (priq = 1-(1H-pyrrol-2-yl) isoquinoline). At the same time, modification of the carbene ligand, via introduction of an –F group, in (fpmi)$_2$Ir(dmpypz), only gave a small shift of emission energy (PL λ_{max} = 455 nm from 466 nm).

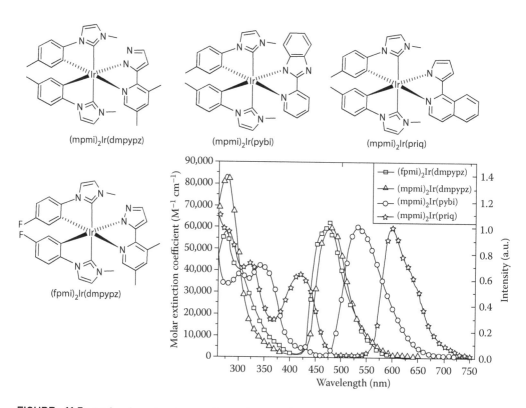

FIGURE 11.7 Molecular structures, absorption, and emission spectra of (mpmi)$_2$Ir(dmpypz), (mpmi)$_2$Ir(pybi), (mpmi)$_2$Ir(priq), and (mpmi)$_2$Ir(dmpypz) in dichloromethane at room temperature. (Adapted with permission from Lu, K.-Y. et al. 2011. Wide-range color tuning of iridium biscarbene complexes from blue to red by different N^N ligands: An alternative route for adjusting the emission colors. *Advanced Materials* 23(42):4933–4937. Copyright 2011 WILEY-VCH Verlag GmbH & Co. KGaA, Weinheim.)

11.2.1.3 Blue Emitters—Complications and Current Status

While green and red phosphorescent OLEDs have reached the market, realization of high performance phosphorescent blue-emitting OLEDs is an ongoing challenge. Although high efficiency blue OLEDs have been demonstrated, their major drawbacks are long-term stability and color quality. For example, efficiency close to the theoretical limit was achieved after elaborate device optimization for an OLED incorporating iridium(III) bis[4,6-di-fluorophenyl)-pyridinato-N,C$^{2'}$]picolinate (FIrpic) as the blue emitting phosphor (Adachi et al., 2001; Holmes et al., 2003; Sasabe et al., 2008). However, the emission chromaticity of the device, with Commission Internationale de L'Eclairage (CIE) coordinates of (0.16, 0.40), does not meet the requirement for the blue color for full-color displays. A true-blue emitter suitable for display applications should exhibit CIE coordinates of (0.14, 0.08). Furthermore, while fluorination of cyclometallating ligands is a common approach to blue-shift the emission of Ir (III) complexes (e.g., FIrpic), this generally makes a complex less thermally and electrochemically robust, which in turn leads to shorter device operational lifetimes.

Device performance is largely governed by the photophysical behavior of the phosphorescent emitter. Compared to green and red cyclometallated Ir(III) complexes, blue emitters have wide energy gaps, high triplet energies and generally exhibit lower emission efficiency and poor photostability due to efficient nonradiative deactivation pathways. Thorough photophysical studies have been performed in order to gain a better understanding of the excited state properties of blue emitters and to provide guidelines for improving the luminescent performance of these materials (Haneder et al., 2008; Li et al., 2005a,b; Sajoto et al., 2005, 2009).

Temperature-dependent luminescent studies of blue phosphorescent cyclometallated Ir(III) complexes showed strong evidence of the presence of a thermally accessible nonemissive (NR) state (Scheme 11.3a) (Sajoto et al., 2009). The nonradiative deactivation in Ir(C^N)$_3$ or (C^N)$_2$Ir(L^X) occurs via five-coordinate species, which is formed by breaking of an Ir–N bond and has triplet metal-centered (^{3}MC) character. For example, no measurable emission from Ir(ppz)$_3$ is observed at room temperature (Sajoto et al., 2005), but emission intensity increases upon cooling when this deactivation pathway is suppressed. This result provides several strategies to improve luminescent efficiency. One approach involves the use of rigid ligands that would prevent metal–ligand bond from

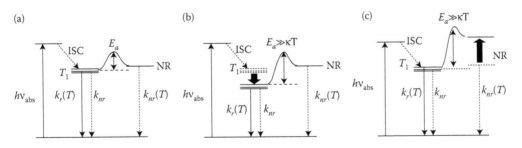

SCHEME 11.3 Schematic representation of radiative and nonradiative deactivation pathways in cyclometallated Ir(III) complexes. (Adapted with permission from Sajoto, T. et al. 2009. Temperature Dependence of Blue Phosphorescent Cyclometalated Ir(III) Complexes. *Journal of the American Chemical Society* 131(28):9813–9822. Copyright 2009 American Chemical Society.)

Chapter 11

dissociation. Another method is to raise the height of the barrier needed to thermally deactivate the complex.

One way to raise the height of the barrier is by lowering the energy of the triplet state (Scheme 11.3b). This has been accomplished by using 1-[(9,9-dimethyl-2-fluorenyl)]pyrazolyl ligand (flz), which is a derivative of ppz ligand where π-system is expanded with a fluorenyl group (Figure 11.5, entry 2) and therefore it has a ³LC state with lower energy (Sajoto et al., 2005). At room temperature the *fac*-Ir(flz)₃ exhibits sky-blue phosphorescence at 480 nm and high quantum yield of 38% and lifetime of 37 μs. The estimated radiative rate is 10^4 s^{-1}, which is an order of magnitude lower than that reported for green ppy-based emitters. The increase of the optical energy gap in blue phosphors due to the lowered HOMO energy level leads to the destabilization of the ¹MLCT state relative to the ³LC state (Scheme 11.2). The larger separation between these two states results in a decreased ¹MLCT character of the lowest excited state and consequently a lower radiative rate.

Another method to suppress nonradiative pathways is to increase the energy of the NR state (Scheme 11.3c). This state is principally consists of metal–ligand antibonding orbitals, so this can be achieved by using high-field ligands that form strong bonds with the metal and therefore shift the metal–ligand antibonding orbitals to higher energy. N-heterocyclic carbenes (NHC) are known to form strong bonds with transition metals (Jacobsen et al., 2009). Cyclometallating bidentate monoanionic ligands bearing an NHC group (C^C:, where the carbene moiety is a neutral two electron donor) have been employed to prepare blue iridium(III) phosphors (Figure 11.8) (Chang et al., 2008; Sajoto et al., 2005).

The first example of a tris-cyclometallated iridium carbene complex, Ir(C^C:)₃, was prepared by Lappert et al. in 1980s; however its photophysical properties were not reported (Hitchcock et al., 1982). Later, 1-phenyl-3-methyl-imidazolin-2-ylidene (pmi) and 1-phenyl-3-methyl-benzimidazolin-2-ylidene (pmb) were used to prepare tris-cyclometallated complexes (Sajoto et al., 2005). Similar to the Ir(C^N)₃ analogs, the lowest excited state of Ir(C^C:)₃ is ³LC with partial MLCT character. The *mer-* and *fac*-isomers of Ir(pmi)₃ and Ir(pmb)₃ show intense structured emission at 77 K in the near-UV part of the spectrum (E_{0-0} = 380 nm) with lifetimes ranging between 2 and 7 μs and, unlike Ir(ppz)₃, also emit at room temperature in solution. However, the quantum yields are

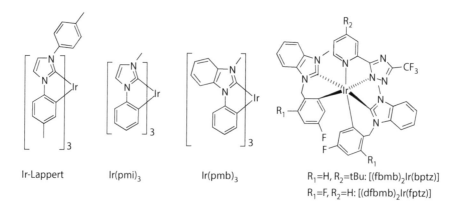

Ir-Lappert   Ir(pmi)₃   Ir(pmb)₃   R₁=H, R₂=tBu: [(fbmb)₂Ir(bptz)]
R₁=F, R₂=H: [(dfbmb)₂Ir(fptz)]

FIGURE 11.8 Molecular structures of selected NHC-Ir(III) complexes.

low (0.002–0.05). This suggests the nonemissive excited state for these complexes has been destabilized relative to the similar state in Ir(ppz)$_3$. The deep blue OLED with *fac*-Ir(pmb)$_3$ (EL λ_{max} = 389 nm) and *mer*-Ir(pmb)$_3$ (EL λ_{max} = 395 nm) doped in the wide energy gap host *p*-bis(triphenylsilyl)benzene (UGH2) have been fabricated (Holmes et al., 2005). The CIE coordinates of (x = 0.17, y = 0.06) were closer to the true blue color necessary for displays than demonstrated by a FIrpic-based OLED. Peak EQEs were 2.6% and 5.8% for *fac*- and *mer*-isomers respectively.

Further examples of NHC-based blue Ir(III) complexes are the heteroleptic complexes (fbmb)$_2$Ir(bptz) and (dfbmb)$_2$Ir(fptz) (Figure 11.8) (Chang et al., 2008). The high triplet energy of nonconjugated C^C: chelating ligand makes the pyridine-triazolate the chromophoric ligand. Efficient blue (PL λ_{max} = 460 nm) phosphorescence with quantum yields of 22% and 73% and lifetimes of 0.219 μs and 0.378 μs in CH$_2$Cl$_2$ were recorded for (fbmb)$_2$Ir(bptz) and (dfbmb)$_2$Ir(fptz) respectively. An OLED fabricated with (dfbmb)$_2$Ir(fptz) as a dopant had CIE coordinates of (x = 0.16, y = 0.13) and maximum external quantum efficiency and power efficiencies of 6.0% and 4.0 lm/W respectively that dropped to 2.7% and 0.9 lm/W at higher current densities.

Another set of ligands that have been employed in the search for efficient true blue phosphors are phosphine-based ligands (Chiu et al., 2009a,b,c; Hung et al., 2009). An interesting example of a blue Ir(III) complex bearing a tridentate dicyclometallated ancillary phosphite ligand (P^C$_2$) was reported by Lin and coworkers (Lin et al., 2011). They prepared a set of complexes Ir(P^C$_2$)(L)(N^N), where L is a monodentate phosphine and N^N is a monoanionic substituted pyridine-triazolate ligand (Figure 11.9) and analyzed the effect of the L ligand on the luminescent properties of the complexes. In a series of Ir(P^C$_2$)(L)(bptz) (bptz = 3-*tert*-butyl-5-(2-pyridyl)triazolate) complexes

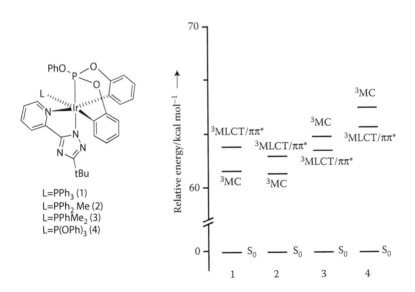

FIGURE 11.9 Molecular structures of complexes Ir(P^C$_2$)(L)(bptz) and energy level diagram obtained from theoretical calculations. (Adapted with permission from Lin, C.-H. et al. 2011. Iridium(III) complexes of a dicyclometalated phosphite tripod ligand: Strategy to achieve blue phosphorescence without fluorine substituents and fabrication of OLEDs. *Angewandte Chemie-International Edition* 50(14):3182–3186. Copyright 2011 Wiley-VCH Verlag GmbH & Co. KGaA, Weinheim.)

the P–Ir bond strength decreases in the order of $P(OPh)_3 > PPhMe_2 > PPh_2Me > PPh_3$ and the solution quantum yields of the respective complexes follow the same order: 87%, 45%, 34%, and 7%. These two characteristics correlate well with the calculated energy difference between the emitting $^3LC/MLCT$ state and the nonradiative ^3MC state. $Ir(P^{\wedge}C_2)(P(OPh)_3(bptz)$ has the strongest P–Ir bond, the highest lying nonradiative state and thus the highest luminescent quantum yield. The electron donor strength of phosphines follow the order $PPhMe_2 > PPh_2Me > PPh_3 > P(OPh)_3$ and correlates well with the radiative rates of the $Ir(P^{\wedge}C_2)(L)(bptz)$ complexes: 3.15×10^4 s^{-1}, 3.06×10^4 s^{-1}, 2.80×10^4 s^{-1}, 1.95×10^4 s^{-1}. This corresponds to a decrease in the MLCT character of the emissive state due to destabilization of the ^1MLCT state relative to the ^3LC state. An OLED fabricated with $Ir(P^{\wedge}C_2)(PPh_2Me)(bptz)$ as a dopant showed high external quantum efficiency and power efficiencies of 11.0% and 16.7 lm/W, with CIE coordinates of ($x = 0.179$, $y = 0.286$).

11.2.2   Pt(II)–Based Emitters

Pt(II) complexes have gained a great deal of attention due to their rich photophysical properties and large potential for various applications. Efficient spin-orbit coupling, induced by third row transition metal Pt, facilitates fast intersystem crossing to the triplet excited state. Thus, Pt(II) can exhibit efficient phosphorescence under ambient conditions. The first triplet emitting complex that demonstrated high efficiency EL was platinum(II) octaethylporphyrin (PtOEP) (Baldo et al., 1998). At room temperature PtOEP exhibits red phosphorescence (PL $\lambda_{max} = 650$ nm), with a quantum yield of 50% in both solution and in a polystyrene film, and a lifetime of 90 μs (Papkovsky, 1995). The EL devices with device structure ITO/NPD/Alq$_3$-PtOEP/Alq$_3$/Mg-Ag were prepared at different doping concentrations and for the first time demonstrated that high efficiency can be achieved with phosphorescent emitters (Figure 11.10) (Baldo et al.,

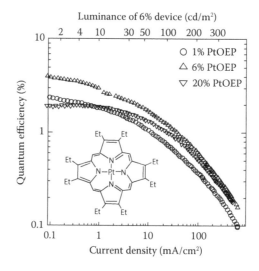

FIGURE 11.10   Current–luminance–efficiency curves of PtOEP-based OLED with different doping levels. (Reproduced from Baldo, M.A. et al. 1998. *Nature* 395(6698):151–154.)

1998). The optimal doping level of 6% gave a maximum external quantum efficiency of 4% at low current densities (0.1 mA/cm^2). However, as current increased, efficiency dropped by more than 50% to <2% at 10 mA/cm^2. The cause of this efficiency roll-off is a triplet-triplet annihilation, which can be efficient in case of PtOEP, as it has relatively long radiative lifetime (Baldo et al., 2000). By optimizing the host material and device architecture, OLEDs incorporating PtOEP dopants have been reported with external quantum efficiencies as high as 9% (Jabbour et al., 2002). This successful result incited interest in platinum-based (and other phosphorescent) materials for OLED applications.

To design a dopant that would satisfy all requirements it is important to understand its specific photophysical behavior that determines its excited state properties. Pt(II) center has a d^8 electronic configuration and a strong preference to adopt a square planar geometry. In this coordination environment, the unoccupied $d_{x^2-y^2}$ orbital is strongly antibonding. When an electron is promoted to this orbital upon excitation, a complex will undergo a tetragonal (D_{2d}) distortion. The resulting d–d state will provide an efficient nonradiative relaxation pathway. To achieve efficient luminescence from Pt(II) complexes, the emissive state (LC, MLCT, etc.) should be the lowest excited state and lie low enough relative to the nonradiative metal-centered (MC) state to impede thermal activation of the nonradiative channel. This can be accomplished by utilizing ligands with low-lying excited states (e.g., porphyrins) that lower the energy of the emitting state relative to the nonemissive MC state. Another approach involves using strong ligand-field ligands (acetylides, cyclometallating ligands) to raise the energy of the MC state. Cyclometallating ligands (C^N) can form Pt(II) complexes that have high emission quantum yields in solution at room temperature. The strong ligand field influence of the carbon atom of the phenyl ring and π-accepting ability of the aromatic rings help minimize nonradiative processes by raising the energy of the d–d state and thereby provide stability to a complex.

Due to their planar geometry, Pt(II) complexes are prone to intermolecular stacking interactions. This can have a significant impact on their photophysical properties. The tendency to form aggregates depends on the steric demands of the ligands. Such interactions involve formation of excimers (Kunkely and Vogler, 1990; Pettijohn et al., 1998; Wan et al., 1991) or metal–metal bound oligomers (Bailey et al., 1995; Buchner et al., 1999; Lai et al., 2002; Miskowski and Houlding, 1991; Yip et al., 1993) and may cause self-quenching at high concentrations (Chan et al., 1992; Connick et al., 1999; Ma et al., 2005). An excimer is a dimer that only exists in the excited state and dissociates in the ground state. By contrast, metal–metal oligomers are stable in the ground state and have weak Pt$\cdots$Pt bonds (3–3.5 Å) due to interaction of d_z^2 orbitals perpendicular to the plane of the complex. Emission from oligomeric structures is attributed to a metal-metal-to-ligand charge transfer (MMLCT) ($d\sigma^* \rightarrow \pi^*$). Both excimers and oligomers usually exhibit broad red-shifted emission compared to monomer emission.

11.2.2.1 Pt(II) Complexes with Bidentate Ligands

Bis-cyclometallated complexes, both homoleptic Pt(C^N)$_2$ and heteroleptic (C^N) Pt(C^N′), have been reported (Williams, 2007) (Figure 11.11). Photophysical properties of complexes based on 2-phenylpyridine (ppy) and 2-(2′-thienyl)pyridine (thpy) ligands and their derivatives have been thoroughly studied since their first report in the 1980s (Barigelletti et al., 1988; Chassot and von Zelewsky, 1987; Chassot et al., 1984; Maestri et al., 1985). The homoleptic complexes Pt(ppy)$_2$ and Pt(thpy)$_2$ (Figure 11.11) did not

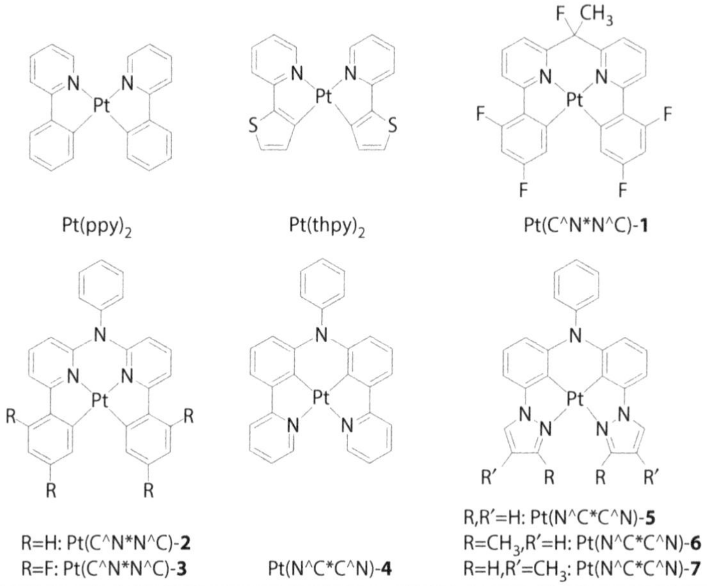

| | Pt(ppy)$_2$ | Pt(thpy)$_2$ | Pt(C^N*N^C)-**1** |

R=H: Pt(C^N*N^C)-**2**
R=F: Pt(C^N*N^C)-**3**

Pt(N^C*C^N)-**4**

R,R′=H: Pt(N^C*C^N)-**5**
R=CH$_3$,R′=H: Pt(N^C*C^N)-**6**
R=H,R′=CH$_3$: Pt(N^C*C^N)-**7**

Complex	PL $\lambda_{max(77\,K)}$, nm	PL $\lambda_{max(298\,K)}$, nm	$\tau_{(298\,K)}$, µs	Φ (%)	References
Pt(ppy)$_2$	491[a]	n/e	n/e	n/e	Maestri et al. (1985)
Pt(thpy)$_2$	570[a]	578	2.2	30	Maestri et al. (1985)
Pt(C^N*N^C)-**1**	–	492,520	0.32	–	Feng et al. (2009)
Pt(C^N*N^C)-**2**	501,539	512,548	7.6	74	Vezzu et al. (2010)
Pt(C^N*N^C)-**3**	481,517	488,523	11.4	75	Vezzu et al. (2010)
Pt(N^C*C^N)-**4**	594,637	613	7.6	14	Vezzu et al. (2010)
Pt(N^C*C^N)-**5**	477,510	484,512	4.9	56	Vezzu et al. (2010)
Pt(N^C*C^N)-**6**	462,491	474	3.4	37	Vezzu et al. (2010)
Pt(N^C*C^N)-7	481,513	486,516	5.7	63	Vezzu et al. (2010)

FIGURE 11.11   Photophysical properties of selected cyclometalated Pt(II) complexes in solution. [a]High-energy peak; n/e = nonemissive.

prove to be well suited for OLED applications. Pt(ppy)$_2$ is nonemissive in solution at room temperature due to efficient d–d deactivation. Pt(thpy)$_2$ and Pt(thpy-SiMe$_3$)$_2$ have rather high solution quantum yields of 35%, but are thermally unstable. Nevertheless, spin-cast OLEDs based on Pt(thpy)$_2$ and Pt(thpy-SiMe$_3$)$_2$ (thpy-SiMe$_2$ = 2-(5-trimethylsilanyl-thiophen-2-yl)-pyridine) were reported to achieve an external quantum efficiency of 5.4% and 11.5% respectively (Cocchi et al., 2004a,b).

The coordination geometry strongly influences emission energy, quantum yield and lifetime. The properties of bis-cyclometallated complexes can be dramatically improved with introduction of a rigid linker between the two C^N ligands (Figure 11.11) (Feng et al., 2009; Vezzu et al., 2010). This suppresses the tetrahedral distortion and makes the molecule more rigid. In this case, formation of bis-cyclometallated complexes with various configurations are possible, for example, Pt(C^N*N^C), Pt(N^C*C^N),

Pt(C^N*C^N). These complexes emit from the mixed ^{3}LC/MLCT state and their phosphorescent efficiencies are among the highest reported for cyclometallated Pt(II) complexes. For example, Pt(C^N*N^C)-**2** (Figure 11.11), where C^N = ppy, has structured green emission (PL λ_{max} = 512, 548 nm), a quantum yield of 74% and a lifetime of 7.6 µs, while Pt(ppy)$_2$ is effectively nonemissive at room temperature. A different isomer Pt(N^C*C^N) shows broader red emission (λ_{max} = 613 nm), a similar lifetime of 7.6 µs, but much lower efficiency of 14%. Pt(C^N*N^C), (C^N = ppy) was chosen for OLED fabrication. After optimization, the best performing device showed 14.7% external efficiency at a current density of 0.01 mA/cm^2 (10.6% at 10 mA/cm^2) (Vezzu et al., 2010).

Heteroleptic (C^N)Pt(LX) complexes offer a high degree of tunability, as their properties can be independently adjusted. This was demonstrated for a series of (C^N)Pt(O^O) complexes, where O^O is a β-diketonate ligand, that is, acetyl acetonate (acac) or its derivative dipivolylmethane (dpm) (Figure 11.12) (Brooks et al., 2002). In this work, the

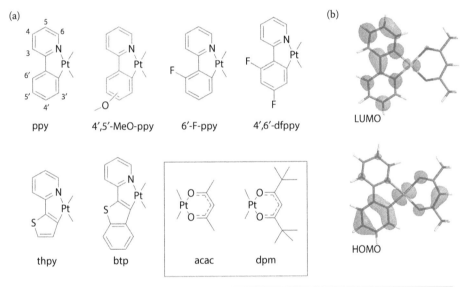

C^N	O^O	PL $\lambda_{max\ (77\ K)}$, nm	PL $\lambda_{max\ (298\ K)}$, nm	$\tau_{(298\ K)}$, µs	Φ (%)
ppy	acac	480	486	2.6	15
ppy	dpm	477	–	–	–
5'-MeO-ppy	dpm	525	–	–	–
4'-MeO-ppy	dpm	480	490	7.4	20
4',6'-dfppy	dpm	458	466	<1.0	2
6'-F-ppy	dpm	468	476	<1.0	6
thpy	dpm	550	575	4.5	11
btp	acac	600	–	–	–

FIGURE 11.12 (a) Molecular structures and photophysical properties of heteroleptic (C^N)Pt(O^O) complexes in 2-methyltetrahydrofuran. (b) HOMO and LUMO orbitals for (ppy)Pt(acac). (Reprinted with permission from Brooks, J. et al. 2002. Synthesis and characterization of phosphorescent cyclometalated platinum complexes. *Inorganic Chemistry* 41(12):3055–3066. Copyright 2002 American Chemical Society.)

emission energy of the complexes was tuned throughout visible spectrum. The lowest excited state was assigned as a mixed ^{3}LC/MLCT state. The result of DFT calculations for (ppy)Pt(acac) showed that its HOMO is distributed among the phenyl moiety, Pt and acac, while the LUMO is predominantly localized on the ppy ligand. Thus, introduction of substituents on the phenyl ring should alter the HOMO, and substitution of the pyridyl moiety should mainly affect the LUMO. The position of substitution is important, as the electron density in the HOMO is centered at the 5′-position of the phenyl ring and nodes exist at the 4′- and 6′-positions. In fact, an electron-donating methoxy group at the 5′-position gives a bathochromic shift: (5′-MeO-ppy)Pt(dpm) has λ_{max} = 525 nm compared to 486 nm for (ppy)Pt(dpm), while (4′-MeO-ppy)Pt(dpm) emits at 480 nm (at 77 K). Electron-withdrawing fluorine groups gives blue-shifted emission: λ_{max} = 458 nm for (4′,6′-dfppy)Pt(dpm) and 468 nm for (6′-F-ppy)Pt(dpm). Other color tuning strategies involve increasing π-conjugation of the C^N ligand, which generally causes a red shift of the emission, and changing the nature of C^N ligand. Orange-red emission was achieved using ligands with low triplet energies, that is, 550 nm for (thpy)Pt(dpm) and 600 nm for (btp)Pt(acac). Many of these complexes are emissive in solution at room temperature with moderate efficiencies (up to 25%).

The ability of (C^N)Pt(O^O) complexes to form excimers at high concentrations was exploited in white OLEDs (WOLEDs) (Adamovich et al., 2002; Cocchi et al., 2009, 2010, 2007a; D'Andrade et al., 2002; Williams et al., 2007; Yang et al., 2008). The broad orange-red (PL λ_{max} = 570 nm) excimer emission of (4′,6′-dfppy)Pt(acac) was combined with the blue monomer emission of FIrpic (PL λ_{max} = 470 nm) in a codoped emissive layer to realize white EL with CIE coordinates (0.35, 0.43) (D'Andrade et al., 2002). The device structure was further simplified when white light was achieved from balanced monomer/excimer emission from a single dopant (Adamovich et al., 2002). Such a simplified structure is advantageous over most common WOLEDs that combine three emitters (red, green, and blue) to achieve white light. It is much easier to control charge recombination and it eliminates the problem of different degradation rates for each dopant that may affect color quality. Figure 11.13 shows the emission spectra for thin films of

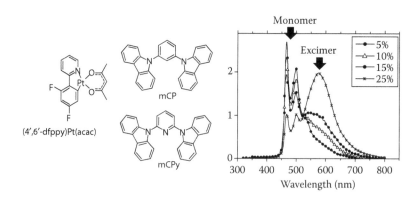

FIGURE 11.13 Molecular structures of (4′,6′-dfppy)Pt(acac), mCP and mCPy and emission spectra for thin films of mCP doped with 5–25 wt% of (4′,6′-dfppy)Pt(acac). (Adapted from Adamovich, V. et al. 2002. High efficiency single dopant white electrophosphorescent light emitting diodes. *New Journal of Chemistry* 26(9):1171–1178, with permission of the Royal Society of Chemistry (RSC) on behalf of the European Society for Photobiology, the European Photochemistry Association and the RSC.)

(4′,6′-dfppy)Pt(acac) doped at various concentrations (5–25 wt%) into an mCP host. In this device, balanced monomer/excimer emission was observed at doping concentration of approximately 15 wt%. Doping levels in the range of 15–20 wt% showed the highest color rendering indexes (CRI) of 68–73. Concentrations between 4–10 wt% showed CIE coordinates closest to white (0.33, 0.33), while at higher concentrations of 20–30 wt% the CIE coordinates were close to those found in incandescent lamps (*ca.* 0.41, 0.41). Based on these results, the optimal doping level of 16 wt% of (4′,6′-dfppy)Pt(acac) in mCP host was chosen for use in the WOLED. The device with a configuration ITO/NPD/Ir(ppz)$_3$/mCP:(4′,6′-dfppy)Pt(acac) (16 wt%)/BCP/Alq$_3$/LiF/Al emitted white light from a single dopant luminescent layer with the maximum external quantum efficiency of 6.4% (12.2 lm/W, 17.0 cd/A) at 1 cd/m^2 and 4.3% (8.1 lm/W, 11.3 cd/A) at 500 cd/m^2, CIE coordinates of (0.36, 0.44) and a CRI of 67. Jabbour and coworkers have optimized the device architecture of a single dopant (4′,6′-dfppy)Pt(acac)-based WOLED and achieved nearly 100% internal quantum efficiency (Williams et al., 2007). They utilized poly(3,4-ethylenedioxythiophene) poly(styrenesulfonate) (PEDOT–PSS) and poly(N-vinylcarbazole) (PVK) instead of NPD as a hole transporting layer (HTL) and 2,6-bis(N-carbazolyl)pyridine (mCPy) host instead of mCP to improve charge transport properties and hole-electron balance in the emissive layer. The device with the configuration of ITO/PEDOT/PVK/mCPy:(4′,6′-dfppy)Pt(acac) (12 wt%)/BCP/LiF/Al showed maximum external quantum efficiency of 15.9% (12.6 lm/W, 37.8 Cd/A) at 500 Cd/m^2, CIE coordinates of (0.46, 0.47) and a CRI of 69.

11.2.2.2 Pt(II) Complexes with Tridentate Ligands

Tridentate ligands can provide additional rigidity to square planar Pt(II) complexes compared to bidentate ligands and potentially improve luminescent performance of the complexes. These complexes offer broad possibilities for structural modifications; therefore they have large potential for tuning of excited state properties. Each constituent moiety of the tridentate ligand can be changed independently by introducing substituents with different electronic and steric properties or by employing different heterocycles (Figure 11.14). A monodentate ligand (anionic or neutral) completes the four-coordinate square planar coordination environment around the Pt center and can be varied independently, thus providing an additional modification site. Several types of tridentate ligands have been explored. Monoanionic cyclometallating ligands N^N^C and N^C^N, in general, give better efficiencies than neutral terpyridine ligands (Aldridge et al., 1994; Arena et al., 1998; Bailey et al., 1995), owing to their high ligand-field strength.

Che and coworkers have extensively studied the photophysical properties of [Pt(N^N^C)X] complexes based on 6-phenyl-2,2′-bipyridine ligand (Hphbpy) (Lai and Che, 2004; Tong and Che, 2009). Complexes shown in Figure 11.14 illustrate the influence of the substituents and the identity of the tridentate ligand as well as the impact of the monodentate ligand (X) on the emission properties. The latter has a noticeable effect on the emission energy. The parent [Pt(phbpy)Cl] shows an emission spectrum peaking at 565 nm in dichloromethane solution at room temperature (Cheung et al., 1996). When the chloride ligand is substituted by strong-field acetylide ligand, the emission wavelength in fluid solution varies from 560 nm to 630 nm depending on the identity of the substituent of the acetylide ligand (–C≡C–R) (Lu et al., 2004). The emissive state

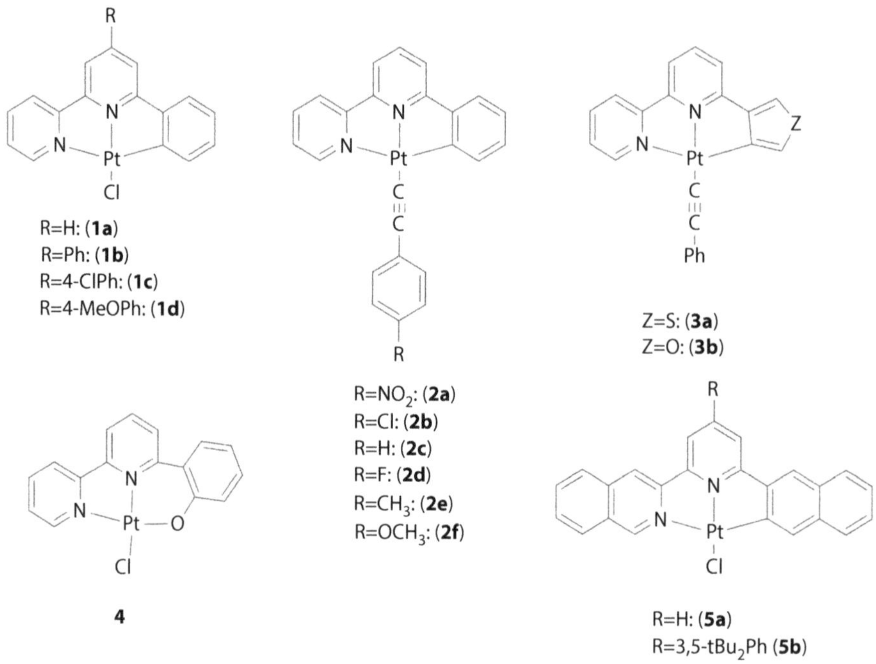

Complex	PL λ_{max} (nm)	τ (μS)	Φ (%)	References
[Pt(N^N^C)Cl]-**1a**	565	0.51	2.5	Lai et al. (1999)
[Pt(N^N^C)Cl]-**1b**	564	0.60	5.2	Lai et al. (1999)
[Pt(N^N^C)Cl]-**1c**	568	0.52	5.4	Lai et al. (1999)
[Pt(N^N^C)Cl]-**1d**	563	0.72	6.8	Lai et al. (1999)
[Pt(N^N^C)X]-**2a**	560	0.9	8.0	Lu et al. (2004)
[Pt(N^N^C)X]-**2b**	578	0.5	8.0	Lu et al. (2004)
[Pt(N^N^C)X]-**2c**	582	0.4	4.0	Lu et al. (2004)
[Pt(N^N^C)X]-**2d**	585	0.3	3.0	Lu et al. (2004)
[Pt(N^N^C)X]-**2e**	600	0.2	2.0	Lu et al. (2004)
[Pt(N^N^C)X]-**2f**	630	≤ 0.1	≤ 0.1	Lu et al. (2004)
[Pt(N^N^C)X]-**3a**	615, 660 (sh)	1.0	5	Lu et al. (2004)
[Pt(N^N^C)X]-**3b**	≈ 640	≤ 0.1	≤ 0.1	Lu et al. (2004)
[Pt(N^N^C)Cl]-**4**	606 (in DMF)	–	–	Kwok et al. (2005)
[Pt(N^N^C)Cl]-**5a**	529,566,595	6	20	Kui et al. (2007)
[Pt(N^N^C)Cl]-**5b**	533,569	7	68	Kui et al. (2007)

FIGURE 11.14   Molecular structures and photophysical properties of selected Pt(II) complexes bearing 6-phenyl-2,2'-bipyridine ligands and its derivatives. Measured in CH_2Cl_2 unless otherwise stated.

in these complexes was assigned as ^{3}MLCT (Pt → π^* (N^N^C)). However, when the acetylenic ligand-localized states ^{3}LC ($^3(\pi \rightarrow \pi)^*$) become accessible (e.g., in extended acetylides) vibronically structured emission is observed in fluid solution indicating the direct involvement of the acetylide ligand in the emission process. Modification of the tridentate ligand by introduction of different substituents at the 4-position of the central pyridyl ring leads to a modest change in the emission maximum (≤5 nm) (Lai et al., 1999; Lu et al., 2004). Larger shifts were observed when the phenyl moiety of the tridentate ligand was replaced with a furyl (λ_{max} = 640 nm), thienyl (λ_{max} = 615 nm), or phenoxy (λ_{max} = 606 nm) group (Kwok et al., 2005) and upon extension of the conjugation of the N^N^C ligand (Kui et al., 2007). Expansion of the π-system also helped to improve the luminescence efficiency of [Pt(N^N^C)X] complexes from less than 10% up to 68% in fluid solution. Due to their thermal stability these complexes are sublimable and thus were utilized as OLED dopants. The yellow-emitting device (EL λ_{max} = 540 nm) that incorporated the highly luminescent complex [Pt(N^N^C)X]-**5a** (Figure 11.14), showed a maximum luminance of 63000 cd/m^2 and a current efficiency of 12.5 cd/A at a current density of 1.8 mA/cm^2 (Kui et al., 2007).

The Williams group has focused on tridentate cyclometallating N^C^N ligands, namely 1,3-di(2-pyridyl)benzene (Hdpyb) and its derivatives (Williams, 2009). Although these ligands are similar to the Hphbpy derivatives discussed above, the coordination mode (N^C^N vs. N^N^C) has a large impact on the luminescent properties of Pt complexes (Figure 11.15). These [Pt(N^C^N)X] complexes exhibit exceptionally high room temperature quantum yields compared to other cyclometallated Pt(II) complexes (Farley et al., 2005; Williams et al., 2003). For example, the [Pt(dpyb)Cl] complex has a quantum yield of 60% in dichloromethane; that is an order of magnitude higher than that of [Pt(phbpy)Cl]. Efficient luminescence of [Pt(N^C^N)X] complexes is attributed to the higher ligand-field strength of dpyb ligands compared to phbpy. This is supported by crystallographic analyses, that revealed that the Pt–C bond in [Pt(N^C^N)X] is shorter than in [Pt(N^N^C)X]. Thus the energy of the nonemissive d–d states is raised in [Pt(N^C^N)X] complexes, therefore the thermally activated nonradiative pathways are efficiently suppressed, leading to higher emission quantum yields. In contrast to the [Pt(N^N^C)X] complexes reported by Che, for which emission energies are almost insensitive to substitution in the 4-position of the central pyridyl ring (Lai et al., 1999), the emission maxima of [Pt(N^C^N)X] complexes can be tuned throughout the visible spectrum by the introduction of substituents with different electronic properties at the central phenyl ring. Various electron donating groups have been employed that led to red shifts of emission maxima. Correlation of the emission energy with the donor ability of substituents was observed (Figure 11.15). The lowest excited state of these complexes was assigned as mixed ^{3}LC/MLCT based on DFT calculations (Sotoyama et al., 2005a), which is consistent with high radiative rates and the structured emission spectra obtained experimentally in solution at low concentrations (≤10^{-5}M). At higher concentrations broad red-shifted emission was observed and was attributed to the formation of excimers, common for Pt(II) complexes.

Highly efficient OLEDs based on [Pt(N^C^N)Cl] complexes were demonstrated (Cocchi et al., 2009, 2010, 2007a,b,c, 2008; Sotoyama et al., 2005b). Their characteristics are among the best achieved with Pt-based phosphors. Color tuning can be

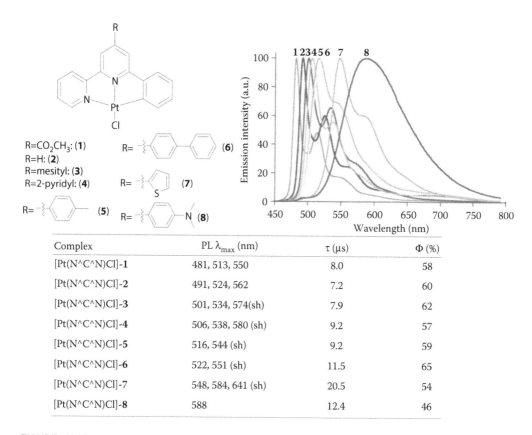

Complex	PL λ_{max} (nm)	τ (μs)	Φ (%)
[Pt(N^C^N)Cl]-**1**	481, 513, 550	8.0	58
[Pt(N^C^N)Cl]-**2**	491, 524, 562	7.2	60
[Pt(N^C^N)Cl]-**3**	501, 534, 574(sh)	7.9	62
[Pt(N^C^N)Cl]-**4**	506, 538, 580 (sh)	9.2	57
[Pt(N^C^N)Cl]-**5**	516, 544 (sh)	9.2	59
[Pt(N^C^N)Cl]-**6**	522, 551 (sh)	11.5	65
[Pt(N^C^N)Cl]-**7**	548, 584, 641 (sh)	20.5	54
[Pt(N^C^N)Cl]-**8**	588	12.4	46

FIGURE 11.15 Molecular structure, emission spectra and photophysical data (dichloromethane, 295 K, C = 10^{-5} M) for [(N^C^N)PtCl] complexes. (Adapted with permission from Farley, S.J. et al. 2005. Controlling emission energy, self-quenching, and excimer formation in highly luminescent N^C^N-coordinated platinum(II) complexes. *Inorganic Chemistry* 44(26):9690–9703. Copyright 2005 American Chemical Society.)

achieved via judicious adjustment of dopant concentration. Figure 11.16 shows the PL spectra of a film of complex [Pt(N^C^N)Cl]-**1** (platinum(II)[methyl-3,5-di(2-pyridyl)benzoate]chloride) doped at 5% in a host matrix (only monomer emission is observed, PL λ_{max} = 490 nm) and 100% evaporated neat film of [Pt(N^C^N)Cl]-**1** (excimer emission, PL λ_{max} = 680 nm). Devices of the configuration ITO/TPD:PC/CBP/CBP:OXA:[Pt(N^C^N)Cl]/OXA/Ca using complexes [Pt(N^C^N)Cl]-**1-6** at low (6 wt%) doping concentration exhibited only monomer emission (Cocchi et al., 2007b). They reached an external quantum efficiency of 16%, luminous efficiency of 40 cd/A and maximum brightness of 12100 cd/m². On the other hand, when neat films of these phosphors were used as emissive layers in the devices having the same configuration, their EL spectra exhibited red/near infrared (NIR) excimer emission centered at 705–720 nm (Cocchi et al., 2007c). Excellent performance for NIR OLEDs has been achieved with external quantum efficiencies up to 14.5% (Cocchi et al., 2008). At intermediate concentrations (10–20 wt%) white light was obtained from balanced monomer/excimer emission of a single dopant (Figure 11.16) (Cocchi et al., 2007a, 2010). Devices reported by Williams and coworkers combine the advantages of single

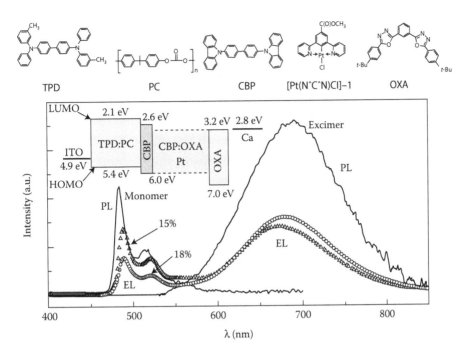

FIGURE 11.16 PL spectra of doped at 5 wt% (monomer emission) and a neat (100 wt%, excimer emission) film of complex [Pt(N^C^N)Cl]-**1** (solid lines). EL spectra at 15 wt% (Δ line) and 18 wt% (○ line) doping concentration of complex [Pt(N^C^N)Cl]-**1** of the OLED shown in the inset. (Reprinted with permission from Cocchi, M. et al. 2007a. Single-dopant organic white electrophosphorescent diodes with very high efficiency and its reduced current density roll-off. *Applied Physics Letters* 90(16): 163508/1–163508/3. Copyright 2007, American Institute of Physics.)

dopant white OLEDs already discussed in this chapter, very high quantum efficiencies and minimal efficiency roll-off at high current densities. The OLED incorporating complex [Pt(N^C^N)Cl]-**1** doped at 15 wt% (Figure 11.16) exhibits a maximum external quantum efficiency of 15.5% at 10 cd/m² and only a small drop to 11.5% at 1000 cd/m² (Cocchi et al., 2007a). The CIE chromaticity coordinates are (0.43; 0.43), which is close to those of an incandescent lamp. The same group later reported EQE up to 18.3% from a single dopant white OLED using a [Pt(dpyb)Cl] derivative (Cocchi et al., 2009).

11.2.2.3 Pt(II) Porphyrins

Pt(II) porphyrins due to their unique photophysical properties have proven to be potentially good dopants for NIR OLEDs, that are of great interest for night-vision displays and sensors. Pt(II) porphyrins demonstrate NIR OLED efficiencies up to 9% (Graham et al., 2011; Sun et al., 2007). Pt(II) porphyrins are known to emit in the low energy part of the visible spectrum. For example, earlier we discussed the PtOEP-based OLED that gives deep-red EL (650 nm) (Baldo et al., 1998). Emission energies can be further red-shifted to the NIR by introduction of bulky groups to the meso positions of the porphyrin core with β-substituted pyrroles and by increasing the π-conjugation of the porphyrin macrocycle (Figure 11.17) (Finikova et al., 2005; Rozhkov et al., 2003). This strategy was successfully applied to obtain a highly efficient NIR emitter

Chapter 11

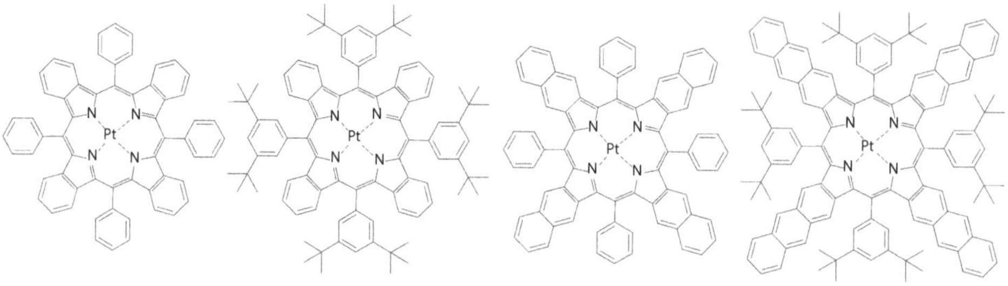

Pt(TPBP)	Pt(Ar$_4$TBP)			Pt(TPTNP)	Pt(Ar$_4$TAP)
Pt(II) porphyrin	PL λ_{max} (nm)	τ (μs)	Φ (%)		References
Pt(TPBP)	771	53	70[a]		Borek et al. (2007)
Pt(Ar$_4$TBP)	772	32	33		Sommer et al. (2011)
Pt(TPTNP)	891	12.7	15		Sommer et al. (2011)
Pt(Ar$_4$TAP)	1022	3.2	8.0		Sommer et al. (2011)

[a]Determined from the ratio of lifetimes at 298 and 77 K.

FIGURE 11.17 Molecular structures and photophysical properties of selected Pt(II) porphyrins in solution.

Pt(II)-5,10,15,20-tetraphenyltetrabenzoporphyrin, [Pt(TPBP)] (Borek et al., 2007). This complex showed a narrow phosphorescence band centered at 765 nm with a radiative lifetime of 53 μs and quantum efficiency of 70% (determined from the ratio of the lifetimes at 298 K and 77 K). Schanze and coworkers further extended the π-system of the porphyrin core and shifted the emission to 891 nm for Pt(II)-5,10,15,20-tetraphenyltetranaphthoporphyrin [Pt(TPTNP)] and to 1022 nm for Pt(II)-5,10,15,20-(3,5-di-*tert*-butylphenyl)tetraanthroporphyrin [Pt(Ar$_4$TAP)] (Sommer et al., 2009, 2011). A decrease of solution PL quantum yield with extension of the π-system was observed. While the emission efficiency of [Pt(TPBP)] was 70%, it decreased to 15% for [Pt(TPTNP)] (lifetime 12.7 μs) and to 8% for [Pt(Ar$_4$TAP)] (lifetime 3.2 μs). A [Pt(TPBP)]-based OLED with a device structure of ITO/NPD/Alq$_3$:[Pt(TPBP)](6 wt%)/Alq$_3$/LiF/Al gave a peak EQE of 6.3% at 0.1 mA/cm^2, and an operational lifetime of greater than 1000 h to 90% of initial efficiency at 40 mA/cm^2 (Borek et al., 2007). The device efficiency reached 8.5% when BCP was used as an exciton blocking and electron transport layer (Sun et al., 2007). An external quantum efficiency of 9.2% was reported for an OLED based on [Pt(Ar$_4$TPBP)], a derivative of [Pt(TPBP)] bearing *tert*-butyl substituents at the 3,5-positions on the *meso*-phenyl groups (Graham et al., 2011).

11.3 Phosphorescent Materials for OLEDs Based on Abundant Metals

The advantages of OLED technology over existing display and lighting technologies are apparent. However, cost is a major challenge for the current generation of OLEDs to overcome in order to become more commercially viable. To date, the best emitters for OLEDs are cyclometallated complexes of expensive metals (Ir and Pt). It is desirable to reduce the cost of materials, especially for large-scale applications (e.g., lighting), without compromising device performance. A potential solution is to develop inexpensive phosphorescent materials based on abundant metals. In search for inexpensive phosphors for OLEDs a great deal of attention has focused on Cu(I)-based materials. Although copper is a first row transition metal and does not possess a strong "heavy atom" effect, it has been shown that Cu(I) complexes can exhibit efficient phosphorescence at room temperature.

11.3.1 Cationic Cu(I) Complexes

The Cu(I) metal ion possesses a $3d^{10}$-electronic configuration and has a preference to form 4-coordinate tetrahedral complexes. Since the first report of room temperature phosphorescence from Cu(I) compounds by McMillin (Blaskie and McMillin, 1980), research has focused principally on cationic complexes $[Cu(N^\wedge N)_2]^+$, where $N^\wedge N$ denotes a chelating bis-imine ligand, typically a substituted phenanthroline (Figure 11.18). The photophysical properties of these Cu(I) complexes have been thoroughly investigated (Lavie-Cambot et al., 2008; McMillin and McNett, 1998; Scaltrito et al., 2000). The absorption spectra of these complexes display LC bands in the UV and weaker MLCT transitions throughout the whole visible region (up to 700 nm). Low-energy MLCT states occur due to low oxidation potential of Cu(I) and easily accessible low-lying empty π^* orbitals of phenanthroline ligands. Consequently, $[Cu(N^\wedge N)_2]^+$ complexes exhibit broad orange-red MLCT emission.

As discussed above, low-lying metal-centered d–d exited states in d^6 (Ir(III)) or d^8 (Pt(II)) complexes can provide efficient pathways for nonradiative decay. The complete filling of d-orbitals in Cu(I) complexes eliminates the possibility of d–d electronic transitions. Yet, their emission efficiency in solution is generally low due to nonradiative deactivation pathways caused by structural reorganization and exciplex formation. PL quantum yields in dichloromethane at room temperature usually do not exceed 0.01 and emission lifetimes are in the range of 70–1000 ns. In coordinating solvents these complexes are effectively nonemissive. The quenching processes in 4-coordinate cationic Cu(I) complexes have been well established (Chen et al., 2002, 2003; Everly and McMillin, 1989; Stacy and McMillin, 1990; Vorontsov et al., 2009). The proposed mechanism is shown in Scheme 11.4. Upon MLCT excitation, the tetrahedral d^{10} Cu(I) center is effectively oxidized to d^9 Cu(II). The d^9 Cu(II) metal center undergoes a Jahn–Teller flattening distortion. This opens an additional coordination site and a 5-coordinate exciplex may be formed upon coordination of nucleophiles such as solvent molecules or counterions. These two processes, significant structural change between the ground and excited states and additional ligand pick up, promote nonradiative decay and significantly quench emission. The generally accepted approach to alleviate this problem and to maximize radiative decay is to increase the steric bulk of ligands, in particular, by using 2,9-substituted phenanthroline derivatives or bulky phosphine ligands (Accorsi et al., 2010; Cuttell et al., 2002; Gothard et al., 2012; Green et al., 2009).

The replacement of one bis-imine ligand in $[Cu(N^\wedge N)_2]^+$ with a bulky phosphine ligand led to higher luminescence efficiency, longer excited state lifetimes and also changed the emission color (Cuttell et al., 2002). The most commonly used phosphines are: triphenylphosphine (PPh_3), 1,2-bis(diphenylphosphino)benzene (dppb) and bis[2-(diphenylphosphino)phenyl]ether (POP) (Kuang et al., 2002; McCormick et al., 2006;

Chapter 11

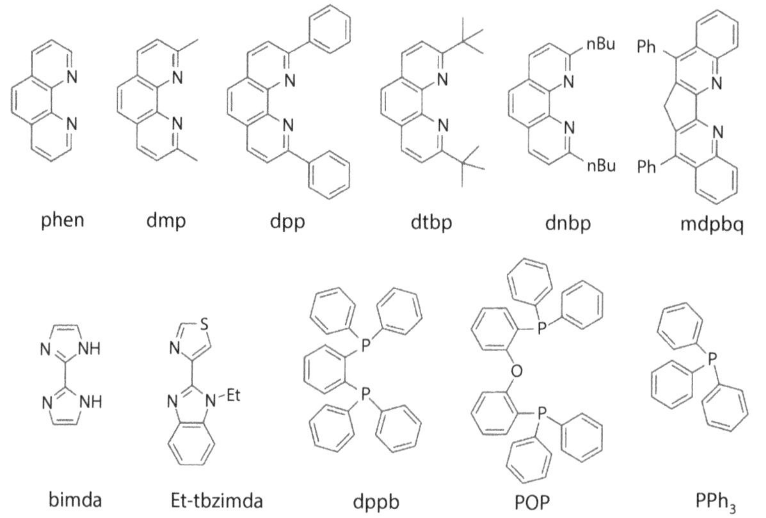

FIGURE 11.18 Ligand definitions and photophysical data (CH$_2$Cl$_2$, RT) for selected cationic Cu(I) complexes.

Entry	Complex	PL λ_{max} (nm)	τ (μs)	Φ (%)	References
1	[Cu(dmp)$_2$]BF$_4$	730	0.09	0.04	Eggleston et al. (1997)
2	[Cu(dbp)$_2$]BF$_4$	725	0.15	0.09	Eggleston et al. (1997)
3	[Cu(dpp)$_2$]PF$_6$	690	0.24	0.087	Miller et al. (1999a)
4	[Cu(dtbp)$_2$][B(C$_6$H$_5$)$_4$]	599	3.26	5.6	Green et al. (2009)
5	[Cu(dmp)(dtbp)]PF$_6$	646	0.73	1.0	Miller et al. (1999b)
6	[Cu(phen)(POP)]BF$_4$	700	0.19	0.18	Cuttell et al. (2002)
7	[Cu(dmp)(POP)]BF$_4$	570	14.3	15	Cuttell et al. (2002)
8	[Cu(dnbp)(POP)]BF$_4$	560	16.1	16	Cuttell et al. (2002)
9	[Cu(dmp)(PPh$_3$)$_2$]BF$_4$[a]	560	0.33	0.14	Rader et al. (1981)
10	[Cu(mdpbq)(PPh$_3$)$_2$]BF$_4$	625	6.9	10	Zhang et al. (2007)
11	[Cu(mdpbq)(POP)]BF$_4$	631	0.75	0.76	Zhang et al. (2007)
12	[Cu(bimda)(POP)]BF$_4$[b]	470	2.1	8.0	Zhang et al. (2009)
13	[Cu(Et-tbzimda)(POP)]BF$_4$[b]	525	16.1	34	Zhang et al. (2009)
10	[Cu(dppb)(POP)]BF$_4$	494	2.44	2.0	Moudam et al. (2007)

[a]In methanol. [b]In PMMA-doped films.

Moudam et al., 2007; Rader et al., 1981; Smith et al., 2010). Photophysicial properties for selected Cu(I)-phosphine complexes are summarized in Figure 11.18. The choice of both bis-imine and phosphine ligands in heteroleptic [Cu(N^N)(P^P)]$^+$ complexes is crucial. The highly emissive [Cu(dmp)(POP)]$^+$ (dmp = 2,9-dimethyl-1,10-phenanthroline) shows broad emission (λ_{max} = 570 nm) with a quantum yield of 15% and 14.3 μs excited state lifetime in degassed dichloromethane. By contrast, [Cu(phen)(POP)]$^+$, bearing unsubstituted phenanthroline, has weak (Φ = 0.18%) red-shifted (λ_{max} = 700 nm) emission with a shorter (0.19 μs) lifetime. A similar effect was observed for [Cu(dmp)(PPh$_3$)]$^+$ where a chelating POP ligand is replaced with two PPh$_3$ ligands. The emitting state in

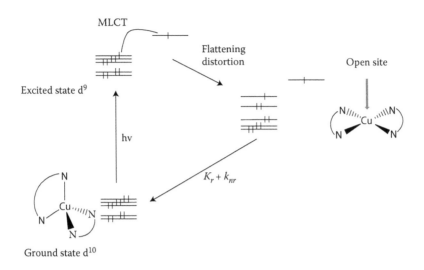

SCHEME 11.4 Excited state (Jablonski) diagram for four-coordinate cationic [Cu(N^N)₂]⁺ complexes.

these mixed-ligand Cu(I) complexes is assigned as ³MLCT. Similar to [Cu(N^N)₂]⁺ complexes, [Cu(N^N)(P^P)]⁺ complexes are also prone to solvent-induced exciplex quenching (Palmer and McMillin, 1987), which is less pronounced for bulkier phosphine ligands, that is, POP (Cuttell et al., 2002; Kuang et al., 2002). However, the combination of steric constraints and electron-withdrawing character of the phosphine ligand leads to an increase in the MLCT energy compared to [Cu(N^N)₂]⁺ complexes. As a result, nonradiative decay is reduced and the emission is blue-shifted.

In view of their promising photoluminescent properties, heteroleptic [Cu(N^N)(P^P)]⁺ complexes have been used in electroluminescent devices (Che et al., 2006; Moudam et al., 2007; Si et al., 2008; Su et al., 2006; Wada et al., 2012; Zhang et al., 2012, 2009, 2007, 2004). These complexes exhibit much higher quantum yields in a film than in solution due to the lack of solvent-induced exciplex quenching. For example, the PL efficiency of [Cu(dnbp)(POP)]BF₄ (dnbp = 2,9-di-n-butyl-1,10-phenanthroline) at room temperature in degassed dichloromethane and 20 wt% doped PMMA film are 16% and 69% respectively (Zhang et al., 2004). The emission maximum in a film is blue-shifted (519 nm) compared to solution (560 nm). Based on its high PL quantum yield in a PMMA film, [Cu(dnbp)(POP)]BF₄ was utilized as a phosphorescent dopant in a green OLED. When [Cu(dnbp)(POP)]BF₄ was doped in commonly used host materials, the PL efficiencies of the films varied depending on the triplet energy of the host materials (Zhang et al., 2012). In TAZ (E_T = 2.6 eV) the PL quantum yield was 16%, while in high triplet energy hosts, that is, 2,6-dicarbazolo-1,5-pyridine (PYD2) (E_T = 2.93 eV), it reached 56%. It was found that the triplet energy of green-emitting [Cu(dnbp)(POP)]BF₄ is 2.72 eV, which is higher than that of Ir(ppy)₃ and blue-emitting FIrpic. Thus, a high triplet energy host is required in order to achieve a high efficiency OLED with this Cu(I) complex. A device comprised of [Cu(dnbp)(POP)]BF₄ doped in a PVK host at 16 wt% exhibited a current efficiency of 10.5 cd/A (Zhang et al., 2004). When PYD2 was used as the host and bis(2-(diphenylphosphino)phenyl)ether oxide (DPEPO) (E_T = 3 eV) as the electron blocking layer, EQEs as high as 15% (49.5 Cd/A) were realized (Zhang et al., 2012). A nondoped

Chapter 11

single layer (ITO/[Cu(dnbp)(POP)]BF$_4$/metal cathode) LEC-type device (light-emitting electrochemical cell) achieved efficiencies up to 56 cd/A, corresponding to an EQE of 16% (Zhang et al., 2006).

By extending the conjugation of the bis-imine ligand of the Cu(I)-dopant complex, Zhang et al. reported the first red-emitting Cu(I)-based OLEDs (Zhang et al., 2007). Complexes [Cu(mdpbq)(PPh$_3$)$_2$]BF$_4$ and [Cu(mdpbq)(POP)$_2$]BF$_4$ (mdpbq = 3,3′-methylen-4,4′-diphenyl-2,2′-biquinoline) (Figure 11.18, entries 10,11) showed efficient PL in 20 wt% doped PMMA film with quantum yields of 0.56 and 0.43 and emission maxima of 606 nm and 617 nm, respectively. After optimization, the OLED with a configuration ITO/PEDOT/TCCz:[Cu(mdpbq)(POP)](BF$_4$)(15 wt%)/TPBI/LiF/Al (TCCz = N-(4-(carbazol-9-yl)phenyl)-3,6-bis(carbazol-9-yl)carbazole; TPBI = 1,3,5-tris-(N-phenylbenzimidazol-2-yl)benzene) exhibited a current efficiency of 6.4 cd/A and an external quantum efficiency of 4.5% at 1.0 mA/cm^2. Su and coworkers were able to shift the emission of [Cu(N^N)(P^P)]$^+$ complexes into the blue-green region of the spectrum (Zhang et al., 2009). Utilization of diimine ligands with high-lying π^* orbitals: 1H,1′H-[2,2′]biimidazole (bimda) and 1-ethyl-2-thiazol-4-yl-1H-benzoimidazole (Et-tbzimda) (Figure 11.18, entries 12,13) selectively increased the LUMO energy level and did not affect the HOMO. The blue-emitting OLED (EL λ_{max} = 480 nm) with [Cu(bimda)(POP)]BF$_4$ doped at 23 wt% into CBP, showed a maximum brightness of 2850 cd/m^2 and low efficiency roll-off at high current density. The efficiency was 1.47 cd/A at low current and only dropped by 10% at 100 mA/cm^2. The green-emitting [Cu(Et-tbzimda)(POP)]BF$_4$-based device (EL λ_{max} = 532 nm) had a maximum brightness of 2320 cd/m^2 and higher efficiency of 2.35 cd/A due to higher PL quantum yield of [Cu(Et-tbzimda)(POP)]BF$_4$ 34% in 5 wt% doped PMMA film versus 8% for [Cu(Bimda)(POP)](BF$_4$). However, the longer PL lifetime (16.1 vs. 2.1 μs) leads to a significant efficiency drop, up to 72% of its maximum at 100 mA/cm^2 (Zhang et al., 2009).

11.3.2   Neutral Cu(I) Complexes

The successful applications of cationic Cu(I) complexes as phosphorescent emitters in OLEDs stimulated recent interest in neutral Cu(I) complexes. Unlike cationic analogs, the photophysical properties of mononuclear neutral Cu(I) compounds have not been well investigated. Nonetheless, current results clearly indicate that the development of charge-neutral Cu(I) complexes is crucial for achieving high phosphorescent radiative rates from these materials and thus for the realization of highly efficient OLEDs that would be able to compete with Ir-based devices.

Following the successful utilization of phosphine ligands to improve photoluminescence in cationic Cu(I) complexes, neutral analogs often possess phosphine ligands as well. Various chelating monoanionic ligands complete the tetrahedral coordination environment (Crestani et al., 2011; Czerwieniec et al., 2011; Hsu et al., 2011; Manbeck et al., 2011; Miller et al., 2007). Peters and coworkers used substituted amidophosphine ligands (^{R}PN) and obtained neutral (^{R}PN)Cu(phosphine) complexes with remarkable luminescent properties (Figure 11.19) (Miller et al., 2007). These compounds have long phosphorescent lifetimes (16–150 μs) in benzene at RT and quantum efficiencies ranging from 16% to 70%, which are among the highest achieved for Cu(I) complexes in solution. Their emission spectra are broad and featureless with maxima ranging from 480 to 552 nm. Interestingly,

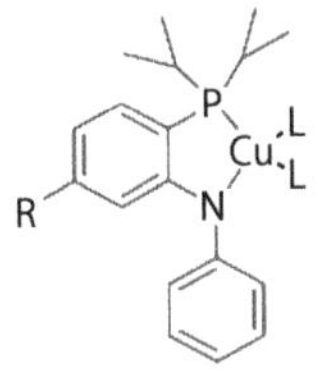

Complex	PL λ_{max} (nm)	Φ (%)	τ (μs)
[PN]Cu(PPh$_3$)$_2$	504	56	20.2(1)
[PN]Cu(PMe$_3$)$_2$	497	21	22.3(7)
[PN]Cu(dppe)$_2$	534	32	16.3(3)
[^{Me}PN]Cu(PPh$_3$)$_2$	498	70	6.7(1)
[^{CF_3}PN]Cu(PPh$_3$)$_2$	552	16	150(3)

R=H: [PN]Cu(L)$_2$
R=Me: [^{Me}PN]Cu(L)$_2$
R=CF$_3$: [^{CF_3}PN]Cu(L)$_2$

FIGURE 11.19 Molecular structures and photophysical properties of neutral (^{R}PN)Cu(phosphine) complexes.

the calculated HOMO and LUMO of (^{R}PN)Cu(phosphine) have very little metal character and are localized on the amidophosphine ligand, suggesting the presence of an ILCT transition. The contribution from orbitals on copper appears in HOMO-1 and HOMO-2. Thus, the lowest excited state may have some mixed-in MLCT character. The importance of participation of metal d-orbitals to achieve effective SOC and to induce efficient phosphorescence in organometallic compounds was discussed earlier in this chapter. This was also demonstrated based on the combination of experimental and computational results in a recent study of isoleptic d^{10}-complexes (Hsu et al., 2011). The photophysical properties of a series of neutral [NN]M(I)[PP] complexes (NN = functionalized 2-pyridyl pyrrolide; PP = POP or PPh$_3$; M(I) = Cu(I), Ag(I), Au(I)) were systematically investigated. One would expect that the radiative rates would decrease from the third row gold(I) to first row copper(I) complexes due to weakening of the "heavy-atom" effect. Interestingly, while the Cu(I) complex showed efficient phosphorescence in solution at RT with a quantum yield and excited state lifetime of 34% and 23.9 μs, respectively, an isostructural Ag(I) complex showed very weak dual emission—fluorescence (0.75%, 176 ps) and phosphorescence (0.95%, 138 μs) and the Au(I) analog displayed weak phosphorescence (3.7%, 124 μs). Based on these data, the radiative rates of these isostructural Cu(I), Ag(I), and Au(I) complexes are $1.4 \cdot 10^4$, $6.88 \cdot 10^1$, and $2.98 \cdot 10^2$ s^{-1} respectively. Theoretical calculations reveal that the lowest excited state of the Cu(I) complex is a triplet LC with mixed in singlet and triplet MLCT character, while the lowest excited states of Ag(I) and Au(I) complexes are pure triplet LC, with no contribution from MLCT (Scheme 11.5). Thus, it was shown that the direct contribution from a metal d orbital in the lowest excited state is more important than the "external" heavy-atom effect in order to attain fast intersystem crossing rates and therefore high phosphorescent radiative rates. This result reveals the potential of a "light" first-row Cu(I) metal for OLED applications.

Several examples of efficient OLEDs containing neutral Cu(I) complexes have been reported (Deaton et al., 2010; Liu et al., 2011; Zhang et al., 2012). Liu et al. demonstrated a new approach for utilizing metal-based phosphors in OLEDs, that involves *in situ* formation of a Cu(I) phosphor by codeposition of CuI and 3,5-bis(carbazol-9-yl)pyridine (mCPy) (Liu et al., 2011). PL quantum yields of codeposited films were as high as 64% depending on the molar ratio of CuI and mCPy. It was determined that the emitting species in CuI:mCPy thin films is the dimeric complex [CuI(mCPy)$_2$]$_2$. OLEDs made using codeposited CuI:mCPy films as emissive layers exhibited green EL at 530 nm and reached maximum luminance of 9700 cd/m^2 and EQE of 4.4%.

Chapter 11

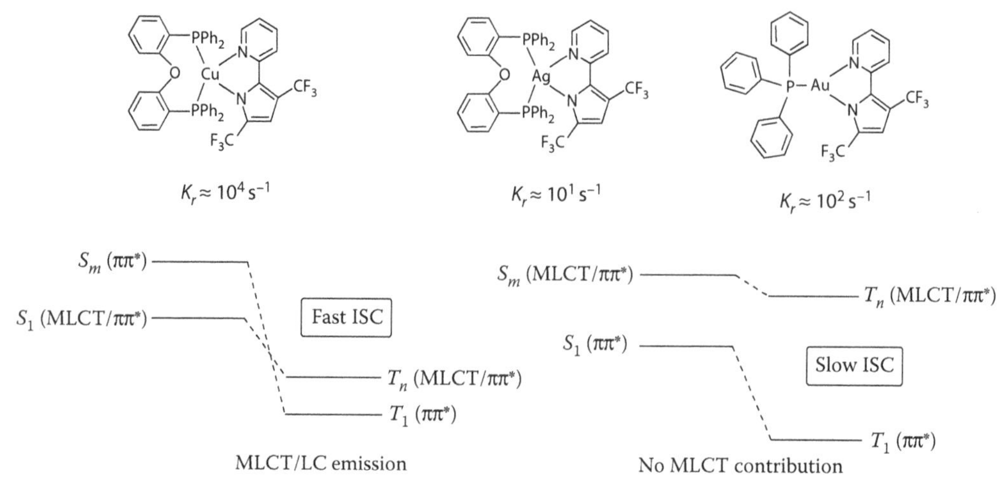

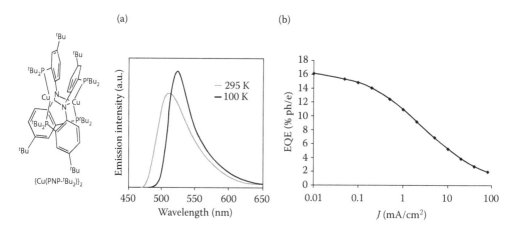

$K_r \approx 10^4\,s^{-1}$ $K_r \approx 10^1\,s^{-1}$ $K_r \approx 10^2\,s^{-1}$

S_m ($\pi\pi^*$) — S_m (MLCT/$\pi\pi^*$) — T_n (MLCT/$\pi\pi^*$)

S_1 (MLCT/$\pi\pi^*$) —   Fast ISC S_1 ($\pi\pi^*$) —   Slow ISC

— T_n (MLCT/$\pi\pi^*$)

— T_1 ($\pi\pi^*$)   — T_1 ($\pi\pi^*$)

MLCT/LC emission   No MLCT contribution

SCHEME 11.5 Diagram representing relative order of the lowest excited states for a series of isostructural Cu(I), Ag(I) and Au(I) complexes. (Adapted with permission from Hsu, C.-W. et al. 2011. Systematic investigation of the metal–structure–photophysics relationship of emissive d(10)-complexes of Group 11 elements: The prospect of application in organic light emitting devices. *Journal of the American Chemical Society* 133(31):12085–12099. Copyright 2011 American Chemical Society.)

A highly efficient OLED was realized by Eastman Kodak using the neutral bis(bis(diisobutylphenylphosphino)amido) dicopper(I) complex [Cu(PNP-^{t}Bu)]$_2$ (Figure 11.20) (Deaton et al., 2010; Harkins and Peters, 2005; Harkins et al., 2008). This complex has a very high PL quantum yield of 57% in solution at RT and lifetime of 11.5 µs. Variable temperature photophysical studies suggest that [Cu(PNP-^{t}Bu)]$_2$ exhibits thermally activated (E-type) delayed fluorescence at RT. Such behavior,

(a)   (b)

FIGURE 11.20 Molecular structure of [Cu(PNP-^{t}Bu)]$_2$ complex, (a) its emission spectra recorded at 100 K and 295 K, and (b) J-EQE curve for OLED fabricated using this complex as an emitter. (Reprinted with permission from Deaton, J.C. et al. 2010. E-type delayed fluorescence of a phosphine-supported Cu$_2$(µ-NAr$_2$)$_2$ diamond core: Harvesting singlet and triplet excitons in OLEDs. *Journal of the American Chemical Society* 132(27):9499–9508. Copyright 2010 American Chemical Society.)

resulting from a small S_1-T_1 energy gap, is quite common for Cu(I) complexes (Blasse and McMillin, 1980; Breddels et al., 1982; Czerwieniec et al., 2011; Kirchhoff et al., 1983). If E-type delayed fluorescence occurs in a system, then a blue shift of the emission maximum and increase in the emission lifetime is observed experimentally as temperature increases. The excited state lifetime of a [Cu(PNP-^{t}Bu)]$_2$ film (doped at 1 wt% into 1,1-bis(4-(N,N-di-p-tolylamino)phenyl)cyclohexane (TAPC)) at 80 K is 336 µs and corresponds to purely triplet emission. Such a long lifetime of a phosphorescent dopant is a drawback for OLED application, because it will lead to efficiency roll-off at high current density due to saturation effects. At RT, the emission lifetime of the [Cu(PNP-^{t}Bu)]$_2$ film is reduced to 11.5 µs due to thermal equilibration of the S_1 and T_1 states that lie approximately 786 cm^{-1} apart. An OLED fabricated with a luminescent layer composed of CBP/TAPC(25%)/[Cu(PNP-^{t}Bu)]$_2$(0.2%) had a maximum external quantum efficiency of 16.1% (47.5 cd/A) at 0.01 mA/cm^2 which dropped to 10.9% (32 cd/A) at 1 mA/cm^2. This performance is among the best achieved with Cu(I)-based devices.

11.3.3 Three-Coordinate Cu(I) Complexes

Although the d^{10} Cu(I) center has a strong preference to form four-coordinate tetrahedral complexes, it is possible to obtain mononuclear three-coordinate Cu(I) compounds; however examples of luminescent complexes are rare. Unlike four-coordinate tetrahedral Cu(I) compounds that undergo a flattening distortion and exciplex quenching in the excited state, the three-coordinate geometry eliminates the possibility of a flattening distortion in the excited state, though exciplex formation is still possible. Therefore, undesirable nonradiative pathways can potentially be minimized in three-coordinate complexes. Recent reports of luminescent three-coordinate Cu(I) complexes suggest that these compounds can exhibit efficient phosphorescence at RT and are attractive candidates for OLED applications (Hashimoto et al., 2011; Krylova et al., 2010; Lotito and Peters, 2010).

The choice of ligands is crucial in order to stabilize three-coordinate Cu(I) structure. Peters and Lotito reported a series of brightly phosphorescent Cu(I) arylamidophosphine complexes [PP]Cu[N], where [PP] is a bidentate phosphine ligand or two monodentate phosphines and [N] is a terminal diphenylamido ligand (Figure 11.21a) (Lotito and Peters, 2010). Solution quantum yields in methylcyclohexane range from 11% to 24% and excited state lifetimes are short and do not vary substantially within the series (1.7–3.17 µs) with the exception of [N] = carbazole (11.7 µs). The calculated (DFT) HOMO is predominantly localized on diphenylamide with little contribution from copper and the LUMO is phosphine-based. This suggests that emission properties can be tuned by variation of both ligands. When the two PPh$_3$ groups in green-emitting (521 nm) (Ph$_3$P)$_2$Cu(NPh$_2$) were replaced with one chelating POP ligand in (POP)Cu(NPh$_2$), a 42 nm red-shift of the spectrum (563 nm) was observed. On the other hand, replacement of NPh$_2$ with carbazole gave a blue-emitting (461 nm) complex (Ph$_3$P)$_2$Cu(cbz). Introduction of electron-donating (–CH$_3$) or electron-withdrawing substituents (–F) in the para position of the phenyl groups of the diphenylamido ligand or on the phosphine did not lead to such drastic shifts, but allowed for fine tuning of the emission color with shifts up to 25 nm.

Chapter 11

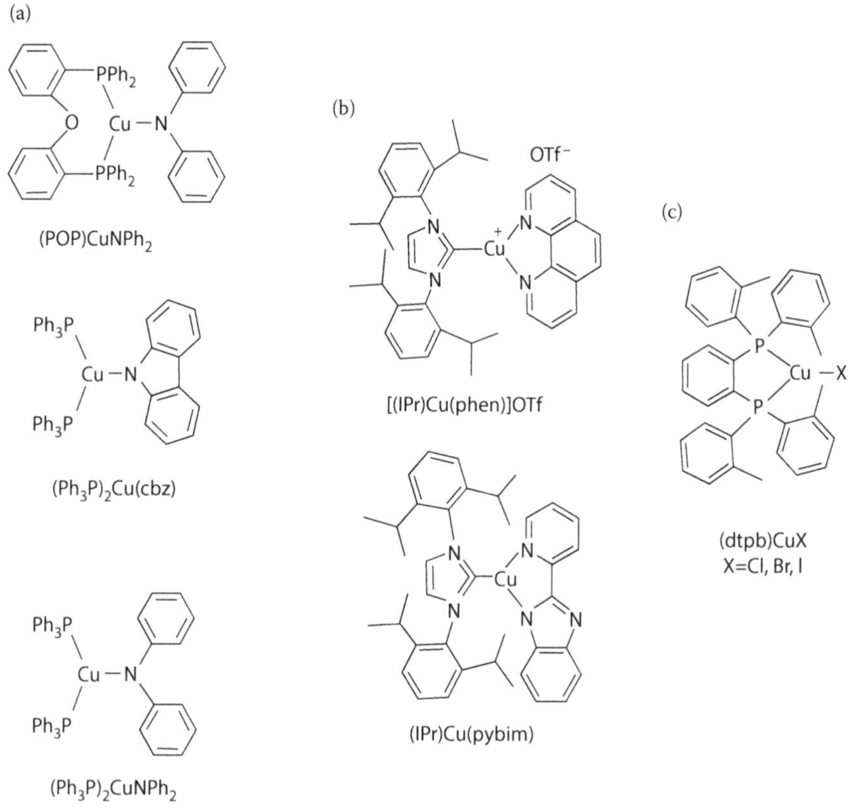

FIGURE 11.21   Molecular structures for the three-coordinate Cu(I)-complexes.

In yet another example, a three-coordinate Cu(I) structure was stabilized with the NHC ligand (Figure 11.21b) (Krylova et al., 2010). Cationic [(IPr)Cu(phen)] OTf (IPr = 1,3-bis(2,6-diisopropylphenyl)imidazol-2-ylidene) and neutral (IPr) Cu(pybim) (pybim = 2-(2-pyridyl)benzimidazole) complexes were obtained. Both complexes are phosphorescent air-stable solids and the neutral complex can be sublimed, which is highly desirable for OLED applications. Emission spectra recorded at 77 K in 2-methyltetrahydrofuran glass display a broad, featureless band centered at 630 nm for [(IPr)Cu(phen)]OTf and 555 nm for (IPr)Cu(pybim). Room temperature solution emission is relatively weak, <0.5%, in degassed CH_2Cl_2. However, in the solid state these complexes are brighter emitters with quantum yields and lifetimes of 2.6% and 1.2 μs for [(IPr)Cu(phen)]OTf and 58% and 33 μs for (IPr)Cu(pybim). Emission enhancement in rigid media suggests that molecular distortions in the excited state may be responsible for quenching in solution. The type of distortion is not yet clear (as the flattening distortion is not possible for the three-coordinate geometry). Barakat et al. have described a Jahn–Teller induced, Y- to T-shape distortion in the excited state of three-coordinate, trigonal planar Au(I) complexes (Barakat et al., 2003). A similar type of distortion may possibly occur in three-coordinate Cu(I) complexes.

An impressive example of utilization of three-coordinate Cu(I) complexes in OLEDs was demonstrated by Hashimoto et al. (2011). They were able to isolate a series of three-coordinate complexes (dtpb)CuX (dtpb = 1,2-bis(o-ditolylphosphino)benzene, X = Cl, Br, I)

(Figure 11.21c) through the use of the bulky dtpb ligands. These complexes phosphoresce in the green region of the spectrum (517–534 nm) and are highly emissive both in dichloromethane solution ($\Phi = 43\%$–60%) and when doped into a mCP film (57%–68%). Excited state lifetimes are relatively short ranging from 3.2 to 6.5 µs. Therefore, the triplet radiative rates of these complexes are up to $9 \cdot 10^5$ s^{-1}, which is comparable to the radiative rate of Ir(ppy)$_3$. The fabricated OLED with (dtpb)CuBr doped into mCP at 10 wt% as an emitting layer showed a maximum external quantum efficiency of 21.3% and current efficiency of 65.3 cd/A, records for OLEDs fabricated with Cu(I) complexes. However significant efficiency roll-off was observed at higher current densities. Nonetheless, this example demonstrates the high potential of three-coordinate Cu(I) complexes for OLED applications.

11.4 Conclusion

Phosphorescent emitters are the key to high efficiency OLEDs. Unlike fluorescent OLEDs, for which the internal quantum efficiency is limited to 25%, OLEDs based on phosphorescent emitters can reach 100% internal efficiency ($\approx$25% external quantum efficiency) due to harvesting of both singlet and triplet excitons that are generated in the electroluminescent process. The choice of phosphorescent dopant for OLEDs is challenging. In order to achieve high performance it is crucial for a triplet dopant to have a short radiative lifetime, high luminance efficiency, low nonradiative rates and be photo- and thermally stable under OLED operating conditions. Organometallic complexes, in particular cyclometalated iridium and platinum-based complexes have proven to be ideal candidates for OLED application. In light of global interest in OLEDs over the last 15 years a great deal of research has focused on photophysical properties of these materials. This has led to a deep understanding of their excited state properties and provided strategies to tune the emission energies and phosphorescence efficiencies in a predictable manner. Although some challenges still remain, cyclometalated iridium and platinum complexes are the most efficient phosphors for OLED yet developed.

To date phosphorescent materials for OLEDs have relied mainly on heavy metal complexes as triplet emitters, utilizing principally iridium or platinum to induce efficient spin-orbit coupling. It was shown that Cu(I) complexes can exhibit efficient phosphorescence at room temperature in both solution and the solid state and some have been used to fabricate OLEDs with external quantum efficiencies exceeding 15%. Although the photophysical and chemical properties of Cu(I)-based phosphors have not been as well-investigated as their third row counterparts, recent results clearly indicate that copper(I) complexes can potentially provide an inexpensive and environmentally friendly alternative to traditionally utilized phosphorescent materials based on heavy metal complexes.

References

Accorsi, G., N. Armaroli, C. Duhayon, A. Saquet, B. Delavaux-Nicot, R. Welter, O. Moudam, M. Holler, and J.-F. Nierengarten. 2010. Synthesis and photophysical properties of copper(I) complexes obtained from 1,10-Phenanthroline ligands with increasingly bulky 2,9-substituents. *European Journal of Inorganic Chemistry* 2010(1):164–173.

Adachi, C., M.A. Baldo, S.R. Forrest, and M.E. Thompson. 2000. High-efficiency organic electrophosphorescent devices with tris(2-phenylpyridine)iridium doped into electron-transporting materials. *Applied Physics Letters* 77(6):904–906.

Adachi, C., R.C. Kwong, P. Djurovich, V. Adamovich, M.A. Baldo, M.E. Thompson, and S.R. Forrest. 2001. Endothermic energy transfer: A mechanism for generating very efficient high-energy phosphorescent emission in organic materials. *Applied Physics Letters* 79(13):2082–2084.

Adamovich, V., J. Brooks, A. Tamayo, A.M. Alexander, P.I. Djurovich, B.W. D'Andrade, C. Adachi, S.R. Forrest, and M.E. Thompson. 2002. High efficiency single dopant white electrophosphorescent light emitting diodes. *New Journal of Chemistry* 26(9):1171–1178.

Aldridge, T.K., E.M. Stacy, and D.R. McMillin. 1994. Studies of the room-temperature absorption and emission-spectra of [Pt(Trpy)X]$^+$ systems. *Inorganic Chemistry* 33(4):722–727.

Arena, G., G. Calogero, S. Campagna, L.M. Scolaro, V. Ricevuto, and R. Romeo. 1998. Synthesis, characterization, absorption spectra, and luminescence properties of organometallic platinum(II) terpyridine complexes. *Inorganic Chemistry* 37(11):2763–2769.

Bailey, J.A., M.G. Hill, R.E. Marsh, V.M. Miskowski, W.P. Schaefer, and H.B. Gray. 1995. Electronic spectroscopy of chloro(Terpyridine) platinum(II). *Inorganic Chemistry* 34(18):4591–4599.

Baldo, M.A., C. Adachi, and S.R. Forrest. 2000. Transient analysis of organic electrophosphorescence. II. Transient analysis of triplet-triplet annihilation. *Physical Review B* 62(16):10967–10977.

Baldo, M.A., S. Lamansky, P.E. Burrows, M.E. Thompson, and S.R. Forrest. 1999a. Very high-efficiency green organic light-emitting devices based on electrophosphorescence. *Applied Physics Letters* 75(1):4–6.

Baldo, M.A., D.F. O'Brien, M.E. Thompson, and S.R. Forrest. 1999b. Excitonic singlet-triplet ratio in a semiconducting organic thin film. *Physical Review B* 60(20):14422–14428.

Baldo, M.A., D.F. O'Brien, Y. You, A. Shoustikov, S. Sibley, M.E. Thompson, and S.R. Forrest. 1998. Highly efficient phosphorescent emission from organic electroluminescent devices. *Nature* 395(6698):151–154.

Barakat, K.A., T.R. Cundari, and M.A. Omary. 2003. Jahn-Teller distortion in the phosphorescent excited state of three-coordinate Au(I) phosphine complexes. *Journal of the American Chemical Society* 125(47):14228–14229.

Baranoff, E., J.-H. Yum, M. Graetzel, and M.K. Nazeeruddin. 2009. Cyclometallated iridium complexes for conversion of light into electricity and electricity into light. *Journal of Organometallic Chemistry* 694(17):2661–2670.

Barigelletti, F., D. Sandrini, M. Maestri, V. Balzani, A. Von Zelewsky, L. Chassot, P. Jolliet, and U. Maeder. 1988. Temperature-dependence of the luminescence of cyclometalated Palladium(II), Rhodium(III), Platinum(II), and Platinum(IV) complexes. *Inorganic Chemistry* 27(20):3644–3647.

Blaskie, M.W. and D.R. McMillin. 1980. Photostudies of Copper(I) systems.6. Room-temperature emission and quenching studies of [Cu(Dmp)$_2$]$^+$. *Inorganic Chemistry* 19(11):3519–3522.

Blasse, G. and D.R. McMillin. 1980. Luminescence of Bis (Triphenylphosphine) Phenanthroline Copper (I). *Chemical Physics Letters* 70(1):1–3.

Blumstengel, S. and R. Dorsinville. 1999. Phosphorescent emission from polymeric light-emitting diodes doped with chrysene-d(12). *Japanese Journal of Applied Physics Part 2-Letters* 38(4A):L403–L405.

Borek, C., K. Hanson, P.I. Djurovich, M.E. Thompson, K. Aznavour, R. Bau, Y. Sun, S.R. Forrest, J. Brooks, L. Michalski, and J. Brown. 2007. Highly efficient, near-infrared electrophosphorescence from a Pt-metalloporphyrin complex. *Angewandte Chemie-International Edition* 46(7):1109–1112.

Breddels, P.A., P.A.M. Berdowski, and G. Blasse. 1982. Luminescence of Some Cu(I) Complexes. *Journal of the Chemical Society-Faraday Transaction II* 78:595–601.

Brooks, J., Y. Babayan, S. Lamansky, P.I. Djurovich, I. Tsyba, R. Bau, and M.E. Thompson. 2002. Synthesis and characterization of phosphorescent cyclometalated platinum complexes. *Inorganic Chemistry* 41(12):3055–3066.

Buchner, R., C.T. Cunningham, J.S. Field, R.J. Haines, D.R. McMillin, and G.C. Summerton. 1999. Luminescence properties of salts of the [Pt(4'Ph-terpy)Cl]$^+$ chromophore: Crystal structure of the red form of [Pt(4'Ph-terpy)Cl]BF$_4$ (4'Ph-terpy = 4'-phenyl-2,2':6',2''-terpyridine). *Journal of the Chemical Society-Dalton Transactions* (5):711–717.

Burroughes, J.H., D.D.C. Bradley, A.R. Brown, R.N. Marks, K. Mackay, R.H. Friend, P.L. Burns, and A.B. Holmes. 1990. Light-emitting-diodes based on conjugated polymers. *Nature* 347(6293):539–541.

Chan, C.W., C.M. Che, M.C. Cheng, and Y. Wang. 1992. Spectroscopy, photophysical properties, and x-ray crystal structure of platinum(II) complexes of quaterpyridine. *Inorganic Chemistry* 31(23):4874–4878.

Chang, C.-F., Y.-M. Cheng, Y. Chi, Y.-C. Chiu, C.-C. Lin, G.-H. Lee, P.-T. Chou, C.-C. Chen, C.-H. Chang, and C.-C. Wu. 2008. Highly efficient blue-emitting iridium(III) carbene complexes and phosphorescent OLEDs. *Angewandte Chemie-International Edition* 47(24):4542–4545.

Chang, C.-J., C.-H. Yang, K. Chen, Y. Chi, C.-F. Shu, M.-L. Ho, Y.-S. Yeh, and P.-T. Chou. 2007. Color tuning associated with heteroleptic cyclometalated Ir(III) complexes: influence of the ancillary ligand. *Dalton Transactions* (19):1881–1890.

Chassot, L., E. Muller, and A. Von Zelewsky. 1984. Cis-Bis(2-Phenylpyridine)Platinum(II) (CBPPP): A simple molecular platinum compound. *Inorganic Chemistry* 23(25):4249–4253.

Chassot, L. and A. Von Zelewsky. 1987. Cyclometalated complexes of platinum(II)—Homoleptic compounds with aromatic C,N ligands. *Inorganic Chemistry* 26(17):2814–2818.

Che, G., Z. Su, W. Li, B. Chu, M. Li, Z. Hu, and Z. Zhang. 2006. Highly efficient and color-tuning electrophosphorescent devices based on Cu-I complex. *Applied Physics Letters* 89(10):103511/1–103511/3.

Chen, L.X., G. Jennings, T. Liu, D.J. Gosztola, J.P. Hessler, D.V. Scaltrito, and G.J. Meyer. 2002. Rapid excited-state structural reorganization captured by pulsed X-rays. *Journal of the American Chemical Society* 124(36):10861–10867.

Chen, L.X., G.B. Shaw, I. Novozhilova, T. Liu, G. Jennings, K. Attenkofer, G.J. Meyer, and P. Coppens. 2003. MLCT state structure and dynamics of a copper(I) diimine complex characterized by pump-probe X-ray and laser spectroscopies and DFT calculations. *Journal of the American Chemical Society* 125(23):7022–7034.

Cheung, T.C., K.K. Cheung, S.M. Peng, and C.M. Che. 1996. Photoluminescent cyclometallated diplatinum(II,II) complexes: Photophysical properties and crystal structures of [PtL(PPh$_3$)]ClO$_4$ and [Pt$_2$L$_2$(μ-dppm)][ClO$_4$]$_2$ (HL = 6-phenyl-2,2'-bipyridine, dppm = Ph$_2$PCH$_2$PPh$_2$). *Journal of the Chemical Society-Dalton Transactions* (8):1645–1651.

Chiu, Y.-C., Y. Chi, J.-Y. Hung, Y.-M. Cheng, Y.-C. Yu, M.-W. Chung, G.-H. Lee et al. 2009a. Blue to true-blue phosphorescent Ir(III) complexes bearing a nonconjugated ancillary phosphine chelate: Strategic synthesis, photophysics, and device integration. *Acs Applied Materials and Interfaces* 1(2):433–442.

Chiu, Y.-C., J.-Y. Hung, Y. Chi, C.-C. Chen, C.-H. Chang, C.-C. Wu, Y.-M. Cheng, Y.-C. Yu, G.-H. Lee, and P.-T. Chou. 2009b. En route to high external quantum efficiency (~12%), organic true-blue-light-emitting diodes employing novel design of iridium(III) phosphors. *Advanced Materials* 21(21):2221–2225.

Chiu, Y.-C., C.-H. Lin, J.-Y. Hung, Y. Chi, Y.-M. Cheng, K.-W. Wang, M.-W. Chung, G.-H. Lee, and P.-T. Chou. 2009c. Authentic-blue phosphorescent iridium(III) complexes bearing both hydride and benzyl diphenylphosphine; control of the emission efficiency by ligand coordination geometry. *Inorganic Chemistry* 48(17):8164–8172.

Cocchi, M., V. Fattori, D. Virgili, C. Sabatini, P. Di Marco, M. Maestri, and J. Kalinowski. 2004a. Highly efficient organic electrophosphorescent light-emitting diodes with a reduced quantum efficiency roll off at large current densities. *Applied Physics Letters* 84(7):1052–1054.

Cocchi, M., J. Kalinowski, V. Fattori, J.A.G. Williams, and L. Murphy. 2009. Color-variable highly efficient organic electrophosphorescent diodes manipulating molecular exciton and excimer emissions. *Applied Physics Letters* 94(7):073309/1–073309/3.

Cocchi, M., J. Kalinowski, L. Murphy, J.A.G. Williams, and V. Fattori. 2010. Mixing of molecular exciton and excimer phosphorescence to tune color and efficiency of organic LEDs. *Organic Electronics* 11(3):388–396.

Cocchi, M., J. Kalinowski, D. Virgili, V. Fattori, S. Develay, and J.A.G. Williams. 2007a. Single-dopant organic white electrophosphorescent diodes with very high efficiency and its reduced current density roll-off. *Applied Physics Letters* 90(16):163508/1–163508/3.

Cocchi, M., J. Kalinowski, D. Virgili, and J.A.G. Williams. 2008. Excimer-based red/near-infrared organic light-emitting diodes with very high quantum efficiency. *Applied Physics Letters* 92(11):113302/1–113302/3.

Cocchi, M., D. Virgili, V. Fattori, D.L. Rochester, and J.A.G. Williams. 2007b. N^C^N-coordinated platinum(II) complexes as phosphorescent emitters in high-performance organic light-emitting devices. *Advanced Functional Materials* 17(2):285–289.

Cocchi, M., D. Virgili, V. Fattori, J.A.G. Williams, and J. Kalinowski. 2007c. Highly efficient near-infrared organic excimer electrophosphorescent diodes. *Applied Physics Letters* 90(2):023506/1–023506/3.

Cocchi, M., D. Virgili, C. Sabatini, V. Fattori, P. Di Marco, M. Maestri, and J. Kalinowski. 2004b. Highly efficient organic electroluminescent devices based on cyclometallated platinum complexes as new phosphorescent emitters. *Synthetic Metals* 147(1–3):253–256.

Connick, W.B., D. Geiger, and R. Eisenberg. 1999. Excited-state self-quenching reactions of square planar platinum(II) diimine complexes in room-temperature fluid solution. *Inorganic Chemistry* 38(14):3264–3265.

Crestani, M.G., G.F. Manbeck, W.W. Brennessel, T.M. McCormick, and R. Eisenberg. 2011. Synthesis and characterization of neutral luminescent diphosphine pyrrole- and indole-aldimine copper(I) complexes. *Inorganic Chemistry* 50(15):7172–7188.

Cuttell, D.G., S.M. Kuang, P.E. Fanwick, D.R. McMillin, and R.A. Walton. 2002. Simple Cu(I) complexes with unprecedented excited-state lifetimes. *Journal of the American Chemical Society* 124(1):6–7.

Czerwieniec, R., J. Yu, and H. Yersin. 2011. Blue-light emission of Cu(I) complexes and singlet harvesting. *Inorganic Chemistry* 50(17):8293–8301.

D'Andrade, B.W., J. Brooks, V. Adamovich, M.E. Thompson, and S.R. Forrest. 2002. White light emission using triplet excimers in electrophosphorescent organic light-emitting devices. *Advanced Materials* 14(15):1032–1036.

Deaton, J.C., S.C. Switalski, D.Y. Kondakov, R.H. Young, T.D. Pawlik, D.J. Giesen, S.B. Harkins, A.J.M. Miller, S.F. Mickenberg, and J.C. Peters. 2010. E-type delayed fluorescence of a phosphine-supported $Cu_2(\mu-NAr_2)_2$ diamond core: Harvesting singlet and triplet excitons in OLEDs. *Journal of the American Chemical Society* 132(27):9499–9508.

Dedeian, K., P.I. Djurovich, F.O. Garces, G. Carlson, and R.J. Watts. 1991. A new synthetic route to the preparation of a series of strong photoreducing agents: Fac Tris-Ortho-Metalated complexes of Iridium(III) with substituted 2-Phenylpyridines. *Inorganic Chemistry* 30(8):1685–1687.

Djurovich, P.I. and M.E. Thompson. 2008. Cyclometallated organoiridium complexes as emitters in electrophosphorescent devices. *Highly Efficient OLEDs with Phosphorescent Materials*, Yersin, H., ed. WILEY-VCH Verlag GmbH & Co. KGaA, Weinheim, 131–161.

Eggleston, M.K., D.R. McMillin, K.S. Koenig, and A.J. Pallenberg. 1997. Steric effects in the ground states of $Cu(NN)^{2+}$ systems. *Inorganic Chemistry* 36(2):172–176.

Endo, A., K. Suzuki, T. Yoshihara, S. Tobita, M. Yahiro, and C. Adachi. 2008. Measurement of photoluminescence efficiency of Ir(III) phenylpyridine derivatives in solution and solid-state films. *Chemical Physics Letters* 460(1–3):155–157.

Everly, R.M. and D.R. McMillin. 1989. Concentration-dependent lifetimes of $Cu(NN)^{2+}$ systems: Exciplex quenching from the ion pair state. *Photochemistry and Photobiology* 50(6):711–716.

Farley, S.J., D.L. Rochester, A.L. Thompson, J.A.K. Howard, and J.A.G. Williams. 2005. Controlling emission energy, self-quenching, and excimer formation in highly luminescent $N^\wedge C^\wedge N$-coordinated platinum(II) complexes. *Inorganic Chemistry* 44(26):9690–9703.

Feng, K., C. Zuniga, Y.-D. Zhang, D. Kim, S. Barlow, S.R. Mardert, J.L. Bredas, and M. Weck. 2009. Norbornene-based copolymers containing platinum complexes and bis(carbazolyl)benzene groups in their side-chains. *Macromolecules* 42(18):6855–6864.

Finikova, O.S., S.E. Aleshchenkov, R.P. Brinas, A.V. Cheprakov, P.J. Carroll, and S.A. Vinogradov. 2005. Synthesis of symmetrical tetraaryltetranaphtho[2,3]porphyrins. *Journal of Organic Chemistry* 70(12):4617–4628.

Flamigni, L., A. Barbieri, C. Sabatini, B. Ventura, and F. Barigelletti. 2007. Photochemistry and photophysics of coordination compounds: Iridium. *Topics in Current Chemistry* 281:143–203.

Gothard, N.A., M.W. Mara, J. Huang, J.M. Szarko, B. Rolczynski, J.V. Lockard, and L.X. Chen. 2012. Strong steric hindrance effect on excited state structural dynamics of Cu(I) diimine complexes. *Journal of Physical Chemistry A* 116(9):1984–1992.

Graham, K.R., Y. Yang, J.R. Sommer, A.H. Shelton, K.S. Schanze, J. Xue, and J.R. Reynolds. 2011. Extended conjugation platinum(II) porphyrins for use in near-infrared emitting organic light emitting diodes. *Chemistry of Materials* 23(24):5305–5312.

Green, O., B.A. Gandhi, and J.N. Burstyn. 2009. Photophysical characteristics and reactivity of bis(2,9-di-tert-butyl-1,10-phenanthroline) copper(I). *Inorganic Chemistry* 48(13):5704–5714.

Grushin, V.V., N. Herron, D.D. LeCloux, W.J. Marshall, V.A. Petrov, and Y. Wang. 2001. New, efficient electroluminescent materials based on organometallic Ir complexes. *Chemical Communications* (16):1494–1495.

Haneder, S., E. Da Como, J. Feldmann, J.M. Lupton, C. Lennartz, P. Erk, E. Fuchs et al. 2008. Controlling the radiative rate of deep-blue electrophosphorescent organometallic complexes by singlet-triplet gap engineering. *Advanced Materials* 20(17):3325–3330.

Harkins, S.B., N.P. Mankad, A.J.M. Miller, R.K. Szilagyi, and J.C. Peters. 2008. Probing the electronic structures of $[Cu_2(\mu-XR_2)]^{n+}$ diamond cores as a function of the bridging X atom (X = N or P) and charge (n = 0, 1, 2). *Journal of the American Chemical Society* 130(11):3478–3485.

Harkins, S.B. and J.C. Peters. 2005. A highly emissive Cu_2N_2 diamond core complex supported by a $[PNP]^-$ ligand. *Journal of the American Chemical Society* 127(7):2030–2031.

Hashimoto, M., S. Igawa, M. Yashima, I. Kawata, M. Hoshino, and M. Osawa. 2011. Highly efficient green organic light-emitting diodes containing luminescent three-coordinate copper(I) complexes. *Journal of the American Chemical Society* 133(27):10348–10351.

Hay, P.J. 2002. Theoretical studies of the ground and excited electronic states in cyclometalated phenylpyridine Ir(III) complexes using density functional theory. *Journal of Physical Chemistry A* 106(8):1634–1641.

Hisamatsu, Y. and S. Aoki. 2011. Design and synthesis of blue-emitting cyclometalated iridium(III) complexes based on regioselective functionalization. *European Journal of Inorganic Chemistry* 2011(35):5360–5369.

Hitchcock, P.B., M.F. Lappert, and P. Terreros. 1982. Synthesis of homoleptic tris(organo-chelate)iridium(III) complexes by spontaneous ortho-metallation of electron-rich olefin-derived N,N′-diarylcarbene ligands and the x-ray structures of tris(ortho metalated-carbene)iridium(III) complexes. *Journal of Organometallic Chemistry* 239(2):C26–C30.

Hofbeck, T. and H. Yersin. 2010. The triplet state of fac-Ir(ppy)$_3$. *Inorganic Chemistry* 49(20):9290–9299.

Holmes, R.J., S.R. Forrest, T. Sajoto, A. Tamayo, P.I. Djurovich, M.E. Thompson, J. Brooks et al. 2005. Saturated deep blue organic electrophosphorescence using a fluorine-free emitter. *Applied Physics Letters* 87(24):243507/1243507/3.

Holmes, R.J., S.R. Forrest, Y.J. Tung, R.C. Kwong, J.J. Brown, S. Garon, and M.E. Thompson. 2003. Blue organic electrophosphorescence using exothermic host-guest energy transfer. *Applied Physics Letters* 82(15):2422–2424.

Hoshino, S. and H. Suzuki. 1996. Electroluminescence from triplet excited states of benzophenone. *Applied Physics Letters* 69(2):224–226.

Hsu, C.-W., C.-C. Lin, M.-W. Chung, Y. Chi, G.-H. Lee, P.-T. Chou, C.-H. Chang, and P.-Y. Chen. 2011. Systematic investigation of the metal-structure-photophysics relationship of emissive d(10)-complexes of group 11 elements: The prospect of application in organic light emitting devices. *Journal of the American Chemical Society* 133(31):12085–12099.

Hung, J.-Y., Y. Chi, I.H. Pai, Y.-C. Yu, G.-H. Lee, P.-T. Chou, K.-T. Wong, C.-C. Chen, and C.-C. Wu. 2009. Blue-emitting Ir(III) phosphors with ancillary 4,6-difluorobenzyl diphenylphosphine based cyclometalate. *Dalton Transactions* (33):6472–6475.

Jabbour, G.E., J.F. Wang, and N. Peyghambarian. 2002. High-efficiency organic electrophosphorescent devices through balance of charge injection. *Applied Physics Letters* 80(11):2026–2028.

Jacobsen, H., A. Correa, A. Poater, C. Costabile, and L. Cavallo. 2009. Understanding the M-(NHC) (NHC = N-heterocyclic carbene) bond. *Coordination Chemistry Reviews* 253(5–6):687–703.

Karatsu, T., E. Ito, S. Yagai, and A. Kitamura. 2006. Radiative and nonradiative processes of meridional and facial isomers of heteroleptic iridium-trischelate complexes. *Chemical Physics Letters* 424(4–6):353–357.

Karatsu, T., T. Nakamura, S. Yagai, A. Kitamura, K. Yamaguchi, Y. Matsushima, T. Iwata, Y. Hori, and T. Hagiwara. 2003. Photochemical mer -> fac one-way isomerization of phosphorescent material. Studies by time resolved spectroscopy for Tris[2-(4′,6′-difluorophenyl)pyridine]iridium(III) in solution. *Chemistry Letters* 32(10):886–887.

Kawamura, Y., K. Goushi, J. Brooks, J.J. Brown, H. Sasabe, and C. Adachi. 2005. 100% phosphorescence quantum efficiency of Ir(III) complexes in organic semiconductor films. *Applied Physics Letters* 86(7):071104/1–071104/3.

King, K.A., P.J. Spellane, and R.J. Watts. 1985. Excited-state properties of a triply ortho-metalated iridium(III) complex. *Journal of the American Chemical Society* 107(5):1431–1432.

Kirchhoff, J.R., R.E. Gamache, M.W. Blaskie, A.A. Delpaggio, R.K. Lengel, and D.R. McMillin. 1983. Temperature-dependence of luminescence from Cu(NN)$^{2+}$ systems in fluid solution: Evidence for the participation of two excited-states. *Inorganic Chemistry* 22(17):2380–2384.

Komada, Y., S. Yamauchi, and N. Hirota. 1986. Phosphorescence and zero-field optically detected magnetic-resonance studies of the lowest excited triplet-states of organometallic diimine complexes. 1. [Rh(bpy)$_3$]$^{3+}$ and [Rh(phen)$_3$]$^{3+}$. *Journal of Physical Chemistry* 90(24):6425–6430.

Krylova, V.A., P.I. Djurovich, M.T. Whited, and M.E. Thompson. 2010. Synthesis and characterization of phosphorescent three-coordinate Cu(I)-NHC complexes. *Chemical Communications* 46(36):6696–6698.

Kuang, S.M., D.G. Cuttell, D.R. McMillin, P.E. Fanwick, and R.A. Walton. 2002. Synthesis and structural characterization of Cu(I) and that Ni(II) complexes that contain the bis[2-(diphenylphosphino)phenyl] ether ligand. Novel emission properties for the Cu(I) species. *Inorganic Chemistry* 41(12):3313–3322.

Kui, S.C.F., I.H.T. Sham, C.C.C. Cheung, C.-W. Ma, B. Yan, N. Zhu, C.-M. Che, and W.-F. Fu. 2007. Platinum(II) complexes with π-conjugated, naphthyl-substituted, cyclometalated ligands (RC-N-N): Structures and photo- and electroluminescence. *Chemistry-a European Journal* 13(2):417–435.

Kunkely, H. and A. Vogler. 1990. Photoluminescence of [PtII(4,7-Diphenyl-1,10-Phenanthroline)(CN)$_2$] in Solution. *Journal of the American Chemical Society* 112(14):5625–5627.

Kwok, C.C., H.M.Y. Ngai, S.C. Chan, I.H.T. Sham, C.M. Che, and N.Y. Zhu. 2005. [(O^N^N)PtX] complexes as a new class of light-emitting materials for electrophosphorescent devices. *Inorganic Chemistry* 44(13):4442–4444.

Kwon, T.H., H.S. Cho, M.K. Kim, J.W. Kim, J.J. Kim, K.H. Lee, S.J. Park et al. 2005. Color tuning of cyclometalated iridium complexes through modification of phenylpyrazole derivatives and ancillary ligand based on ab initio calculations. *Organometallics* 24(7):1578–1585.

Lai, S.W., M.C.W. Chan, T.C. Cheung, S.M. Peng, and C.M. Che. 1999. Probing d^8-d^8 Interactions in luminescent mono- and binuclear cyclometalated platinum(II) complexes of 6-phenyl-2,2'-bipyridines. *Inorganic Chemistry* 38(18):4046–4055.

Lai, S.W. and C.M. Che. 2004. Luminescent cyclometalated diimine platinum(II) complexes: Photophysical studies and applications. *Topics in Current Chemistry* 241:27–63.

Lai, S.W., H.W. Lam, W. Lu, K.K. Cheung, and C.M. Che. 2002. Observation of low-energy metal-metal-to-ligand charge transfer absorption and emission: Electronic spectroscopy of cyclometalated platinum(II) complexes with isocyanide ligands. *Organometallics* 21(1):226–234.

Lamansky, S., P. Djurovich, D. Murphy, F. Abdel-Razzaq, R. Kwong, I. Tsyba, M. Bortz, B. Mui, R. Bau, and M.E. Thompson. 2001a. Synthesis and characterization of phosphorescent cyclometalated iridium complexes. *Inorganic Chemistry* 40(7):1704–1711.

Lamansky, S., P. Djurovich, D. Murphy, F. Abdel-Razzaq, H.E. Lee, C. Adachi, P.E. Burrows, S.R. Forrest, and M.E. Thompson. 2001b. Highly phosphorescent bis-cyclometalated iridium complexes: Synthesis, photophysical characterization, and use in organic light emitting diodes. *Journal of the American Chemical Society* 123(18):4304–4312.

Lavie-Cambot, A., M. Cantuel, Y. Leydet, G. Jonusauskas, D.M. Bassani, and N.D. McClenaghan. 2008. Improving the photophysical properties of copper(I) bis(phenanthroline) complexes. *Coordination Chemistry Reviews* 252(23–24):2572–2584.

Lever, A.B.P. 1984. *Inorganic Electronic Spectroscopy*. Elsevier, Amsterdam; New York.

Li, C.L., Y.J. Su, Y.T. Tao, P.T. Chou, C.H. Chien, C.C. Cheng, and R.S. Liu. 2005a. Yellow and red electrophosphors based on linkage isomers of phenylisoquinolinyliridium complexes: Distinct differences in photophysical and electroluminescence properties. *Advanced Functional Materials* 15(3):387–395.

Li, J., P.I. Djurovich, B.D. Alleyne, M. Yousufuddin, N.N. Ho, J.C. Thomas, J.C. Peters, R. Bau, and M.E. Thompson. 2005b. Synthetic control of excited-state properties in cyclometalated Ir(III) complexes using ancillary ligands. *Inorganic Chemistry* 44(6):1713–1727.

Lin, C.-H., Y.-Y. Chang, J.-Y. Hung, C.-Y. Lin, Y. Chi, M.-W. Chung, C.-L. Lin et al. 2011. Iridium(III) Complexes of a dicyclometalated phosphite tripod ligand: Strategy to achieve blue phosphorescence without fluorine substituents and fabrication of OLEDs. *Angewandte Chemie-International Edition* 50(14):3182–3186.

Liu, Z., M.F. Qayyum, C. Wu, M.T. Whited, P.I. Djurovich, K.O. Hodgson, B. Hedman, E.I. Solomon, and M.E. Thompson. 2011. A codeposition route to CuI-Pyridine coordination complexes for organic light-emitting diodes. *Journal of the American Chemical Society* 133(11):3700–3703.

Lotito, K.J. and J.C. Peters. 2010. Efficient luminescence from easily prepared three-coordinate copper(I) arylamidophosphines. *Chemical Communications* 46(21):3690–3692.

Lu, K.-Y., H.-H. Chou, C.-H. Hsieh, Y.-H.O. Yang, H.-R. Tsai, H.-Y. Tsai, L.-C. Hsu, C.-Y. Chen, I.C. Chen, and C.-H. Cheng. 2011. Wide-range color tuning of iridium biscarbene complexes from blue to red by different N^N ligands: An alternative route for adjusting the emission colors. *Advanced Materials* 23(42):4933–4937.

Lu, W., B.X. Mi, M.C.W. Chan, Z. Hui, C.M. Che, N.Y. Zhu, and S.T. Lee. 2004. Light-emitting tridentate cyclometalated platinum(II) complexes containing σ-alkynyl auxiliaries: Tuning of photo- and electrophosphorescence. *Journal of the American Chemical Society* 126(15):4958–4971.

Ma, B.W., P.I. Djurovich, and M.E. Thompson. 2005. Excimer and electron transfer quenching studies of a cyclometalated platinum complex. *Coordination Chemistry Reviews* 249(13–14):1501–1510.

Maestri, M., D. Sandrini, V. Balzani, L. Chassot, P. Jolliet, and A. Von Zelewsky. 1985. Luminescence of ortho-metalated platinum(II) complexes. *Chemical Physics Letters* 122(4):375–379.

Manbeck, G.F., W.W. Brennessel, and R. Eisenberg. 2011. Photoluminescent copper(I) complexes with amido-triazolato ligands. *Inorganic Chemistry* 50(8):3431–3441.

McCormick, T., W.L. Jia, and S.N. Wang. 2006. Phosphorescent Cu(I) complexes of 2-(2'-pyridylbenzimidazolyl)benzene: Impact of phosphine ancillary ligands on electronic and photophysical properties of the Cu(I) complexes. *Inorganic Chemistry* 45(1):147–155.

McGlynn, S.P., T. Azumi, and M. Kinoshita. 1969. *Molecular Spectroscopy of the Triplet State*. Prentice-Hall, Englewood Cliffs, HJ.

McMillin, D.R. and K.M. McNett. 1998. Photoprocesses of copper complexes that bind to DNA. *Chemical Reviews* 98(3):1201–1219.

Miller, M.T., P.K. Gantzel, and T.B. Karpishin. 1999a. Effects of sterics and electronic delocalization on the photophysical, structural, and electrochemical properties of 2,9-disubstituted 1,10-phenanthroline copper(I) complexes. *Inorganic Chemistry* 38(14):3414–3422.

Miller, M.T., P.K. Gantzel, and T.B. Karpishin. 1999b. A highly emissive heteroleptic copper(I) bis(phenanthroline) complex: [Cu(dbp)(dmp)]$^+$ (dbp = 2,9-di-tert-butyl-1,10-phenanthroline; dmp = 2,9-dimethyl-1,10-phenanthroline). *Journal of the American Chemical Society* 121(17):4292–4293.

Miller, A.J.M., J.L. Dempsey, and J.C. Peters. 2007. Long-lived and efficient emission from mononuclear amidophosphine complexes of copper. *Inorganic Chemistry* 46(18):7244–7246.

Miskowski, V.M. and V.H. Houlding. 1991. Electronic spectra and photophysics of platinum(II) complexes with α-diimine ligands: Solid state effects. 2. metal-metal interaction in double salts and linear chains. *Inorganic Chemistry* 30(23):4446–4452.

Moudam, O., A. Kaeser, B. Delavaux-Nicot, C. Duhayon, M. Holler, G. Accorsi, N. Armaroli et al. 2007. Electrophosphorescent homo- and heteroleptic copper(I) complexes prepared from various bis-phosphine ligands. *Chemical Communications* (29):3077–3079.

Nonoyama, M. 1974. [Benzo[H]Quinolin-10-yl-N]iridium(III) Complexes. *Bulletin of the Chemical Society of Japan* 47(3):767–768.

Nozaki, K. 2006. Theoretical studies on photophysical properties and mechanism of phosphorescence in [fac-Ir(2-phenylpyridine)$_3$]. *Journal of the Chinese Chemical Society* 53(1):101–112.

O'Brien, D.F., M.A. Baldo, M.E. Thompson, and S.R. Forrest. 1999. Improved energy transfer in electrophosphorescent devices. *Applied Physics Letters* 74(3):442–444.

Palmer, C.E.A. and D.R. McMillin. 1987. Singlets, Triplets, and exciplexes: Complex, temperature-dependent emissions from Cu(dmp)(PPh$_3$)$^{2+}$ and Cu(phen)(PPh$_3$)$^{2+}$ in solution. *Inorganic Chemistry* 26(23):3837–3840.

Papkovsky, D.B. 1995. New oxygen sensors and their application to biosensing. *Sensors and Actuators B-Chemical* 29(1–3):213–218.

Paulose, B., D.K. Rayabarapu, J.P. Duan, and C.H. Cheng. 2004. First examples of alkenyl pyridines as organic ligands for phosphorescent iridium complexes. *Advanced Materials* 16(22):2003–2007.

Pettijohn, C.N., E.B. Jochnowitz, B. Chuong, J.K. Nagle, and A. Vogler. 1998. Luminescent excimers and exciplexes of PtII compounds. *Coordination Chemistry Reviews* 171:85–92.

Rader, R.A., D.R. McMillin, M.T. Buckner, T.G. Matthews, D.J. Casadonte, R.K. Lengel, S.B. Whittaker, L.M. Darmon, and F.E. Lytle. 1981. Photostudies of [Cu(bpy)(PPh$_3$)$_2$]$^+$, [Cu(phen)(PPh$_3$)$_2$]$^+$, and [Cu(dmp)(PPh$_3$)$_2$]$^+$ in solution and in rigid, low-temperature glasses: Simultaneous multiple emissions from intraligand and charge-transfer states. *Journal of the American Chemical Society* 103(19):5906–5912.

Ragni, R., E.A. Plummer, K. Brunner, J.W. Hofstraat, F. Babudri, G.M. Farinola, F. Naso, and L. De Cola. 2006. Blue emitting iridium complexes: Synthesis, photophysics and phosphorescent devices. *Journal of Materials Chemistry* 16(12):1161–1170.

Rausch, A.F., H.H.H. Homeier, and H. Yersin. 2010. Organometallic Pt(II) and Ir(III) Triplet emitters for OLED applications and the role of spin-orbit coupling: A study based on high-resolution optical spectroscopy. *Topics in Organometallic Chemistry* 29:193–235.

Rozhkov, V.V., M. Khajehpour, and S.A. Vinogradov. 2003. Luminescent Zn and Pd tetranaphthaloporphyrins. *Inorganic Chemistry* 42(14):4253–4255.

Sajoto, T., P.I. Djurovich, A. Tamayo, M. Yousufuddin, R. Bau, M.E. Thompson, R.J. Holmes, and S.R. Forrest. 2005. Blue and near-UV phosphorescence from iridium complexes with cyclometalated pyrazolyl or N-heterocyclic carbene ligands. *Inorganic Chemistry* 44(22):7992–8003.

Sajoto, T., P.I. Djurovich, A.B. Tamayo, J. Oxgaard, W.A. Goddard, III, and M.E. Thompson. 2009. Temperature dependence of blue phosphorescent cyclometalated Ir(III) complexes. *Journal of the American Chemical Society* 131(28):9813–9822.

Sasabe, H., E. Gonmori, T. Chiba, Y.-J. Li, D. Tanaka, S.-J. Su, T. Takeda, Y.-J. Pu, K.-i. Nakayama, and J. Kido. 2008. Wide-energy-gap electron-transport materials containing 3,5-dipyridylphenyl moieties for an ultra high efficiency blue organic light-emitting device. *Chemistry of Materials* 20(19):5951–5953.

Scaltrito, D.V., D.W. Thompson, J.A. O'Callaghan, and G.J. Meyer. 2000. MLCT excited states of cuprous bisphenanthroline coordination compounds. *Coordination Chemistry Reviews* 208:243–266.

Segal, M., M.A. Baldo, R.J. Holmes, S.R. Forrest, and Z.G. Soos. 2003. Excitonic singlet-triplet ratios in molecular and polymeric organic materials. *Physical Review B* 68(7):075211/1–075211/14.

Si, Z., J. Li, B. Li, S. Liu, and W. Li. 2008. Bright electrophosphorescent devices based on sterically hindered spacer-containing Cu(I) complex. *Journal of Luminescence* 128(8):1303–1306.

Smith, C.S., C.W. Branham, B.J. Marquardt, and K.R. Mann. 2010. Oxygen Gas Sensing by Luminescence Quenching in Crystals of Cu(xantphos)(phen)$^+$ Complexes. *Journal of the American Chemical Society* 132(40):14079–14085.

Chapter 11

Sommer, J.R., R.T. Farley, K.R. Graham, Y. Yang, J.R. Reynolds, J. Xue, and K.S. Schanze. 2009. Efficient near-infrared polymer and organic light-emitting diodes based on electrophosphorescence from (Tetraphenyltetranaphtho[2,3]porphyrin)-platinum(II). *Acs Applied Materials and Interfaces* 1(2):274–278.

Sommer, J.R., A.H. Shelton, A. Parthasarathy, I. Ghiviriga, J.R. Reynolds, and K.S. Schanze. 2011. Photophysical properties of near-infrared phosphorescent π-extended platinum porphyrins. *Chemistry of Materials* 23(24):5296–5304.

Sotoyama, W., T. Satoh, H. Sato, A. Matsuura, and N. Sawatari. 2005a. Excited states of phosphorescent platinum(II) complexes containing N^C^N-coordinating tridentate ligands: Spectroscopic investigations and time-dependent density functional theory calculations. *Journal of Physical Chemistry A* 109(43):9760–9766.

Sotoyama, W., T. Satoh, N. Sawatari, and H. Inoue. 2005b. Efficient organic light-emitting diodes with phosphorescent platinum complexes containing N,C,N-coordinating tridentate ligand. *Applied Physics Letters* 86(15):153505/1–153505/3.

Stacy, E.M. and D.R. McMillin. 1990. Inorganic exciplexes revealed by temperature-dependent quenching studies. *Inorganic Chemistry* 29(3):393–396.

Su, Y.J., H.L. Huang, C.L. Li, C.H. Chien, Y.T. Tao, P.T. Chou, S. Datta, and R.S. Liu. 2003. Highly efficient red electrophosphorescent devices based on iridium isoquinoline complexes: Remarkable external quantum efficiency over a wide range of current. *Advanced Materials* 15(11):884–888.

Su, Z.S., G.B. Che, W.L. Li, W.M. Su, M.T. Li, B. Chu, B. Li, Z.Q. Zhang, and Z.Z. Hu. 2006. White-electrophosphorescent devices based on copper complexes using 2-(4-biphenylyl)-5-(4-tert-butyl-phenyl)-1,3,4-oxadiazole as chromaticity-tuning layer. *Applied Physics Letters* 88(21):213508/1–213508/3.

Sun, Y., C. Borek, K. Hanson, P.I. Djurovich, M.E. Thompson, J. Brooks, J.J. Brown, and S.R. Forrest. 2007. Photophysics of Pt-porphyrin electrophosphorescent devices emitting in the near infrared. *Applied Physics Letters* 90(21):213503/1–213503/3.

Tamayo, A.B., B.D. Alleyne, P.I. Djurovich, S. Lamansky, I. Tsyba, N.N. Ho, R. Bau, and M.E. Thompson. 2003. Synthesis and characterization of facial and meridional tris-cyclometalated iridium(III) complexes. *Journal of the American Chemical Society* 125(24):7377–7387.

Tanaka, D., H. Sasabe, Y.-J. Li, S.-J. Su, T. Takeda, and J. Kido. 2007. Ultra high efficiency green organic light-emitting devices. *Japanese Journal of Applied Physics Part 2-Letters and Express Letters* 46(1–3):L10–L12.

Tang, C.W. and S.A. Vanslyke. 1987. Organic electroluminescent diodes. *Applied Physics Letters* 51(12):913–915.

Tong, G.S.-M. and C.-M. Che. 2009. Emissive or nonemissive? A theoretical analysis of the phosphorescence efficiencies of cyclometalated platinum(II) complexes. *Chemistry-a European Journal* 15(29):7225–7237.

Tsuboyama, A., H. Iwawaki, M. Furugori, T. Mukaide, J, Kamatani, S. Igawa, T. Moriyama, S. Miura, T. Takiguchi, S. Okada, M. Hoshino, and K.Ueno. 2003. Homoleptic cyclometalated iridium complexes with highly efficient red phosphorescence and application to organic light-emitting diode. *Journal of the American Chemical Society* 125(42):12971–12979.

Turro, N.J., V. Ramamurthy, and J.C. Scaiano. 2009. *Principles of Molecular Photochemistry: An Introduction.* University Science Books, Sausalito,CA.

Vanhelmont, F.W.M., H.U. Gudel, M. Fortsch, and H.B. Burgi. 1997. Synthesis, crystal structure, high-resolution optical spectroscopy, and extended Huckel calculations for [Re(CO)$_4$(thpy)] (thpy$^-$ = 2-(2-thienyl)pyridinate). Comparison with related cyclometalated complexes. *Inorganic Chemistry* 36(24):5512–5517.

Vezzu, D.A.K., J.C. Deaton, J.S. Jones, L. Bartolotti, C.F. Harris, A.P. Marchetti, M. Kondakova, R.D. Pike, and S. Huo. 2010. Highly luminescent tetradentate bis-cyclometalated platinum complexes: Design, synthesis, structure, photophysics, and electroluminescence application. *Inorganic Chemistry* 49(11):5107–5119.

Vlcek, A. Jr. and S. Zalis. 2007. Modeling of charge-transfer transitions and excited states in d^6 transition metal complexes by DFT techniques. *Coordination Chemistry Reviews* 251(3–4):258–287.

Vorontsov, I.I., T. Graber, A.Y. Kovalevsky, I.V. Novozhilova, M. Gembicky, Y.-S. Chen, and P. Coppens. 2009. Capturing and analyzing the excited-state structure of a Cu(I) phenanthroline complex by time-resolved diffraction and theoretical calculations. *Journal of the American Chemical Society* 131(18):6566–6573.

Wada, A., Q. Zhang, T. Yasuda, I. Takasu, S. Enomoto, and C. Adachi. 2012. Efficient luminescence from a copper(I) complex doped in organic light-emitting diodes by suppressing C-H vibrational quenching. *Chemical Communications* 48(43):5340–5342.

Wan, K.T., C.M. Che, and K.C. Cho. 1991. Inorganic excimer: Spectroscopy, photoredox properties and excimeric emission of Dicyano(4,4'-Di-Tert-Butyl-2,2'-Bipyridine)Platinum(II). *Journal of the Chemical Society-Dalton Transactions* (4):1077–1080.

Williams, E.L., K. Haavisto, J. Li, and G.E. Jabbour. 2007. Excimer-based white phosphorescent organic light emitting diodes with nearly 100% internal quantum efficiency. *Advanced Materials* 19(2):197–202.

Williams, J.A.G. 2007. Photochemistry and photophysics of coordination compounds: Platinum. *Topics in Current Chemistry* 281:205–268.

Williams, J.A.G. 2009. The coordination chemistry of dipyridylbenzene: N-deficient terpyridine or panacea for brightly luminescent metal complexes? *Chemical Society Reviews* 38(6):1783–1801.

Williams, J.A.G., A. Beeby, E.S. Davies, J.A. Weinstein, and C. Wilson. 2003. An alternative route to highly luminescent platinum(II) complexes: Cyclometalation with N^C^N-coordinating dipyridylbenzene ligands. *Inorganic Chemistry* 42(26):8609–8611.

Yang, C.H., K.H. Fang, C.H. Chen, and I.W. Sun. 2004a. High efficiency mer-iridium complexes for organic light-emitting diodes. *Chemical Communications* (19):2232–2233.

Yang, C.H., C.C. Tai, and I.W. Sun. 2004b. Synthesis of a high-efficiency red phosphorescent emitter for organic light-emitting diodes. *Journal of Materials Chemistry* 14(6):947–950.

Yang, C.-H., C.-H. Chen, and I.W. Sun. 2006. Bathochromic effect of trifluoromethyl-substituted 2-naphthalen-1-yl-pyridine ligands in color tuning of iridium complexes. *Polyhedron* 25(12):2407–2414.

Yang, X., Z. Wang, S. Madakuni, J. Li, and G.E. Jabbour. 2008. Highly efficient excimer-based white phosphorescent devices with improved power efficiency and color rendering index. *Applied Physics Letters* 93(19):193305/1–193305/3.

Yersin, H. and W. Humbs. 1999. Spatial extensions of excited states of metal complexes. Tunability by chemical variation. *Inorganic Chemistry* 38(25):5820–5831.

Yip, H.K., L.K. Cheng, K.K. Cheung, and C.M. Che. 1993. Luminescent Platinum(II) Complexes. Electronic Spectroscopy of Platinum(II) Complexes of 2,2′:6′,2″-Terpyridine (terpy) and p-Substituted Phenylterpyridines and Crystal Structure of [Pt(terpy)Cl][CF$_3$SO$_3$]. *Journal of the Chemical Society-Dalton Transactions* (19):2933–2938.

You, Y.M. and S.Y. Park. 2005. Inter-ligand energy transfer and related emission change in the cyclometalated heteroleptic iridium complex: Facile and efficient color tuning over the whole visible range by the ancillary ligand structure. *Journal of the American Chemical Society* 127(36):12438–12439.

Zhang, L., B. Li, and Z. Su. 2009. Realization of high-energy emission from [Cu(N-N)(P-P)]$^+$ complexes for organic light-emitting diode applications. *Journal of Physical Chemistry C* 113(31):13968–13973.

Zhang, Q., J. Ding, Y. Cheng, L. Wang, Z. Xie, X. Jing, and F. Wang. 2007. Novel heteroleptic Cu-I complexes with tunable emission color for efficient phosphorescent light-emitting diodes. *Advanced Functional Materials* 17(15):2983–2990.

Zhang, Q., T. Komino, S. Huang, S. Matsunami, K. Goushi, and C. Adachi. 2012. Triplet exciton confinement in green organic light-emitting diodes containing luminescent charge-transfer Cu(I) complexes. *Advanced Functional Materials* 22(11):2327–2336.

Zhang, Q.S., Q.G. Zhou, Y.X. Cheng, L.X. Wang, D.G. Ma, X.B. Jing, and F.S. Wang. 2004. Highly efficient green phosphorescent organic light-emitting diodes based on Cu-I complexes. *Advanced Materials* 16(5):432–436.

Zhang, Q.S., Q.G. Zhou, Y.X. Cheng, L.X. Wang, D.G. Ma, X.B. Jing, and F.S. Wang. 2006. Highly efficient electroluminescence from green-light-emitting electrochemical cells based on Cu-I complexes. *Advanced Functional Materials* 16(9):1203–1208.

Chapter 11

Devices and Processing

12. Microcavity Effects and Light Extraction Enhancement

Min-Hao Michael Lu

12.1 Introduction

The practical utility of an organic light-emitting diode (OLED) is largely proportional to the light emitted out of the device. Previous chapters of this book dealt with charge injection, charge transport, and exciton formation in OLEDs that were all essential steps of the complete electroluminescence process. In this chapter, we examine light emission after excitons are formed on the lumiphore, so as to answer two questions: what fraction of these excitons results in light generated within the OLED? And, what fraction becomes useful light emitted out of the OLED?

The foundation of our understanding lies in the realization that the lumiphores—the emitting molecular species—exist in a medium bounded on both sides by reflective surfaces/interfaces—the reflective cathode above, and the glass/indium tin oxide (ITO) interface below in a typical bottom-emitting OLED. The optical medium is usually composed of organic layers with a total thickness of several hundred nanometers; hence, the OLED stack is referred to as a *microcavity*. While the reflectivity of the cathode, be it aluminum (Al) or silver (Ag), is well above 90%, the ITO/glass interface is not very reflective owing to the

Chapter 12

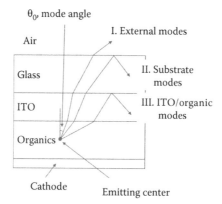

FIGURE 12.1   Three visible radiative modes: (I) external modes, (II) substrate modes, and (III) ITO/organic modes. Also shown is the mode angle θ_0 in the organic layer. (Adapted from Lu, M.-H. and Sturm, J. C. 2001. *Appl. Phys. Lett.* 78: 1927–1929.)

small difference in the indices of refraction: $n_{sub} = 1.5$ and $n_{ITO} \sim 1.8$–1.9 (Figure 12.1); thus the OLED microcavity is sometimes referred to as a *weak* microcavity. Nonetheless, microcavity design can noticeably influence both emission color and intensity.

This chapter is organized as follows: Section 12.2 investigates the emission modes in the OLED (including quantum mechanical (QM) effects, which are relevant because the microcavity thickness is less than wavelengths in the visible range), the implications of these modes on calculated internal quantum efficiency and recombination efficiency, and practical designs for color tuning. Section 12.3 discusses optical extraction techniques that increase the amount of light emitted out of the OLED microcavity. Section 12.4 summarizes and concludes this chapter.

12.2   Microcavity Effects

The basic structure of the OLED microcavity is depicted in Figure 12.1: You can see from the figure (from top to bottom) that it is composed of a reflective cathode, a stack of organic layers with an index of refraction, n_{org}, of approximately 1.7–1.8, an ITO anode (see Chapter 4) with index, n_{ITO}, of 1.8–1.9, and glass substrate with index, n_{sub}, of 1.51. The lumiphore is understood to be located within the organic stack. Understanding of microcavity effects is critically important because it allows a more accurate determination of the internal quantum efficiency, η_{EL}^{int}, the number of internally emitted photons per injected charge carrier. η_{EL}^{int} is not directly observable but is related to the observable external quantum efficiency, η_{EL}^{ext}, the number of externally emitted photons per injected charge carrier via the following equation:

$$\eta_{EL}^{ext} = \eta_{oc}\eta_{EL}^{int} \tag{12.1}$$

where η_{oc} is the outcoupling efficiency. The internal quantum efficiency is in turn determined by

$$\eta_{EL}^{int} = \gamma_{PL} \tag{12.2}$$

where γ is the recombination efficiency or the number of emissive lumiphore excitons created for each injected charge carrier, and η_{PL} is the PL efficiency of the lumiphore in the OLED microcavity. Here γ is a catch-all term that may include the host exciton recombination efficiency and host-dopant energy transfer efficiency in some devices and the efficiency of charge capture on the dopant in others; it further incorporates the singlet–triplet branching ratio in fluorescent OLEDs (see Chapter 6 for a detailed discussion of OLED fundamentals). In Equation 12.2, we have made the assumption that whether the emissive lumiphore exciton is created by photons or electrically has no bearing on how it will decay. It can be seen from Equation 12.1 that any estimation of the ultimate internal device efficiency requires an accurate determination of η_{oc}; furthermore, understanding of the factors that contribute to η_{oc} is instrumental in its enhancement to maximize the external quantum efficiency.

The lumiphore can be modeled classically as an oscillating dipole, or quantum mechanically as a two-level system whose radiative transition rate, k_r, between the excited and ground states is given by Fermi's golden rule, which is given by

$$k_r = \frac{2\pi}{\hbar} \sum_m |\langle m | \bar{\mu} \cdot \bar{E}(\bar{k}, z) | n \rangle|^2 \delta(E_n - E_m - h\upsilon) \tag{12.3}$$

where $\bar{\mu}$ is the dipole moment and $\bar{E}(\bar{k}, z)$ is the electrical field for mode $\bar{k}$ at the location of the dipole; E_m and E_n are the energies of the initial and final states; and $h\upsilon$ is the energy of the emitted photon. The PL quantum efficiency, η_{PL}, of the lumiphore in free space is expressed as

$$\eta_{PL}^0 = \frac{k_r^0}{k_r^0 + k_{nr}^0} \tag{12.4}$$

where the superscript 0 denotes the free-space value; and k_{nr} denotes the nonradiative transition rate that includes intersystem crossing. In practice, the free space PL quantum efficiency is often approximated by the value measured from a dilute solution. If η_{PL}^0 is close to unity as in many highly efficient emitters, then k_{nr}^0 is negligibly small. There is some evidence that the thick film PL efficiency is the same as that in free space (Garbuzov et al. 1996) but the PL efficiency of the same lumiphore in an OLED microcavity, η_{PL}, can be quite different, given by the equation

$$\eta_{PL} = \frac{k_r}{k_r + k_{nr}} \tag{12.5}$$

Here, k_{nr} differs from k_{nr}^0 only when there are extra nonradiative decay pathways specific to the thin film environment: quenching impurities or defects, concentration quenching, and so on. If we limit our consideration to a properly deposited film with a reasonable concentration of efficient dopants, we may neglect any difference in a term that should be small to begin with. On the other hand, k_r is in general very different from the free-space value.

Chapter 12

In a microcavity, the radiative transition rate is composed of contributions from several different modes: external modes that are the useful light emitted out of the OLED (denoted by the subscript "ext"); substrate waveguided modes (denoted by the subscript "sub"); ITO and organic layer waveguided modes (denoted by the subscript "IO"); and finally energy transfer to the cathode (denoted by the subscript "cat"), as given by the equation

$$k_r = k_{ext} + k_{sub} + k_{IO} + k_{cat} \tag{12.6}$$

The first three modes are illustrated in Figure 12.1. They arise from the typical refractive indices encountered in an OLED and the ensuing total internal reflections (TIRs) (Figure 12.1), and can be described by the equation

$$\sin\theta_0 = \begin{cases} 0\cdots n_{air}/n_{org} & \text{external modes} \\ 1/n_{org}\ldots n_{sub}/n_{org} & \text{substrate modes} \\ n_{sub}/n_{org}\cdots 1 & \text{ITO/organic modes} \end{cases} \tag{12.7}$$

where θ_0 is the mode angle (the angle of propagation in the organic layer as defined in Figure 12.1); n_{air}, n_{org}, and n_{sub} are the index of refraction of air, organic layer, and substrate, respectively. Equation 12.7 is nothing more than a description of TIR within the layers by Snell's law in a more general parameterization, where θ_0 may be complex and values of $\sin\theta_0$ from 1 to ∞ are associated with the energy transfer to the cathode, k_{cat} (Chance et al. 1975a,b, 1978). This last term has been identified with energy transferred to the metal surface plasmon polaritons (SPPs) and other surface loss modes (Ziebarth and McGehee 2005).

One way to conceptualize k_{cat} is to consider an oscillation dipole in front of a metallic mirror. An image dipole of opposite orientation is created on the other side of the metallic mirror and serves to dampen radiation from the original dipole. Field lines can be drawn from one dipole to the other, bisected by the metallic mirror, giving rise to surface charge density oscillations. Their associated electromagnetic fields are called SPP waves, which are absorbed by the metal as they propagate. k_{cat} is indubitably radiative in nature but, unlike the three other radiative modes, no actual photon is emitted. Rather we can think of it as the absorption of a "virtual photon." It is in this vein that we revisit Equation 12.5: for an efficient emitter with $k_{nr} \approx 0$, η_{PL} in an OLED microcavity may still be close to unity by this definition, but because of the presence of k_{cat}, the ratio of visible photons emitted to emissive lumiphore excitons may be significantly reduced.

Thus, this leads to the following relationship:

$$\eta_{oc} = \frac{k_{ext}}{k_r} = \frac{k_{ext}}{k_{ext} + k_{sub} + k_{IO} + k_{cat}} \tag{12.8}$$

in a typical bottom-emitting OLED. Therefore outcoupling enhancement techniques, as we will see in Section 12.3, focus on converting k_{sub}, k_{IO}, and k_{cat} into external modes, or on suppressing these rates in the first place.

12.2.1   Ray Optics and Half-Space Dipole Models

Early investigations of light outcoupling in OLEDs were solely focused on the first three "visible" radiative modes. One of the first attempts used a ray optics model that assumed the lumiphore to be an isotropic emitter where the luminous flux is given by (Greenham et al. 1994)

$$F = 2\pi \int I(\theta_0) \sin\theta_0 d\theta_0 \tag{12.9}$$

where the luminance intensity $I(\theta_0) = I_0$ for a Lambertian emitter and θ_0 is the mode angle in the organic layer (Figure 12.1). Contributions to the external, substrate, and ITO/organic modes are arrived at by setting the limits of integration according to Equation 12.7. When only considering the visible radiative modes, the following equation is obtained:

$$\eta_{oc,visible} = \frac{k_{ext}}{k_{ext} + k_{sub} + k_{IO}^2} \approx \frac{1}{2n_{org}} \quad \text{for ray optics model} \tag{12.10}$$

Substituting typical n_{org} values of 1.7–1.8, $\eta_{oc,visible}$ is only in the range of 15%–20% (Madigan et al. 2000). By the same token, nearly 50% of the emission is into the ITO/organic modes owing to the large solid angle these modes occupied (Figure 12.2) (Lu and Sturm 2002).

In polymer OLEDs, it is commonly assumed that the emissive excitons have a preferred orientation because of a long polymer backbone lying in the in-plane direction. A model was constructed where the exciton is represented by a dipole located in a semi-infinite region of organic material on top of a semi-infinite metal cathode (Kim et al. 2000). The outcoupling efficiency was shown to depend on both the dipole–cathode distance and the dipole orientation. The optimal dipole–cathode distance was shown to be at various antinodes when optical interference with the metal mirror was taken into account. For a lossless metal mirror, the first antinode is a quarter wavelength away. $\eta_{oc,visible}$ at the

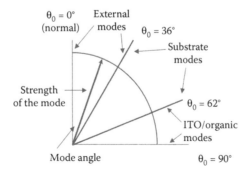

FIGURE 12.2   Radial plot of mode strength versus mode angle. Ray optics model: critical angles for an isotropic point source in the OLED microcavity, calculated based on $n_{org} = 1.71$ and $n_{glass} = 1.51$. (Adapted from Lu, M.-H. and Sturm, J. C. 2001. *Appl. Phys. Lett.* 78: 1927–1929.)

Chapter 12

optimal location was calculated as $0.75/n_{org}^2$ for an isotropic dipole and as much as $1.2/n_{org}^2$ for an in-plane dipole (Lu 2002). In either case, the outcoupling efficiency is much larger than the $1/2n_{org}^2$ predicted by ray optics because the ITO/organic modes were suppressed relative to the external modes in the half-space dipole model. We will return to this point when discussing the quantum mechanical (QM) model in Section 12.2.2.

The half-space dipole model was a step in the right direction as it recognized the optical interference between the lumiphore and the cathode, although the half-space was a gross simplification of the actual microcavity. Ray optics should be understood as the limit where the wavelength $\hat{y}$ of emitted light is infinitesimally small compared with cavity thickness when phase correlation is not a concern and intensities are additive arithmetically.

12.2.2 Quantum Mechanical Microcavity Theory

A QM microcavity model was developed by Bulović et al. (1998). In the QM treatment, the electromagnetic field in the layered microcavity is represented by the sum of eigenmodes of the cavity; the radiating molecule is modeled as a dipole; and the transition rate for each mode is given by Fermi's golden rule. It has the advantage that the modes are computed separately. The QM model is incomplete in that it does not address the energy transfer from the dipole to the metal electrodes, k_{cat}. However, it was shown by Chance et al. (1975a,b, 1978) that this energy transfer arises exclusively from the near field of the dipole; so it does not affect the shape of the normalized far-field intensity patterns. In the model developed by Bulović et al., the QM microcavity treatment is augmented by the Green's function analysis, which is more convenient in computing the total rate of energy loss and the dipole lifetime in layered media.

In the QM approach, Equation 12.3 is summed over all dipole moment $\bar{\mu}$ and electrical field $\bar{E}(\bar{k},z)$. Following the method developed by Ujihara (1975) and extended by Deppe (Deppe and Lei 1990), an imaginary upper boundary is placed at distance L_Z away from the ITO/glass interface. When all the modes are solved for, L_Z is let to tend to infinity. The advantage of this approach is that all modes become normalizable. Since the glass substrate is thick, the transmission from glass to air is handled by ray optics (Greenham et al. 1994, Madigan et al. 2000). On the other hand, the ITO/organic layers are thin when compared with the visible wavelengths; so the ITO/organic modes are discrete. The wave vectors, $\bar{k}$, for these modes are described by transcendental equations that are solved numerically for both the transverse-electric (TE) and transverse-magnetic (TM) polarizations (Hecht 1998).

The electric field is determined by the microcavity structure as given by

$$\bar{E}_{\bar{k}}^{TE} = \bar{A}(\bar{k})\sin^2(k_{0z}l)\hat{x}$$
$$\bar{E}_{\bar{k}}^{TM} = \bar{B}(\bar{k})\cos^2\theta_0\sin^2(k_{0z}l)\hat{y} + \bar{C}(\bar{k})\sin^2\theta_0\cos^2(k_{0z}l)\hat{z}$$

(12.11)

where $\bar{A}(\bar{k})$, $\bar{B}(\bar{k})$, and $\bar{C}(\bar{k})$ are functions of material constants, and $\bar{k}$; k_{0z} is the $\hat{z}$ component of the wave vector in the emitting layer; and l is the dipole–cathode distance.

Here, the cathode is taken to be a perfect reflector. For spontaneous emission, the electric fields are normalized such that the energy in each mode is equal to that of a single photon. Near an antinode in the electric field where the $\sin^2(k_{0z}l)$ terms are maximized, the TE and first half of the TM radiation, both of which arise from the in-plane component of the electric field, are maximized. Near a node in the electric field, the second half of the TM radiation, which arises from the normal component of the electric field, is maximized, leading to a large in-plane TM component. We call the two components in the TM radiation the antinodal and nodal contribution, respectively.

Owing to the high density of modes in the external and substrate modes, the sum in Equation 12.10 may be replaced by a three-dimensional (3D) integral: $\Sigma_{\vec{k}} \Rightarrow \int(\text{mode volume})^{-1} d\vec{k}$. On the other hand, the ITO/organic waveguide can support at most a couple of modes. The sum in Equation 12.10 is then transformed into a two-dimensional (2D) integral in the $k_x - k_y$ plane, and a discreet sum in k_z.

The last point is highly consequential, given the thickness range of typical organic and ITO layers are such that they form a very thin 2D slab waveguide relative to visible wavelength. In most devices, no more than a handful of k_zs are allowed. Therefore, there are generally fewer ITO/organic modes in the QM model than predicted by the ray optics model. It is even possible to construct structures with thin ITO and organic layers where no ITO/organic mode exists at all.

12.2.3   High-Index Substrates and Numerical Results

Figure 12.3 illustrates the radiation pattern near the peak of the Alq_3 spectrum ($\lambda = 524$ nm) in a PVK/Alq_3 bilayer OLED, where the QM microcavity results are in stark contrast to the isotropic radiation of the ray optics model. The external and substrate modes are a continuum because the distance of observation and the thickness of the glass substrate are much greater than the wavelength in question. The combined ITO/organic layer thickness is on the order of a half wavelength; so the modes there, if

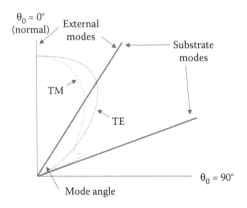

FIGURE 12.3   Radial plot of modal strength as a function of mode angle in Alq_3 at $\lambda = 524$ nm. OLED structure: standard glass/100 nm ITO/40 nm PVK/80 nm Alq_3/Mg:Ag. The exciton is at the PVK/Alq_3 interface. Solid line: TE mode; dashed line: TM mode. The external and substrate modes are a continuum. There are no ITO/organic modes. The cutoff wavelength is ~452 nm for TE modes and ~440 nm for TM modes. (Adapted from Lu, M.-H. and Sturm, J. C. 2001. *Appl. Phys. Lett.* 78: 1927–1929.)

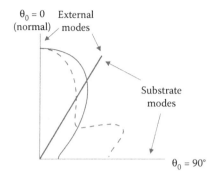

FIGURE 12.4　Radial plot of modal strength as a function of mode angle in Alq₃ at $\lambda = 524$ nm in the same OLED as in Figure 12.3 except on high-index substrate. Solid line: TE mode; dashed line: TM mode. (Adapted from Lu, M.-H. and Sturm, J. C. 2001. *Appl. Phys. Lett.* 78: 1927–1929.)

they exist, are discrete. In this particular example, no ITO/organic mode exist for either TE or TM radiation (Lu and Sturm 2001).

The refractive index of soda lime and borosilicate glasses are about 1.5. However, there are glasses of much higher index (SCHOTT 2012). For this discussion, "high index" is defined as higher than the index of the organic layer where the emission emanates. Using a high-index substrate eliminates TIR at the ITO/glass interface, which means there are no longer any modes confined to the ITO/organic layers. Instead, they are absorbed into the continuum of substrate modes (Figure 12.4). When η_{PL} is close to unity, a high-index substrate does not increase the external quantum efficiency without additional light extraction measures owing to the TIR at the glass/air interface. However, as can be seen in Section 12.3, it is much easier to extract substrate modes than to extract ITO/organic modes; therefore, a high-index substrate combined with optical extraction techniques can drastically improve device external quantum efficiency.

Some concrete examples of modal distribution as a function of OLED structure are shown in Figure 12.5 for a hypothetical OLED with a model Alq/NPD structure. The thickness of the ITO layer affects the modal distribution in two ways. First, it alters the ITO/organic modes by changing the combined thickness of the ITO/organic layer, and second, it alters the external and substrate modes through interference effects. It is possible to have an ITO layer so thin that no mode exists for most of the visible spectrum. In Figure 12.5, the distribution of light emission is calculated for OLEDs with 100 and 200 nm thick ITO layers. It is clear from the figure that the emission into the ITO/organic modes in the OLED with the thinner ITO layer is drastically suppressed, because the cutoff wavelength is only slightly above the low end of the visible spectrum. On the other hand, the cutoff is above the peak emission wavelength of Alq₃ in the OLED with the 200 nm ITO layer, resulting in much stronger ITO/organic modes. The absolute value of the emission in the external and substrate modes is moderately affected by the interference effects, but not enough to prevent the proportion of external emission from increasing with decreasing ITO layer thickness. According to the calculations shown in Figure 12.5, as much as ~52% of the light could be emitted externally in the planar device with 100 nm ITO. For devices on high-index substrates, there is no change in the external emission because it is not dependent upon the index of the intervening layers.

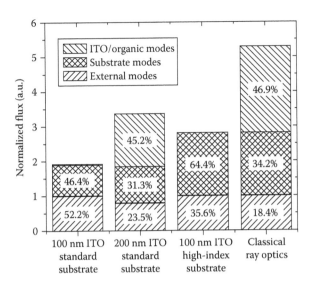

FIGURE 12.5 Calculated distribution of emission into external, substrate and ITO/organic modes for several OLED structures (glass substrate/ITO/40 nm PVK/80 nm Alq$_3$/cathode). The fluxes into the external modes of the sample with 80 nm Alq$_3$ on standard substrates and the ray optics model are normalized to 1. (Adapted from Lu, M.-H. and Sturm, J. C. 2001. *Appl. Phys. Lett.* 78: 1927–1929.)

However, the total emission is different because all modes in the OLED on high-index substrates are continuum modes, whereas the ITO/organic modes in OLEDs on standard substrates are suppressed.

The numerical results from the QM model are very different from the ray optics model, which underestimates the outcoupling efficiency $\eta_{oc,visible}$. Nonetheless, many researchers continue to use $1/2n_{org}^2$ or even 20% as an estimate for the fraction of externally emitted visible radiation.

12.2.4 Microcavity Design

Dodabalapur et al. (1994) demonstrated the ability to control the photoluminescence of Alq in a cavity composed of quartz substrate/(SiO$_2$/Si$_x$N$_y$ quarter wave stack)/polyimide/diamine/Alq$_3$/Al to such an extent that red, green, and blue emission could be obtained (Figure 12.6). The quarter wave stack combined with the Al mirror gave the cavity a higher Q-factor, which made this degree of control possible. A standard OLED microcavity does not have a high Q-factor because the organic/ITO interface is mostly transmissive. Nonetheless, the microcavity effect remains a critical element of device optimization because small alterations to the luminescence spectra can have a large impact on important quality metrics for both display and lighting: color saturation, color rendering, chromaticity angular dependence, and so on.

From an application standpoint, the most frequent use of an OLED optical model is to predict the intensity and spectra as a function of the far-field viewing angle and layer structure. These properties are associated with the visible radiation of the external, substrate, and ITO/organic modes, the analysis of which does not require full QM microcavity model. Instead, a simpler optical model based on a plane wave expansion of the dipole field adequately captures the behavior of the lumiphore in the microcavity.

Chapter 12

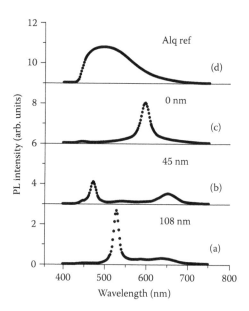

FIGURE 12.6 Photoluminescence spectra of Alq/diamine microcavity devices with different thicknesses of transparent polyimide filler: (a) 108, (b) 45, and (c) 0 nm. Also shown in (d) is the spectrum from a sample without microcavity effects. The thicknesses of the Alq and diamine are 60 MI, and the pump wavelength is 380 nm. (Adapted from Dodabalapur, A. et al. 1994. *Appl Phys Lett.* 64: 2488–2490.)

12.2.5 Wave Optics Model

In this section, we show how the wave optics model is derived. Figure 12.7 shows the setup of the microcavity. The treatment is similar to that of a Fabry–Perot resonator with the distinction that the plane waves originate from within the cavity (Yariv 1991). For each far-field viewing angle θ_{ff} in air, there is an associated mode angle θ_0 according to Snell's law $n_{air} \sin \theta_{ff} = n_{org} \sin \theta_0$ and a forward and backward propagating wave in the organic layer (solid and dash lines in Figure 12.7). Let t, r, t', and r' be the complex transmittance and reflectance at the organic/ITO interface and the organic/metal interface, respectively, successive bounces coherently interfere as shown in the equation

$$E_{ff}\left(\theta_{ff},\lambda\right) = E_0(\theta_0,\lambda)\left[\begin{array}{l}\left(\sum t + trr'e^{i\delta} + tr^2r'^2e^{i2\delta} + \cdots\right) + \\ \left(\sum tr'e^{i\varphi} + tr'e^{i\varphi}rr'e^{i\delta} + tr'e^{i\varphi}r^2r'^2e^{i2\delta} + \cdots\right)\end{array}\right]$$

$$= E_0(\theta_0,\lambda)t\frac{1+r'e^{i\varphi}}{1-rr'e^{i\delta}} \tag{12.12}$$

where $\delta = 4\pi n_{org}d\cos\theta_0/\lambda$ and $\varphi = 4\pi n_{org}l\cos\theta_0/\lambda$ are two phase factors. d and l are the thickness of the organic layer and the emitter–cathode distance, respectively. The polarization and angle-dependent transmittances and reflectances can be solved using a standard transfer matrix formalism (Hecht 1998). For each λ and θ_0, a cavity modification function $G(\theta_0, \lambda)$ relates the far-field intensity to the free-space intensity in the equation

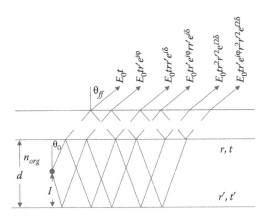

FIGURE 12.7 Microcavity setup for the wave optics model. Successive bounces of forward (solid lines) and backward (dashed lines) propagating waves of the same far-field angle coherently interfere. r and t are the complex reflectance and transmittance from the emitting layer to the final destination layer above the device. r' and t' are the complex reflectance and transmittance from the emitting layer toward the opposite direction.

$$G(\theta_0, \lambda) = \frac{I_{ff}(\theta_{ff}, \lambda)}{I_0(\theta_0, \lambda)} = \frac{n_{air} \cos\theta_{ff}}{n_{org} \cos\theta_0} \left| t \frac{1 + r'e^{i\varphi}}{1 - rr'e^{i\delta}} \right|^2 \quad \text{for TE polarization}$$

$$= \frac{n_{air} \cos\theta_0}{n_{org} \cos\theta_{ff}} \left| t \frac{1 + r'e^{i\varphi}}{1 - rr'e^{i\delta}} \right|^2 \quad \text{for TM polarization} \qquad (12.13)$$

The factors in front of $|\cdots|^2$ account for energy conservation across interfaces. Equation 12.13 takes into account polarization and off-normal observation angles. However, the plane wave expansion ignores the near field that gives rise to the SPP modes. A simple remedy is to multiply $G(\theta_0, \lambda)$ by τ_{cav}/τ_0, where τ_{cav} is the lifetime of the lumiphore in the microcavity and τ_0 is the free-space lifetime (Jordan et al. 1996, Lin et al. 2005, Wu et al. 2005). One may have to resort to a Green's function method to accurately calculate τ_{cav} (Bulović et al. 1998, Chance et al. 1975a,b, 1978, Lu and Sturm 2001) but in practice τ_{cav} varies much more slowly than $G(\theta_0, \lambda)$ versus layer thickness and refractive index; so it is reasonable to ignore this term for device optimization among a set of similar structures. The spectral power distribution of an OLED at any given far-field angle is given simply by

$$S(\theta_{ff}, \lambda) = G[\arcsin(1/n_{org} \sin\theta_{ff}), \lambda] S_0(\lambda) \qquad (12.14)$$

with $S_0(\lambda)$ being the intrinsic spectral power distribution in free space. This treatment is easily extended to more complex stratified media, for example, where additional layers need to be defined, and to transparent and top-emission OLEDs for which similar expressions can be derived for emission through the cathode.

12.2.6 Examples of Color and Intensity Tuning by Microcavity Design

This section will highlight two examples of color and intensity tuning by microcavity effects in OLEDs. As can be seen in the derivation for Equation 12.13, the reflectivity of the anode plays a strong role in the overall emission characteristics of a bottom-emitting OLED. Distributed Bragg reflectors (DBRs) are made of periodic layers of varying refractive index. In the first example, three pairs of TiO_2/SiO_2 that have an estimated reflectivity of 80% were inserted between the ITO and the glass substrate of a bilayer OLED with a dye-doped Alq_3 as the emitter (Figure 12.8) (Jordan et al. 1996). The DBRs were tuned to the noncavity emission peak at 545 nm of a first OLED and indeed a large normal direction luminance enhancement was observed along with a small shift in peak wavelength as observation angle increases (Figure 12.9a). There was also a narrowing of the spectrum—in other words, color saturation increased. Other OLEDs were fabricated where the emissive dopant was changed to one with peak emission centered at 580 nm but with the same device structure. As can be seen in Figure 12.9b, marked shifts in peak wavelengths were observed as a function of observation angle. In this device, maximum luminance occurred at ~40° off the normal direction. This shift in maxima could be understood in the context of the minimization of the denominator in Equation 12.12.

Some general conclusions can be drawn from this work. First, it is possible to use microcavity effects to increase color saturation. Despite a color shift with angle, it may still be desirable in certain display applications. However, it is unlikely that a single device architecture is suitable for the entire visible spectrum and red, green, and blue subpixels will each require their own cavity design.

Metals such as Al and Ag offer high reflectivity across a wide spectrum. In particular, Ag has a reflectivity of 90+% with low absorption. Thus OLEDs with thin Ag electrodes are good candidates for studying microcavity enhancements. Wu et al. (2005) constructed top-emitting, C545T-doped Alq_3 OLEDs with 80 nm Ag as the anode and thin Ag or Ag/TeO_2 as the cathode, where the cathode reflectance was controlled by varying the thicknesses of Ag and TeO_2. The effect of the microcavity can be seen in Figure 12.10. The overall enhancement was a factor of 2. The angular color shift was much smaller than with DBRs because while the phase terms in Equation 12.12 remain, the angular dependence of the reflectance was much weaker for Ag/TeO_2 than for the DBR stack. For this reason, and for simplicity of its processing, the

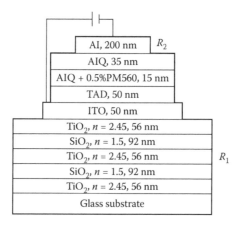

FIGURE 12.8 Schematic of the microcavity organic electroluminescent device. The aluminum electrode has reflectivity $R_2 = 0.90$ and the quarter-wave dielectric stack has reflectivity $R_2 = 0.80$ at 550 nm. (Adapted from Jordan, R. H. et al. 1996. *Appl. Phys. Lett.* 69: 1997–1999.)

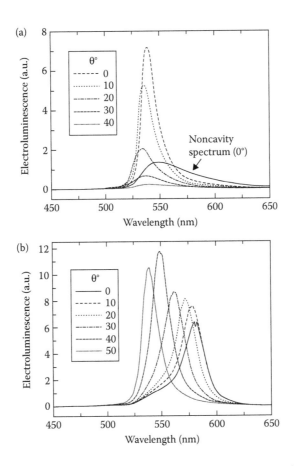

FIGURE 12.9 (a) Electroluminescence spectra measured at 0°, 10°, 20°, 30°, and 40° from the surface normal of a microcavity organic device with normal incidence optical mode located at 545 nm. The 0° spectrum full width at half-maximum is 20 nm. The solid line is the noncavity electroluminescence spectrum measured at 0° from a LED on the same substrate. (b) Electroluminescence spectra measured at 0°, 10°, 20°, 30°, 40°, and 50° from the surface normal of a microcavity organic device with normal incidence optical mode located at 580 nm. Spectral variation with angle is detectable by the human eye. (Adapted from Jordan, R. H. et al. 1996. *Appl. Phys. Lett.* 69: 1997–1999.)

thin metal/dielectric electrode design is preferable over the more complicated DBR approach. Nonetheless, the resonance condition still depends heavily on layer thicknesses such that distinct cavity structures optimized separately for red, green, and blue OLEDs are still required. A common cavity structure suitable for the entire visible range remains elusive.

12.3 Optical Extraction Enhancement

The outcoupling efficiency determines the portion of the internally generated light that becomes useful external emission. As ever more efficient materials and device structures are developed, optical extraction is becoming the bottleneck that limits the ultimate efficacy of OLEDs. Equation 12.8, revisited here, gives the outcoupling efficiency, η_{oc}, for a typical planar OLED:

Chapter 12

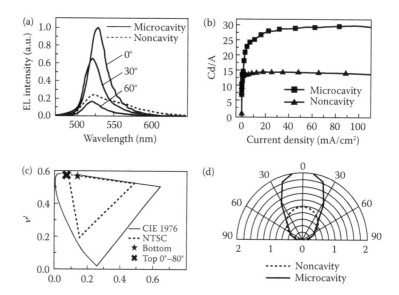

FIGURE 12.10   (a) Measured EL spectra with relative intensities at 0°, 30°, and 60° off the surface normal of the microcavity device, and at 0° of the noncavity device. (b) cd/A efficiencies for both devices. (c) 1976 CIE coordinates of 0°–80° EL of the microcavity device and 0° EL of the noncavity device. (d) Polar plots of measured EL intensities (normalized to the 0° intensity of the noncavity device) for both devices. (Adapted from Lin, C.-L.; Lin, H.-W.; and Wu, C.-C. 2005. *Appl. Phys. Lett.* 87: 021101.)

$$\eta_{oc} = \frac{k_{ext}}{k_r} = \frac{k_{ext}}{k_{ext} + k_{sub} + k_{IO} + k_{cat}} \tag{12.8}$$

Almost all of the approaches to enhance optical extraction can be described by

$$\eta_{oc} = \frac{k_{ext} + p_1 k_{sub} + p_2 k_{IO} + p_3 k_{cat}}{k_{ext} + q_1 k_{sub} + q_2 k_{IO} + q_3 k_{cat}} \tag{12.15}$$

where both the ps and qs are between 0 and 1. In a bottom-emitting device, the ps default to 0 and the qs to 1. The ps represent efforts to convert nonexternal modes into external modes, thus increasing η_{oc} by increasing the numerator in Equation 12.8; whereas the qs represent efforts to suppress nonexternal modes, thus increasing η_{oc} by decreasing the denominator in Equation 12.8. For the rest of this chapter, we define the outcoupling efficiency in two ways: as a percentage of the original light output and as a multiplier of the original light output. Thus, 50% improvement refers to 50% more light extracted into external modes than without any enhancement, while 1.5× also refers to 50% more light extracted.

Before delving further, it behooves us to examine the numerical results from a full QM microcavity model so as to have a basic understanding of the magnitude of each of these modes as well as the amount of enhancement to be expected from optical extraction techniques. Figure 12.11 from Meerheim et al. (2010) includes calculated modal distributions for a red bottom-emitting phosphorescent OLED with conductivity-doped

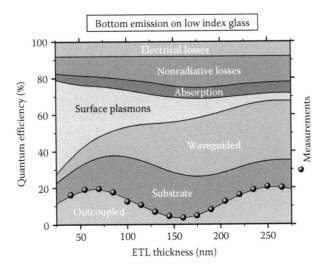

FIGURE 12.11 Measured EQE at 1.51 mA/cm² of the red bottom OLEDs as a function of the n-layer thickness and comparison to simulation results. At this current density, the highest EQE of 21% (at 330 cd/m² brightness) is reached in the second maximum. The figure also shows the distribution of all loss channels in the devices. (Adapted from Meerheim R. et al. 2010. *Appl. Phys. Lett.* 97: 253305.)

hole and electron transport layers (HTL and ETL), and Ag cathode. In addition to the outcoupled (k_{ext}), substrate (k_{sub}), waveguided (k_{IO}), and surface plasmon (k_{cat}) modes, Figure 12.11 also includes absorption (small, fairly constant, mostly from ITO), nonradiative modes (k_{nr}), and electrical losses (*IR* and Stoke losses independent of any of the other modes; these losses are normally not considered in microcavity models). Figure 12.12 from Lu (2010) presents a similar plot of transition rates (as opposed to modal fraction in the previous figure) of the radiative modes as a function of ETL thickness in a bottom-emitting OLED with ITO anode and Al cathode. Figure 12.12 does not include the nonradiative mode, which makes a small, fixed contribution to the total transition rate in efficient phosphorescent devices (Lu 2010). While the modal distribution is a function of device architecture and material choices including the cathode, results in Figures 12.11 and 12.12 from two independent studies show excellent agreement and ought to be generally applicable to bottom-emitting OLEDs on standard glass substrates.

One key takeaway from these numerical results is that the external modes represent about 15%–20% of the total radiative modes at a typical ETL thickness around 50 nm. Interestingly, this figure is precisely what is given by ray optics when only the external, substrate, and ITO/organic waveguided modes are considered—a coincidence that has no doubt contributed to its lasting appeal despite its erroneous premise and failure to predict any wavelength and layer thickness dependencies.

According to both figures, at a typical ETL thickness of 50 nm, extracting all substrate modes would increase the outcoupling efficiency by approximately 80%, with an additional 50% gain possible, were the ITO/organic waveguided modes also completely extracted. Extraction techniques, of course, cannot be 100% effective; so the realizable gain is usually lower. It is also important to note that the peak in the external modes does not coincide with the peak of the combined external and substrate or combined

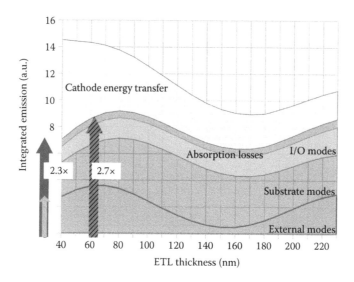

FIGURE 12.12 Calculated transition rates for radiative modes including absorption losses as a function of ETL thickness. The external modes of a control device with a 50 nm ETL is indicated by the gray arrow. The black arrow indicates a maximum enhancement factor of 2.3× with the same device structure. The diagonal lines arrow indicates a maximum enhancement factor of 2.7× with an 80 nm ETL optimized for the combined external, substrate, and ITO/organic waveguided modes. (Adapted from Lu, M. 2010. Presented at *DOE SSL R&D Workshop*, Raleigh, NC, February 3.)

external, substrate, and ITO/organic waveguided modes, which suggest the need to perform separate optimizations for control and extraction-enhanced devices. Figure 12.12 puts this into perspective by highlighting the up to 2.7× extraction gain possible by optimizing extraction for the first maxima in combined modes at an ETL thickness around 80 nm.

Optical extraction is arguably more important in applications in lighting where the total luminous efficacy is a crucial figure of merit than in displays where there is still doubt whether enhanced extraction comes at the expense of reduced resolution. Of course, in displays, enhanced extraction can reduce the power consumption and thereby extend battery life.

Throughout the literature and in this chapter, the enhancement factor—that is, the amount of luminous flux increase versus a control device—is a key figure of merit for evaluating any extraction technique. Measured properly, it provides practitioners—that is, panel manufacturers—an easy means to estimate their ultimate panel efficacy. However, care must be taken to ensure the reported factor is meaningful. There are several simple rules to observe: (1) both the final and control device forward 180° luminous flux must be measured rather than estimated from normal direction luminance values because the intensity distributions are likely quite different. (2) The ratio is ideally taken from an optimized final device and an optimized control device—*in general, these two structures are not the same*. If, however, the ratio is taken with an optimized control device and a final device of the same structure, the resultant enhancement factor may be regarded as a lower bound. (3) A common pitfall is to optimize the enhancement factor as a function of the device structure. That ratio has no bearing to any practical panel one wants to produce. (4) Finally, enhancement factors are not simply additive for the simple reason that the same light cannot be extracted twice. Combining multiple extraction approaches demands thoughtful analysis.

12.3.1 Extracting Substrate Waveguided Modes

Substrate waveguided modes are those with mode angle $\sin\theta_0 = 1/n_{org} \ldots n_{sub}/n_{org}$. They are confined at the substrate/air interface by TIR and easily extracted by any means that disrupts the planar nature of the substrate waveguide. This disruption can be achieved by either modification of the substrate/air interface (also referred to as the backside of the substrate as opposed to the side where OLED is deposited) or by incorporating scattering centers into the substrate itself. Using the convention established in Equation 12.15, the approaches discussed in this section aim to increase p_1 as much as possible, with the other ps and qs at their default values.

One of the first attempts at backside patterning was to attach a small lens directly above the OLED as demonstrated by Madigan et al. (2000), where a small lens of radius (ρ) 3.4 mm was placed above an OLED of 1.75 mm diameter (Figure 12.13). A total luminous flux enhancement factor of 2.0× was observed, with the caveat that it was unclear whether the baseline device was optimized. The enhancement factor increases monotonically with the collection angle of the lens, labeled $\theta_{sub,max}$. Even if an array of lenses was used instead of a single lens, rays of larger angle could only be collected after additional bounces that entail greater losses.

It ought to be clear from the earlier-mentioned discussion on collection angles that the enhancement factor from backside patterning tends to suffer as the OLED size increases, especially as the feature size is to remain small to maintain the overall thin form factor. If the restriction on feature size is relaxed, a "macro" extractor can be used. For example, a large acrylic block several centimeters thick was used with a large-area OLED (25 cm²) and an enhancement factor of 1.7–1.75× was observed (Levermore et al. 2011). On the basis of the numerical results presented in Figures 12.12 and 12.13, 1.8× is likely the upper limit for a properly optimized baseline device using substrate mode extraction techniques alone. For reasons of reducing cost and simplicity of processing, lamination of a microstructured film or direct patterning of the substrate is preferred, and many examples can be found in the literature.

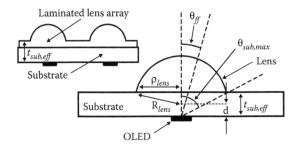

FIGURE 12.13 Small lens attached above the OLED for extraction enhancement. $\theta_{sub,max}$ is the maximum collection angle of the lens. Inset: Spherical features implemented as a plastic lens array laminated to a planar substrate. (Adapted from Madigan, C. F.; Lu, M.-H.; and Sturm, J. C. 2000. *Appl. Phys. Lett.* 76: 1650–1652.)

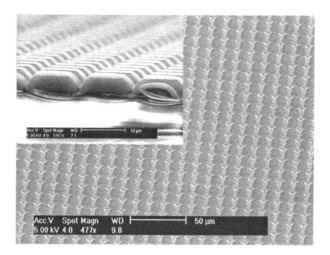

FIGURE 12.14 SEM of a molded PDMS MLA. (Adapted from Möller, S. and Forrest, S. R. 2002. *J. Appl. Phys.* 91: 3324–3327.)

Möller and Forrest (2002) demonstrated an enhancement factor of 1.5× using a molded polydimethylsiloxane (PDMS) microlens array (MLA) (Figure 12.14). Lin et al. (2007) examined OLEDs with laminated commercial optical films (Figure 12.15): a diffuser (Taiwan Keiwa Inc., BS-702) and a brightness enhancement film (3M, Vikuti™ BEF II 90/50), and found integrated luminance increases by factors of 1.42× and 1.08×, respectively. Other approaches include incorporating silica microspheres(Li et al. 2007, Yamasaki et al. 2000) or simply sandblasting the backside of the glass substrate (Chen and Kwok 2010).

Another method to disrupt the substrate waveguide is to introduce volumetric scattering into the substrate or, alternatively, to laminate onto the substrate a film with embedded scattering centers, as illustrated in Figure 12.16. Optically, a volumetric scattering film is functionally equivalent to a film with random surface features, both scatter the substrate light randomly, where forward scattered light exits in the device and the backward scattered light has a chance to be reflected or scattered into an external mode. In practice, volumetric scattering may be easier to control than a surface film by

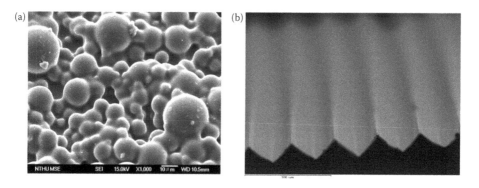

FIGURE 12.15 Microphotographs of (a) the BS-702, tilt angle: 45° and (b) the BEF II 90/50, tilt angle: 90°. (Adapted from Lin, H.-Y. et al. 2007. *Opt. Commun.* 275: 464–469.)

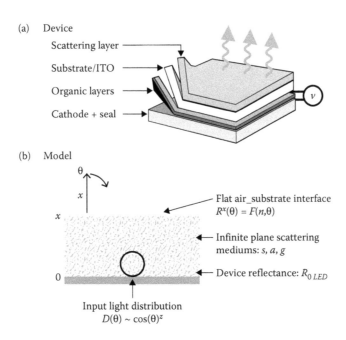

(a) Device

Scattering layer

Substrate/ITO

Organic layers

Cathode + seal

v

(b) Model

θ

x

x

0

Flat air_substrate interface
$R^x(\theta) = F(n,\theta)$

Infinite plane scattering
mediums: s, a, g

Device reflectance: $R_{0\,LED}$

Input light distribution
$D(\theta) \sim \cos(\theta)^z$

FIGURE 12.16   (a) Schematic of an OLED device and (b) the idealized model geometry and model input parameters. (Adapted from Shiang, J. J. and Duggal, A. R. 2004. *J. Appl. Phys.* 95: 2880–2888.)

adjusting particle loading, shape, and size (Shiang et al. 2004). In addition, a flat volumetric scattering layer may be more easily integrated elsewhere in the device, such as between the substrate and ITO.

Shiang applied Mie scattering theory to develop a quantitative model with two key parameters to characterize scattering (Shiang 2004). The first, g, is the asymmetry factor that describes the expectation value of $\cos(\theta - \theta')$, which is a measure of the deflection from the incidence angle, θ, and exit angle, θ', for each scattering event. The second, S, is the scatterance, which is the probability that a scattering event will occur as the ray traverses the substrate. It is a product of the scattering cross section, the particle concentration, and the total substrate thickness. $g = 1$ implies that scattering does not deflect the trajectory of the ray, whereas $g = 0$ corresponds to an exit angle totally uncorrelated with the initial ray angle. Numerical results pointed to a preference to higher g, where the maximum extraction was relatively independent of S. Thus, to maximize design flexibility and tolerance, larger grained scattering particles that exhibit a higher value of g are preferred. This result is expected to be of general validity in external as well as internal volumetric scattering layers. In a companion article (Shiang 2004), ZrO_2 power ($d_{50} = 0.6$ mm) or "cool white" phosphor particles ($d_{50} = 50$ μm) were mixed into a poly dimethyl silicone (PDMS, $n = 1.41$) matrix to create volumetric scattering films. Enhancement factors of up to 1.4×, closed to the predicted maximum, were observed. The results were confirmed in later work based on polymer light-emitting diodes (LEDs) (Bathelt et al. 2007).

Rather than dispersing high-index particles in a low-index matrix material, Nakamura et al. (2006) did the opposite: dispersing silicone particles ($n = 1.43$) in a high-refractive index (HRI, $n = 1.72$) polymer resin. The resultant diffusive HRI film was then used as a substrate adjacent to the ITO anode (Figure 12.17). This approach combines volumetric

Chapter 12

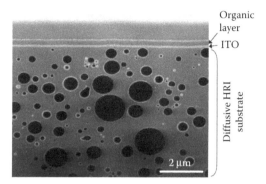

FIGURE 12.17 Cross-sectional scanning ion microscopy (SIM) image of the organic EL device with a diffusive HRI substrate. (Adapted from Nakamura, T. et al. 2006. *Opt. Review* 13: 104–110.)

scattering with the benefit of a high-index substrate, which will be discussed further in the next section. Unfortunately, the amount of luminous flux enhancement was not reported for this work.

In summary, methods for converting substrate modes into external modes are important not simply because they are nearly a necessity in OLED lighting panels, but also because techniques to harness other modes tend to convert them into substrate modes first before being extracted by one of the aforementioned approaches. Both surface and volume scattering can effectively convert substrate waveguided light into externally emitted light, with a highest practical enhancement factor around 1.4× based on an optimized large-area control device. Slightly higher extraction, up to an enhancement factor of 1.5×, can be achieved using a MLA because it refracts light and backscattering is not as significant. For the same reason, MLAs tend to preserve the chromaticity angular dependence of the control device. A possible improvement may be to get the best of both worlds by introducing a small degree of scattering into an MLA film.

12.3.2 Low-Index Aerogel and Top-Emission OLEDs

Tsutsui et al. (2001) reported enhanced outcoupling efficiency in OLEDs with a layer of silica aerogel ($n = 1.03$) inserted between ITO and the substrate (Figure 12.18). The aerogel layer was prepared by the sol–gel method with supercritical drying, and had an index of refraction very close to that of air. Unlike in a regular device where the refractive indexes follow the trend $n_{org} > n_{sub} > n_{air}$, the aerogel layer causes rays of mode angle θ_0, where $n_{air}/n_{org} > \sin\theta_0 > n_{sub}/n_{org}$, that would have existed in the substrate to be confined to the ITO/organic layers as well. The subwavelength thickness of the ITO/organic layers forces all modes in them to become discretized and suppressed. Applying Equation 12.15, q_1 will be 0 while the other ps and qs are at their default values. Tsutsui et al. reported an enhancement factor of 1.8×, although the Alq$_3$ baseline device may not have been fully optimized. The authors reported an external quantum efficiency of 0.765%, while Alq$_3$ devices of external quantum efficiency exceeding 1% were routinely made at that time (Hung et al. 1997).

It was noted that the optical configuration of a top-emitting OLED was almost identical to the device on aerogel, except with a thin semitransparent cathode layer as

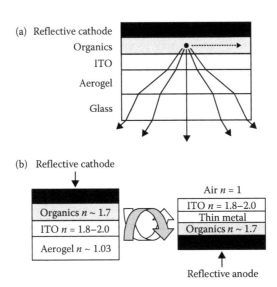

FIGURE 12.18 (a) Light emission in an OLED on aerogel/glass substrate. The waveguided modes (dashed line) in the ITO/organic layer are suppressed. (Adapted from Tsutsui, T. et al. 2001. *Adv. Mater.* 13: 1149–1152.) (b) Comparison of the OLED in part (a) with a TEOLED: neglecting the thin metal film in TOLED and the difference between the refractive indices of aerogel and air, the two structures are identical. (Adapted from Lu, M.-H. et al. 2002. *Appl. Phys. Lett.* 81: 3921–3923.)

illustrated in Figure 12.18 (Lu 2002). This was experimentally confirmed when a top-emitting OLED was shown to emit 20% more photons over a control device with the same layer structure, despite the extra absorption in the thin metal cathode. Figure 12.19 illustrates this graphically by plotting the calculated modal distribution for a top-emitting phosphorescent OLED (Meerheim et al. 2010). Note the absence of the

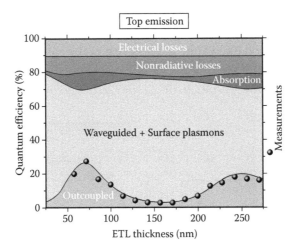

FIGURE 12.19 Measured EQE at 0.74 mA/cm² of the red top OLEDs as a function of the n-layer thickness and comparison to simulation results. At the current density, the highest EQE of 27% (at 176 cd/m² brightness) is obtained for a first-order cavity. The figure also shows the distribution of all loss channels in the devices. Waveguided and plasmonic losses are not distinguished owing to the complex modal cavity structure. (Adapted from Meerheim R. et al. 2010. *Appl. Phys. Lett.* 97: 253305.)

substrate modes and the amalgamated waveguide and surface plasmon modes. The first peak in outcoupled (external) modes may represent a 30%–50% increase versus a bottom-emitting control device.

12.3.3 Extracting ITO/Organic Waveguided Modes

Compared with the substrate modes, ITO/organic waveguided modes are far less accessible. Several approaches have been tried to harvest them: (1) using a substrate with an index higher than the emitting layer (alternatively, reducing index of the emitting layer to below that of the substrate), thereby combining the ITO/organic waveguided modes with substrate modes, which can be harvested as discussed in Section 12.3.1; (2) disrupting the planar nature of the ITO/organic waveguide, which involves introducing scattering centers or reflecting surfaces internal to the device, that is, between the substrate and the cathode; and (3) employing photonic bandgap structures within the device.

12.3.4 High–Index Substrates

When the refractive index of the substrate is higher than that of the emitting layer, there is no longer TIR at the ITO/substrate interface; so there are no longer any modes confined exclusively within the ITO/organic layers. The modes in the substrate still cannot escape if the substrate acts as a planar waveguide, but as shown in Section 12.3.1, there is no shortage of ways to harvest the modes in the substrate once they arrive there. One of the best examples of the use of a high-index substrate used the combination of a high-efficiency phosphorescent OLED architecture with a shaped high-index substrate (Figure 12.20) to produce a then record-breaking white OLED at an efficacy of 90 lm/W at 1000 cd/m² (Reineke et al. 2009). The combined high-index substrate ($n = 1.78$) and the hemispherical lens resulted in an enhancement factor of 2.4×, among the highest reported and close to the maximum predicted by numerical modeling as discussed at the start of Section 12.3.

Even when the index of the substrate is below that of the emitting layer, any increase in the index will still contribute to decreasing the ITO/organic waveguided modes by increasing the critical angle for TIR at the ITO/substrate interface. Once the index of

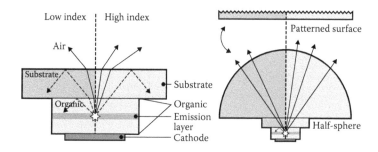

FIGURE 12.20 The left panel shows a cross section of an OLED to illustrate the light propagation. Solid lines indicate modes escaping the device to the forward hemisphere; dashed lines represent trapped modes. The right panel shows how a large half-sphere and a patterned surface can be applied to increase light outcoupling. (Adapted from Reineke, S. et al. 2009. *Nature* 459: 234–239.)

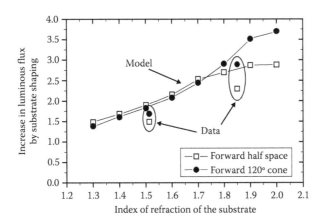

FIGURE 12.21   Predicted increases in luminous flux emitted in the forward half-plane and the forward 120° cone as a function of the index of refraction of the substrate, assuming complete conversion of substrate modes into external modes. OLEDs have the structure: substrate/100 nm ITO/40 nm PVK/80 nm Alq$_3$/cathode. (Adapted from Lu, M.-H. and Sturm, J. C. 2002. *Appl. Phys.* 91: 595–604.)

the substrate is above that of the emitter layer, any further increase will contribute to a forward collimation of rays in the substrate, which should increase the collection efficiency of surface scattering features or reduce the number of bounces in a volumetric scattering layer. Therefore, extraction enhancement for shaped substrates is a monotonically increasing function of the substrate index, with a transition point where the substrate index matches that of the emitting layer, after which the slope of increase flattens (Figure 12.21).

The drawbacks for high-index substrates, both glass and polymer, are the high cost and general lack of availability. In addition, many high-index glasses have high absorption coefficients in the blue region or are very brittle. Meanwhile, the alternative—lowering the index of the emitting layer—is also fraught with challenges. For these reasons, there are currently no ready solutions in high-index substrates to support commercial OLED products.

12.3.5   Internal Scattering, Reflecting, and Photonic Crystal Structures

A number of internal light extraction enhancement structures have been investigated. They can be differentiated into three categories: scattering, reflecting, and photonic crystal (PC) structures.

One of the first internal scattering layers was proposed by researchers at Eastman Kodak (Tyan et al. 2008, Tyan 2011). As shown in Figure 12.22, a morphologically flat but optically inhomogeneous layer was inserted between the substrate and the ITO. The flat surface was necessary to ensure a high device yield. Its location meant almost all visible light passes through it before entering the substrate. In combination with an external light extraction structure on the backside of the glass substrate, the internal scattering layer resulted in an enhancement factor of 2.28× in a white OLED (Tyan et al. 2008).

In addition to work from the Kodak group, several other internal scattering layers have been proposed. They used a diverse range of material systems and processing

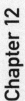

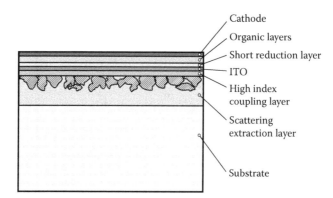

FIGURE 12.22   Cross-sectional diagram showing an internal scattering layer between the substrate and ITO. This device also features a short reduction layer. (Adapted from Tyan, Y.-S. 2010. OLEDs at 150 LPW what are the barriers. Presented at *DOE SSL R&D Workshop*, Raleigh, February 3.)

techniques, but with the common purpose of inserting a flat scattering layer between the substrate and ITO. Hong et al. (2010) investigated spontaneously formed nanofaceted MgO ($n = 1.73$) planarized with ZrO_2 ($n = 1.84$) as the scattering layer (Figure 12.23). Although the use of the pyramidal shape was well reasoned, the small index difference between the two materials as well as the nanometer length scale of the pyramids did not

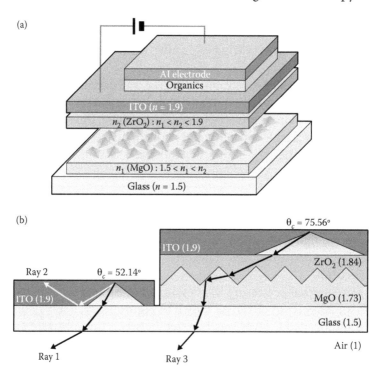

FIGURE 12.23   (a) Schematic of bottom-emitting OLEDs with the embedded refractive index modulation layer (RIML) at the glass/ITO interface. The RIML is composed of electron-beam-evaporated MgO and MgO/ZrO_2. (b) Schematic explanation of the mechanism for improving device outcoupling efficiency with a microfacet-structured RIML. Diagrams are not to scale. (Adapted from Hong, K. et al. 2010. *Adv. Mater.* 22: 4890–4894.)

Nanoparticle + Polymer matrix

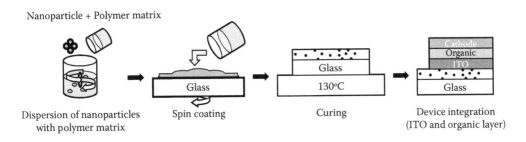

| Dispersion of nanoparticles with polymer matrix | Spin coating | Curing | Device integration (ITO and organic layer) |

FIGURE 12.24 Process flow for preparing the internal scattering layer with nanoparticles and subsequent device fabrication by depositing ITO, organic layers, and cathode. (Adapted from Chang, H.-W. et al. 2011. *J. SID* 19: 196–204.)

lead to strong scattering effects. As a result, the maximum enhancement factor achieved was only 1.19×.

Chang et al. (2011) took another approach to realizing the scattering layer. By fabricating the scattering films with nanoparticles (TiO_2 nanoparticles with d = 100, 400 nm) dispersed in a polymer matrix (transparent photoresist) with smooth enough surfaces for subsequent device integration (Figure 12.24). The total scattering layer thickness was 1.4 μm. The largest enhancement factor reported was 1.96× for a film with the highest loading (0.24 g/cc) of the 400 nm TiO_2 particles. Interestingly, this film had the lowest transmission (71%, with diffuse transmission ~52%), and the highest haze (73%) among the formulations examined. The chromaticity angular dependence was, if anything, weaker in the device with the scattering layer, indicating fairly complete mode mixing and a lack of subwavelength periodicity.

Instead of a random scattering layer in close proximity to the ITO/organic layers, it is possible to place a grid within a large-area OLED, where the edges of the grid act as well-controlled reflecting surfaces for ITO/organic waveguided light. Sun and Forrest (2008) deposited a low-index grid atop the ITO before further deposition of the organic layers and the cathode. Thus waveguided light entering the low-index region sideways could be scattered into the substrate and further extracted via a MLA (Figure 12.25). In this approach, the grid opening was small enough to ensure that absorption in the ITO/organic was kept to a minimum even though a smaller grid opening implied a smaller total aperture. SiO_2 (n = 1.45) was used as the grid material and a grid width of 1 μm and grid opening of 6 μm were used. When combined with an external MLA, a total enhancement factor of 2.3× was observed.

Koh et al. (2010) also applied a grid to a large-area OLED but in a different structure that eliminated the aperture trade-off. First, an ITO grid was realized by photolithographic patterning and etching, leaving a well-controlled tapered edge. Next, a high-conductivity poly(3,4-ethylenedioxythiophene): poly(styrene sulfonate) (PEDOT:PSS) layer was uniformly coated. The combination of ITO and PEDOT:PSS acted as the anode where the patterned ITO grid supplied both the lateral conductivity and the desired reflecting facets. As indicated in Figure 12.26, waveguided modes propagating in either ITO or the organic layer were easily redirected into the substrate. In combination with an external MLA, an enhancement factor of 2.29× was observed. This approach has the advantage of not imposing an additional process step because the ITO needs to be patterned in most

Chapter 12

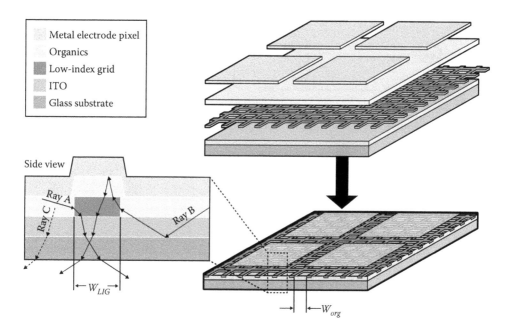

FIGURE 12.25 Schematic of the OLED with the embedded low-index grid (LIG) in the organic layers. The LIG is first patterned on the ITO on a glass substrate. The organic layers and the metal electrode are then subsequently deposited onto the substrate. The pixel size of the metal electrode is one order of magnitude larger than the period of the grid. Inset: enlarged side view of the OLED structure and the mechanism that leads to outcoupling of waveguided light. Diagrams are not to scale. (Adapted from Sun, Y. and Forrest, S. R. 2008. *Nat. Photonics.* 2: 483–487.)

practical applications, and the aperture of the device is not impacted. One key drawback is the need for a solution-processed PEDOT:PSS layer, which may be incompatible with some processing approaches. The reflection at the ITO, organic/PEDOT:PSS interface hinges on their index differential (n_{ITO} = 1.8, $n_{PEDOT:PSS}$ = 1.42, n_{org} = 1.75). Although high-conductivity, vacuum thermal evaporation (VTE)-compatible materials may be available, it is unclear whether they will be in the necessary index range.

A third category of internal enhancement structures are based on PCs. A 2D SiO$_2$ (n = 1.45)/SiN$_x$ (n = 1.93) PCs inserted between the ITO and glass substrate were demonstrated to increase the extraction efficiency by 55% (Do et al. 2004). An improvement by the same group used spin-on-glass (n = 1.28)/SiN$_x$ 2D PCs (Figure 12.27) and as much as 85% improvement was reported on account of the greater index differential (Kim et al. 2006). Waveguided modes can easily access the PC that acts as a Bragg grating, which scatters the modes whose wavenumber satisfies the momentum conservation condition shown in the following equation:

$$k_{para} = k_0 \sin\theta = k_{wg} \pm mk_G \tag{12.16}$$

where k_0 is the wavenumber in free space; k_{para} the in-plane component; k_{wg} the waveguide mode; $k_G = 2\pi/L$ the wavenumber associated with the grating, where L is the

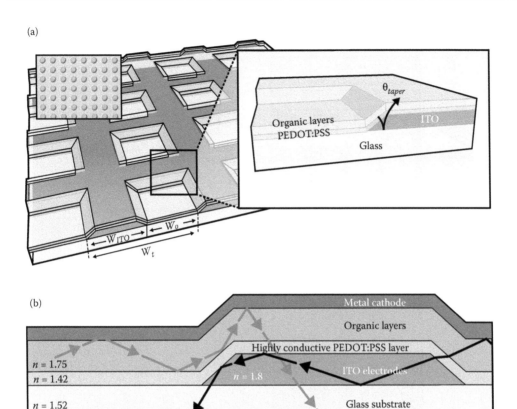

FIGURE 12.26 (a) Proposed electrode structure consisting of patterned ITO and coated high-conductivity PEDOT:PSS (Baytron PH500) layer. The size of the patterned ITO openings is approximately 3 mm wide and the openings repeat in a square lattice format with a spatial period W_t of 6 mm. The taper angle at the edge of the patterned ITO layer is defined as θ_{taper}. Inset in the left corner: optical image of patterned ITO. (b) Representative cases that convert light trapped in organic/ITO layers into outcoupled mode. Rays confined in the organic/ITO layers are reflected at the interface with a low-index PEDOT:PSS layer and change their direction, being extracted into the air. Diagrams are not to scale. (Adapted from Koh, T.-W. et al. 2010. *Adv. Mater.* 22: 1849–1853.)

periodicity of the grating; θ is the mode angle; and m is an integer (Koo et al. 2010). In the wavelength range of interest, m is typically 1 or 2, implying strong angular and wavelength dependence in extraction enhancement (Figure 12.28). This pattern is common in extraction enhancement, based on periodic structures of subwavelength feature size and generally not desirable in practical applications. Moreover, realizing such fine features typically requires laser interference photolithography, which adds to process cost.

Of the three categories of internal extraction structures, the internal scattering layer has the best combination of potential gain in enhancement factor, process simplicity,

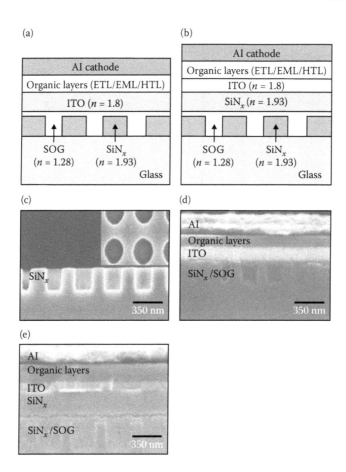

FIGURE 12.27 Schematic of a modified OLED employing 2D PC slab: (a) 2D SiN$_x$/SOG PC layer (300 nm)/SOG-overcoated layer (20 nm)/ITO (150 nm)/organics (145 nm) (type II) and (b) 2D SiN$_x$/SOG PC layer (300 nm)/SOG-overcoated layer (20 nm)/SiN$_x$ high-index layer (260 nm)/ITO (80 nm)/organics (145 nm) (type III). (c) Field-emission SEM images of the top view of the 2D high-index material nanohole pattern (×50,000): period of 350 nm, hole diameter of 250 nm, and height of 300 nm. Focused ion beam SEM images of the cross-sectional view of PC OLEDs inserted with (d) a flatter 2D SiN$_x$/SOG PC layer (×40,000) and (e) a high-index (SiN$_x$)-coated 2D SiN$_x$/SOG PC layer (×55,000). (Adapted from Kim, Y.-C. et al. 2006. *Appl. Phys. Lett.* 89: 173502.)

and cost-effectiveness. It is likely that some versions of it will be found in most commercial OLED lighting panels in the future.

12.3.6 Extracting and/or Suppressing Metal Surface Plasmon Polariton Modes

Of all the modes in the OLED, the SPP modes have proved the most difficult to extract. Various studies have placed its contribution to the total decay rate at anywhere between about 30% to over 40%, a figure so large as to be impossible to overlook. Indeed, whether and how much of the SPP modes can be extracted has become the biggest open question when projecting the ultimate realizable efficacy of an OLED lighting panel (Figure 12.29) (Bardsley Consulting et al. 2012).

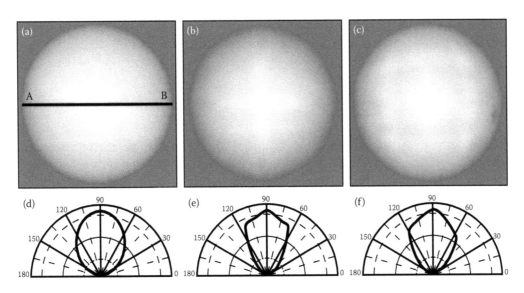

FIGURE 12.28 Measured 2D far-field intensity profiles of (a) a conventional OLED, (b) a PC OLED ($L = 350$ nm), and (c) a PC OLED ($L = 500$ nm). Intensity profiles along the horizontal line (A–B) of the 2D far-field intensity profiles for (d) a conventional OLED, (e) a PC OLED ($L = 350$ nm), and (f) a PC OLED s $L = 500$ nm. (Adapted from Do, Y. R. et al. 2004. *J. Appl. Phys.* 96: 7629–7636.)

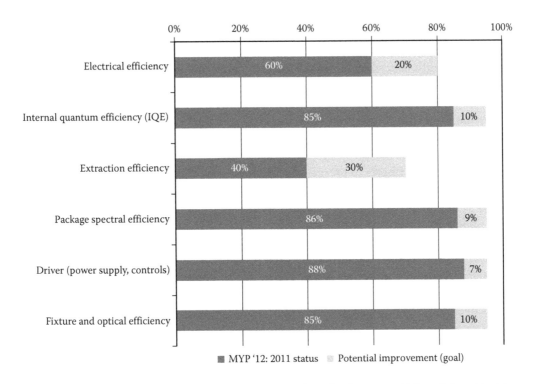

FIGURE 12.29 OLED panel and luminaire loss channels and efficiencies. (Adapted from Bardsley Consulting et al. 2012. Solid-State Lighting Research and Development: Multi-Year Program Plan. Prepared for Lighting Research and Development, Building Technologies Program, U.S. Department of Energy.) *Note*: Assumptions for target figures: CCT: 2580–3710 CRI > 85, 10,000 lm/m².

Chapter 12

SPP modes are the in-plane electromagnetic waves bound to charge density oscillations at the metal/dielectric interface. They have the dispersion relation

$$k_{SP} = k_0 \sqrt{\frac{\varepsilon_m \varepsilon_d}{\varepsilon_m + \varepsilon_d}} \tag{12.17}$$

where k_0 is the free-space wavenumber; ε_m and ε_d are the relative permittivities of the metal and dielectric, respectively. Denoting the complex permittivity and wavenumber of the metal by denoting the real part with primes and the imaginary part with double primes, Equation 12.17 simplifies to

$$k_{SP} = k'_{SP} + ik''_{SP} = k_0 \left(\frac{\varepsilon'_m \varepsilon_d}{\varepsilon'_m + \varepsilon_d} \right)^{1/2} + ik_0 \left(\frac{\varepsilon'_m \varepsilon_d}{\varepsilon'_m + \varepsilon_d} \right)^{3/2} \frac{\varepsilon''_m}{2\varepsilon'^2_m} \tag{12.18}$$

with the assumption that $|\varepsilon'_m| \gg \varepsilon''_m$ and neglecting any absorption in the dielectric. Barnes et al. (2003) contains an excellent discussion on the length scales relevant to plasmonic optics: "There are three characteristic length scales that are important for SP (surface plasmon)-based photonics in addition to that of the associated light. The propagation length of the SP mode, δ_{SP}, is usually dictated by loss in the metal. For a relatively absorbing metal such as aluminum, the propagation length is 2 μm at a wavelength of 500 nm. For a low loss metal such as silver, the length at the same wavelength is increased to 20 μm. By moving to a slightly longer wavelength, such as 1.55 mm, the propagation length is further increased toward 1 mm. The propagation length sets the upper size limit for any photonic circuit, based on SPs. The decay length in the dielectric material, δ_d, is typically on the order of half the wavelength of light involved and dictates the maximum height of any individual features, and thus components, that might be used to control SPs (p. 827)." Thus we have a general guideline on the feature size and material constants necessary for effective scattering of the surface plasmon modes.

Three different strategies to extract the SPP modes have been investigated: (1) index coupling, (2) prism coupling, and (3) grating coupling, which will be addressed individually as follows (Frischeisen et al. 2011).

The index coupling technique was demonstrated by Andrew and Barnes (2004). Optically pumped emission from a green dye (Alq_3) excited an orange dye (R6G) across a thin silver layer, where the excitation energy was transmitted through the silver layer via SPP modes (Andrew and Barnes 2004). Figure 12.30 illustrates the experimental setup as well as the electric and magnetic fields in each layer. Several groups demonstrated this concept in electrically pumped devices. Figure 12.31 shows another example where green OLEDs were fabricated with thin silver cathodes, and red-doped emitting layers were deposited above the cathode to absorb the SPP-mediated energy transfer. The resultant spectra included a mixture of green and red emission (Tien et al. 2010). While this process had sensitivity to silver thickness as to be expected, the amount of energy transmitted was in the 5%–7% range, a small fraction of what is lost to the SPP modes in a regular device. For a similar structure, Celebi et al. (2007) calculated a maximum transfer efficiency of 6% from Alq_3 to R6G through a modified Green's function formalism.

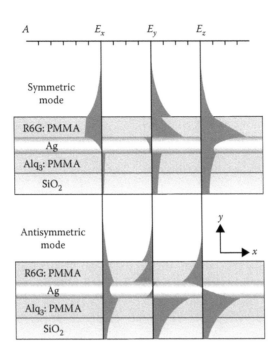

FIGURE 12.30 Coupled SPP modes supported by a dielectric-clad thin silver film. Schematic sample cross sections superposed with field profiles associated with the symmetric (top) and antisymmetric (bottom) coupled SPPs (calculated for maximum Alq$_3$ emission, $\lambda = 520$ nm) illustrating how the fields span the silver film. For clarity, the antisymmetric SPP fields are expanded by a factor of 2. In all calculations, a 60-nm-thick silver film (with complex optical permittivity ($\varepsilon = -9.39 + 0.78i$) is bounded by 80-nm-thick PMMA layers ($\varepsilon = 2.22 + 0i$), supported by a semi-infinite silica substrate ($\varepsilon = 2.12 + 0i$). (Adapted from Andrew, P. and Barnes, W. L. 2004. *Science* 306: 1002–1005.)

The second technique (called prism coupling) is based on the so-called reverse Kretschmann configuration (Figure 12.32). Similar to the index coupling technique, dye emission-induced plasmon modes propagate across a thin silver layer (about 50 nm). Rather than exciting another dye molecule, the plasmon mode emits into a silica substrate. The emission is subsequently released into air by a prism index matched to the

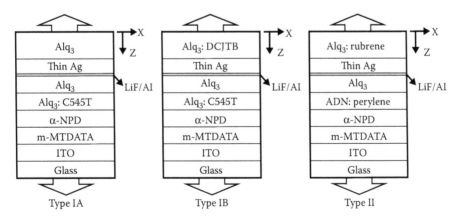

FIGURE 12.31 Schematic device structures under investigation. (Adapted from Tien, K.-C. et al. 2010. *Org. Electron.* 11: 397–406.)

Chapter 12

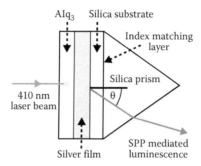

FIGURE 12.32 Schematic of experimental system. The silica substrate supports a silver film 18–91 nm thick. 30 nm of Alq_3 is evaporated on top of the silver and the substrate is index matched to a silica prism. (Adapted from Winter, G. and Barnes, W. L. 2006. *Appl. Phys. Lett.* 88: 051109.)

silica substrate. Hence this mechanism is referred to as "plasmon-mediated emission" as opposed to "plasmon-mediated transmission." Figure 12.33 shows the luminescence spectra through a silver layer and in a control device without the silver. The striking peak in the TM spectra is the clearest evidence of SPP-mode-mediated emission rather than a case of straightforward transmission (Winter and Barnes 2006). While both plasmon-mediated transmission and emission expand the boundaries of nanophotonics, it is unclear how much of the power lost to SPP modes they recover. The fact that they only work for specific wavelengths and angles limits their practical application in displays and lighting.

The third and perhaps most useful technique is to scatter the SPP modes by imposing a periodic structure onto the organic/metal interface such that the subsequent Bragg scattering recovers some of the SPP modes. Hobson et al. (2002) made a periodic "buckles" microstructure by laser interference photolithography. The microstructure was preserved through all layers of the OLED (Figure 12.34). Bragg scattering of the

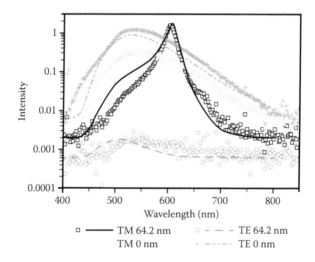

FIGURE 12.33 Experimental and theoretical TM and TE polarized luminescence spectra collected through 64.2 nm silver, and emission spectrum of dye through no silver. All spectra displayed were collected or simulated at a polar emission angle of $\theta = 63.4°$. (Density of data points reduced for clarity.) (Adapted from Winter, G. and Barnes, W. L. 2006. *Appl. Phys. Lett.* 88: 051109.)

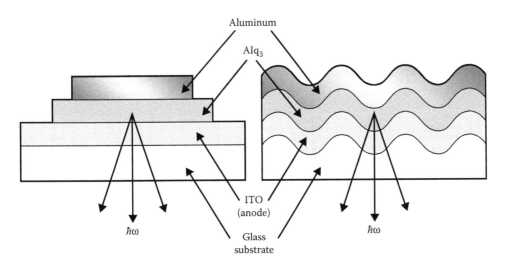

FIGURE 12.34 Schematic cross section of planar (left) and microstructured (right) OLEDs. Actual period 290 nm, depth 30 nm. (Adapted from Hobson, P. A. et al. 2002. *Adv. Mater.* 14, 1393–1396.)

SPP modes follows exactly the same form as Equation 12.16 save for the substitution of k_{SP} for k_{wg} as shown in the following equation:

$$k_{para} = k_0 \sin\theta = k_{SP} \pm m k_G \qquad (12.19)$$

Only a handful of wavelengths are scattered with a grating period of 290 nm, corresponding to the few ms that provide a solution to Equation 12.19. Hobson and Barnes contended that a 2D microstructure can recover ~60% of the power associated with SPP modes but unfortunately did not provide an integrated measurement of the device on "buckles" substrate (Hobson et al. 2002).

Combining the first and third approaches, Feng et al. (2008) demonstrated that a corrugated cathode would increase the efficiency of SPP-mediated energy transfer over a flat device. A one-dimensional (1D) grating was introduced into a silica substrate using a holographic lithography technique. The grating period was 400 nm and the groove depth was 50 nm. Excitation was provided by electroluminescence from an OLED doped with a fluorescent blue dye and the acceptor was a red dye. Energy transfer in the corrugated device was shown to be greater by a factor of 10 over a flat device. Unfortunately, the absolute transfer efficiency was not reported.

Koo eliminated the need for high-resolution lithography by taking advantage of the spontaneously formed "buckles" pattern when Al was evaporated on to a PDMS substrate at 100°C and subsequently allowed to relax (Koo et al. 2010). The "buckled" Al was used as a mold to transfer the pattern to another PDMS layer, on top of which OLEDs were fabricated (Figure 12.35). Rather than a periodic grating, AFM images revealed randomly oriented structures with a somewhat consistent peak-to-peak spacing around 410 nm (Figure 12.36). The authors reported maximum current efficiency and power efficiency increase by factors of 2.2× and 2.9× at 2000 cd/m², however, it is inherently difficult to compare these devices with control devices because the "buckles" structure

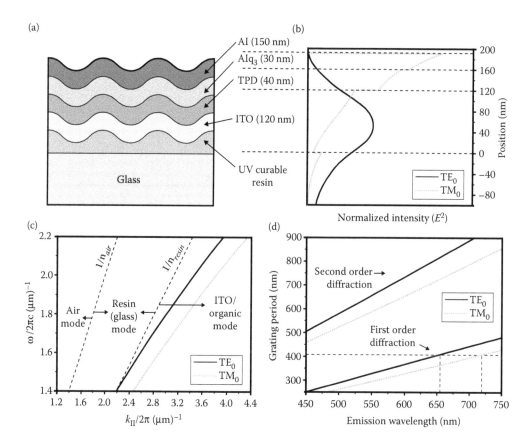

(a)

Al (150 nm)
Alq₃ (30 nm)
TPD (40 nm)
ITO (120 nm)
Glass
UV curable resin

(b)

Normalized intensity (E^2)

Position (nm)

— TE$_0$
····· TM$_0$

(c)

$\omega/2\pi c\ (\mu m)^{-1}$

$1/n_{air}$ $1/n_{resin}$

Air mode   Resin (glass) mode   ITO/ organic mode

— TE$_0$
····· TM$_0$

$k_{\parallel}/2\pi\ (\mu m)^{-1}$

(d)

Grating period (nm)

Second order → diffraction

First order diffraction

— TE$_0$
····· TM$_0$

Emission wavelength (nm)

FIGURE 12.35 Characterization of waveguide modes by the transfer matrix method. (a) Schematic of an OLED with buckles. The buckled PDMS was transferred to UV-curable resin by an imprinting technique. An OLED device was subsequently fabricated on the corrugated resin. AFM (not shown) images confirmed that the corrugated structure was well preserved on the surface of the OLED device. (b) Electric field intensity normalized by the total power for TE$_0$ (black line) and TM$_0$ (dotted line) modes at $l_0 = 525$ nm. (c) Dispersion curves of the TE$_0$ (black line) and TM$_0$ (dotted line) modes confined in the ITO/organic layer. (d) Relation between the outcoupled emission wavelength in the normal direction and the grating period for the TE$_0$ (black line) and TM$_0$ (dotted line) modes plotted using the calculated propagation vector k_{wg} and the Bragg grating equation. First- and second-order diffractions were considered. The dashed horizontal line corresponds to the peak wavelength of the buckling pattern used as the grating, $\lambda_{peak} = 410$ nm. Thus, the maximum outcoupling efficiencies for the TE$_0$ and TM$_0$ modes are expected to be found at emission wavelengths of ~655 and 720 nm, respectively, which are far from the main emission wavelength of the emitting layer, ~525 nm. Additionally, we expect the second-order diffractions of the modes to contribute to the outcoupled emission over all wavelengths. (Adapted from Koo, W. H. et al. 2010. *Nat. Photonics* 4: 222–226.)

permeating through the entire device is expected to alter the injection and transport properties of all layers. Some chromatic shifts in emission spectra were observed in the device on "buckles" but the amount was much less than other SPP mode extraction method mentioned earlier. Intensity distribution of the "buckles" devices on was close to Lambertian, as were the control devices. Both were likely due to the random nature of the "buckles" pattern.

One can conclude from this experience that an ideal device to capture the SPP modes would be one with flat organic layers to preserve the electrical characteristics and randomly roughed cathode to scatter the SPP modes. A recent paper by Novaled AG

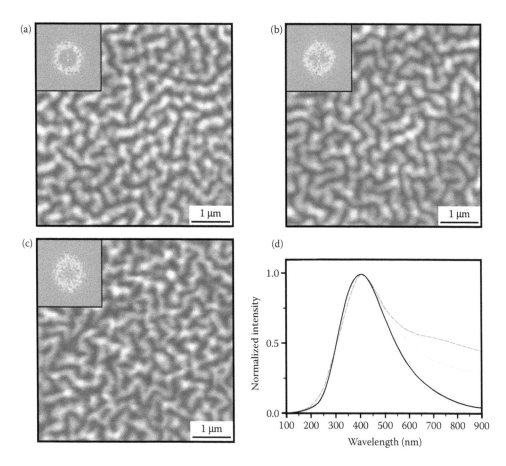

FIGURE 12.36 AFM analysis of buckling patterns. Dimensions 5 μm × 5 μm. (a) Buckled structure formed by a 10-nm-thick aluminum layer. (b, c), Buckled structures formed by deposition of a 10-nm-thick aluminum layer twice and three times, respectively. Resin layers imprinted with a buckled PDMS replica were used for measurement. Inset: FFT patterns of each image. (d) Power spectra from FFTs as a function of wavelength for buckled patterns obtained with deposition of a 10-nm-thick aluminum layer once (black line), twice (dotted line), and three times (dashed line). (Adapted from Koo, W. H. et al. 2010. *Nat. Photon.* 4: 222–226.)

demonstrated a device just like that (Löser et al. 2012). It was reported that a proprietary outcoupling material, NET-61, formed crystallites when deposited by evaporation. When this material was inserted into the ETL, the roughened surface was reproduced in subsequent layers all the way to the cathode (Figure 12.37). While the enhancement factor achieved was only 1.8×, this approach remains the most promising among SPP mode extraction techniques.

Since none of the SPP mode extraction methodologies has been wholly satisfactory, it is logical to look for alternatives. As can be seen in Equation 12.15, suppressing SPP modes, or reducing q_3, is also a valid way to increase the outcoupling efficiency. The amount of coupling into SPP modes decreases monotonically with increasing dipole–metal distance; so increasing the ETL thickness to locate the dipole near the second antinode would lower the power loss to SPP modes to a certain degree. Note that it is assumed that a thick, conductivity-doped ETL that decouples optical interference and carrier transport will be used (Walzer et al. 2007). However, it should also be noted that

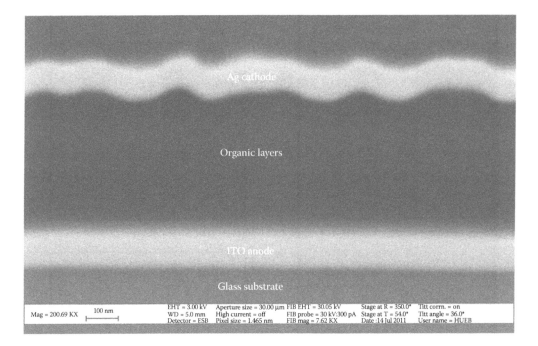

FIGURE 12.37 The evaporated NET-61 forms crystallites on the underlying layer. The morphology and therefore the scattering properties can be controlled by changing evaporation parameters such as nominal layer thickness and evaporation rate. An SEM cross section image of a device with NET-61 is shown. (Adapted from Löser, F.; Romainczyk, T.; and Rothe, C. 2012. *J. Photonic Energy* 2: 021207.)

SPP modes for aluminum cathode drops more slowly with increasing dipole–metal distance than for silver cathode (Figures 12.12 and 12.13).

Hong et al. made the astute observation that in Equation 12.17, if the dielectric material has a relative permittivity greater than the negative real part of the relative permittivity of the metal, then the SPP modes are suppressed (Hong et al. 2011). WO_3 ($\varepsilon = 35.2$) fits that description well. Equation 12.17 gives an answer of $k_{SP} = k_0 (0.22 + 4.02i)$ at the WO_3/Ag interface, that is, with an SPP mode extinction coefficient 150 times greater than at the organic/Ag interface. A sandwich structure such as WO_3/Ag/WO_3 (WAW) should have SPP modes suppressed at both interfaces. It was found to have a transmission as high as 92.9% at $\lambda = 510$ rμ, which makes it well suited as a transparent cathode. Somewhat unfortunately, the WAW cathode was compared with a LiF/thin Al cathode in a top-emitting device when Al is quite absorbing (Figure 12.38). A proper comparison should have been made between the total emission from a transparent device with ITO anode and WAW cathode versus a bottom-emitting device with ITO anode and LiF/Al cathode as control, each with its own optimized layers.

In summary, the SPP modes represent up to 40% or more of the total radiation, and preventing energy loss to, or recovering the energy loss from, these modes has become the forefront of current research in light extraction enhancement. There are several promising approaches, namely the cathode with random subwavelength roughness and the WAW cathode, but no one has yet shown a meaningful reduction of energy loss to SPP modes. A breakthrough there could significantly increase the ultimate efficacy OLEDs can achieve.

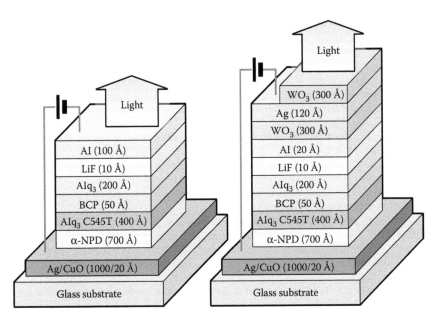

FIGURE 12.38 Schematic of the TOLEDs with different cathode structures (100 Å-thick-Al and WO₃/Ag/WO₃ multilayer). (Adapted from Hong, K. et al. 2011. *J. Phys. Chem.* 115: 3453–3459.)

12.4 Conclusions

In the first part of this chapter, we explored how an exciton decays once formed inside an OLED. Several different microcavity theories were examined, culminating in a QM microcavity theory that, we believe, provides a full view of all the decay pathways in the OLED. A wave optics model that provides a full description of the external emission from an OLED was also presented.

Armed with modal distributions calculated from the QM microcavity theory, we arrived at limits of efficacy improvements from various extraction enhancement techniques that were classified based on the modes they extract. Representative extraction techniques in each category and their respective merits were discussed. Extraction of SPP modes was highlighted as a focus of current research and several promising directions were presented. The understanding of exciton decay mechanism is of both scientific and practical significance. Being able to harness all the radiative modes inside an OLED would realize its full potential as one of the most efficacious light sources devised by man.

References

Andrew, P. and Barnes, W. L. 2004. Energy transfer across a metal film mediated by surface plasmon polaritons. *Science* 306: 1002–1005.

Bardsley Consulting, Navigant Consulting, Inc., Radcliffe Advisors, Inc. et al. 2012. Solid-state lighting research and development: Multi-year program plan. Prepared for Lighting Research and Development, Building Technologies Program, U.S. Department of Energy.

Barnes, W. L.; Dereux, A.; and Ebbesen, T. W. 2003. Surface plasmon subwavelength optics. *Nature* 424: 824–830.

Bathelt, R.; Buchhauser, D.; Gärditz, C. et al. 2007. Light extraction from OLEDs for lighting applications through light scattering. *Org. Electron.* 8: 293–299.

Bulović, V.; Khalfin, V. B; Gu, G. et al. 1998. Weak microcavity effects in organic light-emitting devices. *Phys. Rev. B* 58: 3730–3740.

Celebi, K.; Heidel, T. D.; and Baldo, M. A. 2007. Simplified calculation of dipole energy transport in a multi-layer stack using dyadic Green's functions. *Opt. Express* 15: 1762–1772.

Chance, R. R.; Miller, A. H.; Prock, A. et al. 1975a. Fluorescence and energy transfer near interfaces: The complete and quantitative description of the Eu^{+3}/mirror systems. *J. Chem. Phys.* 63: 1589–1596.

Chance, R. R.; Prock, A.; and Silbey, R. 1975b. Comments on the classical theory of energy transfer. *J. Chem. Phys.* 62: 2245–2253.

Chance, R. R.; Prock, A.; and Silbey, R. 1978. Molecular fluorescence and energy transfer near interfaces. *Adv. Chem. Phys.* 37: 1–65.

Chang, H.-W.; Tien, K.-C.; Hsu, M.-H. et al. 2011. Organic light-emitting devices integrated with internal scattering layers for enhancing optical out-coupling. *J. SID* 19: 196–204.

Chen S. and Kwok, H. S. 2010. Light extraction from organic light-emitting diodes for lighting applications by sand-blasting substrates. *Opt. Express* 18: 37–42.

Deppe, D. G. and Lei, C. 1990. Spontaneous emission from a dipole in a semiconductor microcavity. *J. Appl Phys.* 70: 3443.

Do, Y. R.; Kim, Y.-C.; Song, Y. W. et al. 2004. Enhanced light extraction efficiency from organic light emitting diodes by insertion of a two-dimensional photonic crystal structure. *J. Appl. Phys.* 96: 7629–7636.

Dodabalapur, A.; Rothberg, L. J.; Miller, T. M. et al. 1994. Microcavity effects in organic semiconductors. *Appl. Phys. Lett.* 64: 2488–2490.

Feng, J.; Okamoto, T.; Naraoka, R. et al. 2008. Enhancement of surface plasmon-mediated radiative energy transfer through a corrugated metal cathode in organic light-emitting devices. *Appl. Phys. Lett.* 93: 051106.

Frischeisen, J.; Scholz, B.; Arndt, B. et al. 2011. Strategies for enhanced light extraction from surface plasmons in organic light-emitting diodes. *J. Photonics Energy* 1: 011004.

Garbuzov, D. Z.; Bulović, V.; Burrows, P. E. et al. 1996. Photoluminescence efficiency and absorption of aluminum-tris-quinolate (Alq_3) thin films. *Chem. Phys. Lett.* 249: 433–437.

Greenham, N. C.; Friend, R. H.; and Bradley, D. D. C. 1994. Angular dependence of the emission from a conjugated polymer light-emitting diode: Implications for efficiency calculations. *Adv. Mater.* 6: 491.

Hecht, E. 1998. *Optics*, 3rd edn. Reading, MA: Addison-Wesley.

Hobson, P. A.; Wedge, S.; Wasey, J. A. E. et al. 2002. Surface plasmon mediated emission from organic light-emitting diodes. *Adv. Mater.* 14: 1393–1396.

Hong, K.; Kim, K.; Kim, S. et al. 2011. Optical properties of WO_3/Ag/WO_3 multilayer as transparent cathode in top-emitting organic light emitting diodes. *J. Phys. Chem.* 115: 3453–3459.

Hong, K. and Lee, J.-L. 2011. Review paper: Recent developments in light extraction technologies of organic light emitting diodes. *Electron. Mater. Lett.* 7: 77–91.

Hong, K.; Yu, H. K.; Lee, I. et al. 2010. Enhanced light out-coupling of organic light-emitting diodes: Spontaneously formed nanofacet-structured MgO as a refractive index modulation layer. *Adv. Mater.* 22: 4890–4894.

Hung, L. S.; Tang, C. W.; and Mason, M. G. 1997. Enhanced electron injection in organic electroluminescence devices using an Al/LiF electrode. *Appl. Phys. Lett.* 70: 152–154.

Jordan, R. H.; Rothberg, L. J.; Dodabalapur, A. et al. 1996. Efficiency enhancement of microcavity organic light emitting diodes. *Appl. Phys. Lett.* 69: 1997–1999.

Kim, J.-S.; Ho, P. K. H.; Greenham, N. C. et al. 2000. Electroluminescence emission pattern of organic light-emitting diodes: Implications for device efficiency calculations. *J. Appl. Phys.* 88: 1073–1081.

Kim, Y.-C.; Cho, S.-H.; Song, Y.-S. et al. 2006. Planarized SiN_x/spin-on-glass photonic crystal organic light-emitting diodes. *Appl. Phys. Lett.* 89: 173502.

Koh, T.-W.; Choi, J.-M.; Lee, S. et al. 2010. Optical outcoupling enhancement in organic light-emitting diodes: Highly conductive polymer as a low-index layer on microstructured ITO electrodes. *Adv. Mater.* 22: 1849–1853.

Koo, W. H.; Jeong, S. M.; Araoka, F. et al. 2010. Light extraction from organic light-emitting diodes enhanced by spontaneously formed buckles. *Nat. Photonics* 4: 222–226.

Levermore, P. A.; Dyatkin, A. B.; Elshenawy, Z. M. et al. 2011. Phosphorescent OLEDs: Enabling solid state lighting with lower temperature and longer lifetime. *SID 11 Digest*. 1060–1063.

Li, F; Li, X.; Zhang J. et al. 2007. Enhanced light extraction from organic light-emitting devices by using microcontact printed silica colloidal crystals. *Org. Electron.* 8: 635–639.

Lin, C.-L.; Lin, H.-W.; and Wu, C.-C. 2005. Examining microcavity organic light-emitting devices having two metal mirrors. *Appl. Phys. Lett.* 87: 021101.

Lin, H.-Y.; Lee, J.-H.; Wei, M.-K. et al. 2007. Improvement of the outcoupling efficiency of an organic light-emitting device by attaching microstructured films. *Opt. Commun.* 275: 464–469.

Löser, F.; Romainczyk, T.; and Rothe C. 2012. Improvement of device efficiency in PIN-OLEDs by controlling the charge carrier balance and intrinsic outcoupling methods. *J. Photonic Energy* 2: 021207.

Lu, M. 2010. Outcoupling Efficiency Enhancement in OLEDs. Presented at *DOE SSL R&D Workshop*, Raleigh, NC, February 3.

Lu, M.-H. and Sturm, J. C. 2001. External coupling efficiency in planar organic light-emitting devices. *Appl. Phys. Lett.* 78: 1927–1929.

Lu, M.-H. and Sturm, J. C. 2002. Optimization of external coupling and light emission in organic light-emitting devices: modeling and experiment. *J. Appl. Phys.* 91: 595–604.

Lu, M.-H.; Weaver, M. S.; Zhou, T. X. et al. 2002. High-efficiency top-emitting organic light-emitting devices. *Appl. Phys. Lett.* 81: 3921–3923.

Madigan, C. F.; Lu, M.-H.; and Sturm, J. C. 2000. Improvement of output coupling efficiency of organic light-emitting diodes by backside substrate modification. *Appl. Phys. Lett.* 76: 1650–1652.

Meerheim R.; Furno, M.; Hofmann, S. et al. 2010. Quantification of energy loss mechanisms in organic light-emitting diodes. *Appl. Phys. Lett.* 97: 253305.

Möller, S. and Forrest, S. R. 2002. Improved light out-coupling in organic light emitting diodes employing ordered microlens arrays. *J. Appl. Phys.* 91: 3324–3327.

Nakamura, T.; Fujii, H.; Juni, N. et al. 2006. Enhanced coupling of light from organic electroluminescent device using diffusive particle dispersed high refractive index resin substrate. *Opt. Review* 13: 104–110.

Reineke, S.; Lindner, F.; Schwartz, G. et al. 2009. White organic light-emitting diodes with fluorescent tube efficiency. *Nature* 459: 234–239.

SCHOTT. 2012. *Optical Glass Collection Datasheets*. SCHOTT North America, Inc, Mainz, Germany.

Shiang, J. J. and Duggal, A. R. 2004. Application of radiative transport theory to light extraction from organic light emitting diodes. *J. Appl. Phys.* 95: 2880–2888.

Shiang, J. J.; Faircloth, T. J.; and Duggal, A. R. 2004. Experimental demonstration of increased organic light emitting device output via volumetric light scattering. *J. Appl. Phys.* 95: 2889–2895.

Sun, Y. and Forrest, S. R. 2008. Enhanced light out-coupling of organic light-emitting devices using low-index grids. *Nat. Photonics* 2: 483–487.

Tien, K.-C.; Lin, M.-S.; Lin, Y.-H. et al. 2010. Utilizing surface plasmon polariton mediated energy transfer for tunable double-emitting organic light-emitting devices. *Org. Electron.* 11: 397–406.

Tsutsui, T.; Yahiro, M.; Yokogawa, H. et al. 2001. Doubling coupling-out efficiency in organic light-emitting devices using a thin silica aerogel layer. *Adv. Mater.* 13: 1149–1152.

Tyan, Y.-S. 2010. OLEDs at 150 LPW what are the barriers. Presented at *DOE SSL R&D Workshop*, Raleigh, February 3.

Tyan, Y.-S. 2011. Organic light-emitting-diode lighting overview. *J. Photonic Energy* 1: 011009.

Tyan, Y.-S.; Rao, Y.; Wang, J.-S. et al. 2008. Fluorescent white OLED devices with improved light extraction. *SID 08 Digest*: 935–938.

Ujihara, K. 1975. Quantum theory of a one-dimensional optical cavity with output coupling. Field quantization. *Phys. Rev. A* 12: 148.

Walzer, K.; Maennig, B.; Pfeiffer, M. et al. 2007. Highly efficient organic devices based on electrically doped transport layers. *Chem. Rev.* 107: 1233–1271.

Winter, G. and Barnes, W. L. 2006. Emission of light through thin silver films via near-field coupling to surface plasmon polaritons. *Appl. Phys. Lett.* 88: 051109.

Wu, C.-C.; Chen, C.-W.; Lin, C.-L. et al. 2005. Advanced organic light-emitting devices for enhancing display performances. *J. Disp. Technol.* 1: 248–266.

Yamasaki, T.; Sumioka, K.; and Tsutsui T. 2000. Organic light emitting device with an ordered monolayer of silica microspheres as a scattering medium. *Appl. Phys. Lett.* 76: 1243–1245.

Yariv, A. 1991. *Optical Electronics*, 4th edn. Philadelphia: HRW Saunders.

Ziebarth, J. M. and McGehee, M. D. 2005. A theoretical and experimental investigation of light extraction from polymer light-emitting diodes. *J. Appl. Phys.* 97: 064502-1–7.

13. Device Degradation

Denis Y. Kondakov

13.1 Introduction

Although some observations of electroluminescence in organic films and crystals were reported from 1950s to early 1980s, they were generally considered as a scientific curiosity rather than a promising technology. The invention of the first organic electroluminescent diode structure of potential practical importance was reported by Tang and Van Slyke

Chapter 13

(1987). One revolutionary aspect of the device (named OLED later) was the heterojunction between two amorphous organic insulators. The organic materials were chosen in a specific way to confine the recombination of the majority of injected carriers to the heterojunction region and away from the electrodes. It is noteworthy that despite more than two decades of active research and development in the organic electroluminescence field, modern OLEDs, including polymer-based devices, still almost invariably inherit this feature.

From the very first publication, OLEDs showed efficiencies and drive voltages that were not only much improved in a relative sense (compared with the previous examples of organic electroluminescence), but even the absolute values were order-of-magnitude close to the theoretical limits. Considering the simplicity of the overall device—thin layers of amorphous organics between conductive electrodes—and the potential ease of manufacturing, it had very soon become clear that the useable lifetime of an electroluminescence device represents one of the most important barriers to the commercialization of an OLED technology.

Generally speaking, the lifetime of an OLED depends not only on the elementary device itself but also on the operation and/or storage conditions, such as, for example, applied voltage, temperature, and humidity. The intended application influences useable lifetimes as well: in some cases, loss of half of emission intensity may be tolerable; in others, the maximum allowed loss may be as small as 3%. Changes in emission color, drive voltage, and consumed power exemplify other important characteristics that may limit device lifetime. Nonetheless, OLED lifetimes are most commonly defined in terms of time it takes for the emission intensity to decay to a certain percentage of initial value while the emission-inducing current is unchanged. As there is no commonly accepted standard with respect to the initial value of emission intensity for the lifetime determination, values of initial current density or intensity are normally provided along with the lifetimes. For example, the first reported OLED device showed a half-life (commonly denoted T_{50} or $T_{1/2}$) of ~100 h at 5 mA/cm^2 current density (Tang 1986). Ideally, temperature is to be specified as well. However, it is very rarely reported in practice, which is unfortunate because the ambient temperatures inside lifetime testing systems may vary substantially between different research facilities.

In discussing OLED lifetimes, more general terms such as "stability," "instability," "aging," and "degradation" are frequently used. These terms are normally used to describe OLED lifetimes in a qualitative sense or to refer to a distinctive physical or chemical process affecting or limiting the lifetime.

13.1.1   Historical Perceptions of OLED Instability

Since the earliest publications, OLED stability was considered to be one of the most important characteristics. Aside from the formation of localized defects—dark spots—reducing emissive area and electrical shorting, the steady decrease of luminance typically received the most attention. Initial assessments of OLED lifetimes were performed and reported using T_{50} values obtained at arbitrary chosen current densities, irrespective of device color or efficiency. Not surprisingly, this practice was soon changed to report T_{50} values obtained with initial luminance in 100–200 cd/m^2 range, which was assumed to be close to a useable minimum value for various display applications. Later, the perceived requirements increased by nearly two orders of magnitude, taking into account combined changes in initial luminance (up to 1000 cd/m^2 before various losses associated with display construction) and changes in maximally allowed loss of brightness, for example, T_{90} or even T_{97} instead of T_{50}. Additionally, the lifetime requirements increased along with expansion of intended application scope: For example, the average usage pattern dictates that the small camera displays (which were naturally the earliest to be commercialized) do not need lifetimes of the same magnitude as regular TV sets and computer displays. In essence, for two-and-half decades, the research and development of OLEDs was playing a catch-up game with constantly increasing lifetime requirements.

While the practical considerations and the consequential research and development steadily drove the perceived stabilities of OLEDs lower and the actual stabilities higher, the understanding of the causes and mechanisms underlying OLED degradation developed at a much more uneven pace. Suggestions of oxidation and corrosion of injecting electrodes made in the original publication led to many important studies of the processes involving electrode materials and external agents such as oxygen and moisture (Do et al. 1994). The mechanism of cathode failure was reasonably well understood and, in less than a decade, it had become apparent that the rigorous elimination of reactive agents from the ambient is sufficient to prevent major electrode degradation processes (Burrows et al. 1994).

On the other hand, relatively little attention was directed to regions of OLED devices away from the electrodes. Some reported observations of the processes such as oxidation (Zhang et al. 1995), crystallization (Han et al. 1994), and diffusion (Han et al. 1995) in the bulk of organic layers were later shown to be of minor (if any) importance in well-encapsulated OLED devices with stable and impenetrable electrodes. More detailed discussion of these findings will be given in subsequent sections. In essence, the common perception of the OLED degradation mechanism was that it is primarily electrode related.

However, very significant lifetime differences between OLEDs using different organic materials (Adachi et al. 1995) or even different purity grades of the same material (Sheats et al. 1996) served as a reminder that some (yet unknown) degradation processes may not be restricted to the electrode regions. Although some limited mechanistic suggestions were made in the respective publications, the nature of the degradation processes remained almost completely unclear. Conceptually, the degradation process consists of two distinct components: (i) changes in device composition and (ii) their relationship to changes in luminance or other performance characteristic of interest. Neither component of the bulk degradation process was understood or even generally accepted until 1999, when Popovic et al. presented compelling evidence that one of the "classic" OLED materials—Alq—is unstable in the bulk when exposed to hole current (Aziz et al. 1999). Even though the specific details of what happens on a molecular level were not determined and the proposed action mechanism was limited to generic luminescence quenching, the resulting paradigm shift is hard to overestimate. Suddenly, it had become apparent that an imaginary OLED with perfectly stable injecting contacts and fully isolated from external agents may still have limited lifetime, which is determined by molecular instability of major device component in oxidized state.

In the following years, a number of studies appeared clarifying physics of the degradation process, particularly establishing the importance of nonradiative recombination centers in extinguishing useful electroluminescence and reducing overall efficiency (Kondakov et al. 2003). Although the perception of instability sources shifted from electrodes and external agents to the bulk of organic layers and internally generated charge carriers, the fundamental mechanisms responsible for the change in device composition on the molecular level remained unknown. Furthermore, it had become apparent that the observed instability of oxidized Alq is a fairly uncommon property in OLED materials (Kondakov et al. 2007, Kondakov 2008, Jarikov and Kondakov 2009). Various OLED materials exposed to unipolar currents of electrons or holes showed no appreciable degradation, yet the respective OLED devices had limited lifetimes and exhibited regular signs of accumulation of nonradiative recombination centers and luminescence quenchers as a result of operation. This apparent contradiction was mostly resolved in another paradigm shift, which started recently with learning chemical nature of degradation process and establishing the major importance of chemical reactivity of excited states, which are unavoidably created in the course of normal operation of OLEDs (Kondakov et al. 2007). It is noteworthy that one important but perhaps unfortunate finding was that the degradation process is intimately linked with emission itself. While the external agents can be eliminated completely and the density of charge carriers can be, at least in principle, manipulated independently of emission intensity, the density of the excited states is approximately linearly related to the emission intensity.

13.1.2　Empirical Characteristics of Degradation Process

A large body of research toward developing long-lived and efficient OLEDs has yielded many important empirical observations about OLED degradation. Aside from their obvious value for designing practical devices, the commonly observed empirical characteristics of OLED degradation are also useful to validate mechanistic hypotheses. For example, an early hypothesis of crystallization as a source of luminance loss is inconsistent with observations that luminance loss occurs only under device operation. In the remainder of this section, several potentially useful telltales will be briefly discussed.

13.1.2.1　Luminance Decay Form

From the very first publication, the luminance decays were reported to decelerate gradually during operation, following some distinctly nonlinear function. The exponential function similar to that in first-order chemical kinetics equation does not generally fit the OLED decay. Other simple kinetic equations are also unable to describe typical OLED decays adequately. On the other hand, it is usually possible to find fairly simple mathematical functions to fit luminance decays of a given OLED, especially when only a limited portion of decay—for example, until 50% loss—is considered. Useful functions include stretched exponential—$\exp(-at^b)$, multiexponential—$c\exp(-at) + d\exp(-bt)$, and power function—$at^{-b} - 1$, where a, b, c, and d are fitting parameters. Devices with similar compositions and common fabrication processes can usually be fitted with the same function. Finding the appropriate fitting function and parameters allows predicting the lifetime (e.g., T_{50}) of a device by extrapolating fitted function without actually operating the device until it reaches 50% luminance loss.

Although there is an obvious practical value in fitting, it conveys very little information about the nature of OLED degradation. For example, it is known that the satisfactory fits of experimental luminance decays can be obtained from models based on fundamentally very different mechanisms: first- or zero-order kinetics of accumulation of luminescence quenchers combined with long-range or short-range quenching and exciton migration processes (Kondakov 2010). A trivial loss of emitting centers that is assumed to be nontrivially dependent of their concentration was also shown to be able to mimic experimentally observed decays (Féry et al. 2005). Considering all the necessary assumptions and adjustable parameters, it is perhaps not surprising that the fundamentally dissimilar models are equally capable of fitting the experimental decays. Conversely, observing a satisfactory agreement between some model and an experiment should not be taken as confirmation of validity of a mechanism associated with such model. Overall, the variations in OLED decay shapes as well as the absence of any simple analytic equation that is generally applicable to OLED decay are suggestive of a complex underlying mechanism such as, for example, multistep chemical reactions with feedback inhibition or catalysis.

Aside from the complexity, a typical decay has some other properties that are mechanistically noteworthy. Luminance decays are typically cumulative: the only time the device is subjected to forward current counts with respect to luminance loss. Degradation processes do not appear to extend appreciably into the off-time. This behavior is suggestive of operation-related accumulation (or consumption) of some species inside the device.

The other interesting property of luminance loss is that it is mostly irreversible. Although some deviations from such behavior have been reported, the recoveries of luminance efficiency are usually quite minor (Yahiro et al. 2000). It is, however, unclear whether the reported recoveries are indicative of a partial reversibility of a dominant degradation mechanism or of minor variations in device efficiency prompted by external factors such as reverse bias treatment. Absence of reversibility is suggestive of either chemical reactions or some irreversible physical transformation such as crystallization.

13.1.2.2 Link between Degradation Rate and Current Density

A remarkably general and conceptually important property of luminance decay is its dependence on current rather than on the voltage. Just as the electroluminescence in OLEDs is determined by current, its decay appears to be current driven too. The rate of degradation is a function of current density and, respectively, the extent of degradation is a function of total charge passed through the device. Voltage, on the other hand, does not appear to induce luminance loss directly. It is easy to demonstrate that, for a wide range of applied voltages (or, more pertinently, internal electric fields), there is no luminance decay in the absence of forward current. This does not mean that the applied voltage has no effect on the rate of degradation but rather that it is more appropriate to consider the voltage effect as a perturbation of degradation induced by current.

Initially, the specific form of relationship between luminance decay rate and current density was determined to be approximately linear, which led to classifying the degradation process as "coulombic," that is, determined by the total injected charge (Van Slyke et al. 1996). Because the luminance and current density are also related in an approximately linear manner, the dependence of decay rate on current may be considered as the

Chapter 13

dependence on luminance. The redefined relationship was expressed by making product of T_{50} and initial luminance constant (Popovic and Aziz 2002).

Later, it had become clear that there are substantial and systematic deviations from the linear relationships, usually resulting in decay rates increasing with current densities faster than expected. In some cases, the superlinearity was so great that the decay rates scaled approximately proportionally to luminance (or current density) squared (Sato et al. 1998). For practical purposes of extrapolating device lifetime, the usual lifetime–luminance relationship is usually amended by exponentiating luminance with an additional variable parameter, which is usually called acceleration factor or coefficient. Because of the difficulties associated with measuring lifetimes when current densities are widely varied (low current densities lead to impractically long durations of experiment, whereas high current densities are limited by various destructive phenomena owing to electric field and temperature), it is currently unclear how general is the relationship between lifetime and exponentiated luminance.

The observations of linear and superlinear relationships between the luminance decay rate and current density or luminance provide important clues about degradation mechanism. They suggest that the OLED degradation is driven by the excited states generated by recombining charges. The observed linearity is indicative of the first-order kinetics with respect to concentration of the excited states, and the superlinear relationships suggest some involvement of higher-order kinetics as observed in bimolecular reactions. For comparison, the same observations suggest that the OLED degradation mechanism driven by the reactions of charge carriers without recombination is less likely because their concentrations are not generally expected to scale linearly with current density.

13.1.2.3 Temperature Effects on OLED Degradation

Another generality with respect to OLED degradation is a strong dependence on temperature. As a rule, the operation-induced luminance decay rate monotonically increases with temperature until temperature is kept below a certain, material-dependent limit (Sheats et al. 1996). Under such conditions, significant degradation occurs only when device operates, although minor temperature-induced changes can occur even in the absence of current. Quantitatively, luminance decay rates appear to approximately follow Arrhenius relationship, which is usually interpreted in terms of degradation being a thermally activated process. For example, formal activation energy of the order of 0.3 eV was reported for Alq-based OLED devices (Aziz et al. 2002). Although some attempts to relate these values with properties of OLED materials were made, no compelling evidence of mechanistic connection had been reported so far. In principle, Arrhenius-like behavior is commonly observed for chemical reactions, including very complex reaction mechanisms. However, thermally activated behavior is also quite common for various purely physical processes. Intriguingly, it has been reported that the operation of an ordinary phosphorescent OLED at 20 K led only to ~5× decrease in degradation rate, which formally corresponds to a very low value of activation energy (~3 meV) (D'Andrade et al. 2006). However, with only two data points reported, it is not clear whether Arrhenius treatment is appropriate. For example, it is possible that degradation is dominated by thermally activated mechanism at room temperature, whereas, temperature-independent mechanism overtakes at low temperature. In principle, many barrierless (negligible activation energy) chemical reactions are known, particularly in the field of photochemistry. Overall, the temperature effects do not appear to be particularly informative with respect to OLED degradation mechanisms.

Exceeding the temperature limit typically results in marked and irreversible changes in device characteristics or even a total failure (defined as complete loss of electroluminescence). This type of degradation occurs readily even when device is not operated. The device- and material-specific temperature limit can usually be traced to glass transition temperature of some material in OLED stack (Tokito et al. 1997). A typical OLED material remains a rigid solid in amorphous state when it is within an intended operational range of temperatures. Mobility of molecules of the size of a typical OLED material is essentially nonexistent in that state. Exceeding the glass transition temperature (which, particularly in thin layers, may be somewhat modified by adjacent materials or dopants) results in the material becoming a supercooled liquid and molecular mobility becoming significant factor. Many different physical phenomena such as mixing, phase separation, dewetting, and crystallization become possible in liquid state. Electrical shorts and electrode delaminations frequently occur when temperature limit is exceeded, especially when forward or reverse voltage is applied. Despite the generally detrimental effects of exceeding glass transition temperature, it is noteworthy that it is possible to design OLEDs that gain some advantage from exceeding glass transition temperature: Temporarily liquefied layers were reported as a way to change emission color in a controlled manner (Wu et al. 2003) whereas continually liquid layers (Xu and Adachi 2009) were proposed to markedly increase lifetime of the OLEDs by refreshing emissive layer.

13.1.2.4 Fabrication Process and OLED Stability

It is well known that OLED stability is strongly affected by the various details of device fabrication process (Bohler et al. 1997, Kawaharada et al. 1997, Cheng et al. 2000, Lin et al. 2001, Ikeda et al. 2006, Wang et al. 2007, Fehse et al. 2008, Lee et al. 2009, Noguchi et al. 2010, Mao et al. 2011, Yamamoto et al. 2012). Arguably, that is the most frequent reason why the nominally identical devices produced in different laboratories frequently show markedly different lifetimes. Although some process differences exist, it is instructive to consider device fabrication process in the following stages: substrate preparation (including electrode formation), solution deposition of organic layers, and vacuum deposition of organic and inorganic layers, including counter electrode formation. The surface roughness of the substrate, presence of foreign particles, and other defects within or under the active layers are well known to have detrimental effects on OLED stability owing to electric shorting, dark spot formation, and other localized phenomena. Whenever defect dimensions are comparable to or exceed OLED thickness, the likelihood of formation of a hot spot (less-than-average thickness of organic dielectric) or even a pinhole in a nominally impermeable metal electrode is markedly increased.

In contrast, the spatially uniform and temporally monotonic decay of luminance is affected primarily by the conditions associated with deposition of organic layers. Although vapor devices are often considered as references, it is noteworthy that their stabilities are also strongly dependent on fabrication conditions. Perhaps the most well-known effect is the dependence of the lifetime on pressure maintained in the deposition chamber. Everything else being equal, lower pressure is typically associated with longer lifetimes (Ikeda et al. 2006). Stability of the devices with solution-processed layers

Chapter 13

appears to be strongly affected by annealing, reaching, in same cases, stabilities on par with vapor analogues after appropriate thermal treatment (Mao et al. 2011). Elimination of detrimental volatiles such as water (Fehse et al. 2008, Yamamoto et al. 2012) and film density increases (Lee et al. 2009) were proposed as plausible reasons of lifetime improvements. It has been reported that water vapor is frequently a primary component found in the vacuum chamber. Considering its reactivity toward some OLED materials, most notably Alq, and detrimental properties of the respective reaction products, it is plausible that contamination with water may be a culprit in some cases.

However, many OLED materials (e.g., hydrocarbon hosts and dopants, arylamine hole transport materials and dopants, etc.) are not expected to be reactive with water molecules during deposition. Additionally, small amount of water retained by organic layers after deposition should be eliminated after diffusion to electropositive metal electrode. It is therefore doubtful that water vapor present during vacuum deposition is a common source of OLED instability. The water vapor effects (Ikeda et al. 2006, Yamamoto et al. 2012) may just as well be interpreted in terms of some contaminant that is detrimental for OLED stability (e.g., a halogenated organic material) and with vapor phase concentration naturally changing along with chamber pressure. As an example, minute quantities of reactive species arising from an ion gauge inside the deposition chamber had a marked effect on the internal fixed charge density of organic layers (Noguchi et al. 2010).

OLED stability is also affected by deposition rate and substrate temperature (Cheng et al. 2000, Wang et al. 2007). In addition to the effects of decomposition caused by high deposition rates and temperatures, the lower rates may be more affected by contamination by the putative impurities present in the vacuum chamber as discussed earlier. Even in the absence of changes in chemical composition, differences in surface morphology and density may be sufficient to affect OLED lifetime.

13.1.2.5 Instability Attributable to Materials

Considering a very extensive and continually growing body of research on relationship between OLED materials and device stability, it is next to impossible to provide a meaningful discussion of the subject in this section. Nonetheless, there are several generalities in OLED degradation process with respect to materials that are worth pointing out. In particular, it is noteworthy that the representatives of many classes of organic molecules are invariably detrimental for the device lifetime. For example, halogenated (with a possible exception of fluorine) compounds are capable of reducing device lifetime by orders of magnitude even when present as a minor component (<1%) (Fleissner et al. 2009). As an empirical rule, chemically reactive molecules are usually detrimental for OLED stability. Considering that a low concentration of some reactive material may cause substantial instability, it is sometimes difficult to separate negative effects attributable to a nominal material from the effects of some inadvertently present impurities. As a result, purification is extremely important and sometimes may lead to orders of magnitude improvement in stability (Sheats et al. 1996).

It is noteworthy that the material-related instability does not usually appear to be a result of incompatibility with some other component in the stack but rather an inherent property of a "harmful" material. Nonetheless, the negative effect can often be minimized by preventing such material from participating in chemical reactions by moving it away from reactive species. For example, when a hole transport material with a non-conjugated arylamine moiety is isolated from the recombination zone by a continuous layer of a more stable benzidine derivative, the stability is greatly increased (Kondakov 2008). An alternative approach is to mix material that is detrimental for stability with other material that can effectively compete with physical process, leading to harmful chemistry. Stability improvements resulting from mixing unstable-to-holes Alq with arylamine hole transport materials may serve as an example of this approach (Popovic and Aziz 2002). In either case, it is generally helpful to understand the chemical mechanism behind the instability.

13.2 Operation-Related Changes in Device Structure and Composition

The cumulative and irreversible character of OLED degradation suggests that the structure and composition of an OLED changes during device operation. Aside from the readily identifiable localized defects such as dark spots and edge growth, other changes within an OLED are difficult to detect as a result of device geometry: typically, not more than several hundred of molecular monolayers of organic materials and nearly perfect containment of organic layers between impermeable electrodes. Nonetheless, various methods have been adopted to study processes that occur in degrading OLEDs. Although the distinction may be somewhat arbitrary in a few cases, it is instructive to consider two broad types of operation-related changes: physical and chemical.

13.2.1 Physical Processes

13.2.1.1 Ionic and Molecular Migration

Mass transfer of cathode metal in OLEDs was reported in 1996 as one of the primary processes responsible for formation of visible nonemissive defects (Cumpston and Jensen 1996, Kasim et al. 1997). As expected from an electromigration process in general, the defect growth was linked to current density although it was not clarified whether the specific relationship between the two has coulombic character.

Metal electrodes were also shown to undergo electrochemical dissolution, generating metal ions and creating macroscopic defects—dark spots (Aziz and Xu 1997). The migrations of metal ions originated at the cathode (alkali and alkaline earth metals) and at the anode (indium) were reported by several research groups. Although no systematic studies were reported, the length of migration appears to depend strongly on the nature of ion: alkali metal ions appear to be more prone to diffusion relative to alkaline earths such as Mg (Lee et al. 1999), and the larger yet softer Cs ion migrates more readily than smaller and harder Li (Gerenser et al. 2004). It is likely that the migration is strongly affected by the organic medium as well, but this subject remains unexplored.

It is noteworthy that the several studies report a much higher extent of In migration in degraded OLEDs (Lee et al. 1999, Nguyen et al. 2001). However, it is not clear whether the observed accumulations of In ions in the bulk of organic films were current-driven, coulombic phenomena, which may have been intimately related to luminance decay or, alternatively, induced by electric field or temperature associated with device operation and is only indirectly related to luminance decay. Theoretically, the latter possibility appears considerably more likely. Furthermore, the ion diffusion as primary source of luminance decay is even more unlikely, considering nearly perfect stability of nonoperating OLEDs even if kept at elevated temperature. Field-induced drift of ions as primary source of luminance is similarly unlikely in view of the irreversibility of the luminance decay in contrast to a natural reversibility of ionic drift. A recent study provides experimental evidence that metal ion migration does not play dominate role in luminance decay of OLEDs (Luo et al. 2007).

On the other hand, drift and diffusion of metal ions may still play an important role in voltage changes in an operating device (Shen et al. 2000). As the efficiencies of OLEDs usually show weak dependence on voltage (more precisely, internal electric field), migration of metal ions may be expected to influence luminance decay indirectly.

By analogy to metal migration processes, molecular diffusion can be also envisioned to take place in OLED to result in unintended mixing between layers. For example, mixing at the interface between hole transporting and emissive layer has been reported (Fujijira et al. 1996). It was also suggested that the molecular movement of organic molecules comprising OLED layers can be suppressed by introduction of bulky substituents. In principle, some molecules such as conductivity dopants, which are commonly used in transport layers, can effectively quench electroluminescence upon migration to an emissive zone. However, no direct evidence of such migrations was ever reported, which is understandable while considering that the organic layers are rigid solids under intended conditions of operation. Furthermore, some common OLED architectures should be particularly sensitive to molecular diffusion, yet show no indications that it takes place to any significant degree. For example, there are no operation-related color mixing in white OLEDs, which are frequently designed to include multiple distinct sublayers, each responsible for emission of the part of white spectrum. In general, molecular migration as a significant source of luminance decay is unlikely, considering nearly perfect stability of nonoperating OLEDs.

13.2.1.2 Ferroelectricity and Molecular Orientation

Many OLED materials possess significant dipole moments. It has been reported that the orientation of molecular dipoles in films may not be completely random, resulting in a macroscopic dipole moment and ferroelectric behavior (Berleb et al. 2000, Ito et al. 2002, Kondakov et al. 2003). Although molecular migration in amorphous films is unlikely below glass transition temperature, electric field-induced reorientation of the molecular dipoles may be possible. On the basis of experimental observations of reversible shifts in current–voltage characteristics, molecular dipole reorientation processes have been proposed to occur in operating OLEDs (Zou et al. 1997). In contrast, operation-related luminance decays are mostly irreversible and, therefore, are likely to be unrelated for the most part to molecular orientation changes.

13.2.1.3 Morphology of Device Layers

Morphological changes in OLED layers are often suggested to occur during device operation. However, the majority of the reported studies are about changes in device characteristics that occur upon exceeding glass transition temperature of at least one organic layer (Tokito et al. 1997). Although significant heating can result from the device operation, typical characteristics of operation ($\leq$1000 cd/m^2, <10 V) do not result in internal device temperature being more than 10°C above ambient. Therefore, it is more appropriate to consider morphological changes such as thermal or storage effects as opposed to be caused by device operation on its own accord. It is noteworthy that a systematic study of degradation rate as a function of hole transport material structure showed no correlation between glass transition temperature and lifetime (Adachi et al. 1995).

Morphological changes in the absence of heating have been reported as well. For example, standalone films often undergo crystallization upon storage. Ambient-exposed defect regions of regular OLEDs were also reported to show localized crystallization (Aziz et al. 1998). Here again, general link to OLED operation is unlikely.

13.2.1.4   Excited State Quenching

One of the first suggestions of luminescence quenching as a source of luminance decay in OLEDs came from the studies of photooxidation of films of common EL polymers (Yan et al. 1994, Rothberg and Lovinger 1996) and Alq (Zhang et al. 1995). In-device quenchers were reported soon thereafter (Scott et al. 1996, Aziz et al. 1999) and quenching of luminescence had become one of the most frequently mentioned process responsible for the luminescence decay in operating OLEDs.

In general, luminescence quenching is a process of radiationless energy transfer between the emissive excited state and the quencher, which is a nonemissive molecule or an atom with an accessible excited state. There are several elementary mechanisms of radiation-less energy transfer: the dipole–dipole mechanism (effective for long-range transfer of energy from singlet or triplet excited states to generate singlet excited state of the acceptor molecule) and the electron exchange mechanism (effective at short-range distances for singlet–singlet and triplet–triplet energy transfers). Both processes can result in an excitation migrating tens of nanometers before emitting light or deactivating nonradiatively on a luminescence quencher. The long diffusion lengths result in small concentrations of an effective quencher, as low as a fraction of a percent, being sufficient to extinguish most of the luminescence.

It follows that increasing concentration of quenchers would lead to decreasing luminescence and, vice versa, decreasing luminescence may suggest accumulation of quenchers. As the quenching process is independent of whether the excited states are generated by photoexcitation or electrical excitation, the involvement of quenchers in electroluminescence decay can be readily probed by measuring the photoluminescence efficiency as a function of operation time. In essence, the notion of luminescence quenching provides both a sound hypothesis to explain OLED luminance decay and a method to probe it. It is also noteworthy that the typical OLED structure is very appropriate for nondestructive monitoring of photoluminescence.

Although the quantitative aspects of physics of the elementary energy transfer processes are relatively well understood, the quantitative treatment of the quenching process in OLED layers is typically impossible because of many unknown parameters. For example, by neglecting excited state diffusion via hopping, the determination of a quencher concentration from the extent of photoluminescence efficiency loss requires spectral characterization of the quencher as well as a reasonably accurate profile of its concentrations within the layer of interest, neither of which are generally available. Even the relationship between electroluminescence and photoluminescence efficiency losses is difficult to interpret quantitatively because of the differences in profiles of excited states in electroluminescence and photoluminescence experiments: the electroluminescence is typically concentrated near some interface and distributed unevenly within a nominal emissive layer, whereas the distribution of photoluminescence is more uniform and governed primarily by optical interference effects, which usually results in degradation effects being somewhat attenuated in photoluminescence experiments.

Even limited to qualitative interpretations, the photoluminescence measurements proved to be very effective in mechanistic studies of OLED degradation. Starting from the conclusive demonstration of the first intrinsic degradation mechanism—instability of cation radical of Alq[9]—the accumulation of luminescence quenchers detected by steady-state and time-resolved techniques has been almost invariably shown to occur in operating OLEDs. It is also noteworthy that various complex forms of electroluminescence decay can be readily interpreted in terms of quenching mechanism (Kondakov 2010).

13.2.1.5 Charge Trapping and Energy Level Shifts

As pointed out in the original publication by Tang and Van Slyke (1987), luminance decay in operating OLEDs is usually accompanied by a concomitant voltage rise, which may be attributable to deterioration of one or both injecting electrodes. Although electrode degradation processes were indeed demonstrated to occur in operating OLEDs, they are not the only reason for voltage rise. For example, measuring internal electric fields in operating OLEDs with relatively stable electrodes suggested that the observed voltage rises were unrelated to the electrodes (So and Kondakov 2010). In principle, two other possible causes of voltage rise are an increase in the built-in voltage and generation of new traps away from the electrodes. The changes in built-in voltage should result in current–voltage characteristic shift such that the voltage rise is independent of current density. That is contrary to a typically observed behavior: the voltage rise (i.e., the difference between aged and unaged devices) monotonically increases with current density. Furthermore, a reported case of the direct measurements of built-in voltage on degraded devices showed that not only the absolute magnitude of built-in voltage change attributable to degradation was considerably smaller than the voltage rise but the changes were also in opposite direction—positive voltage rise and negative change in built-in voltage (Kondakov et al. 2003). By elimination, voltage rise in operating OLEDs with stable electrodes is likely to be a manifestation of a build-up of traps.

It should be stressed that, although it is the space charge that is ultimately responsible for the voltage rise, the degradation-related increase in concentration of efficient traps is an essential element here. It is easy to see that the interfacial or space charge densities frequently detected in operating fresh devices cannot be responsible for the voltage rises observed during device degradation and can only have minor influence on luminance decays. In fact, presence of significant density of the carriers of one sign in the recombination zone is usually required for efficient operation of an OLED because it prevents leakage of the carriers of opposite sign to the counter electrode (Albrecht and Bässler 1995). The efficient OLEDs are usually designed to have a marked excess of carriers of one sign (typically, holes) within the recombination zone. Such excess translates to net charge density, which, according to Poisson's equation, corresponds to the difference in electric fields on two sides of recombination zone. It follows that, whenever the electrode deterioration increases the electric field predominantly on one side of the recombination zone, the net charge density within the recombination zone will change. In essence, the charge density is a product rather than a source of voltage rise in devices with unstable electrodes.

In order to explain voltage rise in OLEDs with stable electrodes, one needs to postulate generation of new sites—traps during device operation. The newly created traps are partly filled, resulting in decreasing density of mobile carriers, increasing total density of charges, and, respectively, increasing voltage. Longer operation will lead to a higher concentration of traps, translating in higher concentration of filled traps as well and larger voltage increase. The link between traps and luminance decay is naturally related to nonradiative recombinations involving filled traps: The larger concentration of traps translates into higher probability of recombinations between mobile carriers and filled traps.

Generation of the trapped charge densities during OLED degradation was suggested as early as 1996 based on the observations of the effects of OLED operation with pulsed waveform, which included reverse bias component (Van Slyke et al. 1996). Soon thereafter, trap density was offered as one of the possible interpretations of the experimentally observed differences between photoluminescence and electroluminescence evolutions during aging of Alq-based devices (Popovic et al. 2001). Formation of traps as a result of interaction of Alq with water (Steiger et al. 2001) and photoexcitation (Hashimoto et al. 2003) were reported as well. Detection of filled traps formed as a device degraded in a course of its regular operation was reported in 2003 (Kondakov et al. 2003). The voltammetric technique adopted to detect filled traps also yielded quantitative information about their concentrations. As a result (i) the traps were determined to form gradually during device operation and (ii) the luminance decays were shown to be quantitatively connected to concentrations of filled traps. Operation-induced generation of traps was detected in OLEDs on the basis of various materials, including materials structurally dissimilar to Alq. Filled traps of both signs and various depths were detected as well (Kondakov et al. 2003, 2007, Kondakov 2005, 2008, Renaud and Nguyen 2009, Nakano 2009).

Overall, generation of the substantial densities of traps (comparable to the densities of mobile carriers) during OLED operation appears to be a very common phenomenon. It is noteworthy that the trap generation might explain an early observation of the emission zone shrinkage in degraded OLEDs (Matsumura and Jinde 1997).

13.2.2 Chemical Processes

As discussed in the previous sections, there are many commonly observed physical changes in degraded OLEDs. Taking into consideration the mostly irreversible, coulombic character of the luminance decay as well as its extreme sensitivity to chemical nature of the materials, generation of permanent defects such as deep traps and luminescence quenchers is the most likely to be principal sources of degradation. In the absence of external agents, formation of such defects in amorphous organic materials is difficult to envision without invoking chemical transformations.

It is noteworthy that the potential importance of chemical reactions in OLED operation has been suggested by Rothberg as early as 1996 (Rothberg and Lovinger 1996). He speculated that the two most likely types of chemical reactions are excited-state reactions or electrochemistry. A practically important suggestion that the OLED materials should not have chemically reactive bonds was made a year later (Wakimoto et al. 1997). The first observation of in-device chemistry was reported by Scott et al. (1996). A large degree of oxidation of emissive polymer was observed in the device that had been operated, and ITO electrode was proposed as a source of oxygen. However, follow-up studies showed that indium tin oxide (ITO) is not a significant source of oxygen in operating devices, provided that it is properly fabricated (Cumpston et al. 1997). It was also demonstrated that, although metal electrodes are nominally impermeable, oxygen can migrate through microscopic defects and react with the excited polymer (Tada and Onoda 1999).

The first studies of intrinsic chemical transformations in OLEDs appeared nearly a decade later. The slow rate of advance is undoubtedly due to difficulties associated with the detection of extremely small quantities of products of chemical reactions present in OLEDs. By way of illustration, the experiments simulating degradation by intentional introduction of traps and quenchers showed that even a fraction of percent of a single monolayer is sufficient to extinguish most of electroluminescence. This corresponds to absolute quantities as low as 10^{-12} g/mm^2, which is obviously problematic for various analytical methods. Nonetheless, the structural identification and quantification of the products of operation-induced chemical products are obviously essential elements of understanding the OLED degradation in general and the development of materials and devices with superior stability.

13.2.2.1 Exocyclic C–X Bond Dissociation Reactions

The first intrinsic chemical reaction identified in operating OLEDs was the dissociation of the carbon–nitrogen bonds in CBP (bis-carbazolyl biphenyl) (Kondakov et al. 2007). Although the immediate products of the dissociation—reactive radicals—were detectable only indirectly through increasing density of unpaired electrons, structurally identified stable products contained fragments characteristic of homolytic dissociation of CBP and radical addition reaction to generate stabilized radicals. Quantum chemical calculations indicated that the radical products were deep traps and effective quenchers of luminescence. The quantities of the reacted CBP were order-of-magnitude consistent with the observed losses in luminance.

Based on the studies of CBP photochemistry in solution and in films, it was concluded that the homolytic dissociation of carbon–nitrogen bond occurred in the first singlet excited state. The unimolecular reaction is allowed because of nearly equal carbon–nitrogen bond dissociation energy and the singlet excited state energy. The alternative heterolytic dissociation reactions are unlikely to be based on unfavorable energetics. The first-order kinetics in excited state concentration is consistent with an approximately linear dependence of the rate on intensity of photoexcitation. On the basis of model fitting, a bimolecular reaction between polarons and excited states has been also proposed to occur in another carbazole derivative doped with phosphorescent complex (Giebink et al. 2008, 2009). However, considering many unsupported assumptions about unknown quantities such as spatial distributions of the excited states, polarons, and generated defects within the proposed models, the conclusions appear tentative at best. Ironically, the authors themselves provide (contrary to their conclusions and, perhaps, inadvertently) substantial evidence against the involvement of polar-exciton reactions in luminance decay in a regular OLED. Specifically, the observed absence of luminescence quenching in photodegraded single-carrier devices (Giebink et al. 2009) essentially rules out the process authors study as a dominant degradation mechanism in a regular OLEDs, which, as a rule, show concerted decays of photoluminescence and electroluminescence. From the first principles, the putative nonradiative recombination centers that do not quench photoluminescence of dopant present at 15% concentration appear unrealistic.

Carbon–nitrogen dissociation reactions in operating OLEDs were also shown to involve arylamines, which are typical representatives of hole transport materials (Kondakov 2008, Scholz et al. 2008). As with carbazoles, homolytic dissociations of the excited states were identified as a likely source of operation-induced chemistry. The reactivities of these molecules can be traced to the relatively low bond-dissociation energies. It is noteworthy, that, although no systematic studies have been published thus far, the aromatic molecules with another type of relatively weak exocyclic bond in aromatic molecules—carbon–oxygen bond, for example, ~65 kcal/mol in phenyl methyl ether—appear to cause substantial instability in OLEDs. Halogens, with exception of fluorine, also form relatively weak bonds, which was reflected in instabilities of OLED devices incorporating even low levels of the halogenated molecules (Fleissner et al. 2009). It is noteworthy that the halogenated molecules and aromatic ethers may have additional heterolytic dissociation pathways owing to the presence of good leaving groups—halogen or aryloxy anions.

13.2.2.2 Ligand Dissociations in Organometallic and Coordination Complexes

Ligand dissociation and exchange reactions are particularly important in the OLEDs, considering that the phosphorescent devices invariably utilize organometallic or coordination complexes of transition metals or rare earths as emitters. Even though coordination complexes of main group metals are not as essential, they are also commonly used in OLEDs, starting from the first published OLED device, which included Alq as emitter and electron transport material. Most frequently, complexes used in OLEDs utilize bidentate ligands, which have, at least formally, quite high dissociation energies. However, stepwise dissociations and ligand exchange reactions may overcome energetic restriction attributable to the formal ligand dissociation energies. Considering anionic nature of ligands such as quinolates and acetylacetonates, heterolytic dissociation reactions are also quite likely, making formal dissociation energies less important.

Experimentally, dissociation of one quinolate ligand in Alq during device operation has been proposed based on the detection of Alq fragment with one less quinolate ligand only in aged devices (Scholz et al. 2009a). The product of the exchange reaction between quinolate ligand and phenanthroline derivative present in adjacent layer was also detected only in aged devices. Therefore, the quinolate dissociation appears to be caused by device operation. The initially formed quinolate-less fragment of Alq is coordinatively unsaturated and is likely to form weak complexes with immediate neighbors, which, in case of a strong ligand phenanthroline, is detected directly.

It is noteworthy that the excitation of Alq and related complexes does not result in promotion of an electron to the metal–ligand antibonding orbitals. Bonding and

Chapter 13

antibonding orbitals also do not participate in electrochemical oxidation and reduction reactions. Therefore, the nature of ligand dissociation reaction is unclear and can be envisioned to occur in excited neutral and ion-radical state. Regardless of the mechanism, the elementary dissociation is very likely to be highly reversible. The crucial unknown here is a putative process responsible for preventing the initially dissociated quinolate ligand from recomplexing with Alq fragment. In contrast with the homolytic dissociations of C–X bonds discussed in the previous section, the products of the heterolytic dissociation of Alq are not indiscriminately reactive toward neighboring molecules. Further studies are necessary to elucidate the dissociation mechanism of Alq and structurally related complexes, which have been shown to undergo similar ligand dissociation reactions (Scholz et al. 2007).

Despite having marked differences in photophysical properties and electronic structure when compared with Alq, the organometallic complexes—iridium complexes—appear to undergo formally similar ligand dissociation and exchange processes. The fragments corresponding to formal dissociation of one ligand are frequently detected in aged OLEDs (Meerheim et al. 2008, Scholz et al. 2009b, Moraes et al. 2011a,b). When the adjacent materials are capable of acting as strong ligands for the dissociated iridium complex, the corresponding adducts are detected as well, although it is not clear if these adducts are formed in OLED during operation or during analysis. Just as with the coordinative complexes, the elementary mechanisms of the dissociation itself and the following reactions responsible for irreversibility remain mostly unknown. A notable exception is fragmentation of picolinate ligand, which results in irreversible formation of carbon dioxide (Moraes et al. 2011b).

Although the ligand dissociation reactions appear to be quite common in operating OLEDs, their importance for the luminance decay and voltage rise processes remain unclear because of the insufficient information about quantities and properties of the respective degradation products.

13.2.2.3 Oxidative Coupling Reactions

Couplings of oxidized arylamines and carbazoles are well-known reactions in solution. These reactions involve electrophilic addition of cation radical to electron-rich arylamine followed by the second oxidation and, formally, release of two protons. However, despite the widespread use of the molecules of the same chemical classes, no indications of such reactions have been found in OLEDs. Although the detailed discussion is outside the scope of this chapter, the dielectric nature of OLED layers and the negligible density of counterions are probably the main reasons for the absence of such reactivity in OLED films.

Interestingly, the evidence of oxidative coupling has been found in OLEDs, but it appears to follow completely different reaction path. Instead of electrophilic addition of cation-radical, the carbon–carbon bond-forming coupling appears to involve neutral excited molecules. These reactions were found in the course of investigation of chemical behavior of fully aromatic hydrocarbons on OLEDs. In principle, bond dissociation reactions are highly unlikely for such molecules because they typically lack weak bonds that are comparable to the singlet excited state energies. The other type of chemical reactivity, which will be discussed in the following section, is also unlikely, considering the absence of strong nucleophilic centers, such as sp^2-hybridized nitrogen atom with a lone pair of electrons orthogonal to π-electron system.

The excitation of a typical anthracene host ADN (9,10-bis(2-naphthyl)anthracene) results in an intramolecular cyclization product along with the transfer of various products of hydrogen to neighboring molecules, which can be identified only by an increase in the density of free spins in a degraded sample (Kondakov et al. 2010). Although the complete mechanism remains unknown, the activation energy of the initial cyclization of ADN was calculated to be substantially less than the first excited singlet state energy, which suggests the unimolecular reaction of the neutral excited molecule. As with all other known in-device chemistries, the initial step is likely to be highly reversible (~0.3 eV activation energy for the respective ring-opening reaction) and the overall rate is likely to be controlled by subsequent steps of hydrogen transfer.

Although the chemical structure of the cyclization product was identified only in ADN photo- and OLED degradations, similar accumulations of free-spin densities and losses of luminescence were observed in rubrene films. It is therefore likely that the similar coupling reactions of excited molecules are quite general and may dominate OLED degradation in the absence of faster degradation pathways. In rigid solid films, the intermolecular coupling reactions are also possible although somewhat disadvantaged because of the geometry requirements.

Limited quantitative information about this type of chemistry in operating OLEDs suggests that it is order-of-magnitude significant with respect to the degradation of OLEDs with such hydrocarbons as major components of the emissive layers. Nonetheless, the overall degradation may be dominated by other processes, such as carbon–nitrogen bond dissociation of hole-transporting materials and emissive dopants.

13.2.2.4 Electrophilic Addition Reactions

Although there are many uncertainties about detailed chemical mechanisms of degradation discussed in the preceding sections, their common feature is the central role of the excited states. However, the discussion of known in-device chemistries would be incomplete without mentioning chemical transformation that may occur in the absence of electronic excitation. As briefly mentioned earlier, heterolytic bond dissociations such as dehalogenation might occur in some molecules inadvertently introduced in OLEDs as impurities. Presence of a good leaving group—halogen anion—might cause instability of the respective anion-radicals even in the absence of electronic excitation. Although halogens are normally excluded from OLED materials, the other chemical transformations of ion-radicals may occur for common OLED materials. A good example of such chemistry may be dimerization of phenanthroline derivatives observed in operating device (Scholz et al. 2008). The reaction is likely to occur by addition of electrophilic cation-radical (hole) to strong nucleophilic centers, such as sp^2-hybridized nitrogen atom of phenathroline bearing a lone pair of electrons, which is orthogonal to π-electron system. This type of chemistry is well known in the electrochemical reactions of azines in solution. It is noteworthy that the known instability of Alq with respect to hole current may be a manifestation of the similar chemical reactivity, involving, for example, a free lone pair of quinolate ligand. It is also possible that the same chemistry is responsible for the quinolate ligand dissociation reactions discussed in one of the preceding sections.

13.3 Luminance Loss as Manifestation of Operation-Induced Changes

The preceding sections show that there is a remarkable diversity in physical and chemical processes taking place in operating OLEDs. In all likelihood, these processes are quite general and occur to some extent in a wide variety of OLED types and compositions. Unfortunately, the conclusions about any given process as a focal point of degradation are sometimes determined by the capabilities of a chosen analytical method. In a way, the situation is reminiscent of a classic fable wherein each of six blind men tries to describe a complete elephant after touching different parts of its body.

Notwithstanding potential benefits of a comprehensive description of the processes resulting from OLED operation, OLEDs are practical devices and the degradative processes discussed in the preceding sections need to be understood from the perspective of their influence on luminance loss.

13.3.1 Emission Zone Damage: Nonradiative Recombination Centers and Luminescence Quenchers

As discussed earlier, operation-induced chemical transformations in the emissive layer lead to generation of various molecular fragments—free radicals, coordinatively unsaturated complexes, free ligands, and so on. Some fragments, such as carbon-centered σ-radicals, are known to be very reactive and likely to react with neighboring molecules to produce stabilized π-radicals. Similarly reactive species may be expected to result from hydrogen transfer reactions following oxidative coupling. Regardless of the specific chemical mechanism of ligand dissociation, the products are likely to be ionic or neutral forms of free radicals as well. In the absence of molecular diffusion, the isolated radicals stabilize by reacting with neighboring molecules. The resultant migration of the radical center has very short length, essentially comparable to molecular size. As it is derived from host or dopant molecules, the radical is likely to have its singly occupied molecular orbital within the band gap of the emissive layer and, therefore, act as a deep trap for holes and electrons. Trapping mobile carriers will result in the formation of ionic form of free radical—filled trap, which will compete with mobile (and, possibly, dopant trapped) carriers of the same sign for mobile carrier of opposite sign.

Another consequence of the presence of the unpaired electron is the nonradiative nature of the excited states of such radical species. Fast internal conversion processes, relatively low energies and oscillator strength of electronic transitions are common for radical species. These properties have been confirmed by quantum chemical calculations for several examples of plausible OLED degradation products (Kondakov et al. 2007, Kondakov 2008, 2010). Clearly, charge recombination on these centers would not produce useable light. Similarly, short- or long-range energy transfer from the excited states of the emissive dopant and host to such species would also amount to nonradiative deactivation of the excited states. Even though the ionic form (filled trap) does not have singly occupied molecular orbital, the calculations show that species like that are also likely to be nonemissive and, therefore, effective quenchers of luminescence. Overall, chemical reactions in operating OLEDs lead to generation and accumulation of deep traps—nonradiative recombination centers and luminescence quenchers. It is noteworthy that the chemically dissimilar reaction may result in a qualitatively identical behavior from the perspective of physical methods: combination of increases in density of traps, increases in fixed (deeply trapped) charge density, losses of photoluminescence efficiency, accumulations of permanently present spin density, and so on.

Quantitatively, the deleterious effects of nonradiative recombination and luminescence quenching on OLED luminance strongly depend on the photophysical properties of degradation products and their concentration profiles as well as density profiles of charge carriers and excited states in operating OLED. Because very little information of this kind is available, it is practically impossible to describe luminance loss quantitatively, treating effects of different chemical and physical processes separately. Nonetheless, with some assumptions, it has been shown that the concentrations of the degradation products and the measured densities of filled traps and free spins are order-of-magnitude consistent with luminance losses. Finally, even in the absence of specific information about profiles and spectroscopic properties of degradation products, the relationships between their concentrations and luminance losses can be assessed (Kondakov 2010).

13.3.2   Indirect Mechanisms of Luminance Loss: Remote Interfaces and Layers

Degrading electrodes have been suggested as a possible source of instability as early as 1987 (Tang and Van Slyke 1987). It is plausible that the diffusion of metal ions already present in electrode regions as well as electrode corrosion to generate such ions may result in increases in injection barrier and, respectively, shifts in current–voltage characteristic. Changes in transport layers may have similar effects: for example, conductivities of transport layers may be adversely affected by diffused ions and by chemical reaction induced by the drifting carriers. Overall, it is clear that the substantial and partially reversible increases in drive voltage can result from degradative processes involving electrodes and transport layers. In some cases, even minor decreases in drive voltage on the initial stages of degradation are observed. On the other hand, potential connections between such remote processes and luminance losses are much less understood, except in trivial cases when, for example, metal electrode is losing reflectivity owing to corrosion.

In principle, insufficient carrier injection on one electrode can be expected to result in greatly decreased luminance. In an extreme case of one noninjecting electrode, an OLED becomes a single-carrier device and no electroluminescence is observed because the carriers injected by the other electrode drift across the organic layers and discharge nonradiatively at the noninjecting electrode. Even if the electroluminescence is observed, it is possible that the significant fraction of carriers cross the organic layers without recombining and, again, discharge nonradiatively at the opposite electrode. This behavior is particularly significant for devices without an appropriate heterojunction such as, for example, the original polymer-based light-emitting diodes (Burroughes et al. 1990). Quantitatively, the fraction of carriers undergoing recombination in organic layers rather than leaking to opposite electrode may be denoted as "charge balance factor." Naively, the "balance" may be thought to mean some kind of charge compensation between positive and negative charge densities, where a perfectly "balanced" device would have equal densities of holes and electrons in an electrically neutral recombination zone. However, the efficient OLEDs neither require nor typically have "balanced" charge densities. In fact, modeling (drift-diffusion models with Langevin recombination) shows that it is straightforward to attain >0.99 "charge balance factor" together with a narrow recombination zone even if the density of one carrier exceeds the other one by orders of magnitude.

Despite the "charge balance factor" clearly being a misnomer, it is nonetheless frequently used in a much broader and much less defined sense to refer to any experimentally observed deviation of external quantum efficiencies from a value expected based on photoluminescence quantum efficiency, optical outcoupling, and spin statistics. For example, 15% external quantum efficiency observed in an imaginary phosphorescent OLED utilizing a dopant with 100% photoluminescence quantum yield (measured) and 20% optical outcoupling (predicted) can be interpreted as 0.75 "charge balance factor." However, considering excellent charge confinement in a typical phosphorescent OLED, the 0.75 factor clearly has little to do with carriers leaking to opposite electrodes or with the ratio of hole and electron densities in recombination zone. Instead, the 0.75 factor in this example stems from various exciton loss channels, such as quenching, annihilation, nonradiative recombination, and so on.

Some of the loss channels are affected by carrier densities. For example, because charge carriers are normally ion-radical species with substantial absorbance in visible spectrum, it is not surprising that they have been shown to be responsible for the efficiency lowering (Young et al. 2008). In many types of OLEDs, the concentration of a major carrier in the recombination zone is approximately proportional to drive voltage, which results in an indirect connection between drive voltage and efficiency. The loss channels involving triplet–triplet and triplet–singlet annihilations are primarily affected by exciton concentrations rather than by the carrier densities. However, modeling shows that, with everything else being equal, the drive voltage (or, more precisely, electric field profile in the emissive layer) affects the width of the recombination zone and, therefore, the exciton concentrations and the respective annihilation losses. The width of the recombination zone is also likely to affect carrier densities and optical outcoupling, resulting in additional indirect connections between drive voltage and efficiency.

Existence of the voltage–efficiency connections makes it possible to argue that the operation-induced loss of electroluminescence efficiency might be, to some degree, caused by electrode-deterioration-related voltage change. Because of the great diversity of the chemical and physical degradative processes in operating OLEDs, it is next to impossible to completely separate direct losses discussed in previous sections and indirect ones discussed in the present section. Instead, their relative importance can be probed by, for example, fabricating model devices where only one type of "degradation" occurs. For example, through the careful design of device structure, the connection between drive voltage and efficiency has been demonstrated in OLEDs without any significant leakage of carriers to the opposite electrodes (Kondakov and Young 2010). A set of devices with the systematically varied concentration of the hole-trapping dopant more than 100 nm away from the designated recombination zone was used to vary drive voltage and, thereby, imitate the degradation of injecting contact. Complete absence of the emission component with the distinctive spectrum of the trapping dopant confirmed that holes are effectively confined to the designated recombination zone. Electroluminescence efficiency changes were indeed observed in such experiments supporting the notion that the degradation of the remote interfaces and layers can cause luminance loss through indirect, electric-field-mediated means even in the devices with nearly perfect recombination zone confinement. However, the absolute magnitude of the effect showed that it is only a minor factor in the luminance loss of the actual degrading devices. Given that such experiments were only conducted for a single type of OLEDs, the generality of this behavior is unclear.

It is possible to envision other indirect mechanisms, where degradation of remote layers or interfaces causes luminance loss. For example, it is known that many electron transport materials, such as Alq or phenanthroline derivatives, are unstable when exposed to hole current. Therefore, even a weak hole current that may reach electron transport layer might be sufficient to cause degradation. Typically, the recombination zone is located on the other side of the emissive layer, which means that excitons cannot be significantly quenched by this remote-from-the-recombination-zone region. However, the exciton diffusion process can enable quenching and respective loss of luminance. It is also possible that some degradation products are sufficiently small to actually diffuse across the emissive layer and reach the recombination zone. Unfortunately, such indirect degradation mechanisms remain almost purely speculative in the absence of the detailed studies.

13.4 Sensitivity Factor as a Second-Order Stability Effect

It is usually assumed that the stability of an OLED with a given architecture is determined by the rates of the chemical and/or physical degradative processes. The research publications that focus on the mechanisms of degradative processes often offer some suggestions on how to minimize their rates in order to increase OLED stability. Similarly, in the publications that describe empirically found materials or combinations of materials yielding longer-lived OLEDs, the results are typically interpreted in terms of suppressing some tentative degradative process. By way of illustration, the notion that the hole current damages Alq is logically connected with approaches such as mixing Alq with hole transport materials or manipulating ionization potential of material in adjacent hole transport layer to minimize residency time of hole on Alq molecules and, therefore, suppress the presumed chemical reactions. The deleterious effects of weak bonds in OLED materials are naturally linked to material design as well as to the attempts to eliminate various contaminants in order to eliminate the respective degradative processes. The notions of unwanted ion migration are likewise connected to various electrode modifications and attempt to utilize blocking layers.

Aside from the trivial approaches such as increasing emission area, stacked devices, and so on, the prevalent way of thinking is that the development of longer-lived OLEDs requires suppressing of all unwanted processes. This is a clearly useful approach: it is obvious that decreasing rates of all degradative processes would lead to the improvements in device stability. However, one important and rarely discussed point is that it is incorrect to extend this statement to infer that the stability improvements *require* suppressing the degradative processes. As the quantitative connections between luminance decay and the results of degradative processes are unlikely to be uniformly applicable to different device compositions, the observed changes is that OLED stability may stem not only from the differences in the rates of degradative processes but also from the differences in *sensitivity* of luminance to the products of degradative processes. For example, it may be speculated that two devices with identical initial compositions of the emissive layers and identical concentrations of degradation products after operation may not show identical fraction of luminance loss. Overall, the concept that sensitivity may have a major influence on OLED stability appears plausible, but clearly requires experimental verification.

One way of demonstrating the role "sensitivity factor" plays in OLED stability requires comparing devices with identical compositions of the degrading layers. The composition must also change equally during operation, yet the devices need to have significantly different stabilities because of some factor that affects only the quantitative connection between luminance loss and profiles of degradation products. As it is nearly impossible to ensure identical compositions during degradation in significantly different devices, one experimental study utilized exactly identical devices degraded to the same degree for such comparison (Kondakov and Young 2010). Considering the well-known empirical connection between device voltage and stability, the "strength" of electron injection was chosen as a likely candidate for sensitivity factor. To manipulate injection in the devices that were already fabricated and degraded, the cathode was removed and reformed along with a suitable electron injection layer. The experimental observations confirmed that the major role of sensitivity: in the otherwise identical devices, the modifications of electron injection had a strong effect on the *apparent* extent of degradation. In the alternative experiments, no electrode replacement was carried out. Instead, the degradative processes were imitated by the intentional introduction of quenchers and nonradiative recombination centers during device fabrication. Here again, the "strength" of electron injection was found to markedly affect the sensitivity of luminance to the presence of molecules imitating degradative products.

It should be stressed that electron injection is not expected to be the only "sensitivity factor." In fact, it may not be generally applicable across different OLED types. But it is expected that there are multiple "sensitivity factors" even for a given device structure. Together with the rates of actual degradative processes, these factors ultimately determine device stability. Such effects need to be taken into consideration whenever longer-lived OLED devices are developed. It is also likely that some empirical improvements in OLED stability stem not only from suppressing degradative processes but, at least, partly from decreasing sensitivity. Although this is an almost completely unexplored field, many physical mechanisms to manipulate sensitivity may be envisioned. For example, it is likely that carrier transport in OLEDs is filamentary. The recombination and subsequent degradative processes may therefore occur predominantly on a small fraction of molecules within the nominal recombination zone. It is obvious that changing the proportion of the "active" molecules will likely have a strong effect on device sensitivity to degradative processes.

References

Adachi, C., K. Nagai, and N. Tamoto. 1995. Molecular design of hole transport materials for obtaining high durability in organic electroluminescent diodes. *Appl. Phys. Lett.* 66:2679–81.

Albrecht, U. and H. Bässler. 1995. Efficiency of charge recombination in organic light emitting diodes. *Chem. Phys.* 199:207–14.

Aziz, H., Z. Popovic, and N.-X. Hu. 2002. Organic light emitting devices with enhanced operational stability at elevated temperatures. *Appl. Phys. Lett.* 81:370–2.

Aziz, H., Z. Popovic, and S. Xie. 1998. Humidity-induced crystallization of tris(8-hydroxyquinoline)aluminum layers in organic light-emitting devices. *Appl. Phys. Lett.* 72:756–8.

Aziz, H., Z. D. Popovic, N.-X. Hu, A.-M. Hor, and G. Xu. 1999. Degradation mechanism of small molecule-based organic light-emitting devices. *Science* 283:1900–2.

Aziz, H. and G. Xu. 1997. Electric-field-induced degradation of poly(p-phenylenevinylene) electroluminescent devices. *J. Phys. Chem.* B 101:4009–12.

Berleb, S., W. Brütting, and G. Paasch. 2000. Interfacial charges and electric field distribution in organic hetero-layer light-emitting devices. *Org. Electron.* 1:41.

Bohler, A., S. Dirr, H.-H. Johannes, and W. Kowalsky. 1997. Influence of the process vacuum on the device performance of organic light-emitting diodes. *Synth. Met.* 91:95–7.

Burroughes, J. H., D. D. C. Bradley, A. R. Brown, R. N. Marks, K. Mackay, R. H. Friend, P. L. Burns, and A. B. Holmes. 1990. Light-emitting diodes based on polymers. *Nature* 347:539–41.

Burrows, P. E., V. Bulovic, S. R. Forrest, L. S. Sapochak, D. M. McCarty, and M. E. Thompson. 1994. Reliability and degradation of organic light emitting devices. *Appl. Phys. Lett.* 65:2922–4.

Cheng, L. F., L. S. Liao, W. Y. Lai, X. H. Sun, N. B. Wong, C. S. Lee, and S. T. Lee. 2000. Effect of deposition rate on the morphology, chemistry and electroluminescence of tris-(8-hydroxyqiunoline)aluminum films. *Chem. Phys. Lett.* 319:418–22.

Cumpston, B. H. and K. F. Jensen. 1996. Electromigration of aluminum cathodes in polymer-based electroluminescent devices. *Appl. Phys. Lett.* 69:3941–3.

Cumpston, B. H., I. D. Parker, and K. F. Jensen. 1997. In situ characterization of the oxidative degradation of a polymeric light emitting device. *J. Appl. Phys.* 81:3716–20.

D'Andrade, B. W., J. Esler, and J. J. Brown. 2006. Organic light-emitting device operational stability at cryogenic temperatures. *Synth. Met.* 156:405–8.

Do, L. M., E. M. Han, Y. Niidome, M. Fujihira, T. Kano, S. Yoshida, A. Maeda, and A. J. Ikushima. 1994. Observation of degradation processes of Al electrodes in organic electroluminescence devices by electroluminescence microscopy, atomic force microscopy, scanning electron microscopy, and Auger electron spectroscopy. *J. Appl. Phys.* 76:5118–21.

Fehse, K., R. Meerheim, K. Walzer, K. Leo, W. Lövenich, and A. Elschner. 2008. Lifetime of organic light emitting diodes on polymer anodes. *Appl. Phys. Lett.* 93:083303.

Féry, C., B. Racine, D. Vaufrey, H. Doyeux, and S. Cinà. 2005. Physical mechanism responsible for the stretched exponential decay behavior of aging organic light-emitting diodes. *Appl. Phys. Lett.* 87:213502–3.

Fleissner, A., K. Stegmaier, C. Melzer, H. von Seggern, T. Schwalm, and M. Rehahn. 2009. Residual halide groups in gilch-polymerized poly(p-phenylene-vinylene) and their impact on performance and lifetime of organic light-emitting diodes. *Chem. Mater.* 21:4288–98.

Fujijira, M., L.-M. Do, A. Koike, and E.-M. Han. 1996. Growth of dark spots by interdiffusion across organic layers in organic electroluminescent devices. *Appl. Phys. Lett.* 68:1787–9.

Gerenser, L. J., P. R. Fellinger, C. W. Tang, and L.-S. Liao. 2004. Photoemission investigation of cesium-doped tris(8-hydroxyquinoline) aluminum (alq3) and the effect of dopant diffusion. *Soc. Inf. Disp. Digest* 35:904–7.

Giebink, N. C., B. W. D'Andrade, M. S. Weaver, J. J. Brown, and S. R. Forrest. 2009. Direct evidence for degradation of polaron excited states in organic light emitting diodes. *J. Appl. Phys.* 105:124514–7.

Giebink, N. C., B. W. D'Andrade, M. S. Weaver, P. B. Mackenzie, J. J. Brown, M. E. Thompson, and S. R. Forrest. 2008. Intrinsic luminance loss in phosphorescent small-molecule organic light emitting devices due to bimolecular annihilation reactions. *J. Appl. Phys.* 103:044509–9.

Han, E. M., L. M. Do, Y. Niidome, and M. Fujihira. 1994. Observation of crystallization of vapor-deposited TPD films by AFM and FFM. *Chem. Lett.* 23:969–72.

Han, E. M., L. M. Do, N. Yamamoto, and M. Fujihira. 1995. Study of interfacial degradation of the vapor-deposited bilayer of Alq3/TPD for organic electroluminescent (EL) devices by photoluminescence. *Chem. Lett.* 24:57–8.

Hashimoto, Y., T. Kawai, M. Takada, S. Maeta, M. Hamagaki, and T. Sakakibara. 2003. Trap states of tris-8-(hydroxyquinoline)aluminum degraded by blue laser using thermally stimulated current method. *Jpn. J. Appl. Phys.* 42:5672–5.

Ikeda, T., H. Murata, Y. Kinoshita, J. Shike, Y. Ikeda, and M. Kitano. 2006. Enhanced stability of organic light-emitting devices fabricated under ultra-high vacuum condition. *Chem. Phys. Lett.* 426:111–4.

Chapter 13

Ito, E., N. Hayashi, H. Ishii, N. Matsuie, K. Tsuboi, Y. Ouchi, Y. Harima, K. Yamashita, and K. Seki. 2002. Spontaneous buildup of giant surface potential by vacuum deposition of Alq3 and its removal by visible light irradiation. *J. Appl. Phys.* 92:7306.

Jarikov, V. V. and D. Y. Kondakov. 2009. Studies of the degradation mechanism of organic light-emitting diodes based on tris(8-quinolinolate)aluminum Alq and 2-tert-butyl-9,10-di(2-naphthyl)anthracene TBADN. *J. Appl. Phys.* 105:034905–8.

Kasim, R. K., Y. Chenga, M. Pomerantzap, and R. L. Elsenbaumer. 1997. Investigation of device failure mechanisms in polymer light-emitting diodes. *Synth. Met.* 85:1213–4.

Kawaharada, M., M. Ooishi, T. Saito, and E. Hasegawa. 1997. Nuclei of dark spots in organic EL devices: Detection by DFM and observation of the microstructure by TEM. *Synth. Met.* 91:113–6.

Kondakov, D. Y. 2005. Direct observation of deep electron traps in aged organic light emitting diodes. *J. Appl. Phys.* 97:024503–5.

Kondakov, D. Y. 2008. Role of chemical reactions of arylamine hole transport materials in operational degradation of organic light-emitting diodes. *J. Appl. Phys.* 104:084520–9.

Kondakov, D. Y. 2010. The role of homolytic reactions in the intrinsic degradation of OLEDs. In So, F. (Ed.), *Organic Electronics*, Boca Raton, New York: CRC Press, pp. 211–42.

Kondakov, D. Y, C. T. Brown, T. D. Pawlik, and V. V. Jarikov. 2010. Chemical reactivity of aromatic hydrocarbons and operational degradation of organic light-emitting diodes. *J. Appl. Phys.* 107:024507–8.

Kondakov, D. Y., W. F. Nichols, and W. C. Lenhart. 2007. Operational degradation of organic light-emitting diodes: Mechanism and identification of chemical products. *J. Appl. Phys.* 101:024512–7.

Kondakov, D. Y., J. R. Sandifer, C. W. Tang, and R. H. Young. 2003. Nonradiative recombination centers and electrical aging of organic light-emitting diodes: Direct connection between accumulation of trapped charge and luminance loss. *J. Appl. Phys.* 93:1108–19.

Kondakov, D. Y. and R. H. Young. 2010. Variable sensitivity of organic light-emitting diodes to operation induced chemical degradation: Nature of the antagonistic relationship between lifetime and efficiency. *J. Appl. Phys.* 108: 074513–11.

Lee, S. T., Z. Q. Gao, and L. S. Hung. 1999. Metal diffusion from electrodes in organic light-emitting diodes. *Appl. Phys. Lett.* 75:1404–6.

Lee, T.-W., T. Noh, H.-W. Shin, O. Kwon, J.-J. Park, B.-K. Choi, M.-S. Kim, D. W. Shin, and Y.-R. Kim. 2009. Characteristics of solution-processed small-molecule organic films and light-emitting diodes compared with their vacuum-deposited counterparts. *Adv. Funct. Mater.* 19:1625–30.

Lin, K. L., S. J. Chua, and S. F. Lim. 2001. Influence of electrical stress voltage on cathode degradation of organic light-emitting devices. *J. Appl. Phys.* 90:976–9.

Luo, Y., H. Aziz, Z. D. Popovic, and G. Xu. 2007. Degradation mechanisms in organic light-emitting devices: Metal migration model versus unstable tris(8-hydroxyquinoline) aluminum cationic model. *J. Appl. Phys.* 101:034510–4.

Mao, G., Z. Wua, Q. He, B. Jiao, G. Xu, X. Hou, Z. Chen, and Q. Gong. 2011. Considerable improvement in the stability of solution processed small molecule OLED by annealing. *Appl. Surf. Sci.* 257:7394–7398.

Matsumura, M. and Y. Jinde. 1997. Change of the depth profile of a light-emitting zone in organic EL devices with their degradation. *Synth. Met.* 91:197–8.

Meerheim, R., S. Scholz, S. Olthof, G. Schwartz, S. Reineke, K. Walzer, and K. Leo. 2008. Influence of charge balance and exciton distribution on efficiency and lifetime of phosphorescent organic light-emitting devices. *J. Appl. Phys.* 104:014510–8.

Moraes, I. R., S. Scholz, B. Lüssem, and K. Leo. 2011a. Role of oxygen-bonds in the degradation process of phosphorescent organic light emitting diodes. *Appl. Phys. Lett.* 99:053302–3.

Moraes, I. R., S. Scholz, B. Lüssem, and K. Leo. 2011b. Analysis of chemical degradation mechanism within sky blue phosphorescent organic light emitting diodes by laser-desorption/ionization time-of-flight mass spectrometry. *Org. Electron.* 12:341–7.

Nakano, Y. 2009. Deep-level optical spectroscopy investigation of degradation phenomena in tris(8-hydroxyquinoline) aluminum-based organic light-emitting diodes. *Appl. Phys. Express* 2:092103–3.

Nguyen, T. P., J. Ip, P. Jolinat, and P. Destruel. 2001. XPS and sputtering study of the Alq3/electrode interfaces in organic light emitting diodes. *Appl. Surf. Sci.* 172:75–83.

Noguchi, Y., N. Sato, Y. Miyazaki, and H. Ishii. 2010. Light- and ion-gauge-induced space charges in tris-(8-hydroxyquinolate)aluminum-based organic light-emitting diodes. *Appl. Phys. Lett.* 96:143305–3.

Popovic, Z. D. and H. Aziz. 2002. Reliability and degradation of small molecule-based organic light-emitting devices (OLEDs). *IEEE J. Sel. Top. Quantum Electron.* 8:362–71.

Popovic, Z. D., H. Aziz, N.-X. Hu, A. Ioannidis, and P. N. M. Anjos. 2001. Simultaneous electrolumines-cence and photoluminescence aging studies of tris(8-hydroxyquinoline) aluminum-based organic light-emitting devices. *J. Appl. Phys.* 79:4673–5.

Renaud, C. and T.-P. Nguyen. 2009. Study of trap states in polyspirobifluorene based devices: Influence of aging by electrical stress. *J. Appl. Phys.* 106: 053707–11.

Rothberg, L. J. and A. J. Lovinger. 1996. Status of and prospects for organic electroluminescence. *J. Mater. Res.* 11:3174–87.

Sato, Y., S. Ichinosawa, and H. Kanai. 1998. Operation characteristics and degradation of organic electrolu-minescent devices. *IEEE J. Sel. Top. Quantum Electron.* 4:40–8.

Scholz, S., C. Cortenb, K. Walzera, D. Kuckling, and K. Leo. 2007. Photochemical reactions in organic semi-conductor thin films. *Org. Electron.* 8:709–17.

Scholz, S., B. Lüssem, and K. Leo. 2009a. Chemical changes on the green emitter tris(8-hydroxy-quinolinato)alu-minum during device aging of p-i-n-structured organic light emitting diodes. *Appl. Phys. Lett.* 95:183309–3.

Scholz, S., R. Meerheim, B. Lüssem, and K. Leo. 2009b. Laser desorption/ionization time-of-flight mass spectrometry: A predictive tool for the lifetime of organic light emitting devices. *Appl. Phys. Lett.* 94:043314–3.

Scholz, S., K. Walzer, and K. Leo. 2008. Analysis of complete organic semiconductor devices by laser desorption/ionization time-of-flight mass spectrometry. *Adv. Funct. Mat.* 18:2541–7.

Scott, J. C., J. H. Kaufman, P. J. Brock, R. Dipietro, J. Salem, and J. A. Goitia. 1996. Degradation and failure of MEH-PPV light-emitting diodes. *J. Appl. Phys.* 79:2745–51.

Sheats, J. R., H. Antoniadis, M. Hueschen, W. Leonard, J. Miller, R. Moon, D. Roitman, and A. Stocking. 1996. Organic electroluminescent devices. *Science* 273:884–8.

Shen, J., D. Wang, E. Langlois, W. A. Barrow, P. J. Green, C. W. Tang, and J. Shi. 2000. Degradation mecha-nisms in organic light emitting diodes. *Synth. Met.* 111–112:233–6.

So, F., and D. Y. Kondakov. 2010. Degradation mechanisms in small-molecule and polymer organic light-emitting diodes. *Adv. Mater.* 22:3762–77.

Steiger, J., S. Karg, R. Schmechel, and H. Seggern. 2001. Aging induced traps in organic semiconductors. *Synth. Met.* 122:49–52.

Tada, K. and M. Onoda. 1999. Photoinduced modification of photoluminescent and electroluminescent properties in poly(p-phenylene vinylene) derivative. *J. Appl. Phys.* 86:3134–9.

Tang, C. W. and S. A. Van Slyke. 1987. Organic electroluminescent diodes. *Appl. Phys. Lett.* 51:913–5.

Tokito, S., H. Tanaka, K. Noda, A. Okada, and Y. Taga. 1997. Thermal stability in oligomeric triphenylamine/tris(8-quinolinolato) aluminum electroluminescent devices. *Appl. Phys. Lett.* 70:1929–31.

Van Slyke, S. A., C. H. Chen, and C. W. Tang. 1996. Organic electroluminescent devices with improved sta-bility. *Appl. Phys. Lett.* 69:2160–2.

Wakimoto, T., Y. Yonemoto, and J. Funaki. 1997. Stability characteristics of quinacridone and coumarin molecules as guest dopants in the organic LEDs. *Synth. Met.* 91:15–9.

Wang G. F., X. M. Tao, and H. M. Huang. 2007. Influence of the deposition temperature on the structure and performance of tris(8-hydroxyquinoline) aluminum based flexible organic light-emitting devices. *Appl. Surf. Sci.* 253:4463–6.

Wu, C.-C., C.-W. Chen, and T.-Y. Cho. 2003. Three-color reconfigurable organic light-emitting devices. *Appl. Phys. Lett.* 83:611–3.

Xu, D. and C. Adachi. 2009. Organic light-emitting diode with liquid emitting layer. *Appl. Phys. Lett.* 95:053304–3.

Yahiro, M., D. Zou, and T. Tsutsui. 2000. Recoverable degradation phenomena of quantum efficiency in organic EL devices. *Synth. Met.* 111–112:245–7.

Yamamoto, H., C. Adachi, M. S. Weaver, and J. J. Brown. 2012. Identification of device degradation positions in multi-layered phosphorescent organic light emitting devices using water probes. *Appl. Phys. Lett.* 100:183306–4.

Yan, M., L. J. Rothberg, F. Papadimitrakopoulos, M. E. Galvin, and T. M. Miller. 1994. Defect quenching of conjugated polymer luminescence. *Phys. Rev. Lett.* 73:744–7.

Young, R. H., J. R. Lenhard, D. Y. Kondakov, and T. K. Hatwar. 2008. Luminescence quenching in blue fluo-rescent OLEDs *Soc. Inf. Disp. Digest* 39:705–708.

Zhang, X.-M., K. A. Higginson, and F. Papadimitrakopoulos. 1995. Photoluminescence and electrolumi-nescence quenching in 8-hydroxyquinoline aluminum chelates. *Mater. Res. Soc. Symp. Proc.* 413:43–7.

Zou, D., M. Yahiro, and T. Tsutsui. 1997. Study on the degradation mechanism of organic light-emitting diodes (OLEDs). *Synth. Met.* 91:191–3.

Chapter 13

14. Vapor Deposition Methods and Technologies

Research to Manufacturing

David W. Gotthold

14.1   Introduction

Since the early days of organic light-emitting diode (OLED) development by Tang and VanSlyke at Kodak, vacuum thermal evaporation (VTE) has been the most commonly used deposition process (Tang and Van Slyke 1987). Vacuum thermal deposition is conceptually a very simple process, where pure versions of the desired materials are evaporated from crucibles using thermal energy. Adjusting the temperature of the crucible can control the rate of evaporation, and therefore the rate of deposition. For OLED materials, the evaporation temperature is typically low (<300°C) and the crucibles are resistively heated and controlled using a thermocouple and PID control system. The resulting vapor then

Chapter 14

propagates along a line of sight path to the substrate where it condenses to form a film. This line of sight propagation also enables simple patterning through the use of shadow masking, currently the dominant technique for both lighting and displays.

While the promise of a low-cost printable display technology has been a key driver for interest in OLEDs, vacuum evaporation remains the method currently used for almost all small molecule displays and lighting in production as of 2014.

14.2 Thermal Evaporation Advantages

VTE has several advantages that have led to the dominance of this technology. The first advantage of VTE is the ability to deposit complex, multilayer device structures with almost unlimited combinations of different materials. Vacuum deposition is well suited to forming these complex structures because each layer can be deposited as a pure material without solvents, with precise control of thickness and (with the use of masks) lateral distribution, and with virtually no interactions with the other layers in the structure. Other deposition methods, particularly solution-based methods, typically use solvents to enable printing or coating of the organic materials. While these approaches are typically low cost (solution coating) or simple to pattern (printing) they significantly limit the device design possibilities due to the need for orthogonal solvents. Basically, each successive layer cannot use a solvent that dissolves one of the previous layers. As there are a limited number of potential solvent classes, this significantly limits the total number of device layers. It also limits the types of materials that can be used, as not all are readily soluble. For example, Alq is a classic and historically important OLED material with poor solvent solubility. It was one of the two materials originally used by Tang and VanSlyke and the lack of a good solvent was one of the key reasons they used VTE for their device fabrication approach (Tang and Van Slyke 1987; VanSlyke, S. A. 2012, personal communication).

Second, VTE is one of the cleanest methods for depositing materials for a wide range of applications; cleanliness is critical to achieving high performance and long lifetime OLEDs. This cleanliness arises for two reasons. First, high-purity source material is typically used to prevent degradation during evaporation. Because any material to be deposited using VTE must sublime at appropriate pressures and temperatures, zone sublimation refining can be and is usually used to purify source materials. Second, environmental contamination is absent in the high-vacuum environment used in most VTE systems. Most OLED materials are highly sensitive to oxygen and water vapor and, while inert gas environments can be used to provide some protection, high vacuum is typically the best (Aziz and Popovic 2004; Baldacchini et al. 2005). This vacuum environment not only significantly reduces the potential for contamination species, it also helps to minimize particle contamination during device fabrication. The thin layers (10–100 nm typically) used in standard OLED designs are especially susceptible to particulate damage, which lead to shorts. In the vacuum environment, there are no air currents to carry particulates, and they instead drop to the bottom of the chamber. While this does not completely eliminate particulates (for instance, they can be attracted to the surface electrostatically), vacuum does reduce likelihood of contamination (Ouellette 1997).

A final consideration for VTE, while not quite an advantage per se, is that a vacuum deposition step is typically required for electrode deposition. Standard electrode deposition is usually accomplished using vacuum deposition (either sputtering or thermal evaporation) so that even devices fabricated under atmospheric conditions typically require a vacuum deposition step.

14.2.1 Thermal Evaporation Challenges

Despite all the advantages of VTE described in the previous section, there remain some critical and fundamental disadvantages that have helped limit the penetration of organic devices into more general markets. These disadvantages all revolve around the fact that thermal evaporation of organic materials is difficult to scale up. This is not fundamentally due to the vacuum needs, though the capital costs can present a high barrier to entry, but rather to the organic materials themselves.

The first and most obvious challenge is that most organic materials are very temperature sensitive and rapidly degrade if overheated. This is not generally an issue in low deposition rate research systems where the substrates are small and the rates are low, but for production systems, thermal degradation is usually the limiting factor in scaling both size and throughput. Even though most organic materials are currently thermally purified, they are still large organic molecules and as such have the double (in this case) disadvantage of low vapor pressure and low degradation temperatures. When one compares the typical sublimation temperatures for most relevant organic materials (250–400°C) to the typical degradation temperatures for those materials (300–450°C)—the challenge becomes obvious. Long et al. showed that for typical research deposition rates, Alq has a lifetime orders of magnitude longer than typical deposition times. This is borne out in most OLED work where Alq is considered a very stable and robust material. However, for large area displays or low cost lighting, the deposition rate must be increased by over a factor of 1000 to reduce the cycle times and to cover a larger substrate area. Thus, the overall temperature must be increased to a point where the material begins to degrade after only a few hours. At the high rates needed to achieve acceptable cost targets for OLED-based lighting, the material lifetime is measured in minutes. Table 14.1 shows the relative deposition rates for a nominal 30 nm thick Alq layer on different substrate sizes using typical deposition times.

In order to achieve acceptable system uptimes, it is critical that the temperature of the bulk of the material be reduced and that the evaporative surface area be as large as possible. Specific designs and sources to achieve these simultaneous design criteria will be discussed later in this chapter.

A related challenge is that some materials do in fact start to decompose at temperatures lower than they evaporate, or at least close enough to the evaporation temperature

Table 14.1 Substrate Sizes and Typical Cycle Times for Different Applications Showing the Relative Deposition Rates for a 30 nm Thick Alq Layer

Substrate Class	Size (mm)	Time (s)	Rate (g/h)	Relative Rate
R&D sample	10 × 10	240	0.0021	1
Pilot line	150 × 150	120	0.0142	68
Mobile display	730 × 920	120	1.11	530
Television	2200 × 2400	60	15.1	7200

Chapter 14

that degradation is inevitable. As mentioned previously, currently most, if not all, materials go through a zone sublimation thermal purification process during manufacturing where they are evaporated and then condensed in a thermal gradient to improve purity. Almost by definition, if a material survives this process, it can be deposited using VTE at least at some useful rate. However, as chemical processing becomes more precise and the as-synthesized purity of new materials improves, the need for thermal distillation is reduced. As nonthermal purification improves, materials that cannot be successfully evaporated will become viable device materials for low temperature processing such as solution printing.

A second challenge to VTE is that the material utilization is generally low. This is wasteful of expensive material and, furthermore, the wasted material tends to coat up other parts of the deposition system, leading to shorter uptime and higher particle counts. This challenge arises from the fact that the evaporative flux is extremely difficult to direct. Much work has been done on shaped nozzles to improve the fraction of the evaporated flux that hits the substrate surface with only limited success (Keum et al. 2004; Kim et al. 2008, 2011). The challenge is that without a gas flow to entrain the evaporating molecules, the distribution of molecules leaving the evaporative surface have a profile that can be approximated with a cosine function, shown in Equation 14.1 (Ohring 2001).

$$\frac{d\bar{M}_s}{dA_s} = \frac{\bar{M}_e \cos\phi\cos\theta}{\pi r^2} \tag{14.1}$$

In more complex (real world) sources, where the evaporating surface may be at the bottom of a deep and narrow crucible or come through a nozzle with a high aspect ratio, the flux is better approximated by Equation 14.2, where the exponent n is an experimentally determined fitting factor:

$$\frac{d\bar{M}_s}{dA_s} = \frac{\bar{M}_e(n+1)\cos^n\phi\cos\theta}{2\pi r^2} \tag{14.2}$$

The ideal behavior for the molecular flux would be something collimated like a laser beam, where the flux could be accurately and repeatably directed exactly to the substrate. This corresponds to very high values of the exponent n in Equation 14.2, indicating the flux is highly directional. In practice, though, n can only be varied a small amount, typically between 1 and 3. While increasing n to 2 or 3 is helpful, it still is very small compared to an ideal case, which might someday be possible through the use of baffles and shaped nozzles.

Patterning of thermally evaporated materials, particularly for displays, is done using high precision shadow masks, typically fine metal masks (FMM) to directly pattern the incoming flux during the deposition process (Yotsuya 2005; Kaneko et al. 2006; Matsuura 2012). While this process is relatively straightforward, it is wasteful of material, especially for displays, as most of the evaporated material is deposited onto the mask instead of the substrate and maintaining high precision alignment over large substrates is difficult. This will be covered in more detail in Section 14.6.

Finally, many common organic materials do not melt before evaporating, either because their vapor pressure is sufficiently high at temperatures below their melting point, or because they thermally decompose before melting. Because the evaporating material remains a solid powder, the stability of the flux can be poor for two reasons. First, unlike most liquids, the thermal conductivity of the powder is very low, leading to non-uniform temperature in the source material. Second, the vaporization surface area can change rapidly as the powder depletes, particle sizes change, and particles shift in the deposition boat or crucible. This can be mitigated by adding thermally conductive inert materials, such as ceramic beads, to the powder (CHAOHON TECHNOLOGY CO. 2013).

14.3 Research Scale Deposition

Research scale OLED thermal deposition can be done on systems ranging from simple bell jars with purely manual controls to multichamber systems with fully automated substrate transfer, mask alignment, and transfer and process control. Figure 14.1 shows an example of a part of a research system from Trovato Manufacturing that can be used for material testing and device design. Except for the simplest of systems, typically used to study the deposition process itself or test new materials, most systems have multiple sources to simplify the fabrication of devices containing several different materials. In Figure 14.1, the source flange from a research system contains spaces for 12 thermal boat sources, which will be further discussed in Section 14.4. The panels between the sources prevent cross contamination by preventing material from one source from depositing on another source during normal operation. Also visible in the figure are the quartz crystal monitor (QCM) sensor heads that provide thickness feedback. Depending on the system complexity, layer deposition can be automated using QCM feedback and appropriate control software.

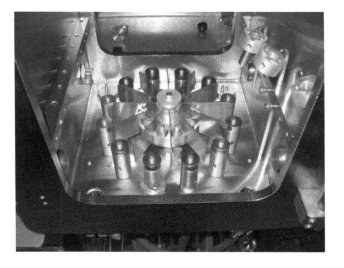

FIGURE 14.1   Source flange for a research system. (Courtesy of Trovato Manufacturing.)

Chapter 14

14.4   Production Scale Deposition Systems

Current production OLED systems are primarily used for display fabrication, and as such are based on traditional LCD motherglass sizes to take advantage of existing process lines. Table 14.2 lists the typical glass sizes that have been used in the LCD industry for each generation of display manufacturing, as well as which sizes are being used for OLED manufacturing. Development work has typically been done on smaller production systems because the equipment is readily available, the capital cost fully depreciated and of declining value for LCD fabrication. Current mobile displays are made on either half or full sheets of 4th generation glass (730 mm × 920 mm) and there are several 8th generation (2200 mm × 2400 mm) systems in development for OLED television production.

Production systems come in two basic configurations, cluster, where the chambers are around a central substrate handling robot, (as shown in the photograph in Figure 14.2

Table 14.2   Typical Glass Substrate Sizes for Different Generations of Displays

Generation	Dimensions	Area (m²)	OLED Application
1	300 mm × 400 mm	0.12	Development
2	400 mm × 500 mm	0.20	Development
3	550 mm × 650 mm	0.36	
4	730 mm × 920 mm	0.67	Mobile displays/lighting
5	1100 mm × 1300 mm	1.43	Lighting/TVs
6	1500 mm × 1800 mm	2.70	
7	1900 mm × 2200 mm	4.18	
8	2200 mm × 2400 mm	5.28	Television
9	2400 mm × 2800 mm	6.72	
10	2850 mm × 3050 mm	8.69	

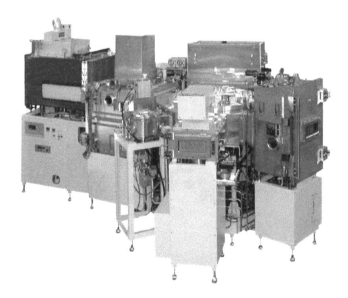

FIGURE 14.2   Cluster deposition system for process development and pilot production. (Courtesy of Tokki Canon.)

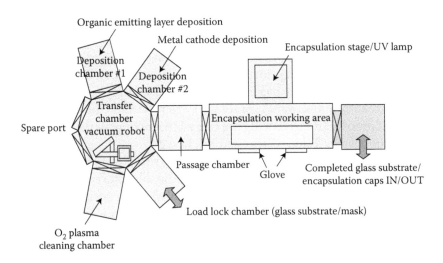

FIGURE 14.3   Cluster production system schematic. (Courtesy of Tokki Canon.)

and the schematic in Figure 14.3) and in-line, where the chambers are in a line and the substrate moves from one end to the other (as shown on the right side of Figure 14.4). Each approach has advantages and disadvantages and the optimal technology often depends on the final product.

Cluster systems use a layout similar to traditional display and semiconductor processing where a central robot is surrounded by an assortment of dedicated process chambers, as shown in Figure 14.3. This approach provides several advantages. First, a cluster tool provides the ability to easily combine multiple process steps into a single system while maintaining a high level of configurability. This enables the system to be optimized for any specific process by adding or removing modules without the need to fundamentally redesign the core hardware. It also enables simple load-balancing, by enabling the addition of multiple process chambers for a slow deposition step. Second, because the chambers are independent and separated by valves, maintenance on an individual chamber can be managed independently without opening the entire system.

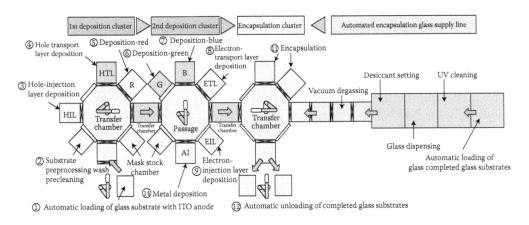

FIGURE 14.4   Hybrid cluster-inline system for higher throughput. (Courtesy of Tokki Canon.)

Third, because the current LCD display industry uses similar equipment for the production of the transistor backplane, the equipment set is well-developed and general purpose components such as substrate cleaning and load locks already exist. However, there are a couple of disadvantages that primarily affect the process cost. First, the cost of the robot and the multiple vacuum management systems, pumps, gauges, valves, and so on, needed add significantly to the overall cost of the system. Second, the transfer from tool to tool requires additional time, reducing the overall throughput of the system. Systems of this type are supplied by Tokki, Ulvac, and K.J. Lesker.

In-line systems use a continuous flow process similar to thin-film photovoltaic and architectural glass coating systems. In these systems, the substrate enters at one end and proceeds through sequential deposition tools, each of which deposits one layer, before emerging fully processed at the end of the line. The primary advantage of this approach is that the throughput to capital expenditure ratio is much better than in a cluster system. The simplified glass transport eliminates the need for the advanced robotic handling, and the straight-line process can reduce the total footprint needed for system installation. The tradeoff for the low cost and high throughput is the lack of configurability, which limits the changes to the device stack that can be made without significant, and expensive, hardware modifications. In addition, the substrate speed is limited by the slowest process. In some cases, for example very thick layers, this can be mitigated by splitting the process into multiple steps, which increases the overall size and cost of the system but enables better cycle time alignment.

Hybrid systems, such as the one shown in Figure 14.4 combine some aspects of in-line (on the right sides) and cluster tools (on the left side) to achieve a balance between configurability and throughput.

14.4.1 Organic Vapor Phase Deposition

An alternate approach to VTE is to use a carrier gas to transport the organic materials from an evaporation zone to a deposition system (Baldo et al. 1998; Shtein et al. 2001; Forrest 2004). This approach has been commercialized by Aixtron AG into a system conceptually similar to those used for Metal-Organic Chemical Vapor Deposition (MOCVD) and uses the ability to easily switch carrier gasses on and off to rapidly control layers (Aixtron 2012). Another advantage of this approach is that very high film thickness uniformity can be achieved across the substrate by careful distribution of the transport gas through a shower-head like distribution system. However, this same distribution can make scaling to very large sizes more difficult as the uniform transport of the carrier gas in a low-vacuum environment becomes challenging as the substrate size increases. Another potential difficulty is that the elevated temperature of the carrier gas needed to prevent premature deposition in the distribution tubing can cause additional material degradation. This can be mitigated by using high gas velocities and reducing transport time as well as by minimizing the heated source volume to reduce the overall time at elevated temperatures.

14.5 Evaporation Source Technology

The key part of any OLED deposition system is the evaporation source itself. For research systems, where the deposition rates are low, uniformity is not a major issue and material utilization is of limited concern, sources can be very simple evaporation cups or boats. However, for production systems, the design of the source is critical to achieving a cost effective and efficient process.

14.5.1 Point Sources

Point sources are conceptually the simplest evaporation sources, consisting typically of an open crucible surrounded by a heating element. However, they can range in practice from a beaker wrapped in resistive wire to a ceramic or refractory metal crucible heated by a CVD grown pyrolytic boron nitride (pBN) and graphite (pG) heater capable of continuous operation at 1500°C in corrosive environments.

For research systems, the typical source is a simple evaporation boat that holds just a few milligrams or grams of material. Figure 14.5 shows a typical source, consisting of a formed refractory metal, with a small source material area and a cover baffle, from a system designed by Angstrom and located at Pacific Northwest National Laboratory. The baffle serves multiple purposes. First, it helps provide a uniform heating environment for the source material, minimizing hot spots that could cause material degradation. Second, by covering the actual source material, it prevents cross-contamination from flakes and/or particles of material falling from the deposition area located above the source. Finally, as described earlier, most OLED source materials are powders and are not particularly stable during evaporation, so the baffle prevents "spitting" of particles and reduces flux variations to some extent.

More advanced sources provide radiative heat to the top surface of the source material using a perforated or slotted heater. Because most organic materials sublime from powder and have very low thermal conductivity, heating from the top limits evaporation to the top surface, generating flux while reducing thermal damage to the bulk of the

FIGURE 14.5 Research sized boat source shown with (a) cover on and with cover removed, (b) to show the interior of source with pockets at both ends for holding material. (Photo: PNNL.)

source material. Kodak developed such a source for their work on OLEDs using a quartz boat to hold the source material and a refractory metal combination cap and heater with perforated holes to allow the flux to escape. A similar approach is used today in many thermal evaporation sources (Witzman et al. 2001; VanSlyke, S. A. 2012, personal communication).

Even with top heating, point sources are fundamentally limited in operation by the thermal damage to the source material. However, since these sources are typically used for research and small-scale development, this is of limited concern.

14.5.2   Linear Boat Sources

For production on larger substrates, achieving the desired uniformity with point sources becomes infeasible due to the high material efficiency penalty and the large chamber size needed (Conroy et al. 2010a,b, 2011). The logical solution is to use a heated crucible configured as a line source (Krug et al. 1993; Witzman et al. 2001). This is exactly the approach taken by several companies, including Tokki, YAS, and Jusung (Krug et al. 1993; OLED Association 2011; Kim et al. 2011). These sources have similar designs to the point sources, but have a much longer evaporation area, providing a line of flux instead of a point. The flux profile from these sources can be simulated as a line of point sources added together. In fact, some boat sources use a series of nozzles instead of a slot to simplify the source design (Keum et al. 2004). This linear flux profile enables uniform deposition across the width of large glass substrates and either the substrate or the glass can be scanned to provide uniform coverage across the entire substrate.

Linear boat sources have similar material limitations as point sources, with the entire mass of source material being heated during normal operation. Like the point sources, it is possible to heat from the top surface and maintain a gradient from the evaporation zone down to the base of the "charge," that is, the source material placed in the boat during a deposition run, but the material will still be exposed to potentially damaging temperatures. Because the sources are designed for production, they contain much more material than point sources, typically 100's of grams (compared to less than a gram for research systems), and are expected to run for longer times, both of which increase the likelihood of material damage. Sources of this type are made by a variety of equipment suppliers, primarily Korean companies such as Sunic, YAS, and Jusung, as well as internally developed sources at the large display manufacturing companies (Kim et al. 2010).

14.5.3   Next Generation Sources

In order to overcome the rate, material lifetime, large area uniformity, and operational uptime challenges that point and boat sources have, different groups have developed a range of novel sources.

Kodak developed an effusion source technology called Variable Injection Source Technology (VIST) (Hamer et al. 2008a,b; Long et al. 2008, 2009). This source was designed to address thermal damage by flash evaporating a small amount of organic material while keeping the bulk cold, and large area deposition by using a distribution nozzle with a carefully tuned flux profile. By evaporating under nonequilibrium conditions, the source enables the coevaporation of materials with different vapor pressure curves, so hosts and dopants can be evaporated from the same source at the same time. They demonstrated the capability for 20 s cycle time and 70% material utilization for an inline 5G deposition system for solid-state lighting (Tyan 2009). However, the source had challenges with reliably feeding organic powders to the flash evaporator, requiring special processing to achieve the necessary consistency (Long et al. 2009). In the end, the source technology was sold when Kodak exited the OLED business in 2010.

Veeco Instruments has developed a series of sources based on their valved source and linear nozzle technologies developed for the MBE and solar markets (Veeco 2009; Campion 2010). By using a valve to modulate the flux, a more stable deposition rate can be achieved, even when using materials that do not have stable evaporation profiles (Kim et al. 2008; Priddy et al. 2008; Conroy et al. 2010a,b). This ability to control the flux response can be extended to enable reloading the source during brief intervals in operations, effectively increasing the source capacity indefinitely without causing additional material damage. This is done by introducing new material into the crucible and using the valve to keep the flux stable as the powder heats up and comes to equilibrium with the crucible temperature. The nozzles of each source are designed to fit closely together, enabling codoping with multiple materials. The utilization of the source is a function of source–substrate spacing and width of the substrate, with efficiencies approaching 90% for 1 m wide glass and 100 mm source-substrate spacing and nearly 40% for a G4 substrate (730 mm × 960 mm) with a 300 mm source–substrate spacing (Gotthold et al. 2011).

Hitachi Zosen has developed a planar source instead of a linear source, in order to deposit over an entire substrate without moving either the source or the substrate, greatly simplifying the system design and mask alignment. A planar distribution manifold spreads the flux across the entire substrate at once, and multiple materials can be codeposited by interleaving multiple distribution manifolds. The manifolds are supplied from a small, separately heated crucible that can be valved off and vented without breaking vacuum, enabling quasi-continuous operation (Fujimoto 2010). Applied Materials has developed a linear nozzle source to enable the glass to be held in a vertical orientation and uses a distribution nozzle to provide a uniform flux (Hoffmann et al. 2011). Ulvac presented a cell, called the g-cell, that uses a carrier gas to transport a small amount of material to a vaporizer zone and then to a linear nozzle for distribution (Waters 2009).

14.6 Process Control and Feedback

Process control and in situ characterization in OLED thermal evaporation is currently relatively immature compared to many other semiconductor processes and this remains a challenge for high volume manufacturing. An example of a high performance OLED lighting stack (in terms of the combination of efficacy and lifetime) is the tandem hybrid white OLED stack shown in Figure 14.6.

In order to produce a product at high yield, layer thickness and composition must be closely controlled. Typically, layer thickness must be maintained to within ±3% of the aim point, or the color, efficiency, voltage, or lifetime will be out of the specifications for the product. The stack in Figure 14.6 shows 11 organic layers. In order for this product to be manufactured with an economically viable yield, each layer must be in control within specification for >99% of the time. If each layer were out of specification 10% of the time, the yield for the full device might be expected drop to 31% (device yield = 0.9^{11}). This low

	Cathode		
Red and yellow emission	n-type contact layer		
	Electron transport layer (ETL)		
	Red emitter (phosphorescent)		
	Yellow emitter (phosphorescent)		
	Hole transport layer (HTL)		
Connector	p-type layer		
	n-type layer		
Blue emission	Electron transport layer (ETL)		
	Blue emitter (fluorescent)		
	Hole transport layer (HTL)		
	Hole injection layer (HIL)		
	Anode		
	Substrate		

FIGURE 14.6   Example of a very high-performance hybrid tandem white OLED architecture with 11 organic layers and two metal/oxide layers.

yield would be difficult to tolerate for displays and would completely eliminate the possibility of cost-effective lighting using thermal evaporation.

14.6.1   Flux Monitoring

The primary technique for flux monitoring in both research and production is a quartz crystal microbalance (QCM) positioned such that the flux from a single evaporation source impinges on the measurement surface. QCMs work by vibrating a small quartz crystal that is exposed to the deposition flux. As material accumulates on the crystal, the resonant frequency shifts. By correlating this shift with the known physical properties of the deposited material, the accumulated mass and deposition rate can be calculated. While this technique works, there are some OLED-specific challenges. First, because organic materials have lower density than most materials the QCMs were developed to measure, the accuracy of the measurements can suffer, especially as the thickness of the accumulated material increases. This is primarily due to the fact that the mechanical properties of organic films are not as well known or as stable with regard to thickness, temperature, and deposition rate as more conventional films such as metals and ceramics.

These issues necessitate the regular replacement of the quartz crystals, which can be done manually or through a reloading system. Manual changes work well for research systems where the deposition chambers are opened regularly and run-to-run variations are less critical. However, this approach is less suitable for production systems where regular system vents cannot be tolerated. Various reloading QCMs are made by the leading sensor manufacturers. For example, Inficon has dual crystal heads (primarily for redundancy) and both 6 and 12 crystal rotatable turret heads designed specifically for long-term operation (Inficon 2013). Colnatec also makes a multicrystal head design with either 12 or 24 crystals specifically for the OLED and thin-film solar industries.

Another path to extending the operational life is to enable heating of the crystal to remove accumulated material. AVL, now PiezoCryst, has commercialized an ortho-phosphate-based crystal that can be heated to temperatures up to 800°C without affecting operation. In addition to their multicrystal heads, Colnatec has also developed a heated sensor head/crystal combination suitable for OLED deposition (Colnatec 2013).

Long et al. have developed a flux monitoring gauge that is incorporated into their VIST source (Long et al. 2008). This gauge, a variation on a Pirani gauge (Ellett 1931) modified to run at low temperature, is capable of providing much faster feedback to enable stable flux control from a variable evaporation material. This is not a direct measurement of the flux being delivered to the substrate, but instead is a measure of the pressure inside the source. Under similar operating conditions and source designs, this measurement will correlate linearly with the delivered flux.

Optical absorption of light that passes through the emitted flux plume also has potential as a measurement tool. This has typically been done using atomic absorption (AA) and is especially useful for separating out the fluxes of different materials that are being codeposited. A major issue with this approach is in the name—this is an atomic absorption phenomenon. As such, it works best with discrete atomic species such as In, Ga, and As. In OLED materials, most of the atomic components are the same (i.e., C, N, H, etc.) and the differences are in the elemental ratios and/or structural. A second major issue with AA is that the evaporation process invariably deposits some material all overso that the viewports used to achieve an optical path will coat up with time. This is an issue for all AA applications, not just for OLEDs, and it requires careful system design and operation to achieve reasonable operating times.

14.6.2 Optical Film Measurement

While QCM measurement can provide a key control signal for the film deposition process, it does not measure the performance of the deposited film. Ellipsometry has been used to characterize deposited films providing a range of useful details, including layer thickness, interface quality, and optical constants (Celii 1997). However, the complexity of the data means that the necessary data fitting and modeling can make real-time feedback difficult.

More recent work has focused on using spectroscopic reflectometry from companies such as LayTec (COMEDD 2013) and k-Space Associates (k-space Associates 2012). Several approaches have been developed for the task of determining the important optical parameters of multilayer thin films. One of the most powerful is phase modulated spectroscopic ellipsometry (PMSE) (Celii 1997; Hartmann et al. 2004; Farahzadi et al. 2010). In this method, the incident light is directed through the deposited layer and a photoelastic modulator mounted in the path of the reflected light measures the change in polarization of the reflected light. The polarization change contains information about the amplitude and phase of the reflected wave as it passes through the different components of the stack. While polarization shifts depend on the thicknesses and optical constants (n and k) of the individual layers, it is not possible to obtain these parameters directly from the ellipsometric angles without intensive spectral modeling and fitting

Chapter 14

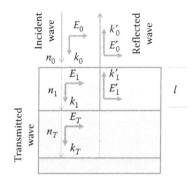

FIGURE 14.7 Normal incidence geometry of spectroscopic reflectance probe for *in-situ* optical constant and thickness measurements. (Courtesy of k-space Associates. 2012. k-Space wins DOE funding for OLED development. Retrieved November 27, 2013, from http://www.k-space.com/k-space-wins-doe-funding-for-oled-development-2012-06-29.)

to many unknowns. For this reason, while PMSE is a useful *ex situ* method, it does not lend itself to *in situ* methods that require real-time analysis of the optical data. A more convenient method is spectroscopic reflectance (SR), which also utilizes phase shifts in a light beam as it propagates through the various film layers. The optical geometry of the measurement is schematically shown in Figure 14.7 at normal incidence (the most favorable situation for *in situ* measurements).

Using SR, one can obtain the required information about the optical constants, the layer thicknesses, and interface roughness by modeling the optical response of the stack at different incident wavelengths, and extracting best fit parameters. Layer thickness precision down to 0.1 nm and refractive index uncertainties at the level of a few percent [k-space Associates] have been demonstrated with this approach for OLED multilayer stacks and efficient routines are readily available for this purpose.

14.7 Patterning

In order to make functional devices, organic films usually have to be patterned in some way. The extent and precision of the patterning depends greatly on the application. Whereas lighting might have centimeter scale pixels with 1 mm resolution, mobile displays can have over 400 pixels/inch (ETNews 2012) with micron resolution requirements as shown in Table 14.3.

Table 14.3 Example Resolutions versus Pixel Size and Typical Size Tolerances

Display Type	PPI	Subpixel (μm)	Tolerance (5%) (μm)
Mobile phone (4.8" 1080p)	459	18×89	1
Mobile phone (4.1" 720p)	358	42×70	2
Tablet (10.1" 1080p)	218	38×116	2
Laptop (15" 1080p)	147	57×173	3
Monitor (27" 1080p)	82	103×311	5
Television (55" 1080p)	40	211×634	10

Unlike traditional semiconductors, postdeposition patterning of organics using etching techniques is generally impractical. While the organic layers can be removed by solvents, controlled removal is difficult, and distinguishing between photoresist and the OLED layers is extremely difficult. The one exception is when the organic layer is not directly patterned, but instead the pixel definition is done with the metal contact layer. This is the approach used in some displays, where the individual pixels are white, and then are processed through a color filter. While this approach does not have quite the efficiency and color purity of direct emission, the lack of patterning makes for a dramatically simpler device fabrication. This approach was originally proposed by Kodak (Hatwar et al. 2004; Hamer et al. 2008) and is the method of choice for some large display manufacturing.

14.7.1 Fine Metal Mask

The most direct method of patterning the organic layers when depositing by vapor deposition is the use of shadow masking with FMM. These are thin foils of Invar™ or similar low thermal expansion metals that are typically photochemically etched to produce the pixel pattern. While FMM patterning is conceptually simple and is currently the method used for most OLED display production, there are several significant challenges.

First, because this is a reductive rather than additive approach (like printing), the material utilization for patterned layers can be quite low, especially for the multiple color emissive layers used in displays. Using the coarse approximation that each primary pixel is made up of a red, green, and blue subpixel, then the maximum material utilization for each emissive layer is only 33%, and in real-world applications, where there is spacing between the pixels and each subpixel, the actual utilization can by much lower, closer to 10%.

Second, the stochastic nature of the flux source limits the resolution to which the sub pixels can be defined. While this is generally not an issue for lighting or larger displays, it is a key limiting factor for achieving high pixel density in smaller displays. Figure 14.8 shows a schematic of the flux patterns through an etched shadow mask illustrating the issue. Until recently, this has limited displays to <200 PPI, but recent advances

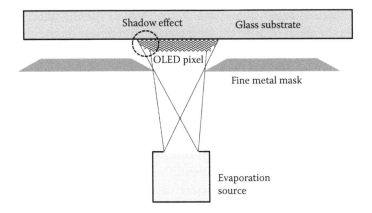

FIGURE 14.8 Schematic of flux pattern through FMM hole. Show beveled edges of a hole, and flux from directly below versus from off angle.

Chapter 14

in pattern layout, mask alignment and flux distribution have enabled higher resolution displays (ETNews 2012).

A third problem with FMM is achieving correct alignment and maintaining that alignment during deposition. For larger patterns, physical alignment methods can be used, where the substrate is held in a mount with the masking system that provides for proper registration. Depending on the size of the substrate, this approach can typically be used for alignment down to less than 0.1 mm (100 μm) (Boroson, M. 2012, personal communication). Where higher accuracy is required, such as in display applications, an optical alignment process similar to that used in LCD mask alignment is typically used. This process, while capable of 1 μm registration on Gen 5 (1100 × 1300 m) substrates, is much more time consuming and can contribute to yield loss, both through errors in registration and through particle generation during the alignment process itself. Once the substrate and mask are aligned, they must be held in alignment during deposition, which is a separate and more difficult issue. It is also clear from the schematic layout in Figure 14.8 that any variation in the mask-substrate spacing could dramatically change the size and edge quality of the masked pattern. In most current processes, the substrate is mounted horizontally and the organic materials are evaporated from below and deposited on the downward facing surface. The orientation limits the number of particulates that could inadvertently land on the substrate before or during evaporation. For smaller substrates, typically <4 G (730 mm × 960 mm), this approach works well, but for larger substrates, maintaining the spacing between the substrate and the mask becomes much more difficult. For this reason, many conceptual designs for larger FMM-based systems have the substrate oriented vertically, or with a small (<5°) tilt so that the mask can rest gently against the substrate. In this configuration, the mask is often dimpled to provide small contact points with the correct spacing when the mask is laid down on the substrate.

Because VTE is, as its name suggests, a high temperature operation, the thermal loads during the deposition process can be significant. The heat load from the source is directed almost entirely at the FMM, causing both local (right above the source) and broad area heating of the FMM. This heating causes thermal expansion and the potential for misalignment. Because the shadow mask is, by definition, blocking the part of the source from the substrate, the two components heat up at different rates. This is the reason that low thermal expansion alloys are one key to the manufacture of FMMs. Figure 14.9 shows the effect of different temperature differences on the alignment capability for FMMs with a range of substrate sizes. This problem is further compounded by the fact that with each process cycle, a new substrate is introduced, but the mask is typically reused, so there are differences in the cumulative heating of each component.

A final challenge to the FMM process is that the material that is blocked by the mask generally accumulates on it, rather than just reevaporating into the chamber. This leads to build up which can slowly clog the holes in the FMM, reducing the size of the deposited features. If the buildup is sufficiently thick, the deposited material can start to flake off, leading to particulate defects. The exact amount varies greatly with material, deposition conditions, and mask details. Regular mask cleanings and exchanges are required to maintain suitable yields. The exact interval of the cleaning and/or exchanges depends on the structure design, the materials being used and the tolerance of the process for particulates.

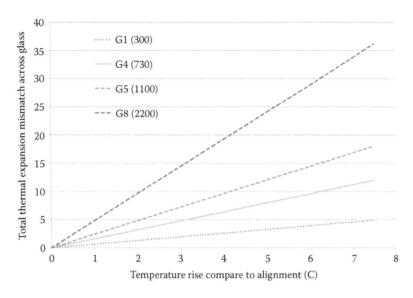

FIGURE 14.9 Alignment mismatch between glass substrate and fine metal mask for different temperature gradients and glass sizes.

A variation on FMM patterning is to use a scanning mask. In this approach, a small, high precision mask is held fixed relative to the source instead of to the substrate. As the substrate is scanned relative to the source, linear patterns or stripes can be produced, which can be used for larger displays or for lighting—where the pixel size can accommodate the lower resolution.

14.7.2   White OLED Plus Color Filter

One approach proposed by Kodak to address many of the shortcomings of FMM patterning for displays is to use individually addressable white pixels matched with a color filter, similar to that used in LCD displays (Hatwar et al. 2004; Hamer et al. 2008). In this approach, the processing is dramatically simplified because only the electrodes need to be patterned, not the organic layers, which completely eliminates the need for shadow masking or post deposition patterning. However, this approach loses some of the efficiency and color purity advantages of direct patterned OLEDs, because white light is generated instead of specific colors. The impact of this can be mitigated by carefully matching the emission colors of the white OLED with the color filters to ensure maximum transmission of the target wavelengths.

14.7.3   Laser–Induced Direct Imaging

Laser-induced direct imaging (LIDI) or laser-induced transfer imaging (LITI) is a novel approach to attempt to address some of the shortcomings of FMM-based patterning (Lamansky et al. 2005). In this process, the organic materials that are to be patterned, typically the emissive layers, are first deposited onto a flexible transparent sacrificial substrate using either traditional thermal evaporation or solution coating.

Chapter 14

This sacrificial substrate is then held in close physical proximity to the device substrate and a laser is used to reevaporate the material, leading to deposition on the primary substrate. Because this transfer process is done using a focused laser beam, the material is reevaporated and deposited only where needed.

This process has been proposed as a method for large area display manufacturing, especially for the red and green subpixels, where performance similar to VTE has been demonstrated. However, challenges with yield and blue emitter deposition have limited the impact to date (Kyoon 2009; ETNews 2012).

14.8   Conclusion

Vacuum thermal deposition, while not without its challenges, continues to be a key technology for the deposition of organic light emitting materials for both research and production. While other technologies, such as direct solution printing, are making progress, VTE remains the highest performance and most flexible deposition method. Key challenges must still be overcome to enable lower cost manufacturing. Continued development in the areas of high capacity large area evaporation sources and advanced process monitoring and control will be needed for OLED VTE to expand to address the needs of the broad display and lighting markets.

References

Aixtron. 2012. *OVPD Mass Production Equipment*. Aixtron, S.E.

Aziz, H. and Z. D. Popovic. 2004. Degradation phenomena in small-molecule organic light-emitting devices. *Chemistry of Materials* 16: 4522–4532.

Baldacchini, G., T. Baldacchini, A. Pace, and R. B. Pode. 2005. Emission intensity and degradation processes of Alq3 films. *Electrochemical and Solid-State Letters* 8: J24–J26.

Baldo, M., M. Deutsch, P. Burrows, H. Gossenberger, M. Gerstenberg, V. Ban et al. 1998. Organic vapor phase deposition. *Advanced Materials* 10: 1505–1514.

Campion, R. P., C. T. Foxon, and R. C. Bresnahan. 2010. Modulated beam mass spectrometer studies of a Mark V Veeco cracker. *Journal of Vacuum Science and Technology B* 28: C3F1.

Celii, F. G., T. B. Harton, and O. F. Phillips. 1997. Characterization of organic thin films for OLEDs using spectroscopic ellipsometry. *Journal of Electronic Materials* 26: 366–371.

CHAOHON TECHNOLOGY CO., L. 2013. OLED/MBE/CIGS-related equipment and instruments. Retrieved November 27, 2013, from http://www.chaohon.com.tw/product_cg7969.html.

Colnatec. 2013. Colnatec tempe datasheet. Retrieved November 27, 2013, from http://colnatec.com/colna_cms/wp-content/uploads/2013/09/Colnatec_Tempe_9-2013b.pdf.

COMEDD. 2013. Flexible OLED from the roll. Fraunhofer COMEDD.

Conroy, C., S. W. Priddy, J. A. Dahlstrom, R. C. Bresnahan, D. W. Gotthold, and J. C. Patrin. 2010a. Linear deposition source. United States Patent Application S20100159132 A1.

Conroy, C., S. W. Priddy, J. A. Dahlstrom, R. C. Bresnahan, D. W. Gotthold, and J. C. Patrin. 2010b. Linear deposition source. WO Patent WO/2010/080,268.

Conroy, C., S. W. Priddy, J. A. Dahlstrom, R. C. Bresnahan, D. W. Gotthold, and J. C. Patrin. 2011. Linear deposition source. WO Patent WO/2011/065,998.

Ellett, A. and R. M. Zabel. 1931. The pirani gauge for the measurement of small changes of pressure. *Physical Review* 37: 1102–1111.

ETNews. 2012. Samsung high resolution OLED display. Retrieved December 21, 2012, from http://www.etnews.com/news/device/device/2621080_1479.html.

Farahzadi, A., M. Beigmohamadi, P. Niyamakom, S. Kremers, N. Meyer, M. Heuken et al. 2010. Characterization of amorphous organic thin films, determination of precise model for spectroscopic ellipsometry measurements. *Applied Surface Science* 256: 6612–6617.

Forrest, S. R. 2004. The path to ubiquitous and low-cost organic electronic appliances on plastic. *Nature* 428: 911–918.

Fujimoto, E. 2010. *OLED Manufacturing System Equipped by Planar Evaporation Source*. SID Display Week, Seattle, WA.

Gotthold, D. W., M. O'Steen, W. Luhman, S. Priddy, C. Counts, and C. Roth. 2011. *Challenges for OLED deposition by vacuum thermal evaporation*. Presented at OLED Materials for Lighting and Display, Minneapolis, MN.

Hamer, J. W. 2008. *Advances towards Realizing Production of Large-Size OLED TVs*. OLEDs World Summit, La Jolla, CA.

Hamer, J. W., A. D. Arnold, M. L. Boroson, M. Itoh, T. K. Hatwar, M. J. Helber et al. 2008. System design for a wide-color-gamut TV-sized AMOLED display. *Journal of the Society for Information Display* 16: 3–14.

Hartmann, E., P. Boher, C. Defranoux, L. Jolivet, and M. O. Martin. 2004. UV-VIS and mid-IR ellipsometer characterization of layers used in OLED devices. *Journal of Luminescence* 110: 407–412.

Hatwar, T., J. Spindler, C. Brown, and M. Ricks. 2004. White devices with color filter arrays. United States Patent US20050147844 A1.

Hoffmann, U., H. Landgraf, M. Campo, S. Keller, and M. Koening. 2011. New concept for in-line OLED manufacturing. 79540Z–79540Z.

Inficon. 2013. RSH-600 Rotary Sensor–INFICON. Retrieved June 30, 2013.

k-space Associates. 2012. k-Space wins DOE funding for OLED development. Retrieved November 27, 2013, from http://www.k-space.com/k-space-wins-doe-funding-for-oled-development-2012-06-29/.

Kaneko, K., Y. Takeda, K. Kobayashi, E. Chin, and K. Murayama. 2006. Method of forming vapor-deposited film and method of manufacturing EL display device. United States Patent US2006/0240669 A1.

Keum, J. H., C. S. Ji, H. M. Kim, and S. T. Namgoong. 2004. Heating crucible and deposition apparatus using the same. United States Patent US8025733 B2.

Kim, H. W., S. Y. Han, H. B. Shim, J. Patrin, R. Bresnahan, C. Conroy et al. 2008. Improvement of material utilization of organic evaporation source for manufacturing large-sized AMOLED devices. *SID Symposium Digest of Technical Papers*. 39: 1450–1453.

Kim, S., K. Jeong, and D. Chi. 2010. Organic deposition methods for AM-OLED by using compact linear nozzle source (CLNS) and hybrid source. *SID Symposium Digest of Technical Papers* 41: 692–694.

Kim, S.-M., K.-H. Jeong, M.-W. Choi, and H.-W. Park. 2011. Multiple nozzle evaporator for vacuum thermal evaporation. United States Patent US7976636 B2.

Krug, T., F. Anderle, A. Feuerstein, E. Sichmann, and W. Buschbeck. 1993. Linear thermal evaporator for vacuum vapor depositing apparatus. United States Patent US5216742 A.

Kyoon, C. H. 2009. New Opportunities of AMOLED in Display World. *OLED World Summit 2009*, San Francisco, CA, Intertech PIRA.

Lamansky, S., T. R. Hoffend Jr., H. Le, V. Jones, M. B. Wolk, and W. A. Tolbert. 2005. Laser induced thermal imaging of vacuum-coated OLED materials. *Proceedings of the SPIE 5937: Organic Light-Emitting Materials and Devices IX*. 593702; doi:10.1117/12.627063.

Long, M., M. L. Boroson, D. R. Freeman, B. E. Koppe, T. W. Palone, and N. P. Redden. 2008. Cost competitive vacuum deposition technology for small molecule OLED manufacturing. *SID Symposium Digest of Technical Papers*. 39: 507–510.

Long, M., B. Koppe, N. Redden, and M. Boroson. 2009. Responsive vacuum deposition technology for cost-effective OLED manufacturing. *SID Symposium Digest of Technical Papers* 40: 943–946.

Long, M. H. 2008. Pressure gauge for organic materials. United States Patent US7322248 B1.

Long, M. H., T. W. Palone, and B. E. Koppe 2009. Controllably feeding organic material in making OLEDs. United States Patent US7625601 B2.

Matsuura, H. Y. 2012. Vapor deposition method and apparatus. United States Patent US8313806 B2.

Ohring, M. 2001. *The Materials Science of Thin Films*, 2nd edition. San Diego, CA: Academic Press.

OLED Association. 2011. OLED displays and lighting–2010/2011 state of the art.

Ouellette, J. 1997. Contamination control in vacuum. *The Industrial Physicist* 3:4.

Priddy, S. W., R. C. Bresnahan, and C. M. Conroy. 2008. Vapor deposition sources and methods. United States Patent US20080173241 A1.

Shtein, M., H. F. Gossenberger, J. B. Benziger, and S. R. Forrest. 2001. Material transport regimes and mechanisms for growth of molecular organic thin films using low-pressure organic vapor phase deposition. *Journal of Applied Physics* 89: 1470–1476.

Tang, C. W. and S. A. VanSlyke. 1987. Organic electroluminescent diodes. *Applied Physics Letters* 51: 913–915.

Chapter 14

Tyan, Y.-S. 2009. OLED architecture: implications for manufacturing. *DOE Solid State Lighting Manufacturing Workshop*. Fairfax, VA.

Veeco Instruments. 2009. *Datasheet–Valved Cracker for Arsenic*. Veeco Instruments.

Waters, D. 2009. Developments in Vacuum Deposition Equipment for OLED. *OLED World Summit 2009*, San Francisco, CA, Intertech PIRA.

Witzman, M. R., R. A. Bradley Jr., C. W. Lantman, and E. R. Cox. 2001. Linear aperture deposition apparatus and coating process. United States Patent US6202591 B1.

Yotsuya, S. 2005. Thin film formation method, thin film formation equipment, method of manufacturing organic electroluminescence device, organic electroluminescence device, and electronic apparatus. United States Patent US20050153472 A1.

15. Solution Deposition Methods and Technologies

Research to Manufacturing

Vsevolod V. Rostovtsev and Curtis R. Fincher

15.1 Introduction

"There's a new TV on the block, and its picture is so amazing, it makes plasma and L.C.D. look like cave drawings," so begins a review of Sony's XEL-1 organic light-emitting diode (OLED) TV published in *The New York Times* in 2008 (Pogue 2008). High contrast ratio, large viewing angle, and low power consumption contained in a light and thin panel make displays based on OLED technology very attractive for TV manufacturers and promise a major improvement in viewing experience for consumers. Several large display manufacturers have recently introduced OLED TVs into the market. Both LG and Samsung are currently offering the new sets for sale on a limited basis.

A typical OLED device consists of several thin layers deposited on top of each other (see Section I of this book for detailed discussions of various aspects of OLED structures). Starting from the anode, these typically include the hole-injection layer (HIL), hole-transport layer (HTL), emissive layer (EML), and electron-transport layer (ETL). Some device architectures incorporate one or several blocking layers, but their utilization is typically limited to vapor-deposited devices. All four organic layers can potentially be deposited out of solution. The thickness of individual layers is typically between 5 and 100 nm. Patterning is used to ensure that at least one layer is no wider than the resolution of the overall display. Thickness uniformity is critical, since changes in layer thickness result in poor color, high voltage, and lower device lifetime.

No matter how attractive this technology is to the consumer, for the displays made with OLED technology to be commercially successful, not only must display performance be outstanding, but cost effectiveness also needs to be competitive with other technologies (such as plasma and LCD). Currently, the cost position of OLED displays suffers in comparison to LCDs with a large-screen,OLEDs currently costing anywhere from 3 to 10 times

Chapter 15

the cost of comparable LCDs. It has been suggested that the cost competitiveness of OLED displays can be improved through the use of solution-based deposition techniques. Not only does solution deposition potentially require lower capital investment, it is also compatible with large area substrates now used in manufacturing (Gen 8 glass, e.g., is 2.5 m wide and 2.8 m tall). Furthermore, it is considerably more efficient in material utilization. Since material costs are projected to be a significant expense in the manufacture of OLEDs, material utilization will be a key performance parameter. Nevertheless, active OLED materials—emitters, hosts, and charge transport materials— are critical to the quality of OLED displays and are projected to grow into a $3 billion market (NanoMarkets 2012).

This overview will not cover drive electronics and design of backplanes that OLED devices are built on (these are very briefly discussed in Chapter 17). Instead, the chapter will describe solution-processing techniques that are used in building multi-layer device structures with organic materials and discuss their applicability for the manufacturing of large-area OLED displays. The structure and properties of the active materials need to be tailored for these processes and we will discuss some of these approaches as well.

15.2   Methods of Printing

Although many methods exist for printing of organic materials onto flat services, only a few are suitable for the creation of patterned multilayer OLED structures. The optimal printing method will need to be able to deliver an OLED material or mixtures of OLED materials as an ink at high throughput rates, with high feature definition, and be able to be printed as several layers on top of each other. These challenges are not easily overcome with traditional printing methods.

Gravure, offset, flexographic, screen, and ink-jet printing are currently the most commonly used techniques in large-scale commercial printing.

A general schematic of the gravure printing process is shown in Figure 15.1.

Gravure printing is quite fast compared to other R2R printing techniques, which makes it attractive for OLED-based lighting manufacturing where the cost per printed unit is a major factor. The main challenge with gravure-printed OLEDs remains their low emission efficiency, but recent advances show promise of improvement. A cartoon schematic of a gravure printing unit is shown in Figure 15.1. The unit consists of four

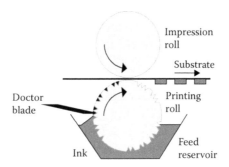

FIGURE 15.1   Schematic illustration of the gravure printing process. (Reproduced from Kopola, P. et al. *Thin Solid Films* **2009**, *517*, 5757–5762.)

basic components: a printing roll, an impression roll, a feed reservoir, and a doctor blade. The printing roll has engraved cells regularly patterned on the surface in which the density of cells, that is, line density, cell depth, and shape can be varied, allowing control of transferred amount of ink on the substrate. The printing roll picks up the reagent ink from the feed reservoir. The doctor blade removes the excess ink to ensure that only the engraved cells are filled with ink. The ink is then transferred to the surface of the substrate forming a wet coating as the substrate moves between the impression roll and the actual printing roll. The characteristics of the printing roll have an influence on the formation of printed film (Kopola et al. 2009).

Flexography is a printing technique that utilizes a flexible relief printing form, referred to as a plate. Ink is transferred to the printing form using an engraved cylinder known as an anilox roll. The surface of the anilox roll is covered with large numbers of finely engraved cells, which are filled with ink from an ink chamber, as shown in Figure 15.2.

The anilox roll is brought into contact with the flexible printing form, thus only allowing transfer of the ink from the anilox roll to those areas where the relief image makes contact. The printing form is then brought into contact with the substrate to complete the printing process. The anilox roll is the primary means of control of the quantity of ink transferred to the plate, and subsequently the print. The volume it can hold is determined by the size and density of the engraved cells. Flexographic printing does not rely on merging of discrete dots that are characteristic of other printing processes, such as screen, inkjet, and conventional gravure. The lack of a continuous line by screen printing, gravure, and inkjet, particularly for fine features, has previously been shown to adversely affect line consistency, caused by pin-holing, cell blocking, and missing dots. Flexographic printing technology offers a solution to these defects.

An example of an offset printing process, in this case, a gravure-based offset printing used for fabrication of organic solid films for applications in electronics, is shown in Figure 15.3.

This version of the offset printing technique relies on the use of a polymer rubber pad or roll to pick up the ink that fills the prepatterned printing plate (the excess ink is removed by doctor-blading before the pick-up roll contacts the printing plate). The printing plate can be made of a variety of materials, including polymers, glass, steel,

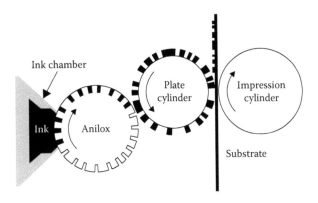

FIGURE 15.2 A schematic of flexographic printing process. (Reproduced from Deganello, D. et al. *Thin Solid Films* **2012**, *520*, 2233–2237.)

Chapter 15

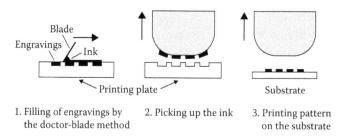

1. Filling of engravings by   2. Picking up the ink   3. Printing pattern
   the doctor-blade method   on the substrate

FIGURE 15.3 A schematic of an example of gravure-based offset printing process used in fabrication of organic films. (Reproduced from Lahti, M.; Leppävuori, S.; Lantto. V. *Applied Surface Science* **1999**, *142(1–4)*, 367–370.)

and so on, defined by the application, and its patterning is done accordingly. Among the advantages of offset printing compared to other methods such as screen printing is resolution, that is, the ability to create finer printing patterns. The high cost of the printing equipment has so far prevented its widespread use for fabrication of organic circuits (Zielke et al. 2005).

Screen printing is one of the most inexpensive and fast methods of depositing patterned organic films. Typically, film thicknesses obtained by this method exceed 0.5 microns; however, significantly thinner films fabricated by screen printing have been demonstrated (Birnstock et al. 2001, Pardo et al. 2000).

The screen printing process, a schematic of which is depicted in Figure 15.4 above was used in fabricating a single layer white OLED that turned on at 10 V and exhibited 9 cd/A efficiency (Lee et al. 2009a).

Among the solution deposition methods, inkjet printing offers the most control over the precision of printed patterns. This control can be achieved without the use of shadow masking; this makes inkjet printing one of the most cost-efficient ways of producing complex patterns. Because of the noncontact nature of this method, the contamination issues that are characteristic of other solution process methods are also minimized.

Existing inkjet printing technologies include continuous printing, thermal inkjet, and piezo-actuated drop-on-demand (DOD) printing heads. Among these, the DOD, shown in Figure 15.5, is the most applicable to the production of organic electronic devices (Tekin et al. 2008).

When voltage is applied to the piezoelectric actuator, the piezoelectric material changes shape, generating pressure pulse that propagates through the capillary and forces the droplet through the nozzle. The dynamic of droplet generation from an inkjet printing head is shown in Figure 15.6.

Among the methods described above, only ink-jet printing is capable of delivering inks to create pixels with the dimensions necessary for state-of-the-art OLED performance (Sirringhaus et al. 2006). The other techniques suffer from poor registration (flexography), high layer thickness (gravure (Lee et al. 2008), offset, and screen printing), and high ink viscosity (offset and screen printing). Not surprisingly, the displays industry has initially focused on ink-jet printing as the preferred method for the production of OLED displays. Although the detailed outcomes and conclusions of these early developmental efforts remain for the most part undisclosed, OLED displays available on the market right now are made by vapor deposition. It is likely that commercially

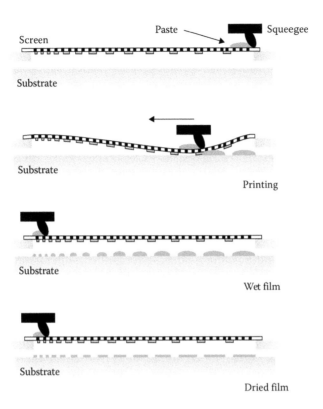

FIGURE 15.4 Schematic of screen printing process. (From Lee, D.-H. et al. *Current Applied Physics* **2009a**, *9*, 161–164.)

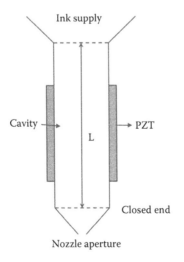

FIGURE 15.5 Piezo-actuated, drop-on-demand (DOD) inkjet print head. (Tekin, E.; Smith, P. J.; Schubert, U. S. *Soft Matter* **2008**, *4*, 703–713. Reproduced by permission of The Royal Society of Chemistry.)

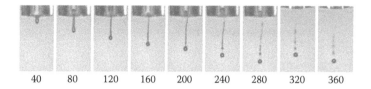

40   80   120   160   200   240   280   320   360

FIGURE 15.6   Droplet generation of a solution of polystyrene in acetophenone as a function of time. The droplet remains attached to the nozzle through a persistent filament. Detachment occurs by formation of a pinch point after 200 ms and formation of secondary satellite droplets after 240 ms. (Reprinted from de Gans, B.-J. et al. *Macromolecular Rapid Communications* **2004**, *25(1)*, 292–296. With permission.)

viable displays could not be manufactured with ink-jet technology due to low through-put times (total accumulated cycle time, or TACT) and yields (mura). It is worth noting that since large-scale commercial printing was optimized for flexible substrates, these methods might find future use in the manufacturing of printable and flexible OLED lighting panels (Kopola et al. 2009, Youn et al. 2012).

Although used widely for the preparation of small-size samples (Duan et al. 2010), spin coating is not suitable for the manufacturing of OLED devices on large backplanes. Spin coating, widely used by the electronics industry for the deposition of photoresist, entails rapid spinning of a substrate combined with the deposition of a liquid ink. The liquid is spread by centrifugal force, while the excess liquid spins off the edge of the substrate, leading to, under ideal conditions, a thin, uniform coating. The thickness of the film depends upon the viscosity of the liquid and the speed of spinning. The solvent used in the ink is often volatile, which upon spinning leads to evaporation and a stable film. Nonuniformity in the final film can arise from particles, "coffee ring" drying, and other defects. Even though many advances have been made in the development of spin-coating techniques for the assembly of multilayer devices, these do not translate well to larger substrate sizes. Nonetheless, spin coating remains a very useful laboratory technique for the initial optimization of materials.

Rapid delivery of inks can be achieved with nozzle printing technology. A nozzle printer simultaneously delivers several continuous streams of ink onto the substrate at high speed. The amount of the delivered material can be controlled through the flow rate of the ink and the scan speed of the printer. This technique is capable of producing thin organic layers suitable for OLED devices. Commercial printing systems have been produced for up to Gen 7 substrate sizes and can be extended to Gen 8 size substrates. The application of this technology to OLED manufacturing has recently been demon-strated by DuPont using a Gen 4 nozzle printer developed by Dai Nippon Screen Co. (DNS) (Flattery 2010). The performance of test displays rivals that of displays made by vacuum deposition. It has been estimated that the cost of OLED manufacturing can be lowered by as much as 60% as compared to vacuum deposition methods.

Deposition of small amounts of organic materials can also be achieved by transfer printing (Kim et al. 2008). In this method, an organic material is solution-coated onto a transfer pad, which is then "stamped" onto the substrate (the display backplane in the case of OLED displays, or the patterned or unpatterned electrode for lighting). However, the amount of the transferred material is too small and it is presently unlikely that this and similar methods would be adopted for large-scale production.

An interesting deposition technique was recently disclosed by Kateeva (Vronsky et al. 2012). In this method, a small amount of organic material is delivered to a print head as a solution, solvent is then dried off, and the remaining solids are heat transferred to the substrate. It remains to be seen whether this composite approach will be able to deliver low TACTs at high substrate sizes with the required reproducibility and uniformity.

Another interesting recently demonstrated process is an electrospray method. This approach was used to solution-deposit blends of polymers and small molecules (Hwang et al. 2012). Chae and coworkers used electrospray technique to deposit 70–160-nm-thick EMLs composed of poly(vinylcarbazole) (PVK), 2-(4-biphenyl)-5-(4-tert-butylphenyl)-1,3,4-oxadiazole (PBD), N,N'-diphenyl-N,N-(bis(3-methylphenyl)-[1,1-biphenyl]-4,4'-diamine) (TPD), and tris(2-(4-tolyl)phenylpyridine)iridium (Ir(ppy)$_3$). The performance of the resulting devices was comparable to that of the devices prepared by spin coating. It will be interesting to see whether this approach could be used to put together multilayer devices.

Multilayer structures can be put together in a semicontinuous fashion when blade-coating is combined with quick drying techniques (Chen et al. 2011, Yeh et al. 2012). Red, green, blue, and white OLED devices can be prepared using this method. Although the work in the cited references was done on small-area substrates, it could potentially be extended to larger areas and roll-to-roll coating.

In some cases, the device architecture may allow the use of common layers. Common layers are those that are shared by all pixels in the display unit and do not require changes in layer thickness from pixel to pixel. In such cases, solution-based coating techniques may be used to cover large areas of the OLED display at high rates. The feature size requirement becomes more relaxed and such printing methods as slot-dye coating can be utilized. Imagine, for example, an electron-transport material that can be used for all three pixels. In such a case, red, green, and blue pixels are all optimized to use an ETL with the same thickness. Or, alternatively, HIL and HTL with the same thickness could potentially be shared by red, green, and blue pixels.

A major outstanding issue to be resolved by any adopted printing technology is that the quality of printed OLED displays will be determined by the uniformity within an individual pixel and/or by pixel-to-pixel uniformity. The dimensions of each pixel are defined by the resolution of the display and the design of its backplane. Layers that comprise each OLED pixel need to be placed in predetermined positions without displacement or overlap between the layers. Inks need to be contained within the designated area during the drying step, remaining so located during subsequent processing steps. Both backplane design and surface modification can help with ink confinement. Differences in surface energies between regions of the backplane can be used as an effective approach for ink containment (Sirringhaus et al. 2006). The amount of ink dispensed, the solvent composition of the ink, as well as duration and temperature regime of the drying step determine the thickness uniformity of layers (Flattery et al. 2010).

The other primary challenge with printing methods is due to the solvent in the ink. During solution deposition of one organic layer on top of another, the solvent penetrates into the underlying layer and leads to swelling or even "wash out" (where the underlying layer is actually removed). Significant mixing along the boundary between the layers can also occur. In some cases, delamination may also be observed. Thus, if

Chapter 15

several consecutive layers of an OLED device are solution-deposited, swelling, mixing, and wash out would obviously lead to inferior device performance. Several approaches are used to prevent these behaviors. Among the most popular are cross-linking, the use of polymeric binders, and the use of orthogonal solvent systems. The rest of this section describes these approaches in more detail.

In the cross-linking approach, a promising polymer, small molecule, or dendrimer is modified with a cross-linking group. Both thermal (styrenes, siloxanes, trifluorovinylethers, benzocyclobutenes) and photochemical (acrylates, cinnamates, chalcones, oxetanes) cross-linkers can be used (Zuniga et al. 2011). The modified material is deposited and then treated thermally or photochemically to generate cross-link bonds. The cross-linked layer becomes insoluble in processing solvents (although there is still possibility of solvent uptake that leads to swelling) and allows solution processing of the next layer. Both the chemical composition and loading of the cross-linking groups need to be carefully optimized since unreacted functional groups may lead to poor device performance. This approach is most commonly used with polymers.

The impact of a printing step on the underlying layer can also be minimized if the materials in the existing layer are not soluble in the solvent or solvent mixture used for printing. This is called using orthogonal solvents. For example, water-soluble poly-(3,4-ethylenedioxythiophene) doped with polystyrenesulfonate (PEDOT:PSS) is commonly used as a HIL and is often deposited from aqueous solutions. The low solubility of PEDOT:PSS in organic solvents allows the use of organic solvent-based inks for the deposition of the next layer, typically a polymer- or small molecule-based HTL. The search for suitable pairs of solvents can be aided by the use of Hansen solubility parameters (Hansen 2007). Once a desired set of solvents is identified, the solubility of the OLED materials needs to be optimized for the particular set—without sacrificing the other properties such that the overall OLED device performance is maintained.

Recent neutron reflectivity work demonstrated that even nonsolvents are capable of changing the surface of the substrate on which they are being coated on and, therefore, alter the interface created between the layers deposited out of orthogonal solvents (Fujii et al. 2009). The exact implications of these results for interpretation of device performance of solution-processed OLED devices remain to be worked out.

Some polymer-based devices can be built by solution methods without the use of cross-linking groups. In such cases, polymer also plays the role of a binder so polymers with very poor charge transport properties (e.g., polystyrene) can also be used (Choi et al. 2011). Meng and coworkers prepared a series of orange OLED devices using poly-(3,4-ethylenedioxythiophene) doped with PEDOT:PSS as the HIL, poly[(9,9-dioctylfluorenyl-2,7-diyl)-*co*-(4,4′-(*N*-(4-*s*-butylphenyl))diphenylamine)] (TFB) as the HTL, 1,3,5-tris(*N*-phenylbenzimidazol-2-yl)benzene (TPBI) as the ETL, and a blend of the orange emitter, TPD, and PBD in PVK as the EMLs (Huang et al. 2010). All four layers were deposited out of the solution by blade coating, followed by spin coating and annealing (annealing was omitted for the ETL). A current efficiency as high as 20 cd/A was achieved for some devices, although with high turn-on voltages. No lifetime data was reported in the paper.

In another study, the same group prepared several blue OLED devices based on a pyrene emitter using all-solution methods and provided a comparison with evaporated devices made out of the same materials (You et al. 2009). Only one layer, the HIL, contained a polymer; all the other layers were composed of small molecules. The authors used blade coating to deposit consecutive layers and avoided spin coating to prevent significant mixing of layers. The solution-processed devices did not perform as well as evaporated ones. This was attributed to differences in the morphology of individual layers, particularly crystallization of individual materials after the annealing step. It is reasonable to project that optimization of the molecular structure of small molecules will lead to devices with improved performance.

It is clear that printing and other solution processing methods can be used to fabricate a wide range of device architectures. The wide variety of specific methods, coupled to a wide variety of materials that can be deposited, provides a great deal of flexibility for OLED fabrication. The next section addresses the principles of designing materials for solution-processed OLEDs.

15.3 Materials

A typical approach for the development of solution-processable materials starts by identifying a molecular core during studies of vacuum-deposited devices (Duan et al. 2010). The material is then modified to be soluble and stable (including chemical, photophysical, and structural stability) in an ink formulation. Although improving materials' solubility may seem like an easy task, it takes a lot of effort to maintain a balance between increasing solubility and maintaining photophysical and transport properties. For example, alkyl- and alkoxy- groups are typically used to adjust the solubility and glass transition temperature of small molecules (Duan et al. 2010, Shirota 2007). Both substituents lead to changes in the energy of ground and excited states. The presence of bulky alkyl groups might also alter the morphology of solid thin films and disrupt carrier transport pathways. Although a significant amount of effort has been devoted to the study of film morphology and solution-to-vapor comparisons, the results appear to be material- and process-dependent, having little predictive power (Kim et al. 2008, Lee et al. 2009b). More work is needed in this area.

In contrast to evaporation, the pixel performance of solution-deposited OLEDs depends upon all components of the ink composition. Thus, special attention needs to be paid not only to the solubility and purity of the OLED materials as is the case with thermally evaporated materials, but also to purity of the solvents used for printing. Any impurities carried by the solvent may be extremely difficult or impossible to remove from a printed layer after deposition and drying. However, excess ink can potentially be recycled to further improve cost-effectiveness of the solution process. This is a significant advantage for materials utilization.

As described in the previous section, the interaction between a printed layer and the solvent used for deposition of the next layer during the printing process can also be minimized if the materials in the underlying layer are not soluble in the printing solvent. For instance, using three different solvent systems allow us to create two interfaces (or three layers) from solution. As discussed above, state-of-the-art OLED devices consist of four organic layers (or three interfaces). The most common solvents used to build multilayer OLED devices are water (for water-soluble polymers such as PEDOT:PSS); aromatics such as toluene or xylene (for both HTL and EML layers) and halogenated solvents (chloroform, chlorobenzene, and dichlorobenzene). The vast majority of materials used in OLED research are poorly soluble in polar protic solvents, so additional approaches were developed to improve the solubility of OLED materials in such solvents. The addition of solubilizing groups such as ethyleneglycol ethers, dimethylamino group, phosphonate esters, and sulfonate group has been used in this fashion (Zhong 2011). Since many electron transport materials contain heterocyclic moieties, Jenekhe and coworkers were able to use aqueous formic acid as a solvent, relying on the formation of a weak salt (Earmme et al. 2010, 2012). In principle, it is possible to create two or even three interfaces with just two immiscible solvent systems, but in practice, this is much more difficult to achieve without sacrifices in device performance as many of the polar groups needed to solubilize molecules in polar or protic solvents have a detrimental effect on device performance. Therefore, a combination of the three solvents systems mentioned above (aqueous, aromatic/halogenated, and protic) with a cross-linked polymer layer would provide a way to build a solution-based four-layer OLED device. Because of the potential for significant mixing along the interface between the printed layers (Fujii et al. 2009), the best results will most likely require careful optimization and matching of components both within and between layers.

Materials with a low glass transition temperature (T_g) can undergo a phase transition during the "bake," or drying, step, which is used for solvent removal, forming crystalline domains in the device. This process becomes even more prevalent if the annealing temperature is above T_g. The glass transition temperature can be increased through careful optimization of the molecular structure: introduction of bulky substituents coupled with a decrease in planarity and symmetry of the structure typically lead to materials with a higher T_g (Shirota 2007).

One advantage of solution-processed EMLs is the ability to easily generate charge balance. To achieve the desired charge balance in vapor-deposited devices, many researchers chemically combine hole- and electron-transporting molecules into mixed-conducting or ambipolar hosts (see Chapter 10 for a detailed discussion of ambipolar hosts). This is not necessary with solution-based devices as various cohosts can be dissolved in the printing ink in any desired ratio.

The purity of OLED materials is critical for device performance. The presence of minute amounts of impurities (either reactive or photoactive) results in poor device performance, reducing efficiency and/or lifetime (Becker et al. 2010). However, the addition of some inert molecules (e.g., polystyrene) has no significant impact on device performance. Therefore, special care needs to be taken during characterization and testing of OLED materials in order to make sure that the observed performance is intrinsic to the studied materials. The next section discusses general requirements for OLED performance fabricated using solution processing.

15.4 Performance

Performance specifications for active OLED materials depend on device and backplane architectures. Accordingly, exact numbers vary from one manufacturer to another. Variations in device architecture range from large to subtle, and are common, but in each case require optimization. However, some general comments can be made:

- The color coordinates of the individual red, green and blue pixels should be deep enough to guarantee that the entire range of the sRGB color gamut of current HDTVs can be achieved. This translates to the following International Commission on Illumination color coordinates: red—(0.64, 0.33), green—(0.30, 0.60), blue—(0.15, 0.06). For lighting color requirements, see the detailed discussion in Chapter 16.
- The turn-on voltage should be as close to the blue bandgap as possible to minimize power usage. For the deep blue required for TVs and displays, this is about 2.8 V. For lighting, a lighter blue may be used to generate acceptable white light, leading to a slightly lower turn-on voltage. Nonetheless, the overall voltage of the OLED should be as close to this value as possible.
- High internal quantum efficiency is required, preferably as close to 100% as possible, in order to maximize the current efficiency. This requirement suggests the use of triplet emitters (see Chapters 6 and 11 for a detailed discussion) where possible. However, the development of an efficient, long-lived deep blue triplet emitter system remains a challenge.
- The lifetime must be acceptable for the application. For small displays that are used infrequently or with short product lifetimes, this may not be as long as for, for example, lighting or TVs. Two lifetime metrics are reported in the industry: T_{97} and T_{50}. These refer to the amount of time it takes for the luminance to fall to 97% and 50% of its initial value, respectively. The lifetime requirements for TVs are harder to define, but are probably on the order of hundreds of hours for T_{97} and in tens or even hundreds of thousands of hours for T_{50} at 1000 cd/m². Unlike color, efficiency, and voltage, lifetime numbers are rarely reported in the literature, making direct comparison between different materials and device architectures difficult. Nonetheless, the overall requirement is quite challenging, and more work is required in this area.

Together, these performance requirements translate to significant challenges for deposition processes and materials, requiring continued development to achieve significant market penetration.

15.5 Conclusions

Remarkable progress in the performance of solution-processed OLED devices as well as in methods for their preparation has been made in the past several years. The performance of solution-deposited devices no longer lags behind the performance of devices made by vapor deposition for most of the performance parameters critical for display performance. Furthermore, given the aggressive cost requirements for OLED lighting, some form of solution processing is likely to be part of the manufacturing process for OLED lighting products. Continued work on materials and processing is necessary to achieve the stringent cost and lifetime requirements for OLED lighting. Nonetheless, given the well-documented advantages of solution processing for the large-scale manufacturing of OLED displays and the lack of the performance gap with the vapor-deposited devices, the technology is ready to be applied to commercial display production.

Chapter 15

References

Becker, H.; Bach, I.; Holbach, M.; Schwaiger, J.; Spreitzer, H. Purity of OLED-materials and the implication on device performance. *SID Symposium Digest of Technical Papers.* **2010**, *41*, 39.

Birnstock, J.; Blaessing, J.; Hunze, A.; Scheffel, M.; Stoeßel, M.; Heuser, K.; Wittmann, G.; Woerle, J.; Winnacker, A. Screen-printed passive matrix displays based on light-emitting polymers. *Applied Physics Letters* **2001**, *78*, 3905–3907.

Chen, C.; Chang, H.; Chang, Y.; Chang, B.; Lin, Y.; Jian, P.; Yeh, H. et al. Continuous blade coating for multi-layer large-area organic light-emitting diode and solar cell. *Journal of Applied Physics.* **2011**, *110*, 094501.

Choi, E. Y.; Seo, J. H.; Jin, Y. Y.; Kim, H. M.; Je, J. T.; Kim, Y. K. Solution processed small molecule-based green phosphorescent organic light-emitting diodes using polymer binder. *Journal of Nanoscience and Nanotechnology.* **2011**, *11*, 10737–10739.

de Gans, B.-J.; Kazancioglu, E.; Meyer, W.; Schubert, U. S. Ink-jet printing polymers and polymer libraries using micropipettes. *Macromolecular Rapid Communications* **2004**, *25(1)*, 292–296.

Deganello, D.; Cherry J. A.; Gethin, D. T.; Claypole, T. C. Impact of metered ink volume on reel-to-reel flexographic printed conductive networks for enhanced thin film conductivity. *Thin Solid Films.* **2012**, *520*, 2233–2237.

Duan, L.; Hou, L.; Lee, T.-W.; Qiao, J.; Zhang, D.; Dong, G.; Wang, L.; Qiu, Y. Solution processable small molecules for organic light-emitting diodes. *Journal of Materials Chemistry.* **2010**, *20*, 6392.

Earmme, T.; Ahmed, E.; Jenekhe, S. A. Solution-processed highly efficient blue phosphorescent polymer light-emitting diodes enabled by a new electron transport material. *Advanced Materials.* **2010**, *22*, 4744–4748.

Earmme, T.; Jenekhe, S. A. High-performance multilayered phosphorescent OLEDs by solution-processed commercial electron-transport materials. *Journal of Materials Chemistry.* **2012**, *22*, 4660.

Flattery, D. K.; Fincher, C. R.; LeCloux, D. L.; O'Regan, M. B.; Richard, J. S. Clearing the road to mass production of OLED television. *Information Display.* **2010**, *27(10)*, 8–13.

Fujii, Y.; Atarashi, H.; Hino, M.; Nagamura, T.; Tanaka, K. Interfacial width in polymer bilayer films prepared by double-spin-coating and flotation methods. *ACS Applied Materials and Interfaces.* **2009**, *1*, 1856–1859.

Hansen, C. M. *Hansen Solubility Parameters: A User's Handbook*, 2nd edn. CRC Press, Boca Raton, FL, **2007**.

Huang, S.; Meng, H.; Huang, H.; Chao, T.; Tseng, M.; Chao, Y.; Horng, S. Uniform dispersion of triplet emitters in multi-layer solution-processed organic light-emitting diodes. *Synthetic Metals.* **2010**, *160*, 2393–2396.

Hwang, W.; Xin, G.; Cho, M.; Cho, S. M.; Chae, H.; Cho, S.; Hwang, W. et al. Electrospray deposition of polymer thin films for organic light-emitting diodes. *Nanoscale Research Letters.* **2012**, *7*, 52.

Kim, K.; Huh, S.; Seo, S.; Lee, H. H. Solution-based formation of multilayers of small molecules for organic light emitting diodes. *Applied Physics Letters.* **2008**, *92*, 093307.

Kopola, P.; Tuomikoski, M.; Suhonen, R.; Maaninen, A. Gravure printed organic light emitting diodes for lighting applications. *Thin Solid Films.* **2009**, *517*, 5757–5762.

Lahti, M.; Leppävuori, S.; Lantto. V. Gravure-offset-printing technique for the fabrication of solid films. *Applied Surface Science.* **1999**, *142(1–4)*, 367–370.

Lee, D.-H.; Choi, J. S.; Chae, H.; Chung, C.-H.; Cho, S. M. Highly efficient phosphorescent polymer OLEDs fabricated by screen printing. *Displays.* **2008**, *29*, 436–439.

Lee, D.-H.; Choi, J. S.; Chae, H.; Chung, C.-H.; Cho, S. M. Screen-printed white OLED based on polystyrene as a host polymer. *Current Applied Physics.* **2009a**, *9*, 161–164.

Lee, T.; Noh, T.; Shin, H.; Kwon, O.; Park, J.; Choi, B.; Kim, M.; Shin, D. W.; Kim, Y. Characteristics of solution-processed small-molecule organic films and light-emitting diodes compared with their vacuum-deposited counterparts. *Advanced Functional Materials.* **2009b**, *19*, 1625–1630.

NanoMarkets LLC. *OLED Materials Markets.* **2012**.

Pardo, D. A.; Jabbour, G. E.; Peyghambarian, N. Application of screen printing in the fabrication of organic light-emitting devices. *Advanced Materials* **2000**, *12*, 1249–1252.

Pogue, D. TV images to dazzle the jaded. *The New York Times.* **2008**, p. C1.

Shirota, Y.; Kageyama, H. charge carrier transporting molecular materials and their applications in devices. *Chemical Reviews.* **2007**, *107*, 953–1010.

Sirringhaus, H.; Sele, C. W.; Werte, von, T.; Ramsdale, C. In: *Organic Electronics, Materials, Manufacturing and Applications*, Klauk, H. (Ed.). Wiley-VCH, Weinheim, Germany, **2006**, p. 294.

Tekin, E.; Smith, P. J.; Schubert, U. S. Inject printing as a deposition and patterning tool for polymers and inorganic particles. *Soft Matter.* **2008**, *4*, 703–713.

Vronsky, E.; Gassend, V.; Golda, D.; Kim, H.-S. Control systems and methods for thermal-jet printing. US patent application, US20120056923 A1, application number US 13/219,515, **2012**.

Yeh, H.; Meng, H.; Lin, H.; Chao, T.; Tseng, M.; Zan, H. All-small-molecule efficient white organic light-emitting diodes by multi-layer blade coating. *Organic Electronics.* **2012**, *13*, 914–918.

You, J.; Tseng, S.; Meng, H.; Yen, F.; Lin, I.; Horng, S. All-solution-processed blue small molecular organic light-emitting diodes with multilayer device structure. *Organic Electronics.* **2009**, *10*, 1610–1614.

Youn, H.; Jeon, K.; Shin, S.; Yang, M. All-solution blade–slit coated polymer light-emitting diodes. *Organic Electronics.* **2012**, *13*, 1470–1478.

Zhong, C.; Duan, C.; Huang, F.; Wu, H.; Cao, Y. Materials and devices toward fully solution processable organic light-emitting diodes. *Chemistry of Materials.* **2011**, *23*, 326–340.

Zielke, D.; Hübler, A. C.; Hahn, U.; Brandt, N.; Bartzsch, M.; Fügmann, U.; Fischer, T. Polymer-based organic field-effect transistor using offset printed source/drain structures. *Applied Physics Letters.* **2005**, *87*, 123508.

Zuniga, C. A.; Barlow, S.; Marder, S. R. Approaches to solution-processed multilayer organic light-emitting diodes based on cross-linking. *Chemistry of Materials.* **2011**, *23*, 658–681.

Chapter 15

16. Design Considerations for OLED Lighting

Yuan-Sheng Tyan

16.1 Introduction

Most commercial organic light-emitting diodes (OLED) activity throughout history has focused on displays. Lighting is a relatively new application and focus, which requires different considerations in the device design. First of all, in lighting, color is critical. Most OLED displays use single color pixels to form full color devices and these devices have been commercialized in cells phones and other small devices, and are in the early stages of commercialization for televisions. For general lighting applications, white light is required. There is a strict definition of what constitutes white light and there are stringent requirements regarding its chromaticity, color-rendering ability, color uniformity over angle, and color consistency over lifetime, among other criteria. These color requirements have a very significant impact on OLED device design.

The second crucial consideration in device design for OLED lighting is efficacy. It is the key reason for our interest in OLED lighting, and a larger concern in lighting than in displays. OLEDs need to compete against not just incandescent lamps, but also fluorescent

Chapter 16

lamps, compact fluorescent lamps (CFLs), and LEDs. Fortunately, the nature of lighting grants us a wider repertoire of efficacy optimization methods than is available to display applications. For example, we no longer need to worry about resolution and ambient contrast, which gives us freedom to explore techniques to increase light extraction from the device and allows us to achieve significant improvement in device efficacy. This chapter will provide an overview of light extraction relevant to lighting; see Chapter 12 for a detailed discussion of light extraction.

The third key consideration in OLED lighting design is device longevity, or how long the device lasts before it reaches the end of service life. OLEDs offer a new luminaire, that is, the lighting fixture plus the light source, not a new light bulb. People replace their light bulb frequently but not their luminaires. OLEDs need to deliver a long lifetime even with devices operating at a much higher luminance than in display applications. For instance, 200 cd/m² luminance is bright enough for a display, but for lighting applications, the OLED may need to operate at 2000 cd/m² or brighter. The challenge is that the lifetime of an OLED decreases superlinearly with increasing luminance, which makes the materials and device lifetime challenges even more difficult to meet than for displays.

Finally, even with a high luminance level, OLED panels must be large enough to deliver the lumens of a typical luminaire. There are many challenges to creating large OLED panels. The cost of large equipment to create panels with good uniformity at high yield is only the first challenge. We must also address the issue of electrical current distribution. How do we supply 20 A/m² of current uniformly over a large area through a thin electrode layer having 10 ohms/sq resistance to a device that operates under 3 V? How do we keep the two electrode layers of a large-area device from contacting each other through defects and shorting out the device when these electrode layers are separated by only a fraction of a micrometer? The rest of this chapter will describe these challenges in greater detail, as well as the current status of the technology in meeting these challenges.

16.2 Color

In popular, or even the scientific, literature, the definition of "white light" has been rather liberal. Often, any light with a broad emission band that looks whitish to the eye is called a white light. For lighting applications, however, the definition is much more stringent. For example, the "ENERGY STAR® Program Requirement for Solid State Lighting Luminaires" (ENERGY STAR 2007) has very clear requirements regarding chromaticity (color coordinates), correlated color temperature (CCT), color spatial uniformity, color maintenance, and color rendering index (CRI) (Table 16.1). These requirements are necessary because color quality has a great impact on the public acceptance of lighting technologies. Linear fluorescent lamps and CFLs are several times more efficacious than incandescent lamps. Yet, mostly as a result of their poor perceived color quality, even today, more than 70 years after their introduction, fluorescent lamps are not widely used for residential and high-end commercial applications. Consequently, these stringent color requirements have a great impact on OLED lighting device design. The requirement for chromaticity alone, for example, is very difficult to satisfy because most published high-efficacy OLED devices do not meet this requirement (Tyan 2009).

According to the ENERGY STAR requirement, a luminaire must have its color coordinates fall within one of the tolerance quadrangles in the CIE (International Commission

Table 16.1   ENERGY STAR Program Color Requirements for Solid-State Lighting Luminaires

Luminaire Requirements:

CCT	The luminaire must have one of the following designated CCTs and fall within the 7-step chromaticity quadrangles as defined in the Appendix of the ENERGY STAR® Program Requirement for Solid State Lighting Luminaires (ENERGY STAR 2007)

Nominal CCT[(1)]	CCT (K)
2700 K	2725 ± 145
3000 K	3045 ± 175
3500 K	3465 ± 245
4000 K	3985 ± 275
4500 K	4503 ± 243
5000 K	5028 ± 283
5700 K	5665 ± 355
6500 K	6530 ± 510

Color spatial uniformity	The variation of chromaticity in different directions (i.e., with a change in viewing angle) shall be within 0.004 from the weighted average point on the CIE 1976 (u′, v′) diagram
Color maintenance	The change of chromaticity over the lifetime of the product shall be within 0.007 of the CIE (u′, v′) diagram
CRI	Indoor luminaires shall have a minimum CRI of 75

on Illumination) 1931 chromaticity diagram (Figure 16.1). This requirement document was developed for LED-based luminaires. Since the color specification was developed by ANSI to be as consistent as possible with existing fluorescent lamp standards, the requirement for OLED-based luminaires is expected to be similar. Almost all OLED lighting devices emit white light by combining light from two or more emitters. White light with different color temperatures can be built by selecting appropriate emitters and adjusting the proportion of emission of each emitter. The proportion has to be controlled precisely. To illustrate this and other points discussed in this chapter, we use model white OLED lighting devices constructed from the following reference emitters: SEB-095 from Merck for the blue, the $Ir(ppy)_3$ for the green, and $Ir(MDQ)_2acac$ for the orange–red. The proportion of photons generated by the three emitters to achieve the color coordinates at the center of each quadrangle is listed in Table 16.2. Note that the difference in proportions between the neighboring color temperatures is quite small. The OLED must be able to deliver these precise proportions to produce the correct colors. Some material combinations and some device structures are simply not able to deliver the correct proportions. Some other devices cannot produce the correct proportion of photons if all emitters are at their maximum efficiency. For these devices, some of the component emitters have to have their efficiency detuned in order for the emitted light to meet the chromaticity requirements. These points will be discussed in further detail later in this chapter.

Chapter 16

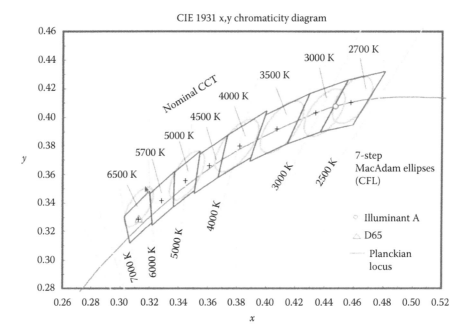

FIGURE 16.1 CIE 1931 chromaticity diagram showing the eight nominal coordinated color temperature (CCT) tolerance quadrangles.

CRI is another measure that has to be met. CRI measures how well a light source displays the color of objects relative to reference sunlight or blackbody emission. Even though some have criticized the accuracy of CRI as a measure of color quality and have proposed other measures, such as the color quality scale (CQS), to replace it (Davis and Ohno 2005), CRI is still the most widely accepted industrial index to measure color quality. For general lighting applications, a value of 80 or higher is normally considered desirable. The ENERGY STAR Requirement only requires a CRI value of 75. The lowered number is to accommodate the LED devices that, because of the spiky spectra of the emitters, have difficulty achieving high CRI numbers. OLED emitters have broader

Table 16.2 Proportion of Photons in Red, Blue, and Green Spectral Regions for White Lights of Different CCT's

CCT	CIE-x	CIE-y	% Photon		
			Blue	Green	Red
2700	0.458	0.410	10.9	21.1	68.0
3000	0.434	0.403	14.2	22.4	63.4
3500	0.407	0.392	18.5	23.1	58.4
4000	0.382	0.380	23.1	23.5	53.4
4500	0.361	0.366	27.5	22.7	49.7
5000	0.345	0.355	31.2	22.2	46.6
5700	0.329	0.342	35.4	20.8	43.8
6500	0.312	0.328	39.8	19.4	40.8

emission bands than LED emitters, and it is therefore easier for OLED devices to deliver good CRI values. The emission bands are, however, generally not broad enough to achieve high CRI numbers if only one or two emitters are used. The requirement for CRI practically limits OLED lighting device design to those employing three or more emitters. Furthermore, the CIE report clearly states that the CRI values are only meaningful if the "chromaticity difference" ($\Delta u'v'$) is less than 5×10^{-3} (CIE 1995). $\Delta u'v'$ is the length of color difference vectors in the CIE 1964 Uniform Color Space between a light source and its reference light source. This statement further enforces the need for light sources to have their color coordinates located near the Planckian Locus. The Planckian Locus is the path tracing the change of color coordinate of the emission of a black body as its temperature changes. It is not uncommon to report CRI numbers for devices that fail to meet the chromaticity requirement. According to CIE, these numbers are not very meaningful.

It is difficult enough to have an OLED lighting device that meets the stringent color requirements; it is much more challenging to have consistency of color for all products in the product line. In LED manufacturing, the devices are sorted into bins, according to their color. This procedure guarantees consistency of color in a bin, but increases production costs significantly. Some people think that since OLED emitters produce light that is characteristic of the emitter molecules, the color will be consistent (Burrows 2009); unfortunately, this is not the case. OLED lighting devices are constructed of multiple emitters; a change in the proportion of light from the individual emitters changes the emission color. Device structures that are less susceptible to color variations are therefore favored. Additionally, OLED devices are multilayer optical thin-film stacks that exhibit microcavity effects. This means that a change in layer thickness also affects the efficacy of the device and the color of the emitted light. It is difficult and expensive to ensure accurate and uniform thicknesses over large areas during manufacturing. For example, the thickness uniformity in the vacuum deposition process is often achieved by moving substrates far away from the sources. This practice sacrifices material utilization efficiency and wastes a lot of the expensive OLED materials. It is therefore very desirable to design OLED devices having inherently large thickness tolerance.

One of the desirable features of OLED lighting is the ability for users to adjust its luminance (dimming). Dimming can save energy and can also change the mood of the illuminated room. Dimming can be accomplished either by adjusting the amount of current used to drive the device or by changing the duty cycle using a pulse-width modulated (PWM) driver. However, OLED devices are constructed of multiple emitters and the proportion of photons from the individual emitters can change as the current level changes. When this happens, the emitted color will change. This happens in some device structure more than in others and efforts have been spent to reduce this color shift. There are, however, no specifications in the ENERGY STAR Requirement on allowable color shifts with dimming. In practice, the incandescent lamps change color temperature when dimmed, and this has not been objectionable to users. In fact, the reduction of color temperature seen with dimmed incandescent lamps might be viewed as desirable by some consumers. In any case, color change is not an issue if PWM drivers are used. Color maintenance through the lifetime of a device might be a more needed feature and is actually demanded by the ENERGY STAR Requirement. To meet the requirement, the individual emitters must have similar aging behavior. There

Chapter 16

are very few reports of aging behavior of white OLEDs. Levermore et al. reported a color shift $\Delta u'v' = 0.008$, just slightly larger than the ENERGY STAR Requirement of less than 0.007, after a device degraded to 64% of its original luminance [Levermore et al.]. Admovich et al. reported a $\Delta u'v' = 0.02$ and 0.006, respectively, for two types of two-unit tandem all-phosphorescent devices (Adamovich et al. 2012). This is an issue that needs to be a key consideration in device design and warrants further study.

Another ENERGY STAR Requirement is spatial uniformity of color. The specification is that the variation of chromaticity in different directions (i.e., with a change of viewing angle) shall be within 0.004 from the weighted average point on the CIE 1976 (u', v') diagram. This is actually quite difficult to achieve for a conventional white OLED device. The various materials in the OLED stack have different optical indices. Refraction takes place at all the interfaces and the optical interference among all light rays can be strong. The interference effect causes emitted light of different wavelengths to have different angular distributions. The color of a broadband white emitter will therefore have a large angular dependence. Figure 16.2a shows the calculated CIE u',v' coordinates as a function of angle for a white OLED device with a simple structure of glass/ITO(100 nm)/HTL(90 nm)/EML-R(1 nm)/EML-G(1 nm)/EML-R(1 nm)/ ETL(60 nm)/Al (Tyan, Y.-S. unpublished results 2014). All organic layers are assumed to have the same dispersive properties as α-NPD. The emission spectra of the B, G, and R emitters are taken to be those of SEB-95, Ir(ppy)$_3$, and Ir(PDQ)$_2$acac. Figure 16.2b shows the calculated $\Delta u'v'$ from the weighted average as a function of angle. This device represents perhaps the best case because the layers are thin and the structure has been optimized for angular dependence. Even here, the calculated angular dependence is too large to meet the ENERGY STAR requirement. In real devices where some of the layers are frequently thickened to reduce shorting, the angular dependence gets even worse. It also gets worse in tandem devices, which have greater total layer thicknesses. In addition, light extraction enhancement schemes utilizing periodical structures, such as photonic crystals (Lee et al. 2003) or periodic microlenses (Yamasaki et al. 2000), can induce some degree of Bragg diffraction, causing additional angular dependence.

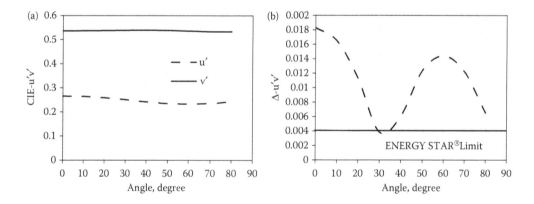

FIGURE 16.2 Calculated results of (a) dependence of CIE-u',v' coordinates on angle, and (b) Δ(u'v') from weighted average on angle for a white OLED device with a structure glass/ITO(100 nm)/HTL(90 nm)/ EML-blue(1 nm)/EML-green(1 nm)/EML-red(1 nm)/ETL(60 nm)/Al. All organic layers are assumed to have the same dispersive properties as α-NPD. The emission spectra of the blue, green, and red emitters are taken to be those of SEB-95, Ir(ppy)$_3$, and Ir(PDQ)$_2$acac.

16.3   Efficacy

Although OLED lighting may offer many interesting features, such as flexibility or transparency, the main reason for the increasing interest in this technology is the energy saving potential of OLED lighting. An OLED has a similar theoretical efficacy limit as that of the inorganic light-emitting diodes (LED). An LED is a point source, however. A LED chip is only about a millimeter in size; all the applied power is concentrated in this small area, and all the light is generated from this small area. Even with aggressive heat sinking, the operating temperature of the LED is high, which leads to significant efficacy and lifetime losses in both the LED chip and the phosphor that converts the blue light into white light. Fixtures are also needed to reduce the glare, which introduces further efficacy loses. Similar situation exists with luminaires made from other light sources as well because they are all point or line sources. An OLED, on the other hand, is a large area light source. The applied power is distributed over the large area; so temperature rise can be minimal. An OLED naturally produces a diffuse light gentle to the eye and hence it suffers none of the optical losses associated with fixtures used to hide the glare. OLED luminaires can therefore have higher operating efficacy than luminaires made from other light sources, including LEDs. This is supported by recent CALiPER studies (DOE 2012). CALiPER is a US Department of Energy program to test the real-life operating performance of commercially available luminaires. Figure 16.3 shows the Round

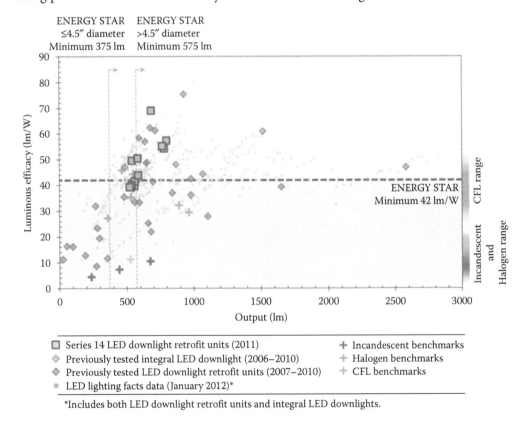

FIGURE 16.3   CALiPER Round 9 test results of commercially available ceiling lights. (DOE CALiPER Program (http://www1.eere.energy.gov/buildings/ssl/about_caliper.html).)

Chapter 16

14 test results of commercially available downlights made from incandescent lamps, CFLs, and LEDs. The report showed that the operating efficacy of incandescent luminaires was about 5–10 lm/W and that of CFLs was about 25–45 lm/W. These values were about 1/3–2/3 of the rated efficacy of the corresponding light sources: 15 lm/W and 65 lm/W, respectively. The efficacy of the LED luminaires covered an even greater range. The worst LED luminaire was as inefficient as those made from the incandescent lamps and the best was measured at about 80 lm/W. In contrast, LED packages of over 254 lm/W efficacy were reported in 2012 (Cree 2012), the year this Round 14 report was published. The wide range of luminaire performance and the big gap between LED package efficacy and LED luminaire efficacy are due to the small size of LED chips, as discussed earlier. The performance of LED-based luminaires depends greatly on the quality of heat dissipation and light distribution. In comparison, the efficacy of OLED luminaires should be similar to that of the OLED panels, and panels of over 100 lm/W efficacy have been reported (Komoda et al. 2012, Tyan et al. 2014, Yamae et al. 2013). Although these were laboratory devices, these results showed the energy-saving potential of the OLED luminaires. The recent rapid progress in the OLED performance suggests that even higher efficacy can be expected in the near future. This section examines the factors controlling the efficacy of OLED devices and their implications for OLED device design.

Figure 16.4 shows a schematic of the light-emitting process in an OLED device. To operate an OLED device, we must provide a voltage to overcome the injection barriers and the resistance of the electrode and the transport layers. A detailed discussion of this process is given in Chapter 6. The remaining net voltage must be high enough to create excitons with energy higher than that of the highest energy photons in the emission spectrum. A fraction η_c of the injected electrons and holes may then combine into excitons. Not all the excitons may emit light, however. Some of the excitons are spin-prohibited to return to the ground via the radiative process. η_s expresses the fraction of excitons that are radiative and can emit light. But even some of these radiative excitons can be lost to nonradiative processes and η_r expresses the fraction of radiative excitons that actually generates light. Furthermore, the high index of refraction of the light emitting materials leads to only a small fraction η_{extr} of the generated light can actually be emitted into the air. The rest suffers total internal reflection; becomes trapped into the surface plasma mode, the organic/TCO mode, and the glass mode; and is eventually absorbed by materials in the OLED device. The product $\eta_c\eta_s\eta_r$ gives the internal

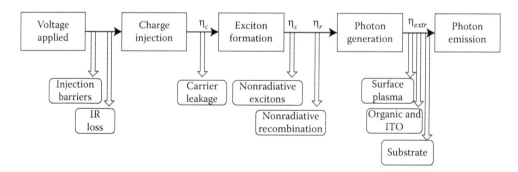

FIGURE 16.4 Schematic of the light emission process in an OLED device.

quantum efficiency (IQE). The product of IQE and η_{extr} gives the external quantum efficiency (EQE), which is actually the only parameter discussed in this paragraph that can be easily measured. The following paragraphs discuss in more detail these parameters and their implication for OLED lighting device design.

16.3.1   Internal Quantum Efficiency

The internal quantum efficiency, or IQE, measures the device's ability to convert the injected carriers to photons. It is the product of three parameters: η_c, η_s, and η_r. η_c is called the charge balance efficiency. It is a measure of the probability of injected carriers to combine with the opposite charge carriers to form excitons. When the charge is not balanced, some of the injected carriers flow though the device and are not participating in the light-generation process. Some common techniques to improve η_c are balancing the electron and hole injection (see Chapter 6) (Wang and Ma 2010); use of bipolar hosts or mixed hosts in the emitting layers (see Chapter 10) (Kondakova et al. 2008a) and the use of exciton, electron, and hole blocking layers (see Chapter 6) (Kijima et al. 1999), and so on. These techniques can be applied to both display and lighting devices.

η_s measures the fraction of excitons that can be radiative and hence light generating. For phosphorescent emitters η_s can approach 100% because both singlet and triplet excitons produce light; while for fluorescent emitters this value is generally thought to be less than 25% because only singlet photons participate in light producing. There has been increasing evidence, however, suggesting that fluorescent emitters can have much higher η_s than 25% (Kondakov 2009, Kondakov et al. 2009). The triplet–triplet annihilation (TTA) process can convert some of the nonradiative triplet excitons into radiative singlet excitons. For devices designed to take advantage of TTA, η_s can approach 40% and, for some special materials, it can be as high as 62.5% (Kawamura et al. 2010). In addition, recently it has been suggested that a properly designed fluorescent-emitting molecule can actually have an IQE equals that of phosphorescent emitters (Adachi 2013). For lighting applications, the stringent color requirement demands that the photons from the component emitters have very specific proportions (Table 16.2). This may force some of the emitters in the devices to operate at less than the maximum efficiency. The theoretical maximum η_s for some OLED lighting device structures is listed in Table 16.3 and discussed below. These values were calculated based on the three model emitters, as discussed before.

All-phosphorescent white devices have been fabricated using multiple dopants in a single emitting layer (Wu et al. 2008), multiple singly doped or doubly doped emitting layers (Komoda et al. 2011, Levermore et al. 2010, Sun and Forrest 2007, Yook and Lee 2008), a set of horizontal segments (Arnold et al. 2008), and tandem structures with multiple emitting units (Loeser et al. 2012, Qi et al. 2008, Wang et al. 2009). Not all the devices can have a theoretical maximum η_s approaching 100%, however. For example, a three-unit 2700 K white tandem device comprising blue-, green-, and red-light-emitting units (Qi et al. 2008) has a η_s maximum of only 49%. This is because the 2700 K spectrum contains only 10.9% and 21.1% blue and green photons, respectively. To get the proper color, only the red-emitting unit can be allowed to emit at the maximum efficiency. The efficiency of the blue and the green emission units have to be tuned way down in order to match the red emitter. For higher color temperature

Chapter 16

Table 16.3   Theoretical Maximum EQE for Some Exemplary Device Structures

CCT	% Photon			All Phosphorescent					Hybrid			
	Blue	Green	Red	Single Multiple-Doped Layer	Exciton Sharing Stacked Layers	Tandem RGB+RGB	Tandem R+G+B	Tandem 2+1	Triplet Harvesting	Exciton Sharing	Tandem B+RG	Tandem B+R+G
2700	10.9	21.1	68	100	100	100	49.0	73.5	84.2	75.4	56.1	49.0
3000	14.2	22.4	63.4	100	100	100	52.6	78.9	87.4	70.1	58.3	52.1
3500	18.5	23.1	58.4	100	100	100	57.1	85.6	92.0	64.3	61.3	55.6
4000	23.1	23.5	53.4	100	100	100	62.4	93.6	97.5	59.1	65.0	59.3
4500	27.5	22.7	49.7	100	100	100	67.0	99.5	90.8	54.8	69.0	48.4
5000	31.2	22.2	46.6	100	100	100	71.5	93.6	80.1	51.7	64.1	42.7
5700	35.4	20.8	43.8	100	100	100	76.1	89.0	70.6	48.5	56.5	37.7
6500	39.8	19.4	40.8	100	100	100	81.7	84.5	62.8	45.6	50.3	33.5

devices, the blue and green units can be allowed to operate more efficiently, and the η_s improves. The device can also be much more efficient if the blue and green emission layers are combined in one single light-emitting unit. In this case, the η_s reaches 73.5% for the 2700 K device. However, if the green and red layers are combined in one unit matching a blue-emitting unit, the η_s can be at most 56.1%, similar to a hybrid tandem device described later. Here, the blue-emitting unit is forced to operate as inefficiently as a fluorescent unit. On the other hand, the η_s can reach 100% if all emitting units are all three-color phosphorescent RGB devices having the proper proportion of photons themselves.

Presently, green and red phosphorescent materials have demonstrated good performance and lifetime, but the blue-emitting materials are lagging behind (Mahon 2011). Because of the difficulty in finding good phosphorescent-blue-emitting systems, many of the current best OLED lighting devices are made using the hybrid structures combining fluorescent blue emitters with phosphorescent green and red emitters (Derks 2012, Haldi 2012, Kang et al. 2011, Mikami and Goto 2012, Sun et al. 2006, Schwartz et al. 2008, Schwartz 2007, Tyan et al. 2009, 2014, Wang and Ma 2010). Fortunately, warm-white lights are often preferred for lighting applications and the blue content in a warm-white light is small (Table 16.2). This reduces the negative impact of using a less-efficient blue fluorescent emitter. Furthermore, fluorescent blue emitters usually require lower operating voltages than phosphorescent blue emitters (Weaver et al. 2012), reducing even further the impact of the lower η_s of the fluorescent blue on efficacy.

There are several approaches to constructing hybrid devices. One approach is to use green and red triplet dopants to harvest the otherwise nonradiative triplet excitons from a fluorescent blue emitting layer (Sun et al. 2006). Several versions of this approach have been presented (Kondakova et al. 2008b, Schwartz 2007, Schwartz et al. 2008, Wang and Ma 2010). The basic principle is to design the device such that the singlet excitons from the blue-emitting layer are not quenched by the triplet dopants and also that the triplet excitons can reach the triplet dopants to be harvested. Using the harvesting approach, η_s can conceivably reach 100% because all singlet and triplet excitons can generate light. However, this high η_s is actually possible for only one particular color temperature because the ratio of the generated blue photons to the green and red photons is 1:3, and this ratio is only approximately true for a particular color temperature. Of all the white emitters constructed using our model emission spectra, the 4000 K white light has the closest photon ratio (Table 16.3). For other color temperatures, either the triplet or the singlet emission needs to be tuned down to have the proper photon ratio. η_s decreases for both higher and lower color temperatures: 84.2% for the 2700 K device and 62.8% for the 6500 K device. These values are still very respectable. However, good white devices meeting the color requirements based on the harvesting scheme have not yet been reported to our knowledge. The challenge for the approach is the limited choice of materials that can make it work.

Another approach for hybrid devices is to vertically stack a fluorescent blue emissive layer with green and red phosphorescent emissive layers (Schwartz et al. 2006). An interlayer is usually inserted between the layers to prevent the quenching of singlet and triplet excitons on the other side. The device is designed such that excitons are generated

Chapter 16

in both the singlet layer and the triplet layers, and hence is referred to as an exciton-sharing device. In the blue emission layer, only the singlet excitons are effective in producing light. Thus, three triplet excitons are wasted for every blue photon generated. This approach is therefore more appropriate for lower color temperature devices requiring fewer blue photons. For a 2700 K device, η_s can reach 75.4%. η_s decreases monotonically for increasing color temperatures (Table 16.3).

The approach that has received increasing attention lately is to use the tandem structure to construct hybrid devices (Derks 2012, Haldi 2012, Kang et al. 2011, Mikami and Goto 2012, Tyan et al. 2009, 2014). In a tandem structure, two or more light emitting units are stacked vertically on top of each other. Take the two-unit tandem device as an example: each light-emitting unit has all of the organic layers and one of the two electrodes. The electrodes between the two emitting units are replaced by a charge generation structure (CGS) (Kido et al. 2003). A CGS is similar to an n+/p+ bi-layer in inorganic semiconductor devices. It is capable of injecting electrons into the electron transport layer on the one side and injecting holes into the hole transport layer on the other. A voltage applied between the two outer electrodes sends an electrical current through both light-emitting units, causing both units to emit light simultaneously. In principle, many units can be stacked (Kido et al. 2003, Liao et al. 2004). Both the luminance and the voltage are the sum total of the all the units in the tandem. There may be only a small improvement in the efficacy, but there are four other advantages of using a tandem structure. First, for the same luminance, the operating current is only a fraction of that of a single-unit device. Since the lifetime decreases superlinearly with increasing drive current, the lifetime of a tandem device is much better than the units making up the tandem. Second, the reduced current also improves current distribution and hence the uniformity of a large area device. Third, the units operate independently. Fluorescent units can be mixed with phosphorescent units without having to worry about exciton quenching. Fourth, if the tandem unit is operating at the same current density as the single unit device, the luminance is higher, which is roughly proportional to the number of units in the device. For the same total luminance output, the device area can be reduced proportionally. Since the device cost is roughly proportional to device area, the use of the tandem can save costs. Of course if we run the tandem OLED at the same current density, we will see no benefit of lifetime improvement. A choice can be made between better lifetime and lower cost.

The η_s, defined as the number of photons generated per electron injected, can be larger than 100% for tandem devices. However, because the voltage of a tandem device is also the sum total of the voltages of all the units, for comparison, a more meaningful value to use is the effective η_s. This is the η_s of the tandem device divided by the number of units in the device. For example, a tandem unit having one phosphorescent unit and one fluorescent unit can have a maximum η_s of 125% if the latter has no TTA contributions. Its effective η_s, however, is only 62.5%. The effective η_s can reach 70% if TTA raises the η_s of the fluorescent unit to 40%. For special materials, η_s can reach 81.3% if TTA raises the η_s of the fluorescent unit to 62.5% (Kondakov et al. 2009b). The color requirements again have an impact on the theoretical maximum η_s that can be expected for each type of device. The calculated values are listed in Table 16.3.

η_r measures the fraction of radiative excitons that actually produces light. In addition to nonradiative annihilation through interaction with other excitons (TTA) or fixed charges, triplet excitons can also drift into low triplet energy neighboring layers and be quenched (Mikami and Goto 2012). Techniques to improve η_r for display devices such as the addition of exciton blockers (Baldo et al. 1999) and the broadening of recombination zones using mixed hosts (Kondakova et al. 2008a) or bipolar hosts (Mi et al. 2010) also work for lighting devices. However, because lighting devices frequently have more than one emitting layer, the management of η_r is somewhat more complicated owing to the need to balance the photon ratios.

16.3.2   Light Extraction Efficiency

η_{extr} (light extraction efficiency) measures the fraction of light generated by the emitting materials that can actually emit into the air and be used for lighting or display applications. Because of the high refractive index of the organic-light-emitting materials, light generated in the OLED devices are mostly trapped as a result of total internal reflection. Roughly only about $1/n^2$, equaling less than 20%, of the generated light actually gets into the air to do useful work, where n is the refractive index of the organic materials and is generally between 1.7 and 1.8. There is a good discussion on the limited applicability of this formula in Chapter 12, wherein a detailed description of light extraction in OLEDs is provided. The light extraction issue affects both lighting and display devices. However, because there are no resolution or image contrasts concerns, there are more opportunities to improve η_{extr} for lighting devices. This topic will be discussed in more detail in Section 16.4.

16.3.3   Lumens per Ampere

The unit for the luminous output of a luminaire is lumens. This is not a purely physical quantity because it involves human eye sensitivity as well. The human eye sensitivity is expressed by the normalized photopic luminousity function $y(\lambda)$, which is highly wavelength dependent. It peaks at about 560 nm with a value of 683 lm/W and falls quickly toward shorter and longer wavelengths. The luminous output of a light source with an emission spectrum $J(\lambda)$ is given by

$$L = 683 \int y(\lambda) J(\lambda)\, d\lambda \tag{16.1}$$

The shape of the emission spectra of an OLED lighting device has, therefore, an impact on the luminous output efficacy and needs to be optimized. The efficacy E of a light source is defined in Equation 16.2:

$$E = L/VI \tag{16.2}$$

where V is the operating voltage, I is the operating current, and L is the luminous output of the device.

Chapter 16

Table 16.4   Calculated LPA Values for White Devices with Different Color Temperatures Using the Reference Spectra of SB–095, Ir(ppy)$_3$, and Ir(PDQ)$_2$acac

	Color Coordinates		LPA	%Photon in Spectrum		
CCT	CIE-x	CIE-y	lm/A	B	G	R
2700	0.457	0.411	619.0	10.9	21.1	68.0
3000	0.433	0.404	624.9	14.2	22.4	63.4
3500	0.406	0.393	626.5	18.5	23.1	58.4
4000	0.381	0.381	626.3	23.1	23.5	53.4
4500	0.360	0.367	619.4	27.5	22.7	49.7
5000	0.344	0.356	614.5	31.2	22.2	46.6
5700	0.328	0.343	603.8	35.4	20.8	43.8
6500	0.311	0.329	593.1	39.8	19.4	40.8

The emitted light $J(\lambda)$ is a flow of photons with the photon flux P, which can be calculated according to

$$P = \int \frac{J(\lambda)}{\left(\dfrac{1.24}{\lambda}\right)} \, d\lambda \tag{16.3}$$

where $(1.24/\lambda)$ gives the energy of the photon with wavelength λ. The external quantum efficiency EQE is calculated as P/I. It is useful to define a parameter LPA (Lumens per Ampere):

$$LPA = L/P = 683 \frac{\int y(\lambda) J(\lambda) \, d\lambda}{\int \dfrac{J(\lambda)}{1.24/\lambda} \, d\lambda} \tag{16.4}$$

which expresses the maximum number of lumens an emission spectrum can deliver if every electron in 1 A of driving current is converted into photons and every photon is emitted into the air. LPA depends only on the emission spectrum, which, in turn, is determined by the selection of the individual emitters. To maximize efficacy, LPA needs to be maximized by selecting the best combination of the individual emitters. The LPA for white lights of different color temperatures constructed using our reference R, G, B spectra is given in Table 16.4. For example, a 3500 K CCT hybrid white comprising the blue, green, and red spectrum in Tyan, Y.-S. (unpublished results 2014) has 627 lm/A of LPA. It is interesting to note that if the spectrum of the green emitter is red shifted by 20 nm or if that of the red emitter is blue shifted by 8 nm, the *LPA* would increase to 663 lm/A. The selection of these emitters is by no means optimized. There are many other combinations of emitters that can deliver significantly higher LPA.

16.3.4   Voltage

Voltage has a much more direct impact on lighting devices than on display devices. In display devices such as AMOLEDs, the OLED voltage is only a fraction of the total drive voltage. In lighting devices, however, the OLED voltage is the denominator of the equation for the efficacy calculation. Any changes in drive voltage reflect proportionally into changes in the efficacy of the device. What is then the theoretical limit on the lowest possible voltage? The fundamental need is to supply enough energy to the emitting photons. OLEDs, however, are broadband-emitting devices. The applied voltage needs to be high enough to emit the highest energy photons in the emission spectrum. For example, the short wavelength edge of the blue, green, and red emitters' emission spectrum in Tyan, Y.-S. (unpublished results 2014) are about 430 nm, 480 nm, and 550 nm, respectively, corresponding to photon energies of 2.88 eV, 2.58 eV, and 2.25 eV. For a single unit white device comprising emission from these three emitters, we need to supply a voltage of at least V_{min} = 2.88 volt. For a two-unit tandem having a blue-emitting unit and a G–R emitting unit, one unit will need to be driven at 2.88 V and the other at 2.58 V. This means a V_{min} of 5.46 V for the device. Similarly, the minimum voltage for the three-unit tandem device with a blue, a green, and a red unit is V_{min} = 7.71 V. Per-unit voltages for the three tandem devices are 2.88 V, 2.73 V, and 2.57 V, respectively. The three-unit tandem device has therefore the lowest effective driving voltage. This consideration comes into play when trying to design the most efficacious devices.

There are other contributors to the device voltage. There is external voltage drop in the electrode layers and connectors, which will be discussed later when we deal with the issues relating to the large area requirements of the lighting devices. Within the device, the injection and the transport of charges are major contributors to the voltage. These contributions can be minimized in well-designed devices by selecting the right materials or by doping, similar to what is normally done in OLED display devices (Wellmann et al. 2005). The blue phosphorescent emitter seems to present a more unique problem, however. It is because it needs wide-energy-gap host and charge transport materials, which can confine the energy and the excitons to the phosphorescent dopant (Ide et al. 2006). These wide-energy-gap host and charge transport materials can make it difficult to inject and transport carriers, resulting in high operating voltages. A state-of-the-art blue phosphorescent device with long lifetime required a 5.6 V driving voltage at 1000 cd/m², which is almost twice the energy for blue photons (Weaver et al. 2012). This high driving voltage needs to be reduced to fully realize the potential of all phosphorescent OLED devices (see Chapter 11 for one approach using conductivity dopants). The high voltage issue can potentially be resolved by using mixed-hosts (Tyan, Y.-S. unpublished results 2014).

16.3.5   Theoretical Limit of Efficacy

The theoretical limit on OLED lighting device efficacy can simply be expressed by Equation 16.5:

$$Efficacy_{max} = LPA/V_{min} \tag{16.5}$$

Take the 2700 K white light (Table 16.4) model device, for example, (Tyan, Y.-S. unpublished results 2014). The calculated LPA is 619 lm/A. Using the V_{min} of 2.88 V corresponding to the highest energy blue photons in the emission spectrum, the *Efficacy*$_{max}$ is calculated to be about 215 lm/W. If the 2700 white light is constructed using a two-unit tandem structure combining a blue-light-emitting unit and a green/red bilayer emitting unit, the LPA becomes 693 lm/A. The higher LPA value is achieved because the IQE of the tandem device can reach 112.2%, assuming the phosphorescent green/red unit reaches the maximum 100% IQE and the blue-light-emitting unit is tuned to 12.2% IQE to have the proper photon ratio for the 2700 K white light. The maximum efficacy calculated is then 127 lm/W, using the V_{min} of 5.46 V calculated in the last section. For 4000 K, the LPA increases to 805 lm/A and the *Efficacy*$_{max}$ to 147 lm/W. This spectrum is by no means optimized. By selecting a different set of blue, green, and red emitters, an LPA of 984.8 lm/A was calculated and the maximum efficacy increased to 184 lm/W (Tyan 2011). In addition, Tyan predicted that *Efficacy*$_{max}$ can be as high as 245 lm/W (Tyan 2014) if a three-unit tandem hybrid structure having a LPA of 1636 lm/A is used and if the voltage of the individual units can reach the low values reported for single-color devices (Novaled 2014). For the all-phosphorescent single-unit device, the calculated *Efficacy*$_{max}$ for the same emission spectrum is 266 lm/W (Tyan 2014).

The best reported efficacy to date for an all-phosphorescent device that nearly meets the Energy Star chromaticity requirement seems to be a two-unit tandem device at 114 lm/W (Yamae et al. 2013) and the best reported hybrid device is a three-unit tandem device at 112 lm/W (Tyan et al. 2014). These values are both significantly lower than their theoretical limits. The EQE of the phosphorescent device was reported to be 99%. Since the IQE of good two-unit tandem phosphorescent devices is believed to be close to 200% (Adachi et al. 2001), the light extraction efficiency is estimated to be around 50%. The effective EQE of the tandem hybrid device was 120%, which was again about 50% of the theoretical maximum IQE value of 240%, assuming the blue emitter had the TTA contribution. Light trapping loss therefore represents the largest loss from the theoretical efficacy. It also represents the largest opportunity for efficacy improvements, as discussed in the next section.

16.4   Extraction Efficiency

This section discusses the implications of light trapping on OLED lighting design. η_{extr} (light extraction efficiency) measures the fraction of light generated by the emitting materials that actually enter into the air and contribute to lighting or display applications. Because of the high refractive index of the organic light-emitting materials, light generated in the OLED devices are mostly trapped owing to total internal reflection at interfaces where light enters from a high-index material into a low index material. It has been a common view that roughly only about $1/n^2$, equaling less than 20%, of the generated light actually enters into the air to do useful work. "n" is the refractive index of the light emitting organic materials and is generally between 1.7 and 1.8. The limits of applicability of this formula are discussed in detail in Chapter 12 of this volume, wherein a more generic discussion of the light extraction issue is presented. In this chapter we focus on the impact of extraction efficiency on device design.

What happens optically in real devices is much more complicated than the simple picture derived earlier, because OLED devices are constructed of many thin-film layers of different optical indices. Interference of refracted light beams from the interfaces between the layers controls the distribution of the light into different modes. There are several published model calculations on light emission in OLEDs (Lin et al. 2006, Meerheim et al. 2010, Nowy et al. 2008, Werner et al. 2008). All models showed that light from the emitting layer is divided into several modes: the air mode is the part that actually emits into the air; the substrate mode is trapped in the substrate; the organic/ITO mode, or sometimes called wave-guided mode, is trapped in the organic layers and in the transparent conductive electrode layer; the surface plasmon-mode is coupled to the surface plasmons at the organic/cathode interface. In addition, there may be absorption and other losses that were included in some of the calculations. For our discussion, we will lump the surface plasmon and the other losses together and call it the mode-4 light. All models agree on the qualitative features of these modes but the quantitative details seem to differ, especially for the mode-4 light. All showed that spacing between the emitting layer and the reflective electrode has the biggest influence on the distribution of light. In particular, the air mode light oscillates strongly with this spacing, as the directly emitted light and the reflected light from the electrode go through the interference cycles. OLED devices are normally optimized by adjusting the layer thicknesses to give the maximum light emission into the air mode. Even with this optimization, however, the emitted light is not much higher than $1/n^2$ and it still constitutes a very small fraction of the total generated light. Enhancing the extraction efficiency to emit more light into the air presents the biggest opportunity for improving the efficiency of OLED devices.

The importance of η_{extr} enhancement has been recognized for some time (Gu et al. 1997, Horikx et al. 1999) and there have been many η_{extr} enhancing schemes reported in the literature. Before reviewing and comparing the effectiveness of these schemes, it is important to point out that "enhancement factor" that many use to report their η_{extr} enhancement results can be very misleading (Krummacher et al. 2006). In many reports, the EQE of the same device before and after the application of the extraction scheme were compared and the ratio was reported as the "enhancement factor." As we mentioned in the previous paragraph, however, the distribution of light in the various modes depends strongly on the device structure. The magnitude of the air mode light can vary over a wide range with device structure. It is easy for an enhancement scheme to show large enhancement factor if the device had little air mode light to begin with. The goal of η_{extr} enhancement, however, is to increase the EQE beyond what is possible by merely optimizing the device layer structure. The control device should therefore be a device that has already been optimized for maximum air mode light, and not merely a device with identical layer structure. Tyan et al. (2008), for example, showed that the EQE of one of their devices with the external extraction structure (EES) was almost 2.7× that of the same device without the EES. The EQE of the best EES device, however, was actually only 1.9× that of the best device without the EES. Unfortunately, this more meaningful procedure of comparing the best devices with and without the extraction enhancement has seldom been followed in the literature. Many studies are based on devices with very poor performance, which makes it even more likely that the control devices were not properly chosen. Their reported enhancement factors may therefore not be accurate indicators of the effectiveness of their enhancement schemes. This is something to keep in mind in the following discussions, and when designing devices for practical application.

The η_{extr} enhancement schemes can be divided into two categories: the external extraction scheme (EES) and the internal extraction scheme (IES). Most lighting devices are bottom-emitting devices with light emitting through a transparent substrate. EES are those applied to the surface of the transparent substrate opposite to the one with the OLED structures. EESs are easier to make because they can be applied after the OLED device is fabricated and the application process can be relatively benign to the OLED device. Since it is located far away from the organic/TCO layers, however, it cannot extract the organic/TCO mode light or the mode-4 light. The enhancement factor is therefore limited. The IESs, on the other hand, are disposed somewhere right next to the OLED layer structure. An IES can be placed between the substrate and the transparent electrode layer, between the electrode layer and the organic layer stack, or between the organic layer stack and the reflecting electrode. It can extract the organic/TCO mode light in addition to the air mode and the substrate mode; therefore its potential enhancement factor is larger than with EES. Since it contacts the OLED layers directly, however, it can damage the device mechanically by introducing shorts or chemically by interacting with the layers. IES schemes are therefore much more difficult to implement.

In addition to these two categories, there are also schemes using the microcavity effect (see Chapter 12). The enhancement due to the microcavity effect usually operates over a very narrow wavelength range, resulting in a sharp peak in the emission spectrum and a strong angle dependence. This is acceptable and may even be desirable for display applications because typically only the normal angle emission of a single color is required. However, for lighting applications where broadband emission with little angular dependence is allowed, this becomes a problem. Shiga et al. (2003) suggested the use of multiwavelength resonant cavities to maintain white emission, but this does not solve the angular dependence problem. Lim et al. (2006) applied a random microlens array to reduce the angular dependence of emission from a green emitting microcavity OLED. The EQE was improved by 80% but the angular dependence, though improved, remained strong. Tyan et al. (2005) showed that by adding a diffusing element to the outside surface of the substrate of a microcavity device with a broadband emitter, broadband emission was obtained with no angular dependence. The EQE for this device was increased by more than 30% over the device with the same emitter but without the microcavity. It was not reported, however, whether there was a net gain in efficiency when compared with a device having the diffusing element only. For lighting applications, the effectiveness of microcavity for light extraction improvement has not been demonstrated.

16.4.1　External Extraction Schemes

Many EES schemes have been reported including roughening the substrate surfaces (Zhou et al. 2011), applying brightness enhancement films (BEF) designed for LCD (Krummacher et al. 2006), using microlens arrays (Yamasaki et al. 2000), and using scattering films (Kawamura et al. 2010, Tyan et al. 2008, Tyan 2009). Because an OLED is a large-area light source with light emitting in all directions and a broadband emitter, the effectiveness of lenses or photonic crystals to direct light is dubious. All these and other EES schemes, however, are capable of reducing the total internal reflection at the substrate–air interface, which causes light trapping in the substrate. EES structures are either applied directly onto the surface of the glass substrate or coated on a flexible foil and then laminated to the outside surface of the substrate. As mentioned previously, the effectiveness of some of these schemes cannot be accurately judged because the control devices were not properly specified. A few reports did have better defined controls and the reported enhancement factors are more meaningful. Krummacher et al. (2006) reported a 27% improvement in EQE by using a BEF designed for LCD displays. Shiang et al. (2004) reported a 40% improvement in EQE using a volume scattering layer. Loebl et al. (2011) also reported a 40% improvement by attaching a scattering foil over the substrate. Nakamura et al. (2004), judging from the reported figure, achieved about 30% improvement. Tyan et al. (2008) demonstrated a 92% improvement using a scattering film coated on the surface of the substrate. The last result was higher than the others but it seemed to be consistent with the extremely high 11% EQE achieved with their fluorescent device. A good white fluorescent device without extraction enhancement typically has an EQE around 6%. The good result was probably achieved through not only an improved EES design but also a device design optimized for EES. It is to be noted that a device without EES has 100% of the air mode light emitted into the air. The application of EES actually reduces the emission of the original air-mode light because the EES structure cannot be 100% transparent to it. What EES does is to extract a fraction of both the air mode and the substrate mode light with roughly equal efficiency. For effective extraction, the device structure needs, therefore, to be designed for maximum air-plus-substrate light. The effectiveness of an EES scheme should then be judged by what fraction of the total air-plus-substrate mode light the scheme can extract.

For measuring the total air-plus-substrate mode light, Nakamura et al. (2004) used a large-area Si photodiode optically coupled to the substrate of the OLED device via an index matching oil. The presence of the index matching oil eliminates the total internal reflection at the substrate–air interface and allows both the air mode and the substrate mode light to enter the photodiode. Their measured ratio of air-plus-substrate mode to air mode varied from 2 to about 5, depending on the thickness of the electron transport layer. The authors estimated that their diffuser film was able to extract 50% of the combined light. Krummacher et al. (2006) used the same photodiode technique in their study. Using a simple green OLED device, they estimated that the maximum substrate mode light was about 1.2× of the maximum air mode light. The BEF used as an EES was able to extract about 63% of the combined light in the best case. On the other hand, the enhancement factor calculated by dividing the EQE of the EES device by the EQE of the same device without EES was as high as 2.25 because the control device in this case had very little air mode light to begin with. The more meaningful factor comparing the best device with EES with the best device without EES was only about 1.27. Sun and Forrest (2008) used a similar approach to measure the air-plus-substrate mode light and showed that, for their particular all-phosphorescent white device, the substrate mode was 1.4× of the air mode and their microlens array was able to extract 70% of the combined light. The enhancement factor was therefore about 1.68 (= 0.7 × 2.4).

Another approach to measure the combined air-plus-substrate light is to attach a large hemispherical lens to the device. If the hemispherical lens is index matched to the substrate and the light-emitting device is at the center of the hemisphere, all emitted

Chapter 16

light arrives at the glass/air interface at normal angle and no light is trapped in the substrate. By integrating the emitted light over all angles, one knows how much air-plus-substrate mode light is present in the original device. Loebl et al. (2011) used this approach and found that their EES in the form of a scattering film extracted about 70% of the air-plus-substrate light for both their two-unit tandem hybrid white device and their three-unit tandem hybrid white device. Their numbers suggested that before the application of EES the substrate-mode light was almost equal to the air-mode light. The observed enhancement factor in this work was about 1.4. Tyan (2011) took a different approach. They used an optical model to calculate the ratio between the air mode light and the air-plus-substrate light, which they could calculate more accurately than the absolute values of these modes, and then used the measured air-mode light to calibrate their calculated results. They estimated that the substrate-mode was about $1.15 \times$ of the air-mode light and their EES was extracting over 84% of the available air-plus-substrate mode light. For their single-unit fluorescent white device, the measured enhancement factor was 1.8. The measured 11.7% EQE for the fluorescent white device seemed to support such high extraction effectiveness. Shiang et al. (2004) used volumetric scattering films comprising scattering particles of ZrO2 or cool white phosphor powders dispersed in polymeric matrix as EES. They reported a 40% increase in light emission and calculated that they were extracting 70% of the substrate-mode light. In summary, it seems common for the EES scheme to extract about 70% of the combined air-plus-substrate light with the best case capable of extracting 84%.

There are other factors to consider in addition to the extraction effectiveness in selecting EES. Manufacturing cost is an obvious important factor. Another important factor for lighting applications is the requirement for color spatial uniformity. The stringent ENERGY STAR Program Requirement leaves very little room for any angular dependence. Schemes that utilize ordered structures such as Bragg reflection grating, BEF, photonic crystals, and ordered-microlens all tend to lead to angular-dependent emission. For practical applications, random features such as random microlens arrays or scattering films might be preferred or even required.

16.4.2   Internal Extraction Schemes

IES needs to be optically at close proximity to the light-emitting layer to be effective in extracting the organic/ITO mode light. Although there are other possibilities, most frequently the IES structure is placed between the glass substrate and the transparent electrode layer. In some cases, it is also applied over the top transparent electrode of a top emitting OLED device. In the former case, the first challenge is to provide a smooth surface to build the OLED structure so that the OLED is not shorted. Lee et al. (2003) constructed a photonic crystal by etching a 200 nm SiO_2 film deposited on the glass substrate to form an array of rods. A thick SiN_x film was then deposited by plasma-enhanced chemical vapor deposition (PE-CVD) over the patterned SiO2 to planarize the surface. They were successful in building the device, but a highly angular-dependent emission was observed as a result of the periodicity of the photonic crystal. Integrated over all angles, an extraction enhancement of 50% was observed. However, the surface of the photonic crystal still showed corrugation of the underlying structure, and this was said to have led to current leakage and the observed lower emission. Kim et al.

(2006) improved the structure by etching a photonic crystal structure into a SiN_x film on the glass substrate and overcoating it with a low-index spin-on-glass layer. An etch-back process was then used to smooth the surface of the spin-on glass. They reported an 85% improvement in emission efficiency in the normal direction. No angular dependent data were provided. Peng et al. (2004) used the underside of a nanoporous anodized aluminum film as the substrate for their OLED device. An aluminum film was coated over a Ti underlayer on a glass carrier. The aluminum film was anodized for form the nanoporous structure, which was then glued to a glass substrate. Etching the Ti underlayer away left the anodized aluminum film on the glass substrate, with the original underside exposed for the subsequent deposition of the OLED. The nanoporous structure acted as a light-scattering IES that increased the angular-integrated light output by about 50%. A somewhat similar idea was used by Tyan et al. (2010). They first coated a scattering layer onto a carrier glass having a smooth surface. Then a UV polymer material was introduced between the scattering layer and the substrate glass. The curing of the UV polymer layer glued the scattering layer to the substrate surface. When the substrate was then separated from the carrier glass, the scattering layer stayed on the substrate. The OLED device was then built on the original underside of the scattering layer, which had been in contact with the smooth surface of the carrier. To further planarize the substrate to ensure the integrity of the OLED device, a high-index smoothing layer was applied over the surface of the scattering layer prior to the deposition of the transparent conductive electrode. A fluorescent white device fabricated this way showed an EQE of 14.5% corresponding to an enhancement factor of 2.3× over the optimized control device (Tyan et al. 2008). A tandem hybrid white OLED device using the same IES was reported to have an EQE of 49.2%, again a 2.3× enhancement factor over the optimized control device without the IES (Tyan et al. 2009). The efficacy of the device was 56 lm/W. Later, using a similar IES but a different device structure, Tyan (2011) reported an improved EQE of 54.8% and efficacy of 66 lm/W at 1000 cd/m^2. As a reminder, the observed EQE value should really be divided by the number of units in the device for making meaningful comparisons. Since the device was a hybrid, the theoretical maximum IQE is somewhere between 125% and 140% depending on whether there was the TTA contribution. The observed EQE was therefore about 44% to 39% of the theoretical maximum IQE. Nakamura et al. (2009) applied a layer of high-index glass loaded with ceramic particles as an IES over a soda-lime glass substrate followed by a high-index smoothing layer. They observed an enhancement factor of 2.1×. Chang et al. (2011) used a nanocomposite scattering layer as IES. They managed to achieve a smooth surface by adjusting the ratio of the matrix and the scattering particles. The device showed an enhancement of 1.96×. However, a low-index matrix was used for the nanocomposite scattering layer and it could likely have caused some internal reflection loss at the ITO/scattering interface. It is difficult to judge how much of the organic/ITO-mode light was actually extracted in the last two cases because the control devices were not well defined and had low performance. Sun and Forrest (2008) applied a low index grid over the ITO electrode layer, which improved the EQE output of their all-phosphorescent white device from 14.7% to 19%, an increase of 32%. The low index grid also shifted some of the organic/ITO-mode light into additional substrate-mode light. Attaching a large photodiode to the device and using index-matching oil, they showed that the substrate-mode light indeed had increased from 21% to 28% because of the low index grid. Attaching a microlens array to

Chapter 16

the device, they were able to extract this light and increase the EQE to 34%, corresponding to about 72% of the combined air-plus-substrate mode light. This final EQE value was about 2.3× that of the control device without these features. The authors suggested that with further improvements the enhancement factor could increase to 3.4×, reaching about 50% EQE. Using a coated IES and a three-unit tandem hybrid white device structure having a fluorescent blue emitting unit and two phosphorescent units, Tyan et al. (2014) achieved an EQE of 120% and an efficacy of 112 lm/W. The overall η_{extr} was 50% assuming full TTA contribution for the fluorescent unit.

Other schemes used high-index substrates to build OLED devices. Using high-index substrates converts all the organic/ITO-mode light into substrate-mode light. One can then use EES schemes to extract all but the mode-4 light. Since high-index substrates are expensive, these schemes may not be practical for use in OLED lighting products. They do help provide insights, however, into the magnitude and the nature of the organic/ITO-mode light. Mikami and Koyanagi (2009) built a green phosphorescent Ir(ppy)$_3$ device over an $n = 2.0$ high-index glass and used a microlens array on the substrate for extraction enhancement. They reported an EQE of 56.9% and an efficacy of over 200 lm/W at low light levels of 1–100 cd/m^2. Again, no angular data were provided and this was single-colored, not a white device. Yamae et al. (2013) reported an all-phosphorescent device prepared on a high-index ($n = 1.78$) polymer substrate. One side of the high-index polymer had patterns for light scattering; the OLED structure was built on the other, smooth side. The patterned side of the high-index polymer was glued to a glass plate, and then a cap glass was glued to the glass plate to encapsulate the device to protect it from moisture. They reported a very high EQE of 42% and an efficacy 87 lm/W for a single unit white device; and an 99% EQE, 114 lm/W efficacy for a double-unit tandem device. The overall η_{extr} was about 50%.

The application of IES reduces the extraction of air mode light but extracts all other internal modes except the mode-4 light with roughly the same efficiency. To really know the effectiveness of IES schemes, we have to find out how much nonmode-4 light there is in the first place. Although modeling efforts have tried to predict this, there are significant disagreements among them on the magnitude of the mode-4 light. Therefore, these ratios must be inferred from various experimental studies. Tyan (2011), on the basis of the comparison between the experimental result discussed earlier and model calculations, deduced that the nonmode-4 light of their fluorescent white device equals about 19.5% EQE, or 78% of the theoretical maximum 25% IQE for the fluorescent device with no TTA contribution. Their IES was able to extract about 75% of the nonmode-4 light. Reineke et al. (2009) fabricated phosphorescent white devices on low- ($n = 1.51$) and high-index ($n = 1.78$) substrates. On the low index substrate without extraction enhancement, the EQE was 13.1%. With the high-index substrate and an index-matching hemisphere, the measured EQE was 34%, representing the magnitude of the total nonmode-4 light. The authors further fabricated two devices with the emitters located at slightly different locations but both near the second interference maximum (see Chapter 12 for a detailed discussion of microcavity effects). Moving the emitter farther away from the metal cathode reduced the surface plasmon losses. With the index matching hemisphere attached to the high-index substrate, the EQE increased to 46% and 44%, respectively, for two devices. These numbers represent the fraction of the nonmode-4 of the total generated light if the IQE were 100%. The white device used for this study, however, showed only

13.1% EQE on the low index substrate without extraction enhancement. In comparison, a well-tuned high efficiency phosphorescent device normally has an EQE higher than 20%. We believe, therefore, that the IQE of Reineke et al.'s device was far less than 100%, maybe no more than 13.1/20 = 65%. The nonmode-4 light would then equal to a fraction of 52% and 68% of the total generated light at the first and second maximum, respectively. With a pyramid array attached to the high-index substrate, they achieved 26% EQE for the device at the first maximum and 34% EQE at the second maximum, suggesting that the pyramid was able to extract about 26/34 = 76%, 34/44 = 77%, and 34/46 = 74%, respectively, of the nonmode-4 light. Meerheim et al. (2010) combined modeling with an experimental approach. They fabricated a series of high-efficiency phosphorescent red devices with varying electron transport layer thickness on a high-index ($n = 1.78$) substrate. Attaching a high-index hemispherical lens using index-matching oil, they measured an EQE of about 47% (judging from their reported figure) near the first interference maximum and 54% near the second maximum. The measured data seemed to fit the model calculation well although there were a couple of adjustable parameters. The model also predicted that the nonmode-4 light was higher with the standard $n = 1.5$ substrates than with the high-index substrate. It started at about 50% EQE near the first maximum and increased monotonically with increasing ETL thickness reaching almost 70% with an ETL thickness of 250 nm. They did not explain, however, why the magnitude of the nonmode-4 light depended on the index of the substrates. Komoda et al. (2012) coupled an all-phosphorescent, two-unit tandem white device to a high-index hemisphere and reported an efficacy of 142 lm/W and an EQE of 123%. The total nonmode-4 light should be at least 61.5% of the total because the IQE was most likely less than 200%. In comparison, the best devices using high-index substrates by the same group had 114 lm/W efficacy and an EQE of 99%, suggesting that they were able to extract about 80% of the available nonmode-4 light to achieve an overall η_{extr} of about 50% (Yamae et al. 2013).

The earlier-mentioned discussions suggest that the best present working schemes can extract about 80% of the nonmode-4 light. There is still a substantial fraction of light that is in the mode-4, however. Mode-4 light comprises mainly of absorption losses and surface plasmon losses. The former can be tackled by carefully selecting materials that do not absorb light in the visible wavelength range. The latter, however, presents a formidable challenge. Various attempts have been made to extract the plasmon mode: using a corrugated substrate (Gifford and Hall 2002), placing a light emitting layer on the other side of a thin Ag cathode layer (Tien and Wu 2007); using a corrugated structure over a thin Ag cathode (Wedge et al. 2004), index coupling, prism coupling, and grating coupling (Frischeisen et al. 2011) (see also Chapter 15 on the extraction of plasmon modes). In all cases the efficiency was very low and the emitted light was highly angular and wavelength dependent. Recently a report by Canzler et al. (2011) was interesting. In this approach, a material that crystallizes upon deposition was deposited right in the n-doped electron transport layer, after the rest of the layers had been deposited and before the deposition of the top cathode layer. The crystallization of this material caused the reflecting cathode on top of it to become corrugated. When applied to a single-unit fluorescent white device, the added structure increased the EQE by about 50% to a value of 12.6%. With an additional EES in the form of a microlens array, the EQE was further increased to 15.7%, representing an 83% increase over the control device. The corresponding overall η_{extr} was almost 63%, assuming the commonly regarded IQE limit of

25%, which is much higher than other reported values. This was later attributed to the extraction of mode-4 light because of the presence of the corrugated reflecting cathode (Murano et al. 2012). This might represent the first successful extraction of mode-4 light in a practical scheme. Applying the same structure and EES to a two-unit tandem device, however, they reported an EQE of 40.5%, representing a η_{extr} of only 32% even assuming no TTA was present in the fluorescent unit.

In summary, good progress has been made in enhancing the extraction efficiency. Practical solutions for white lighting devices are available and nearly all OLED lighting "products" have incorporated some form of external extraction structure. Promising results using IES have been reported, but the challenge of finding practical, low-cost IES solutions remains. At the moment, there appears to be few demonstrated successes extracting the mode-4 light.

It is important to point out that applying light extraction enhancement structure to an OLED device can do much more than enhancing the extraction efficiency:

1. Owing to interference effects in the multilayer optical structure of a typical white OLED, the emitted light always has angular-dependent properties, making it difficult to meet the ENERGY STAR Requirement. The application of an extraction scheme based on a random structure can significantly decrease the angular dependence of the emitted color.
2. The application of an extraction enhancement layer can decrease the dependence of emitted properties on layer thickness. This is because air mode light is the

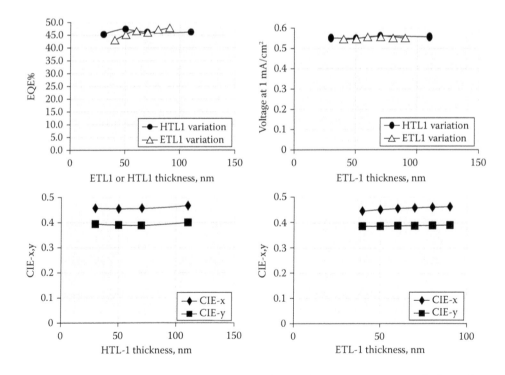

FIGURE 16.5 Dependence of EQE, voltage, and CIE color coordinates on ETL or HTL thickness for two series of OLEDs.

most sensitive to layer thickness. EES extracts both the air and substrate modes; the dependence of the properties of the emitted light on layer thickness is thereby reduced. If IES is used, only the nonmode-4 light is not extracted, the layer thickness dependence is even more reduced. The increased thickness tolerance simplifies manufacturing process, leading to higher manufacturing yields (Figure 16.5).

16.5 OLED Lifetime

In order for OLED lighting products to be positioned in the market place as a new luminaires and not as new light bulbs, OLED panels must last for many years. According to "ENERGY STAR Program Requirements for CFLs Partner Commitments" (ENERGY STAR 2003), to label a 13-year lifetime for residential use, a CFL must have 15,000 h of usable life. This is based on an average daily usage of 3 h per day. For commercial and industrial applications, the daily usage will be a lot longer, perhaps 40,000–50,000 h of usable life will be required.

The definition of lifetime is quite complicated. If an OLED is used for display applications, the lifetime is typically defined as half-life (T50): the time it takes for the display to lose half of its initial luminance. The T50 values are what frequently reported in the literature. For lighting applications, fluorescent, incandescent, and compact fluorescent lamps usually lose about 5%, 15%, and 20%, respectively, of their initial luminance before they burn out. The luminance depreciation for these traditional lamps is not significant and the lifetime is defined as the time it takes for half of the lamps to burn out. LEDs can show much more luminance depreciation before they are expected to burn out; so the lifetime needs to be defined as the time it reaches a certain levels of acceptable luminance depreciation. That level, of course, depends on the application. It has been recommended that, for general lighting applications in office environments, luminance depreciation to 70% of its initial value be used as the definition for useful life (DOE 2009b). No standard has been established for OLEDs, but it is probably a good idea to consider a similar definition for its lifetime as well. Another complication is that the luminance decay rate is a strong function of the luminance level (Wellmann et al. 2005). For most OLEDs, the lifetime seems to follow an inverse power relationship on luminance level as shown in the following equation:

$$T \cdot B^m = \text{Constant} \tag{16.6}$$

where T and B are the lifetime and the luminance level, respectively. The exponent m is between 1.5 and 1.7. This equation means that the brighter the OLED runs, the shorter the lifetime. For $m = 1.6$, the lifetime decreases by a factor of three for every doubling of luminance. Most of the reported lifetime is based on an initial luminance level of 1000 cd/m², a value appropriate for display applications. For lighting, because the cost of OLED panels is area-based, it makes economic sense to operate OLED panels at higher luminance levels, but doing so will decrease the lifetime significantly. It is a compromise that needs to be decided by the market place. As a point of reference, the average output of a 2′ × 2′ or 2′ × 4′ fluorescent troffer is roughly 5000 to 7000 lm/m² (DOE 2009a). The corresponding luminance is about 1500–2200 cd/m², assuming Lambertian light distribution. Operating the OLED at 2000 cd/m² might be a reasonable choice, at least for

Chapter 16

reference comparison purposes. For comparison, the reported lifetime with a beginning luminance of 1000 cd/m² needs to be divided by a factor of 3. For OLEDs, the luminance depreciation usually starts quickly and then slows down gradually. It has been estimated that the T70 is very roughly 1/3 of the T50 value. Again, for comparison, the reported T50 lifetime needs to be divided by another factor of three to convert it to a T70 lifetime. These factors need to be kept in mind in the following discussions.

There has been rapid progress in device lifetime in recent years. This means that OLED lifetime have been getting longer and, to measure the device lifetime, some sort of acceleration methods have to be used. The acceleration can be done either by increasing the luminance level during testing (Wellmann et al. 2005) as discussed in the previous paragraph or through increasing temperature (Aziz et al. 2002, Ishii and Taga 2002). Since the former is easier experimentally, most recent data are based on this technique. In this technique the lifetime is tested at several luminance levels and extrapolated back to the operating level. However, as Birnstock et al. (2008) pointed out, when the luminance level is increased, the temperature is increased as well. The increased degradation rate with increasing luminance is partially due to increased current flow that caused chemical damage and partially due to the increased temperature that increases the damage rate for the same current density. Equation 16.5 may need to be multiplied by another factor to describe the temperature effect that becomes more important at higher luminance levels. Without it the apparent m value gets bigger at high luminance level. Greatly exaggerated lifetime predictions can result when extrapolation is made from the high luminance data (Figure 16.6). To reduce this error, the accelerated lifetime test should be done at luminance levels as close to the operating condition as possible. It is also prudent to watch out for high m values when looking at lifetime predictions from accelerated tests.

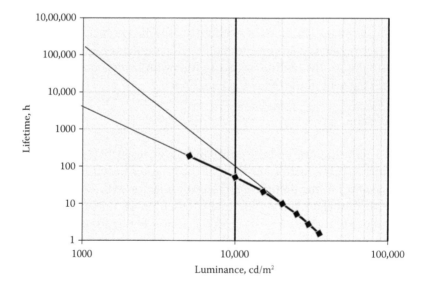

FIGURE 16.6 Predicted lifetime using accelerated test at higher luminance. As luminance value increases, the effect of heating increases, causing accelerated degradation. Greatly exaggerated lifetime predictions can result from this effect.

Komoda et al. (2011) reported a T50 lifetime of 150,000 h, starting at 1000 cd/m^2 for their 56 lm/W tandem hybrid device. Using the factor of 3 approximate relationships between T50 and T70 and between 1000 cd/m^2 and 2000 cd/m^2 initial luminance, the corresponding T70 lifetime for a 2000 cd/m^2 luminance would be about 17000 h. Haldi (2012) reported a T70 lifetime at an initial luminance of 2000 cd/m^2 of about 8000 h for their 60 lm/W tandem hybrid device. Levemore et al. (2011) reported 55,000 h T70 lifetime for an all-phosphorescent device starting at an initial luminance of 1000 cd/m^2, which translates to about 18,000 h lifetime starting at 2000 cd/m^2. These data suggest that current lifetime of OLED lighting device are long enough for residential lighting applications, but more improvements are needed for commercial and industrial applications. Where are these lifetime improvements coming from?

Materials and device improvements will lead to longer lifetime, as they have in recent years (see Chapter 13 for a detailed discussion of degradation processes). Even bigger gain, however, might come from device efficiency improvements. With improved efficacy, a lower drive current is needed for the same luminance and lifetime will improve significantly. The availability of blue phosphorescent system and the success in developing a manufacturable IES have the potential to double the OLED efficacy from their current level. Doubling the efficacy will lead to a ~3× improvement in lifetime and bring the OLED lifetime to a suitable level for commercial and industrial applications. Using a tandem structure with more emitting units will reduce the drive current and markedly improve lifetime as well, although it comes at the expense of a more complicated manufacturing process. Implementing occupancy sensing to turn off the light when there is no one around will reduce the daily usage and extend the useful life. Implementing luminance feedback to increase the drive current when device efficiency decreases will extend the device life as well, since the lifetime is defined by luminance depreciation and not by efficiency depreciation. Finally, the use of T70 to define the lifetime needs to be evaluated carefully. We only have to look at the ceiling in most buildings and offices to see that significant variations exit in the luminance and color of the installed fluorescent troffers.

Luminance depreciation is not the only mode of failure for OLED devices, however. OLEDs are notoriously prone to damage by moisture and need to be protected from the atmosphere (see Chapter 3 for a detailed discussion of encapsulation). The technique that has been used for years for fabricating OLED displays is to encapsulate the OLED between the substrate and a cap made from glass or metal. The perimeter is sealed with low water-permittivity adhesives and the inside is provided with a desiccant. This technique has been proved to be effective (Boroson 2009) and can provide OLED displays with many years of problem-free service. It is rather expensive, however. For larger panels there is also the danger of the cap touching and damaging the OLED owing to flexing. The same concern exists for another encapsulation method that uses a glass frit as the sealant (Li et al. 2003). When properly done, an almost hermetic seal can be made using the glass frit, but there are still technical challenges with this approach, especially if soda-lime glass substrates are involved. Thin-film encapsulation has been studied for years and many materials have been developed that are capable of preventing water penetration over most of the area of the device. The problem has been defect handling. Water penetrating through a single defect can damage the entire device. As the area of the device increases, the probability of having no defects becomes vanishingly small.

Chapter 16

One suggestion to deal with the defect problem is to use atomic layer deposition (Ghosh et al. 2005) to prepare highly conformal thin films as a barrier to moisture penetration. The hope is that the conformal films will wrap around pinholes and particles to prevent shorting. Another, perhaps more successful, defect-handling technique is to utilize alternating layers of organic and inorganic materials to bury the defects and to increase the path length for water to travel (Moro et al. 2004). Some active matrix OLED display products have been reported to use this technology. However, the fabrication equipment is complicated because the organic and inorganic films are made by different processes. The use of high dosage of UV during the fabrication process also tends to damage the OLED device. A low-cost, high-reliability encapsulation method for a large-area OLED lighting panel has yet to be demonstrated.

Another mode of failure is the formation of shorts during operation. Some devices have been found to develop shorting and die abruptly during usage (Blochwitz-Nimoth 2008). The cause is still unknown, but efforts to reduce shorting defects in as-made devices seem also to reduce this problem. There are not enough statistics to know whether this is a serious problem in well-designed mass-produced devices.

16.6 Challenges for Large Area OLEDs

An OLED is a diffuse large-area light source. It produces light that has no glare and is gentle to the eyes. Because the energy is applied uniformly over a large area, the temperature rise during operation is small; so an OLED does not have the heat dissipation problem plaguing inorganic LEDs. These are both attractive features. At the same time, however, it also means that large-area panels are needed to produce enough light for lighting applications. According to a US Department of Energy CALiPER report (DOE 2012), the average luminous output of 4″ and 6″ recess down lights is about 600 lm. An OLED panel operating at 2000 cd/m² luminance produces about 6000 lm/m². It therefore takes about 0.1 m² of OLED panel to produce the lumens of an average recess downlight. In order to make OLED panels that big, several issues must be addressed.

First of all, an OLED is a high-current, low-voltage device. A good OLED device operates at less than 3 V for each light-emitting unit; it is only slightly above the photon energy of the emitted light. The current efficiency for a good single-unit OLED with EES is around 50 cd/A or 150 lm/A. Thus, to produce 600 lm requires more than 4 A of current. The transparent ITO electrode layer has a sheet-resistivity of around 10 ohms/sq, so that the 4 A current produces an IR drop across the ITO that is much too large in comparison with the drive voltage. As we shall see later, this produces significant nonuniformity in the light output over the surface of the panel and causes power losses. Using the tandem architecture lowers the operating current and eases the problem somewhat, but not nearly enough to bring it to an acceptable level.

Next, the OLED structure comprises thin organic films sandwiched between two conducting electrodes. The total thickness of the organic films is only a fraction of a micrometer. Any pinholes or particle defects can cause direct contact between the two electrodes, forming shorting paths. The applied electrical current will flow through these shorting paths and bypass the OLED device. A small shorting path can reduce the light output and a larger shorting path can kill the OLED device completely. Since the distribution of random defects follow the Poisson distribution

$$P(0) = \exp(-N_d \cdot A) \tag{16.7}$$

the probability of defect-free devices P(0) falls exponentially with increasing device area A, where N_d is the defect density and A is the area of the device. If, for example, the yield of making laboratory devices having a typical area of 0.1 cm^2 is 99%, the corresponding yield of making a 0.1 m^2 OLED panel is 3.7×10^{-44}, or essentially zero. Reducing the density of shorting defects is therefore of ultimate importance to the mass production of OLED lighting panels.

Another issue is heating. Even the most efficacious OLED devices could only achieve about 50% of the theoretical maximum EQE (Tyan et al. 2014, Yamae et al. 2013) owing to extraction efficiency loss. In addition, there are Stokes losses because the operating voltage has to deliver the energy of the highest energy photons in the emission spectrum. This means that, for the longer wavelength photons there is excess energy, which is converted to heat. Even with the most efficient OLEDs, most of the applied electrical energy to drive an OLED panel is still lost as heat. This heat energy will increase the OLED panel temperature and shorten the device lifetime. It has been shown that every 10°C increase in temperature shortens the lifetime by a factor of 1.65 (Mahon 2011). The larger the device area, the more difficult it is to dissipate the heat, particularly at the center of the panel. Heat dissipation, although not nearly as much of a problem as in LEDs, still is an important factor in OLED panel design.

16.6.1 IR Drop Due to the Limited Conductivity of the ITO

A one-dimensional schematic of an OLED device is shown in Figure 16.7a and its equivalent circuit in Figure 16.7b (Tyan 2009). Most OLED devices have one transparent anode and one metallic reflecting cathode. The metallic cathode has high electrical conductivity and can be treated as having equal potential. The transparent anode layer, however, has limited conductivity. When current flows along the anode layer, a voltage gradient is developed along the current path. This means points along the current path on the anode are at different potentials and hence the OLED at these points operates at

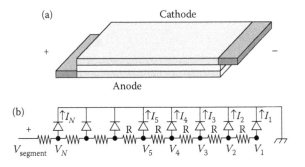

FIGURE 16.7 (a) Schematic of an OLED device and (b) equivalent circuit.

different voltages. Nonuniformity in emission therefore results. The magnitude of this nonuniformity depends on several factors:

1. Size of the OLED
2. Sheet resistivity of the transparent anode layer
3. Magnitude of current flow in the panel and, therefore, factors that affect the current flow such as luminance level, luminance efficiency (whether the device is a single-unit device or a tandem device), and so on
4. Current–voltage relationship of the device: The steeper the IV curve the more non-uniformity results from a given voltage gradient; the IV relationship is usually steeper for more efficacious devices and milder for tandem devices

The effect of these factors has been calculated (Tyan, Y.-S. unpublished results 2014) and the results are presented in Figure 16.8. Figure 16.8a shows the anode sheet resistivity required to give a 20% nonuniformity as a function of OLED panel (segment) size for both a single-unit OLED device (50 cd/A) and a two-unit tandem device (100 cd/A) operating at 2000 cd/m². The nonuniformity is defined as $(L_{max} - L_{min})/L_{max}$, where L_{max}

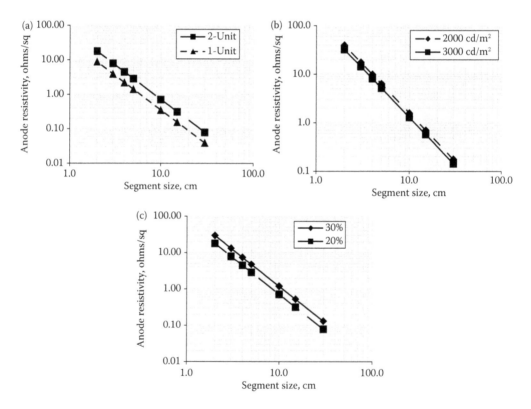

FIGURE 16.8 Calculated transparent anode sheet resistivity requirements for various segment sizes (a) two-unit tandem OLED versus single unit OLED. The OLEDs are operating at 2000 cd/m², with 20% nonuniformity; (b) effect of panel luminance on anode resistivity requirement. The OLED is a two-unit tandem device operating at 2000 cd/m² with 20% nonuniformity; (c) effect of nonuniformity on anode resistivity requirement. The OLED is a two-unit tandem device operating at 2000 cd/m². (Tyan, Y.-S. unpublished results.)

and L_{min} are the luminance level at the brightest and the darkest area of the panel. For a sheet resistivity of 10 ohms/sq, typical of a good-quality ITO, the maximum segment size that can give the required 20% nonuniformity is less than 2 cm for a single-unit device and less than 3 cm for a tandem device. The tandem device has an advantage because it requires less current for the same luminance and also because its current–voltage relationship is less steep. Figure 16.8b shows that as luminance level increases, higher anode conductivity is required, but the relationship is not very strong. Figure 16.8c shows that the anode resistivity requirement can be relaxed significantly for the same segment size if the nonuniformity requirement is relaxed. For a given type of device, the I^2R power loss was found to be independent of segment size and only depends on the allowed nonuniformity. The loss level is found to be quite small. For the two-unit tandem device operating at 2000 cd/m², the power loss is 0.8% and 1.26% for 20% and 30% nonuniformity, respectively.

There are basically two approaches to solve the nonuniformity problem for large OLED panels. One is to add auxiliary electrodes to help conduct the current along the anode. The other is to divide the OLED into small segments and connect them in series monolithically.

In the auxiliary electrode approach, a metal grid is deposited onto the transparent anode to help carry the current normally flowing in the anode layer (Amelung et al. 2008). The exact modeling is complicated and probably is best done using a numerical analysis program. A simpler way to visualize the situation is to mentally start with a uniform layer of metallic conductor and then pile it into the grid, assuming the grid has the same conductance as the uniform metal layer. Figure 16.8a shows that to construct a 30 cm × 30 cm OLED panel running at 2000 cd/m² with 20% nonuniformity, the sheet conductivity of an uniform electrode layer needs to be 0.08 ohms/sq for a two-unit tandem device and 0.04 ohms/sq for a single-unit OLED. The bulk resistivity of silver is 1.6×10^{-6} ohm-cm. It therefore takes about 2×10^{-5} cm thick of pure Ag to deliver the 0.08 ohms/sq sheet resistivity needed for the 30 cm × 30 cm OLED. Assuming this Ag film is piled up to form a grid with 95% aperture, the grid line would have about 20 times the film thickness, or 4 um. A film of this thickness probably needs to be prepared by screen-printing using Ag paste or a similar technique. The fired Ag paste, however, typically has a bulk resistivity 10–100 times higher than that of pure Ag and costs several times that of Ag. This can be rather expensive.

The second approach is to divide the OLED panel into small segments and connect them in series monolithically (Duggal et al. 2003, Tyan 2002). This approach is very similar to what is being done for constructing large-area thin-film solar cells. Both are high-current, low-voltage devices and both need to deal with the IR issue of the electrode layers. As illustrated in Figure 16.9, the transparent layer is first divided into stripes separated by narrow gaps. The organic layers are then deposited over the anode layer, also in the form of long stripes. This can be done either by masking during deposition, or by scribing mechanically or by scribing using laser after the deposition. The metal electrode layer stripes are then applied over the organic stripes. The anode, the organic, and the metal stripes are so positioned such that the metal stripes of one segment can make electrical connection to the anode stripes of the next segment through the gaps in the organic layer. This way a serial connection of the segments can be made. The panel voltage is the sum total of the voltages of all stripes; the panel current is the current of one

Chapter 16

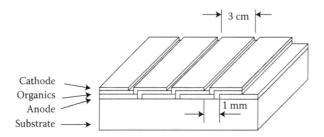

FIGURE 16.9 A monolithic serial OLED lighting device.

single stripe. The high current, low-voltage OLED device is thus converted into a high-voltage, low-current device and the IR problem is reduced. The width of the individual segment is the same as what was calculated in Figure 16.8. For a two-unit tandem device operating at 2000 cd/m² with 20% nonuniformity and 10 ohms/sq anode sheet resistivity, the maximum segment width will be about 3 cm. The 30 cm × 30 cm panel will be composed of 10 segments in series. If the individual two-unit tandem device operates at 6 volt, the panel will operate at 60 V. Assuming the 100 cd/A luminance efficiency, the operating current of the panel will be about 0.18 A. The gaps between the stripes will be dark areas. They may affect the aesthetic of the panel but they waste no power, assuming the contact resistance between the cathode and the anode through the gaps is small. The low operating current will also ease the construction of the driver, especially if the panel voltage matches that of the line voltage. The biggest drawback of this approach is the need to pattern the organic layers and the metal cathode layer. Masking and aligning during the vacuum deposition steps are not always desirable.

16.6.2   Shorting in Large Area OLED Panels

A shorting defect can kill an entire OLED lighting panel. Although there are designs to limit the damage caused by a shorting defect (Blochwitz-Nimoth 2008, Duggal et al. 2003) to a smaller local area, the end result may still not be acceptable aesthetically in a product. Good clean room practice during the manufacturing process is a must, but even that may not be enough. It is therefore desirable in the design of the OLED devices to incorporate some additional safeguards.

One approach practiced in the manufacturing of passive matrix OLEDs is to increase the thickness of the hole transport layers (HTL) to increase the production yields. HTL thickness has a smaller effect on optical tuning than electron transport layer thickness. Furthermore, doping (see Chapter 9) or using a good hole injection layer (see Chapter 5) can reduce the voltage increase from increasing thickness. Following the same principle, the use of tandem structures vertically stacking more than one light-emitting unit also increases the total thickness and reduces the probability of shorts. Since tandem structures have several other benefits, it is probably a better approach than merely increasing the HTL thickness.

Another approach is to use a hole-injection layer coated from the liquid phase (Komoda et al. 2010). Films coated from the liquid phase have the ability to planarize the surface and hide the roughness on the substrate that can cause shorts. Coating from liquid also has the possibility of turning small point defects into line defects or "comets," however. Good clean room practice is even more important in this case.

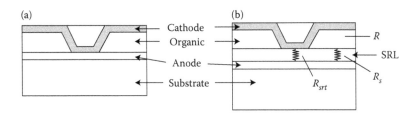

FIGURE 16.10 (a) An OLED device with a shorting defect that allows direct contact between the anode and the cathode. The applied current prefers to flow through the defect, rendering the OLED device nonoperative. (b) An OLED device with a shorting defect and a short reduction layer, SRL. The SRL adds resistance R_{srt} to the shorting defect and R_s to the OLED device. The applied will current flow through the device if $R_{srt} \gg R + R_s$ and the shorting defect no longer affects the operation of the OLED.

Yet another approach is to add a high resistivity layer between the transparent electrode and the hole injection layer (Tyan et al. 2007). Figure 16.10 illustrates how this works. Figure 16.10a shows an OLED device with a shorting defect that allows direct contact between the anode and the cathode. The applied current flows through the defect, rendering the OLED device nonoperative. Figure 16.10b shows the same OLED device with a shorting defect but with a short reduction layer (SRL), added between the bottom electrode layer and the organic layers. The SRL adds resistance:

$$R_{srt} = \rho \frac{t_{srl}}{A_s} \tag{16.8}$$

to the shorting defect and

$$R_s = \rho \frac{t_{srl}}{A} \tag{16.9}$$

to the OLED device, where ρ is the bulk resistivity of the short reduction layer, t_{srl}, A_s, and A are the thickness of the SRL layer, the area of the shorting defect, and the area of the device, respectively. The applied current will no longer flow through the shorting defect if $R_{srt} \gg R + R_s$, where R is the effective resistance of the OLED device given by $R = V/I$, where V is the normal operating voltage and I is the normal operating current of the device. The condition can easily be met because the area of the lighting device A is many orders of magnitude larger than the area of typical shorting defects A_s and the equivalent resistance R of a working device is a very small quantity. The attractiveness of this technique is that no prior knowledge of the existence or the distribution of the shorting defects is needed, and the SRL layer can be prepared in the same equipment that makes the transparent anode layer. It can also be used along with the other two approaches mentioned earlier.

16.6.3   Panel Temperature

Even though device temperature rise during operation is not as serious an issue as for LED's, panel temperature does rise during OLED operation and a raised temperature

Chapter 16

does affect lifetime. Heat dissipation becomes more of an issue as panel size increases. Levermore et al. (2011) reported a surface temperature of less than 30°C for their 49 lm/W, 15 cm × 15 cm, all-phosphorescent OLED panels operating at 3000 cd/m² luminance. In comparison an 8 cm × 8 cm all-fluorescent OLED panel operating at the same luminance had a surface temperature of about 40°C. In both cases, roughly 4 W of electrical power was required to drive the panels, but the more efficient larger panels had more surface to dissipate the heat. This example illustrated the importance of panel efficiency in reducing the operating temperature and hence in improving the lifetime of OLED panels. Even with high efficiency, it is still prudent to pay attention to panel design to facilitate heat dissipation and to avoid local heating. Improving the uniformity of emission by properly designed grid electrodes or serial connecting structure and reducing minor shorts to avoid localization of current flow are examples worth suggesting. Using more thermally conductive substrates to reduce temperature gradient on the substrate is also advisable.

16.7 Conclusion

As this chapter describes, designing OLEDs for lighting applications is quite challenging. OLED lighting must deliver acceptable color over wide angles at low cost and high efficiency, all with a lifetime measured in years. Rapid advances in materials, device structures, light extraction enhancement techniques, and designs have improved OLED lighting performance to a level that makes OLED lighting competitive against other lighting technologies. However, more work must be done to make OLED lighting successful in the marketplace.

References

Adachi, C., M.A. Baldo, M. Thompson, and S.R. Forrest. Nearly 100% internal phosphorescence efficiency in an organic light-emitting device. *J. Appl. Phys.* 90, 2001, 5048.

Adachi, C. Third generation OLED by hyperfluorescence. *SID Symposium Digest of Technical Papers.* 44(1), 2013, 513–514, doi:10.1002/j.2168-0159.2013.tb06257.x.

Adamovich, V.I., P.A. Levermore, X. Xu, A. B. Dyatkin, Z. Elshenawy, M.S. Weaver, and J.J. Brown. High-performance phosphorescent white-stacked organic light-emitting devices for solid-state lighting. *J. Photonics Energy*, 2, 2012, 021202.

Amelung, J., M. Toerker, Y. Tomita, M. Hoffmann, T. Schmitt, Ch. May, U. Todt et al. Large area OLED lighting manufacturing. Presented at *Plastic Electronics 2008*, Berlin, Germany, 2008.

Arnold, M.S., G.J. McGraw, S.R. Forrest, and R.R. Lunt. Direct vapor jet printing of three color segment organic light emitting devices for white light illumination. *Appl. Phys. Lett.* 92, 2008, 053301.

Aziz, H., Z.D. Popovic, and N.-X. Hu. Organic light emitting devices with enhanced operational stability at elevated temperatures. *Appl. Phys. Lett.* 81, 2002, 370.

Baldo, M.A., S. Lamansky, P.E. Burrows, M.E. Thompson, and S.R. Forrest. Very high-efficiency green organic light-emitting devices based on electrophosphorescence. *Appl. Phys. Lett.* 75, 1999, 4–6.

Birnstock, J., T. Canzler, M. Hofmann, A. Lux, S. Murano, P. Wellmann, and A. Werner. PIN OLEDs—Improved structures and materials to enhance device lifetime. *J. Soc. Inf. Disp.* 16, 2008, 221–229.

Blochwitz-Nimoth, J. *Plastic Electronics*, Berlin, October 2008.

Boroson, M. *DOE Manufacturing Workshop*, Vancover, MA, 2009.

Burrows, P.E., *Plastic Electronics Europe 2009*, Dresden, Germany, October 2009.

Canzler, T. W., S. Murano, D. Pavicic, O. Fadhel, C. Rothe, A. Haldi, M. Hofmann, and Q. Huang. Efficiency enhancement in white PIN OLEDs by simple internal outcoupling methods. *SID Symposium Digest of Technical Papers.* 42, 2011, 975–978.

Chang, H.-W., K.-C Tien, M.-H. Hsu, Y.-H. Huang, M.-S. Lin, C.-H. Tsai, and C.-C. Wu. Organic light-emitting devices integrated with internal scattering layers for enhancing optical out-coupling. *J. Soc. Inf. Disp.* 19, 2011, 196–204.

CIE 1995. Method of measuring and specifying colour rendering properties of light sources. CIE Publication 13.3-1995.

Cree, Inc. company news release, April 12, 2012.

Davis, W. and Y. Ohno. Toward an improved color rendering metric. *Proceedings of SPIE,* Vol. 5941, 2005, doi:10.1117/12.615388.

Derks, P. *China International OLED Summit,* Beijing, China 2012.

DOE Solid-State Lighting CALiPER Program Application Summary Report 14: LED Downlight Retrofit Units, March 2012.

DOE: Caliper Summary Report: DOE Solid State Lighting Caliper Program Summary of Results: Round 9 of Product Testing, October 2009a.

DOE Lifetime of white LEDs. Building Technologies, PNNL-SA-50957, US Department of Energy 2009b.

Duggal, A.R., D.F. Foust, W.F. Nealon, and C.M. Heller. Fault-tolerant, scalable organic light-emitting device architecture. *Appl. Phys. Lett.* 82, 2003, 2580.

ENERGY STAR Program Requirements for CFLs Partner Commitments. *Final Version.* 10/30/2003.

ENERGY STAR Requirements for SSL Luminaires, Eligibility Criteria. *Version 1.1.* 2007.

Frischeisen, J., B.J. Scholz, B.J. Arndt, T.D. Schmidt, R. Gehlhaar, C. Adachi, and W. Brutting. Strategies for light extraction from surface plasmons in organic light-emitting diodes. *J. Photonics Energy,* 1, 2011, 011004.

Ghosh, A. P., L. J. Gerenser, C. M. Jarman, and J. E. Fornalik. Thin-film encapsulation of organic light-emitting devices. *Appl. Phys. Lett.* 86, 2005, 223503.

Gifford, D.K. and D.G. Hall. Emission through one of two metal electrodes of an organic light-emitting diode via surface-plasmon cross coupling. *Appl. Phys. Lett.* 80, 2002, 3679.

Gu, G., D. Z. Garbuzov, P. E. Burrows, S. Venkatesh, S. R. Forrest, and M. E. Thompson. High-external-quantum-efficiency organic light-emitting devices. *Opt. Lett.* 22, 1997, 396–398.

Haldi, A. *China International OLED Summit,* Beijing, China 2012.

Horikx, J.J.L., C.T.H.F. Liedenbaum, M.B. Van der Mark, A.J.M. Berntsen, J.J.M. Vleggaar, and H.M.J. Boots. Electroluminescent illumination system with an active layer of a medium having light-scattering properties for flat-panel display devices. *US Patent* 5,955,837, 1999.

Ide, N., T. Komoda, and J. Kido. Organic light-emitting diode (OLED) and its application to lighting devices. *Proc. SPIE 6333, Organic Light Emitting Materials and Devices X,* 63330M, 2006, doi:10.1117/12.683215.

Ishii, M. and Y. Taga. Influence of temperature and drive current on degradation mechanisms in organic light-emitting diodes. *Appl. Phys. Lett.* 80, 2002, 3430.

Kang, M.S., M.K. Joo, J.H. Lee, Y.H. Ham, J.B. Kim, K.S. Moon, and S. Son. Performance of a large-size white OLED for lighting application. *SID Symposium Digest of Technical Papers.* 42, 2011, 972–974.

Kawamura, M., Y. Kawamura, Y. Mizuki, M. Funahashi, H. Kuma, and C. Hosokawa. Highly efficient fluorescent blue OLEDs with efficiency-enhancement layer. *SID Symposium Digest of Technical Papers.* 41, 2010, 560–563.

Kido, J., T. Matsumoto, T. Nakada, J. Endo, K. Mori, N. Kawamura, and A.Yokoi. High efficiency organic EL devices having charge generation layers. *SID Symposium Digest of Technical Papers.* 34, 2003, 964–965.

Kijima, Y., N. Asai, and S. Tamura. A blue organic light emitting diode. *Japan J. Appl. Phys.* 38, 1999, 5274.

Kim, Y.-C., S.-H. Cho, Y.-W. Song, Y.-J. Lee, Y.-H. Lee, and Y. R. Do. Planarized SiN_x/spin-on-glass photonic crystal organic light-emitting diodes. *Appl. Phys. Lett.* 89, 2006, 173502.

Komoda, T., H. Tsuji, N. Ito, T. Nishimori, and N. Ide. High-quality white OLEDs and resource saving fabrication processes for lighting application. *SID Symposium Digest of Technical Papers.* 41, 2010 993–996.

Komoda, T., H. Tsuji, K. Yamae, K. Varutt, Y. Matsuhisa, and N. Ide. High performance white OLEDs for next generation solid state lighting. *SID Symposium Digest of Technical Papers.* 42, 2011, 1056–1059.

Komoda, T., K. Yamae, V. Kittichungchit, H. Tsuji, and N. Ide. Extremely high performance white OLEDs for lighting. *SID Symposium Digest of Technical Papers.* 43, 2012, 610–613.

Kondakov, D. Role of triplet-triplet annihilation in highly efficient fluorescent devices. *J. Soc. Inf. Disp.* 17, 2009, 137–144.

Kondakov, D. Y., T. D. Pawlik, T. K. Hatwar, and J. P. Spindler. Triplet annihilation exceeding spin statistical limit in highly efficient fluorescent organic light-emitting diodes. *J. Appl. Phys.* 106, 2009, 124510.

Kondakova, M.E., T.D. Pawlik, R.H. Young, D.J. Giesen, D.Y. Kondakov, C.T. Brown, J.C. Deaton, J.R. Lenhard, and K.P. Klubek. High-efficiency, low-voltage phosphorescent organic light-emitting diode devices with mixed host. *J. Appl. Phys.* 104, 2008a, 094501.

Chapter 16

Kondakova, M.E., D.J. Giesen, J.C. Deaton, L-S. Liao, T.D. Pawlik, D.Y. Kondakov, M.E. Miller, T.L. Royster, and D.L. Comfort. Highly efficient fluorescent/phosphorescent OLED devices using triplet harvesting. *SID Symposium Digest of Technical Papers*. 39, 2008b, 219–222.

Krummacher, B. C., M. K. Mathai, V. Choong, S. A. Choulis, F. So, and A. Winnacker. General method to evaluate substrate surface modification techniques for light extraction enhancement of organic light emitting diodes. *J. Appl. Phys.* 100, 2006, 054702.

Lee, Y.-J., S.-H. Kim, J. Huh, G.-H. Kim, and Y.-H. Lee. A high-extraction-efficiency nanopatterned organic light-emitting diode. *Appl. Phys. Lett.* 82, 2003, 3779.

Levermore, P. A., V. Adamovich, K. Rajan, W. Yeager, C. Lin, S. Xia, G.S. Kottas et al. Highly efficient phosphorescent OLED lighting panels for solid state lighting. *SID Symposium Digest of Technical Papers*. 41, 2010, 786–789.

Levermore, P.A., A.B. Dyatkin, Z.M. Elshenawy, H.Pang, R.C. Kwong, R. Ma, M.S. Weaver, and J.J. Brown. Phosphorescent OLEDs: Enabling solid state lighting with lower temperature and longer lifetime. *SID Symposium Digest of Technical Papers*. 1, 2011, 1060–1063, doi:10.1889/1.3621006.

Li, C.H., C.C. Chen, J.W. Tsai, C.N. Yeh, and L.-C. Chen. Encapsulation of a display element and method of forming the same. US Patent 20030066311 A1 2003.

Liao, L. S., K. P. Klubek, and C. W. Tang. High-efficiency tandem organic light-emitting diodes. *Appl. Phys. Lett.* 84, 2004, 167.

Lim, J., S.S. Oh, D.Y. Kim, S.H. Cho, I.T. Kim, S.H. Han, H. Takezoe et al. Enhanced out-coupling factor of microcavity organic light-emitting devices with irregular microlens array. *Opt. Express* 14(14), 2006.

Lin, C.-L., T.-Y. Cho, C.-H. Chang, and C.-C. Wu. Enhancing light outcoupling of organic light-emitting devices by locating emitters around the second antinode of the reflective metal electrode. *Appl. Phys. Lett.* 88, 2006, 081114.

Loebl, P., C. Goldmann, V. Van Elsbergen, S. Grabowski, H. Boerner, and D. Bertram. Hybrid white OLEDs for general lighting. *SID Symposium Digest of Technical Papers*. 42, 2011, 979–982.

Loeser, F., T. Romainczyk, C. Rothe, D. Pavicic, A. Haldi, M. Hofmann, S. Murano, T. Canzler, and J. Birnstock., Improvement of device efficiency in PIN-OLEDs by controlling the charge carrier balance and intrinsic outcoupling methods. *J. Photonics Energy*. 2, 2012, 021207.

Mahon, J.K. *OLED World Summit*, San Francisco 2011.

Meerheim, R., M. Furno, S. Hofmann, B. Lüssem, and K. Leo. Quantification of energy loss mechanisms in organic light-emitting diodes. *Appl. Phys. Lett.* 97, 2010, 253305.

Mi, B.X., Z.Q. Gao, Z.J. Liao, W. Huang, and C.H. Chen. Molecular hosts for triplet emitters in organic light-emitting diodes and the corresponding working principle. *Sci. China Chem.* 53, 2010, 1679–1694.

Mikami, A. and T. Koyanagi. High efficiency 200-lm/W green light emitting organic devices prepared on high-index of refraction substrate. *SID Symposium Digest of Technical Papers*. 40, 2009, 907–910.

Mikami, A. and T. Goto. Optical design of enhanced light extraction efficiency in multi-stacked OLEDs coupled with high refractive-index medium and back-cavity structure. *SID Symposium Digest of Technical Papers*. 43, 2012, 683–686.

Moro, L.L., T.A. Krajewski, N.M. Rutherford, O. Philips, R. J. Visser, M.E. Gross, W.D. Bennett, and G.L. Graff. Process and design of a multilayer thin film encapsulation of passive matrix OLED displays. *Proc. SPIE 5214, Organic Light-Emitting Materials and Devices VII*. 83, 2004, doi:10.1117/12.506549.

Murano, S., D. Pavicic, M. Furno, C. Rothe, T. W. Canzler, A. Haldi, F. Löser, O. Fadhel, F. Cardinali, and O. Langguth. Outcoupling enhancement mechanism investigation on highly efficient PIN OLEDs using crystallizing evaporation processed organic outcoupling layers. *SID Symposium Digest of Technical Papers*. 2012, 687–690.

Nakamura, T., N. Tsutsumi, N, Juni, and H. Fujii. Improvement of coupling-out efficiency in organic electroluminescent devices by addition of a diffusive layer. *J. Appl. Phys.* 96, 2004, 6016.

Nakamura, N., N. Fukumoto, F. Sinapi, N. Wada, Y. Aoki, and K. Maeda., Glass substrates for OLED lighting with high out-coupling efficiency. *SID Symposium Digest of Technical Papers*. 40, 2009, 603–606.

Novaled White paper downloaded from Novaled.com. Low operating Voltages/Molecular p and n doping. Accessed March 14, 2014.

Nowy, S., B.C. Krummacher, J. Frischeisen, N.A. Reinke, and W. Brütting, Light extraction and optical loss mechanisms in organic light-emitting diodes: Influence of the emitter quantum efficiency. *J. Appl. Phys.* 104, 2008, 123109.

Peng, H. J., Y. L. Ho, X. J. Yu, and H. S. Kwok. Enhanced coupling of light from organic light emitting diodes using nanoporous films. *J. Appl. Phys.* 96, 2004, 1649.

Qi, X., M. Slootsky, and S. Forrest. Stacked white organic light emitting devices consisting of separate red, green, and blue elements. *Appl. Phys. Lett.* 93, 2008, 193306.

Reineke, S., F. Lindner, G. Schwartz, N. Seidler, K. Walzer, B. Lussem, and K. Leo. White organic light-emitting diodes with fluorescent tube efficiency. *Nature.* 459, 2009, 234.

Schwartz, G., K. Fehse, M. Pfeiffer, K. Walzer, and K. Leo. Highly efficient white organic light emitting diodes comprising an interlayer to separate fluorescent and phosphorescent regions. *Appl. Phys. Lett.* 89, 2006, 083509.

Schwartz, G. Ph D Thesis, Technischen Universitat Dresden, 2007.

Schwartz, G., S. Reineke, K. Walzer, and K. Leo. Reduced efficiency roll-off in high-efficiency hybrid white organic light-emitting diodes. *Appl. Phys. Lett.* 92, 2008, 053311.

Shiang, J.J., T.J. Faircloth, and A.R. Duggal. Experimental demonstration of increased organic light emitting device output via volumetric light scattering. *J. Appl. Phys.* 95, 2004, 2889.

Shiga, T., H. Fujikawa, and Y. Taga. Design of multiwavelength resonant cavities for white organic light-emitting diodes. *J. Appl. Phys.* 93, 2003, 19.

Sun, Y., N.C. Giebink, H. Kanno, B. Ma, M.E. Thompson, and S.R. Forrest. Management of singlet and triplet excitons for efficient white organic light-emitting devices. *Nature.* 440, 2006, 908–12.

Sun, Y. and S. R. Forrest. High-efficiency white organic light emitting devices with three separate phosphorescent emission layers. *Appl. Phys. Lett.* 91, 2007, 263503.

Sun, Y. and S. R. Forrest. Enhanced light out-coupling of organic light-emitting devices using embedded low-index grids. *Nature Photonics* 2, 2008, 483–487.

Tien, K.-C. and C.-C. Wu. Recycling surface plasmon polaritons of OLED for tunable double emission and efficiency enhancement. *SID Symposium Digest of Technical Papers.* 38, 2007, 806–809.

Tyan, Y.-S. OLED apparatus including a series of OLED devices US Patent No. 6693296. 2002.

Tyan, Y.-S. *OLED Lighting Designers Summit*, Boston, MA, 2009.

Tyan, Y.-S. Organic light-emitting-diode lighting overview. *J. Photon. Energy* 1, 2011, 011009. doi:10.1117/1.3529412.

Tyan, Y.-S., OLED Lighting @ First O-Lite, OLED World Summit 2014, Shanghai, China.

Tyan, Y.-S., J. D. Shore, G. Farruggia, and T. Cushman. Broadband-Emitting Microcavity-OLED Device. *SID Symposium Digest of Technical Papers.* 36, 2005, 142–145.

Tyan, Y.-S., J. Farruggia, and T. R. Cushman. Reduction of shorting defects in OLED devices. *SID Symposium Digest of Technical Papers.* 38, 2007, 845–848.

Tyan, Y.-S., Y.Q. Rao, J.-S. Wang, R. Kesel, T. R. Cushman, and W.J. Begley. Fluorescent white OLED devices with improved light extraction. *SID Symposium Digest of Technical Papers.* 39, 2008, 933–936.

Tyan, Y.-S., Y.Q. Rao, X.F. Ren, R. Kesel, T.R. Cushman, W.J. Begley, and N. Bhandari. Tandem hybrid white OLED devices with improved light extraction. *SID Symposium Digest of Technical Papers.* 40, 2009, 895–898.

Tyan, Y.-S., J.-S. Wang, R. Kesel, J. Farruggia, and T. Cushman. Electroluminescent device having improved light output. US Patent #7,851,995, 2010.

Tyan, Y.-S., Y.-X. Shen, J.-J. Peng, L. Zhao, C. Feng, Y.-W. Sui, H. Lu et al. Efficient tandem hybrid white OLEDs for solid state lighting applications. *SID Symposium Digest of Technical Papers.* 45(1), 2014, 675–678, doi:10.1002/j.2168-0159.2014.tb00177.x.

Wang, Q., J. Ding, Z. Zhang, D. Ma, Y. Cheng, L. Wang, and F. Wang. A high-performance tandem white organic light-emitting diode combining highly effective white-units and their interconnection layer. *J. Appl. Phys.* 105, 2009, 076101.

Wang, Q. and D. Ma. Management of charges and excitons for high-performance white organic light-emitting diodes. *Chem. Soc. Rev.* 39, 2010, 2387–2398.

Weaver, M.S., P. Levermore, V. Adamovich, R. Ma, M. Hack, J.J. Brown, and M. Lu. *DOE Solid-State Lighting R&D Workshop*, Atlanta, GA, 2012.

Wedge, S., J. A. E. Wasey, W. L. Barnes, and I. Sage, Coupled surface plasmon-polariton mediated photoluminescence from a top-emitting organic light-emitting structure. *Appl. Phys. Lett.* 85, 2004, 182.

Wellmann, P., M. Hofmann, O. Zeika, A. Werner, J. Birnstock, R. Meerheim, G. He, K. Walzer, M. Pfeiffer, and K. Leo. High-efficiency p-i-n organic light-emitting diodes with long lifetime. *J. Soc. Inf. Disp.* 13, 2005, 393–397.

Werner, A., C. Rothe, U. Denker, D. Pavicic, M. Hofmann, S. Mladenovski, and K. Neyts. The light distribution in OLEDs and ways to increase the light outcoupling efficiency. *SID Symposium Digest of Technical Papers.* 39, 2008, 522–525.

Chapter 16

Wu, H.B., J.H. Zou, F. Liu, L. Wang, A. Mikhailovsky, G.C. Bazan, W. Yang, and Y. Cao. Efficient single active layer electrophosphorescent white polymer light-emitting diodes. *Adv. Mater.* 20, 2008, 696.

Yamae, K., H. Tsuji, V. Kittichungchit, N. Ide, and T. Komoda. Highly efficient white OLEDs with over 100 lm/W for general lighting. *SID Symposium Digest of Technical Papers.* 2013, 916–919.

Yamasaki, T., K. Sumioka, and T. Tsutsui. Organic light-emitting device with an ordered monolayer of silica microspheres as a scattering medium. *Appl. Phys. Lett.* 76, 2000, 1243.

Yook, K. S. and J.Y. Lee. High efficiency and low efficiency roll off in white phosphorescent organic light-emitting diodes by managing host structures. *Appl. Phys. Lett.* 92, 2008, 193308.

Zhou, J., N. Ai, L. Wang, H. Zheng, C. Luo, Z. Jiang, S. Yu, Y. Cao, and J. Wang. Roughening the white OLED substrate's surface through sandblasting to improve the external quantum efficiency. *Org. Electron.* 12, 2011, 648–653.

17. Materials, Processing, and Applications

Daniel J. Gaspar

17.1 Introduction

The first sixteen chapters of this book have described the important materials used in an organic light-emitting diode (OLED), as well as key materials processing technologies used for device fabrication, and principles of device operation and design. This chapter summarizes the outlook for materials, devices, and processing, along with integration challenges posed as well as OLED advantages offered by different applications and technology choices. The pioneering work of numerous researchers in industry, academia, and national laboratories has led to dramatic improvements in the fundamental understanding of the materials, physics, and chemistry of OLEDs. These include improvements in materials lifetime, materials cost, OLED system efficiency, and manufacturing cost. OLEDs also enjoy strong prospects for fantastic and unique product features such as transparency, flexible devices, and exotic form factors. This chapter briefly discusses some of these aspects and ties together the technical factors focused on system design and integration in the context of applications.

OLEDs have the potential to change the way we view information and light our world, but will only do so if companies see a financial benefit in manufacturing OLEDs. OLEDs

Chapter 17

have had significant commercial success in small displays, and now appear to be poised for bigger (literally) success in televisions and lighting. Applications including mobile and handheld device displays (i.e., cellphones, smart watches, tablets, cameras, and other devices), televisions, and automobiles, as well as OLED lighting, including general indoor lighting, outdoor lighting, specialty applications, and automotive lighting, are all current or near-future products. While OLEDs are poised to enter the general lighting market, it has taken much longer than many predictions over the past 10 years due to a combination of factors related to cost and performance. In order to succeed in the lighting marketplace, OLEDs will have to continue to decrease cost, while taking advantage of unique attributes such as thin form factor, lightweight, transparency, flexibility, color tunability, color quality, switching speed, and power consumption, although different attributes will be important for different applications. For the near future, it appears that OLEDs will continue to be relatively expensive, with OLED cost higher than incumbent high-efficiency/performance technologies, particularly inorganic light-emitting diodes (LEDs) and fluorescent lighting for lighting applications, and liquid crystal displays (LCDs), including those with LED backlights for large displays. As OLED performance increases and costs decline, increasing use of OLEDs in applications in televisions, automotive lighting, other outdoor lighting, and general and specialty indoor lighting will become widespread.

17.2 Materials and Components

As has been emphasized throughout this book, the choice of a material for a given purpose within an OLED depends upon the rest of the system. For a detailed discussion of materials choices for lighting design, see Chapter 16. It is impossible to specify materials for a "best" OLED, but generally, the required OLED properties help narrow choices for materials and device design. The question of what materials will be incorporated into commercial products is an important one; one particular market analysis indicates that the global OLED materials market will grow to $3.7B by 2019 (Nanomarkets 2014c). Note that the materials choices and device design discussed in the next section completely neglects many of the manufacturing aspects (covered in Chapters 14 and 15) that are essential to developing processes that deliver high yields, system throughput, and high material utilization rates needed for cost-effective manufacturing. Given the effort spent, and subsequent high costs, to synthesize and purify the OLED materials, it seems a shame to use wasteful processing methods during OLED fabrication.

Within a given system, it is clear that design dictates certain choices. For instance, a transparent OLED obviously requires the use of a transparent substrate and transparent encapsulation. Furthermore, a flexible device dictates certain choices not only in substrate and encapsulation but also in the electrodes and drive circuit. In general, total systems engineering principles provide a useful approach, wherein the interfaces between components are explicitly considered so that the entire system can be optimized. These design questions have been addressed many times over the past 25 years of the modern OLED era. The first comprehensive approach to display design was Gu and Forrest in 1998 (Gu and Forrest 1998) and the principles therein are still useful.

Chapter 16 summarizes the design considerations for lighting. In the rest of this section, we summarize the questions that one should ask to address the materials and processing trade-offs that are inherent in any design.

OLED system design must start with a good understanding of the physical, electrical, and optical fundamentals of the system, as described in Chapters 6, 12, and 16. The optical properties of the system should be considered prior to beginning design, in order to answer questions such as: Will internal and/or external optical extraction be incorporated? What color(s), and what quality of color, need to be incorporated? How stable does the color need to be? Also, what are the optical design requirements for substrate, packaging, controls, and electrodes? What constraints are placed on organic materials, with regard to both composition and layer thickness? This step is usually completed based on the color requirements, and then the energy levels of the emitter dictate the electronic structure of the rest of the materials in the stack. In many cases, strategic alliances or joint development partnerships are necessary in the OLED marketplace. Does the team have access to materials that meet performance, stability, and, potentially, cost requirements, and if not, what partnerships might be available to fill the gap? What deposition approaches are to be used? Are the components (and deposition methods) compatible with later processing steps, and with materials that have already been deposited? What internal structure will be used for the OLED (single stack, tandem, etc.)? Will uniform or graded layers be used? There are many variations on simple 2 or 3 layer devices incorporating blocking layers, energy-level matching layers, conductivity dopants, injection layers, multiple host:emitter layers, emissive layers with multiple emitters, or graded layers. Each layer adds cost to the OLED system, but the trade-offs with performance are not always straightforward. There is no substitute for experience, but a strong grounding of the fundamentals is important to ensure an informed approach to answering these questions.

There are multiple approaches for substrate, packaging, and electrodes, as well. What are the limitations on the mechanical properties of the substrate and encapsulation? In what kinds of environments will the OLED be used? How will contacts be made? What patterning needs to be incorporated? How will current be delivered and controlled? Is an auxiliary anode required? Are thin-film transistors (TFTs) necessary to switch on and off individual pixels? Finally, how will the OLED be integrated into the final product (lighting or display)? These questions should be asked during design, which is usually an iterative process, where the integration takes place over many tests. The information in the preceding chapters provides a sound basis for addressing these questions, including useful survey of small-molecule approaches to the organic materials for most components within the OLED. For the electrodes, materials for the transparent electrode (usually the anode) are covered in Chapter 2. However, the control of the current may be more complex to ensure uniformity and constant light output. For lighting, the question of current spreading was addressed in Section 5 of Chapter 16, and patterning may be relatively simple. However, control of the pixels in a display, where the brightness must be controlled carefully and switched on and off quickly, requires a more complicated control circuit. The next section addresses OLED control circuits, including materials choices and operating principles for the TFTs used in OLED displays, along with the basic design of the control circuit for displays and

Chapter 17

integration into the OLED structure for both active matrix (AMOLEDs) and passive matrix OLEDs (PMOLEDs).

17.2.1 Backplane

OLEDs are current-driven devices, which mean the light output is dependent upon the continuous delivery of electrons and holes to generate excitons in the emissive layer. The light output of an OLED is thus very sensitive to slight variations in current and because the eye is very sensitive to small variations in intensity and color, the current source must be very stable. Other design considerations for the current source include switching speed, process compatibility, power consumption, and possibly characteristics such as transparency and flexibility. In a display, the circuit that translates a signal voltage into an output (voltage for an LCD and current for an OLED) to drive the display is called the backplane.

The backplane for an OLED display is typically deposited onto the substrate first, with the OLED fabricated on top. For a bottom emission device, this implies that either the backplane is transparent or the OLED is deposited offset to the backplane, and the light is emitted through the aperture created thereby (Miura et al. 2011; Nathan et al. 2004). Generally, the most basic circuit to be used in the backplane consists of two TFTs, one of which acts as a switch and the other as the current source. The TFT that delivers the current is usually placed in parallel with a capacitor, often called the storage capacitor, C_{ST}, which provides the charge to quickly generate and hold the gate voltage to the current source TFT constant, so that the transistor operates at a constant current. Such a circuit is designated 2T1C. Depending upon the polarity of the TFTs (p-channel or n-channel) and the integration scheme, the source or drain side of the TFT may be in contact with either anode or cathode (Gu and Forrest 1998; Hildner 2005). For most of the materials in the following sections, the TFT is p-channel type, in contact with the anode and with the OLED on the drain side (with the current flowing through the OLED to the cathode). However, for the metal oxide materials, the channel is n-type, which leads to other acceptable configurations (see, e.g., Park 2013). In the rest of this section, we very briefly describe the operating principles of TFTs, survey the current state of TFT materials and assess suitability for use in controlling OLEDs, and then briefly describe the application of various TFTs to OLED displays.

The objective of the backplane is to quickly deliver a constant current when a signal voltage switches the first TFT into an *on* position, and continue to deliver the current as long as the signal voltage remains at the *on* value. This is achieved by providing a signal—a change in the gate voltage on the switch TFT, which applies a controlled gate voltage, which leads to current flow through the OLED. TFTs operate on the same principles as other field effect transistors, essentially as a voltage-controlled resistor. Briefly, a gate voltage changes the state of the channel from insulating to conducting, and current flows between the source and drain, which are held at different potentials. Standard field effect transistor models can be used to understand TFT behavior. The drain current, I_D, is given in the linear region by Equation 17.1 (after Street 2009), or any elementary textbook on semiconductor devices):

$$I_{D,lin} = C_G \mu \frac{W}{L} [(V_G - V_T)V_D - V_D^2/2] \tag{17.1}$$

and in the saturation region by

$$I_{D,sat} = C_G \mu \frac{W}{2L} [V_G - V_T]^2 \tag{17.2}$$

where C_G is the gate capacitance, μ the mobility, W and L the width and length of the TFT, V_G the gate voltage, V_T the threshold voltage, and V_D the drain voltage. In other words, the amount of current that can be delivered is proportional to the mobility, scaled by the ratio of channel width to length and applied voltage above threshold. The transfer characteristics (top) and I–V curves for an organic TFT (OTFT) are shown in Figure 17.1 (from Park et al. 2007). The linear and saturation regions are clearly shown in the bottom panel.

For an OLED display, the pixel current is about 1–10 μA (Park J.-S. et al. 2009; Street 2009). The size of the TFT (i.e., the cross-sectional area of the channel determines the number of charge carriers) and the mobility determines the amount of current that can be delivered through the current source TFT. For relatively low-mobility materials, the

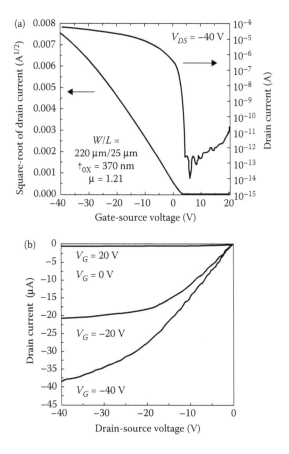

FIGURE 17.1 Characteristic transistor curves showing linear and saturation regions. Obtained from drop-cast OTFT devices. (a) Log and square root of drain current (I_{DS}) as a function of gate-source voltage (V_{DS}) for a drain-source voltage (V_{DS}) of −40 V. (b) Drain current (I_{DS}) as a function of drain-source voltage (V_{DS}) for several values of gate-source voltage (V_{GS}). See Table 17.1 for performance requirements for OLED applications. (From Park, K.P. et al. 2007. *Applied Physics Letters.* 91: 063514, with permission.)

size of the TFT needed to drive an OLED pixel is limited to sizes that are good only for low-resolution displays (i.e., large pixel size); as the channel shrinks with the rest of the TFT dimensions, there comes a point at which the channel cannot carry enough current to drive the pixel. For small OLEDs with high resolution, $\mu > 100$ cm^2 V^{-1} s^{-1} is necessary to deliver high currents and operate at high speed. For high-resolution OLED televisions, this requirement is relaxed, to about $\mu > 30$ cm^2 V^{-1} s^{-1} due to the larger pixel sizes. Stability should be very good, with little change in threshold voltage, V_{TH}, even under bias stress. A change in V_{TH} of as little as 0.1 V can change the pixel brightness by as much as 20% depending on the specifics of the TFT and the OLED (In and Kwon 2009); unfortunately, many materials exhibit V_{TH} changes under bias as we shall see below. Short- and long-range uniformity should be very good, and leakage current very low to minimize power consumption.

Figure 17.2 shows several TFT circuit designs for a variety of candidate TFT materials for OLED displays. Figure 17.2a shows the proposed circuit for the OTFT circuit demonstrated by Zhou et al. (2005). This 2T1C is the simplest circuit devised to generate a constant current. The other panels, (b)–(f), are described below, along with more details on (a).

17.2.1.1   Organic TFTs

Organic TFTs seem to be a natural fit for OLEDS—the processing should be compatible with OLEDs (solution or vapor deposition), and flexibility and transparency are suitable for interesting devices (Sony 2010; Zhou et al. 2005). The mobilities of some organics can exceed a-Si:H (Park et al. 2007; Sakanoue and Sirringhaus 2010) for both solution and vapor-processed devices, which makes them acceptable substitutes for low-resolution displays. This is especially true where extreme flexibility is desired (Sony 2010). The primary challenges in the use of OTFTs to drive OLEDs are related to the high *turn-on* voltage required for TFT performance, passivation, and stability. The practically achievable mobilities are just too low and the voltages too high for commercial use, and the susceptibility to degradation (intrinsic or due to moisture) remains suspect.

17.2.1.2   Amorphous Silicon

The initial material of choice for OLED backplanes was hydrogen-terminated amorphous thin-film silicon (a-Si:H) because it was in widespread use in active matrix LCD displays (AMLCDs), and the maturity of manufacturing technologies developed for AMLCD suggested an inexpensive route to AMOLED manufacturing (Nathan et al. 2004). Although a-Si:H has a relatively low mobility (~1 cm^2 V^{-1} s^{-1}), it has a very good on–off ratio and other performance characteristics, as well as easy integration into displays, which make it well suited for voltage-driven LCD devices. However, a-Si:H TFTs have a fatal flaw when it comes to current-driven OLED devices—under bias, the threshold voltage shifts irreversibly over time, which translates into a change in the current output, and therefore the pixel brightness. Additional TFTs can be incorporated into the circuit to compensate for the threshold voltage shift. Figure 17.2b and c is taken from two attempts to improve the performance of a-Si:H TFTs for OLED displays (Goh et al. 2003; Lee et al. 2005). As is clearly shown, the circuit devised by Goh et al. required two additional TFTs while that devised by Lee et al. required four additional TFTs, for a total of six, in order to compensate for V_T changes. Each TFT consumes additional electrical power and therefore increase power consumption, which is a major drawback especially

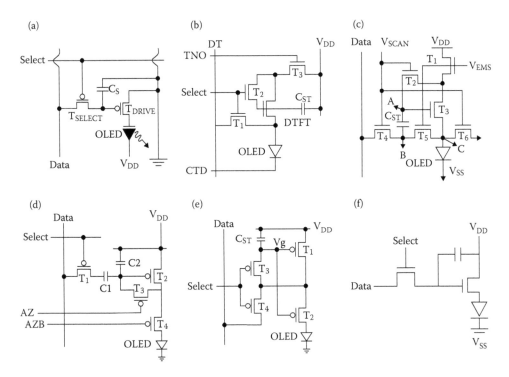

FIGURE 17.2 Circuit diagrams proposed for active matrix OLED display backplane, all based on delivering constant current using at least two TFTs and a storage capacitor: (a) Pentacene OTFT. (After Zhou, L. et al. 2005. *IEEE Electron Device Letters.* 26(9): 640–642, with permission.) (b) Designed for a-Si:H. (After Goh, J.C. et al. 2003. *IEEE Electron Device Letters.* 24(9): 583–585, with permission.) (c) A more complicated version to compensate for threshold voltage degradation using a-Si:H TFTs. (After Lee, J.-H. et al. 2005. *IEEE Electron Device Letters.* 26(12): 897–899, with permission.) (d) Voltage compensation approach to pixel-to-pixel variation in LTPS TFTs. (After Hong, S.-K.; B.-K. Kim; Y.-M. Ha. 2007. *SID Symposium Digest of Technical Papers.* 38: 1366–1369, with permission.) (e) Current voltage compensation approach to pixel-to-pixel variation in LTPS TFTs. (After Hong, S.-K.; B.-K. Kim; Y.-M. Ha. 2007. *SID Symposium Digest of Technical Papers.* 38: 1366–1369, with permission.); note that Hong et al. (2007) also demonstrated a digital driving picture structure and a "global mura compensation" external compensation approach developed by Kodak and implemented in the integrated circuit driver (not shown). (f) Simple two-transistor, one-capacitor control circuit enabled by high-performance metal oxide (IGZO) TFTs. In all cases, V_{DD} is the drive voltage line, V_{SS} is ground, T# is a transistor, C_{ST}/C# is a capacitor, and Data, Select, and other labeled lines are control voltages.

for mobile devices. Despite its early promise, given the low mobility and threshold voltage shift, a-Si:H has been largely abandoned for use in OLEDs.

17.2.1.3 Low-Temperature Polysilicon

An alternative material to a-Si:H is low-temperature polysilicon (LTPS). LTPS was recognized as an alternative early in the history of the development of OLEDs (Stewart et al. 2001), with the caveat that roughness and improvement of low-temperature processes were necessary for successful integration. LTPS has much higher mobility (30–300 cm² V⁻¹ s⁻¹ depending upon the details of its processing), and LTPS TFTs exhibit good on–off ratios, ~10⁶, although they have higher leakage current (which also leads to additional power consumption) (Lin and Chen 2007; Street 2009). Unfortunately, LTPS requires additional processing steps (~5). The Si is deposited by sputtering, and the amorphous film is annealed using an UV excimer laser annealing (ELA) step, sometimes called excimer laser crystallization (Tsai et al. 2008), as well as possibly metal-induced crystallization. The ELA step is compatible with low-temperature processing, but the additional process steps required for LTPS TFTs increase cost substantially. Furthermore, LTPS suffers from variations in mobility (due to different degrees of crystallization) at a length scale that leads to variations in brightness on an OLED display, and may also suffer from threshold voltage shifts (Lin et al. 2012). Figure 17.2d and e shows two examples of compensating LTPS TFT backplane circuit designs. In both cases, four transistors are used (Hong et al. 2007; Lin and Chen 2007) to compensate for variations in OLED brightness. In Figure 17.2d, Lin and Chen (2007) proposed a new circuit using a feedback structure to compensate for decreased light output of the OLED being driven, as well as to correct the threshold voltage shift. In Hong et al. (2007), the authors explored the nonuniformity introduced by the ELA process and compared uncompensated OLED pixels with voltage programming compensation (shown in Figure 17.2e), current programming compensation (not shown), and compensation using a "global mura compensation" method. This method, developed by Kodak, entailed storing an image of the nonuniformities in the integrated circuit (IC) driver, which would then correct the IC output to reduce nonuniformity. Finally, the authors also developed an alternative solid-phase crystallization method, and demonstrated its use in an OLED backplane with much improved uniformity. There are many other examples in the technical and patent literature that focuses on circuit design to overcome these fundamental materials limitations. For some time, LTPS was the preferred material for OLED display backplanes, and was used extensively by Samsung (OLED-Info 2012) and is still in use in some applications.

17.2.1.4 Metal Oxides

Metal oxide TFTs are a relatively recent invention (Hoffman et al. 2003; Nomura et al. 2003), with the first reports using intrinsic ZnO and crystalline indium–gallium–zinc oxide ($InGaO_3(ZnO)_5$/IGZO). In the work of Hoffman, Norris, and Wager, the device used n-type ZnO in the channel, with aluminum–titanium dioxide as the gate insulator. The ZnO TFTs were transparent, although they exhibited photoconductivity under UV exposure, with good on–off ratio (~10^7), threshold voltages from 10 to 20 V, moderate leakage currents, and mobilities up to 2.5 m^2 V^{-1} s^{-1}. The mobility, although comparable to a-Si:H, was not adequate for OLED applications. Nomura took the approach of using a single crystalline film and building the rest of the TFT on top of the IGZO. This device showed excellent mobility due to the absence of grain boundaries, with μ_e ~ 80 cm^2 V^{-1} s^{-1} at room temperature. This device was also a transparent device, had an on–off ratio of ~10^6, low leakage current, and a low threshold voltage (~3 V).

Over the next 10 years, metal oxide TFT technology progressed incredibly quickly (Jeong 2011; Kamiya et al. 2010; Miura et al. 2011; Park J.-S. et al. 2009; Park S.H.K. et al. 2009; Park J.S. et al. 2012; Su et al. 2014), to the point where amorphous IGZO TFTs have demonstrated performance that enables the use of the simple circuit depicted in Figure 17.2f in some cases (Miura et al. 2011). IGZO TFTs have very low leakage current, very low threshold voltage (<1 V), and very high on–off ratio (>10^7) (Jeong 2011; Kamiya et al. 2010; Park J.S. et al. 2012; Su et al. 2014). IGZO TFTs can be quite flexible (Chien et al. 2011; Miura et al. 2011; Park J.-S. et al. 2009) and compatible with low-temperature processing for use in conformable or flexible OLEDs. Uniformity is quite good. IGZO does undergo a voltage threshold shift under positive bias, which may need to be compensated for; see Park (2013) for two examples, 5T2C and 5T1C, of compensation schemes. Furthermore, most metal oxide TFTs are susceptible to degradation due to moisture or interaction with other components such as Cu, so encapsulation and/ or passivation are required. Some metal oxide materials (i.e., ZnO) are photoactive in the UV, and would be susceptible to degradation in outdoor use.

The reduced complexity (no ELA, easy passivation) of the metal oxide TFT technology has enabled reduced capital and manufacturing costs (BNP Paribas 2011; EE Times 2013; Street 2009). This has led to a shift in technology to the point where all major display manufacturers—LCD and OLED—are likely to switch over to metal oxide TFTs by 2014 (EE Times 2014; ID Tech 2014; Nanomarkets 2012). The advent of 4K televisions driven by metal oxide TFT backplanes in 2013 (EE Times 2013), including demonstrations by Sony, Panasonic, and LG, has clearly shown the advantage in performance and processing.

17.2.1.5 Exotic Materials

One of the most exciting new developments in TFT materials is the advent of low-dimensional materials with excellent transport properties. These materials include one- and two-dimensional materials that can be used as the channel in a TFT. For some time, one-dimensional materials such as carbon nanotubes or ZnO nanowires seemed poised to be the newest material to challenge for technological relevance (Street 2009). However, the integration challenges have proven to be formidable. New two-dimensional materials such as graphene (Eda et al. 2008; Lu et al. 2012; Lee et al. 2012) and even newer materials such as WSe_2 (Das et al. 2014) have demonstrated remarkable mobilities (30–300 cm^2 V^{-1} s^{-1}) comparable to metal oxide TFTs, even with few or single-layer devices. Of course, the current is limited by the cross-sectional area of the TFT (see above), but even so the TFT characteristics are impressive. These exotic materials may indeed supplant metal oxide TFTs for use in OLED devices, but that appears to be far off due to the integration challenges over large areas, coupled to the synthetic challenges associated with making relevant amounts of the low-dimensional material at a cost that is acceptable. Nonetheless, given the incredible performance of these materials in lab devices and the flexibility and transparency of some of these devices, it would not surprise the author if these materials enter the marketplace sooner rather than later.

To summarize the properties of the materials covered in this section, Table 17.1 shows the relative performance TFT characteristics of the various materials.

17.2.1.6 Integration Challenges

Integration challenges for passive matrix electrode structures are much more easily overcome, and are covered in detail in Section 16.5.1 of Chapter 16, including approaches for achieving uniform current spreading across large areas. Approaches to overcoming the finite resistance of transparent electrodes include the addition of a metal grid, as shown in Figure 17.3 (Park J.W. et al. 2012). For transparent devices, the metal grid approach must be coupled to the use of a semitransparent electrode, in which the transparent anode would be replaced by a thin layer of Mg:Ag (which has relatively high transparency).

Chapter 17

Table 17.1   Comparison of TFT Performance Requirements for OLEDs, along with Properties of Thin-Film Transistor Materials Classes Proposed for OLED Applications

Material Class	Mobility (cm² V⁻¹ s⁻¹)	Leakage Current	Threshold Voltage	Transparency	Flexibility	Process Compatibility	Stability	Uniformity
OLED requirement[a]	30–100	Variable; <1 pA	Low	Variable	Variable	Excellent	Excellent	Excellent
Metal oxides[b]	10–100	<10 fA	Low	Excellent	Average	Very good	Good	Excellent
LTPS[c]	30–300	~100 fA	Low	Poor	Average	Average	Excellent	Poor
a-Si:H[d]	~0.5–1	~100 fA	Low	Poor	Average	Excellent	Poor	Excellent
Organics[e]	Up to ~1	~1 nA	High	Excellent	Excellent	Average	Poor	Variable
Exotics (graphene, nanowires, etc.)[f]	1–300	<100 pA	Low/variable	Could be excellent	Excellent	Poor	Unknown	Variable

[a] Park (2013) and references therein.

[b] Hoffman et al. (2003); Nomura et al. (2003); Park J.-S. et al. (2009); Park S.H.K. et al. (2009); Street (2009); Kamiya et al. (2010); Chien et al. (2011); Jeong (2011); Miura et al. (2011); Park J.S. et al. (2012); Park (2013); Su et al. (2014).

[c] Stewart et al. (2001); Lin and Chen (2007); Hong et al. (2007); Tsai et al. (2008); In and Kwon (2009); Street (2009); Choi et al. (2010); Lin et al. (2012).

[d] Yang et al. (2000); Goh et al. (2003); Nathan et al. (2004); Lee et al. (2005); Street (2009).

[e] Zhou et al. (2005); Park et al. (2007); Choi et al. (2008); Street (2009); Sakanoue and Sirringhaus (2010); Nomoto et al. (2012).

[f] Eda et al. (2008); Street (2009); Lee et al. (2012); Lu et al. (2012); Das et al. (2014).

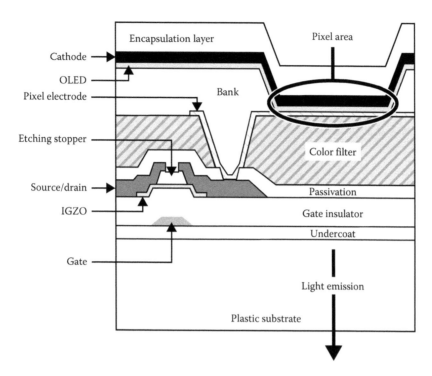

FIGURE 17.3 Auxiliary electrode structure and integration for lighting. Note that the actual current source must deliver large currents, but can be handled by conventional control circuits. (After Park, S.H.K. et al. 2009. *SID Symposium Digest of Technical Papers.* 39(1): 629–632, with permission.)

Integration schemes for AMOLED backplanes have been worked out at some length as well. One drawback is the loss of emissive area by the opaque backplane. Much effort has gone into increasing the aperture area to maximize light output and minimize cost. An example of a flexible device deposited on a plastic substrate from Miura and coworkers is shown in Figure 17.4. In Figure 17.4, the emitting aperture area is shown circled; it is clear that a fair amount of area with expensive organic materials deposited is not emitting light. Of course, increasing the aperture area is one of the key reasons for pursuing solution deposition methods such as printing. Nonetheless, manufacturers have found ways to maximize the emitted light using patterning schemes and other approaches, including using transparent materials (Park S.H.K. et al. 2009).

17.3 Processing and Supply Chain

In order to succeed in the marketplace, OLED products must be built at an acceptable cost and appeal to consumers. OLEDs have many properties that have the potential to differentiate them from other lighting and display products. The success of OLED displays is a strong indicator of the preference consumers have for attributes such as color quality, contrast ratio, switching speed, and so on. Currently, the OLED supply chain spans the globe, although the majority of production capacity is in Asia, with >$10B/year capital investments in production (DOE 2014). The primary contributors to OLED cost are cost of capital (which depends on a variety of factors, including total accumulated

Chapter 17

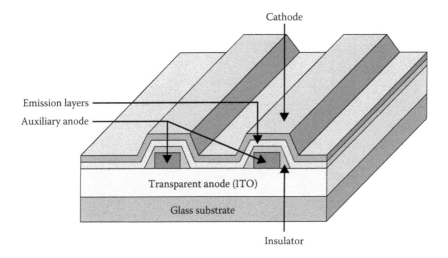

FIGURE 17.4   Schematic showing AMOLED backplane integration. Note the emissive area defined by the electrodes, indicated by the oval. (After Miura, K. et al. 2011. *SID Symposium Digest of Technical Papers.* 42(1): 21–24, with permission.)

cycle time [TACT] and depreciation), overhead (including taxes, shipping and distribution, insurance, product development, subsidies, etc.), patterning (much more cost associated with displays than lighting), substrate (including electrodes and possibly optical outcoupling layers/materials), organic stack (including losses due to lower utilization rates), packaging, and assembly (including product testing), all of which scale with the manufacturing yield. The U.S. Department of Energy Solid State Lighting Program has set targets that, if achieved, would lead to cost-competitive OLED lighting panels within the next 5–10 years. The targets are aggressive, but the program has laid out a research agenda to achieve these targets.

Chapters 14 and 16 discuss vapor and solution approaches to material deposition for OLED fabrication in detail. It is clear from the amount of research, development, and production using both vapor and solution approaches that there is room in the marketplace for both. Each will find applications and practitioners who favor one approach over the other, based on technical needs, cost models, and materials compatibility. Numerous companies in the OLED materials market have developed improved ways to solubilize charge transport or other materials, while the performance enhancements in lifetime and often performance that come from using vapor deposition for some steps appear to be persistent. Still, the performance gap continues to close, and for cost reasons, printing and other solution deposition methods remain quite attractive.

Capitalization remains a significant hurdle for OLED manufacturing. For that reason, many companies continue to seek ways to reduce the scale necessary for profitable OLED production. However, a more likely near-term approach is the conversion of fully depreciated LCD display manufacturing facilities, or the conversion of older OLED lines from display to lighting. This approach is often discussed in the context of the large display manufacturers such as LG and Samsung. For example, a report by BNP Paribas in 2011 (BNP Paribas 2011) predicted the use of Gen8 LCD fabs converted to AMOLED for television applications would reduce the cost by 45% compared to a new

OLED TV fab. In general, requiring the largest production scales for profitability does not lead to rapid introduction and adoption of a new technology. Therefore, it will likely fall to those who have such facilities already (i.e., display manufacturers in Asia) or those who have strong lighting businesses and associated supply chain (i.e., lighting manufacturers in Asia and Europe). For now, startups in Asia, the United States, and Europe are also looking to enter into the OLED fabrication business, but most other players in the marketplace are focused on securing a strong position in the supply chain. Of course, supply chain integration means that a great deal of expertise in OLED design, fabrication, and processing exists outside of the OLED manufacturers, as evidenced by a history of demonstrating highly efficient, long-lived OLEDs by companies such as Novaled and Universal Display Corporation.

17.4   Applications

It was only 9 years from the first demonstration of the heterostructure OLED by Tang and VanSlyke (1987) at Kodak in 1987 and 6 years after the group at Cambridge demonstrated the first polymer devices (Burroughes et al. 1990) to the first commercial PMOLED product. This early product, a monochrome car audio system display with modest resolution (256×64 pixels), was introduced by Pioneer in 1996, and Cambridge Display Technologies (CDT) demonstrated the first light-emitting polymer devices the same year. Two years later, Kodak and Sanyo demonstrated the first full color AMOLEDs, establishing that OLEDs would compete with LCDs in the small and large display markets. This led to a boom, in which it seemed that every company in displays, materials, or electronics planned to manufacture OLEDs, at least 40 by 2003 (BNP Paribas 2011). By 2008, most of these companies had backed out (if they were still in existence). Nonetheless, several key companies persevered and OLEDs are now a consumer favorite in mobile devices, poised to enter the television market in a meaningful way, and potentially finally ready to enter the lighting market.

One indicator of the growth of the OLED industry is the proliferation of market reports and predictions. As the size of the market grows, so too does the number of companies that offer reports with predictions of growth across market segments. For up-to-date information on the OLED industry, markets, and other public reports from industry and other researchers, the website OLED-info.com is a useful resource. In addition to aggregating public media reports of OLED announcements, OLED-info.com provides links to reports available for sale from various companies that assess the technology and markets, including flexible, printed, and component markets (BNP Paribas 2011; DisplaySearch 2010; ID Tech 2014; Markets 2013; Nanomarkets 2012, 2014a,b,c; Navigant 2013; OLED-info 2014a,b,c,d; Transparency 2013; Yole 2014).

17.4.1   Displays

During the 10 years after the first display product (from about 1996 to 2006), remarkable progress was made on OLED displays, as lighting remained a potentially important but more distant market. The number of players in the supply chain and manufacturing landscape increased. Key companies undertook strategic alliances to try to advance the technology and build in supply chain integration. However, all was not rosy, as a number of companies dropped out of the market as the technical barriers made commercialization difficult, including Pioneer. Small monochrome displays and subdisplays on the exterior of MP3 players, mobile phones, and other devices were a key market for early manufacturers such as Sony and RiTdisplay.

By 2003, Kodak introduced the first camera incorporating an AMOLED display, while dramatic improvements in both small molecule and polymer OLEDs efficiency and lifetime were reported. By 2007, Sony announced the first OLED TV, an 11″ device called the XEL-1, but by the end of the year, it looked like OLED televisions were going to take much longer to make it to market; the XEL-1 was not widely available (only a few thousand were produced) and was very expensive. Projections for production by numerous companies were pushed back well beyond 2009. LG has focused their efforts on OLED televisions (as opposed to mobile displays), and as of 2014 offers OLED televisions in most markets, although these are not yet cost-competitive with LCD TVs. At the time this book is written, OLED TVs remain a high-end novelty, although that appears poised for change.

In the meantime, it was the commercial success of AMOLEDs in mobile phones that drove much of the revenue growth and technology development in the industry, particularly due to the efforts of Samsung. In only a few years after shipments of mobile phones incorporating OLED displays began in earnest (late 2007), AMOLED displays became the dominant mobile phone display, with more than 70% of the market, led by Samsung. This remarkable growth coincided with a number of key factors. First, the cost of the OLED display declined rapidly as manufacturing costs were wrung out of production lines. Second, technology for the backplane and the OLED continued to improve dramatically. Third, weight, brightness, viewing angle, and contrast ratio led to consumer preference for OLED displays over LCDs. As power efficiency increased, battery lifetime was also improved, which removed one potential drawback for consumers. One obvious trend in this marketplace is the increasing display size, which is now suitable for high-resolution video on larger phones, tablets, and "phablets." At this point, it is clear that OLED displays are here to stay, and will complete their takeover of the mobile market in the near future.

OLED televisions are finally poised to become commercially viable, with prices dropping rapidly (though still higher than LCD TVs). The introduction of 4K OLED TVs by LG, Panasonic, and Sony has ensured that OLED TVs can be competitive on a performance basis with LED-backlit LCD televisions, in a market where the resolution is driven as much by consumer hype as performance need. Furthermore, the ease with which OLEDs can be made in other than a planar geometry has led to the development of curved TVs by LG, Panasonic, and Samsung. The curved TV is supposed to provide a more immersive viewing experience. Whether this has staying power in the marketplace, or is a gimmick to differentiate OLED TVs from LCD, remains to be seen. Nonetheless, the curved TVs are impressive—the polymer OLED TVs by Panasonic are printed using materials developed by Sumitomo, and can be made to curve in either direction. To summarize, while OLED TV sales have continued to be slow due to high prices (in large part due to yield issues), favorable consumer reviews of performance suggest that OLED TVs will do in the TV market what OLED displays did in the mobile display market, and laptop and desktop displays may follow.

17.4.2　Lighting

General lighting is perhaps the toughest challenge for OLED technology. The price sensitivity of consumers, which implies a very long lifetime if the initial price is higher, makes gaining a foothold in this market very challenging for a new technology that has not come down the experience curve on price. The U.S. Department of Energy Solid State Lighting program estimates that to be competitive in the lighting market, manufacturing costs need to decrease by a factor of 30 (DOE 2014), to $10/klm (klm = 1000 lumens) or $100/m² for a panel emitting light at a brightness of approximately 3000 cd m⁻². This is a big challenge, of course, but this is the cost required for widespread general illumination. OLEDs are entering the market in niche products initially, which will help provide revenue and learning for OLED manufacturers. The horizon for widespread adoption seems to retreat periodically (King 2011), but has not disappeared entirely. In general, OLEDs will find uses initially in applications that require diffuse light (instead of point sources), lower brightness, conformable or curved surfaces, transparent sources, and initially where price sensitivity is reduced.

One example of a market that is small but growing is the automotive market. The global automotive lighting market is expected to be about $29B by 2019 (Marketwatch 2014), with a number of technologies, including tungsten filament, halogen, xenon, LED, and now OLED filling various niches. There are several distinct attributes of automotive applications that make OLEDs particularly well suited for early entry into the automotive lighting market (Nanomarkets 2014a). While headlights require too high a brightness for OLEDs to currently compete, taillights and interior lights are suitable for OLED adoption. For high-end automotive applications, that is, lighting and displays on and in luxury vehicles, performance is paramount. OLEDs are in use in interior displays in several vehicles, while exterior lighting such as taillights and brake lights are in the prototype stage. There are several reasons why automotive manufacturers would be interested in OLEDs, including energy efficiency, package size, and distinctive design options, particularly conformability. Of course, the lifetime and efficiency of red (taillight) and amber (turn signal) OLEDs are relatively good, so long as encapsulation can be maintained, and the conformability of OLEDs lends itself to placement on the curved rear panel. A quick Internet search for pictures of OLEDs used in automotive applications shows a large number of distinctive designs. For instance, in early 2013, Audi rolled out a "3-D OLED" lighting assembly for the A8 taillight called "the swarm," developed in concert with Philips. In fact, as of 2014, there are a number of higher-end vehicles with internal and external lighting components incorporating PMOLEDs and AMOLEDs.

OLEDs have demonstrated great promise for lighting applications, with significant R&D efforts by all major lighting manufacturers around the world. However, as the technical challenges elucidated in this book have made clear, continued difficulty in overcoming these challenges has prevented OLEDs from entering the lighting markets in any significant volume as of the writing of this book. Nonetheless, many analysts believe the time is right for OLEDs to enter the lighting market (Nanomarkets 2014b). Figure 17.5 shows three luminaires from Acuity Brands Lighting incorporating currently available OLED panels. The performance of these panels is quite good from a

Chapter 17

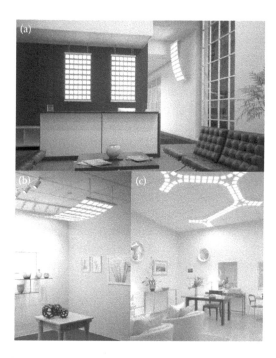

FIGURE 17.5   Pictures of large-area lighting products designed by Acuity Brands Lighting. The luminaire brand names, light output, and number of panels are (a) CANVIS™ vertical: 54 panels, 3885 lumens/luminaire, 46 lm W⁻¹ for 3000 K panels; (b) CANVIS™ horizontal: 66 panels, 4712 lumens/luminaire, 53 lm W⁻¹ for 3000 K panels; and (c) TRILIA™: 96 panels (4 luminaires); 67,560 lumens as shown; 46 lm W⁻¹ for 3000 K panels. All products use the same panels, which are available in three correlated color temperatures (CCT): 3000, 3500, and 4000 K. L70 lifetimes: 40,000 h at 3000 cd m⁻² or 72,000 h at 2000 cd m⁻² (3000 K panel); 30,000 h at 3000 cd m⁻², and 54,000 h at 2000 cd m⁻² (3500 and 4000 K panels). Color specifications: 3000 K panel: CCT = 2972 K; CRI = 89; R9 value = 29; Duv value = 0.0019. 3500 K panel: CCT = 3521 K; CRI = 86; R9 value = 24; Duv value = 0.0013. 4000 K panel: CCT = 3853 K; CRI = 90; R9 value = 41; Duv value = 0.0004. (Photos courtesy of Acuity Brands Lighting, www.acuitybrands.com/oled.)

form factor, color quality, efficiency, and lifetime standpoint. The caption provides detailed performance data.

One potential advantage that OLEDs have over LEDs for general lighting is the relatively warm light generated by OLEDs. It is easier to generate a warm white than a cool white with an OLED with good lifetime and efficiency using excellent phosphorescent green and red emitters, with a smaller amount of blue mixed in, than it is for LEDs. The panels shown in Figure 17.5 are available in three temperatures with correlated color temperatures (CCT) ranging from 3000 to 4000 K. The efficiency reduces slightly as the color temperature rises—in contrast to what is seen with LED lights. In any event, the designs take advantage of the thin form factor of the panels, and in the case of the CANVIS™ in Figure 17.5b, flexible OLED panels. Another interesting set of designs for accent or task lighting is shown in Figure 17.6. These designs, also by Acuity Brands Lighting, demonstrate similar performance to the area lighting designs in Figure 17.5; the NOMI Curve in Figure 17.6 (top panel) also takes advantage of panel flexibility.

These are only some of the commercially available panels and luminaires that are in the marketplace. It does seem that OLED lighting is poised to take off in the next few years, at least for indoor lighting where environmental conditions are not harsh. While

FIGURE 17.6 Pictures of two decorative/specialty indoor lighting products designed by Acuity Brands Lighting. Top panel: NOMI™ Curve; 138 lumens per luminaire; 2 panels; 32 lm W^{-1} as shown. Bottom panel: REVEL™; 321–349 lumens total output; 5 panels; 37 lm W^{-1} for dimmable 3000 K luminaire. Color and lifetime specifications are the same as for Figure 17.5. (Photos courtesy of Acuity Brands Lighting, www.acuitybrands.com/oled.)

it does not seem that OLED lighting is going to fail in the marketplace, harsh, although growth does appear to be fairly slow.

17.5 Summary

This chapter has summarized the materials, device design, and structures that are the basis for high-efficiency OLEDs for displays and lighting applications. The requirements for performance vary by application, but the underlying physical principles are similar. OLED researchers and developers must understand the physics of charge injection, transport, and recombination, the physics of light emission and propagation in planar media (i.e., optical system design), the chemistry and mechanics of materials (organic and inorganic), the interplay between deposition methods and chemistry, the system integration requirements for electrode and electrical circuit design, and the chemistry and materials science of encapsulation. Finally, the OLED developer should also understand both the physics and biology of color—color temperature, color rendering index, and other measures of color quality, and how color interacts with efficiency and other performance criteria within an OLED. We hope that this book has provided a solid basis for this understanding.

OLEDs have the potential to transform the way we display information, design and light our work and living spaces, and generally interact with light. In order to succeed in the marketplace, some of the claims that have been made for OLED technologies must be fulfilled: lighting that is lighter, thinner, flexible, less expensive, brighter, and lasts nearly forever, transparent and/or with programmable colors; or displays with incredible contrast, fast switching speed, with true black and vivid colors, all in a package that is thin, light, and cheap.

While not all of these attributes will be achieved simultaneously, it is clear that seemingly insurmountable technical hurdles—such as lifetime or manufacturing yield—will be overcome by dedicated and clever researchers in industry, academia, and national laboratories around the world. In fact, the huge success of OLEDs in mobile displays and the growing success of OLED televisions, along with an OLED lighting industry poised to take off, point the way to overcoming technical and market barriers. We hope that this book will contribute in some small way, and look forward with great enthusiasm to the continued and future success of OLEDs.

References

BNP Paribas. Available at: http://www.bnppresearch.com/ResearchFiles/13338/Asia%20OLED-Mar11.pdf. 2011. Accessed 10/1/2014.

Burroughes, J.H.; D.D.C. Bradley; A.R. Brown; R.N. Marks; K. Mackay; R.H. Friend et al. Light-emitting diodes based on conjugated polymers. 1990. *Nature*. 347: 539–541.

Chien, C.-W.; C.-H. Wu; Y.-T. Tsai; Y.-C. Kung; C.-Y. Lin; P.-C. Hsu et al. High-performance flexible a-IGZO TFTs adopting stacked electrodes and transparent polyimide-based nanocomposite substrates. 2011. *IEEE Transactions on Electron Devices*. 58(5): 1440–1446.

Choi, J.W.; J.H. Cheon; J.H. Oh; J. Jang; S.-C. Kim; J.-S. Ahn et al. Reduction of off-state currents in silicon on glass thin film transistor by off-state bias stress. 2010. *Electrochemical and Solid State Letters*. 13(7): J85–J87.

Choi, J.W.; S.H. Noh; D.K. Hwang; J.-M. Choi; S.J. Jang; E. Kim et al. Pentacene-based low-leakage memory transistor with dielectric/electrolytic/dielectric polymer layers. 2008. *Electrochemical and Solid State Letters*. 11(3): H47–H50.

Das, S.; R. Gulotty; A.V. Sumant; A. Roelofs. All two-dimensional, flexible, transparent, and thinnest thin film transistor. 2014. *Nano Letters*. 14: 2861–2866.

DisplaySearch. Available at: http://www.displaysearch.com/cps/rde/xchg/displaysearch/hs.xsl/landing_2014_oled_technology_report.asp. 2010. Accessed 10/13/2014.

DOE. Available at: http://apps1.eere.energy.gov/buildings/publications/pdfs/ssl/ssl_mfg_roadmap_aug2014.pdf. 2014. Accessed 9/22/2014.

Eda, G.; G. Fanchini; M. Chhowalla. Large-area ultrathin films of reduced graphene oxide as a transparent and flexible electronic material. 2008. *Nature Nanotechnology*. 3: 270–274.

EE Times. Available at: http://www.eetimes.com/document.asp?doc_id=1319808. 2013. Accessed 10/1/2014.

Goh, J.C.; J. Jang; K.S. Cho; C.K. Kim. A new a-Si:H thin film transistor pixel circuit for active matrix organic light emitting diodes. 2003. *IEEE Electron Device Letters*. 24(9): 583–585.

Gu, G.; S.R. Forrest. Design of flat-panel displays based on organic light-emitting devices. 1998. *IEEE Journal of Selected Topics in Quantum Electronics*. 4(1): 83–99.

Hildner, M.L. OLED displays on plastic. In: *Flexible Flat Panel Displays*, ed. G.P. Crawford. 2005. 285–312. John Wiley and Sons, Chichester, West Sussex, England.

Hoffman, R.L.; B.J. Norris; J.F. Wager. ZnO-based transparent thin-film transistors. 2003. *Applied Physics Letters*. 82(5): 733–735.

Hong, S.-K.; B.-K. Kim; Y.-M. Ha. LTPS technology for improving the uniformity of AMOLEDs. 2007. *SID Symposium Digest of Technical Papers*. 38: 1366–1369.

ID Tech. Available at: https://www.linkedin.com/pulse/article/20140822061819–173774513-metal-oxide-tft-backplanes-for-displays-2014-market-size-technologies-and-2024-forecasts-research-report. 2014. Accessed 10/1/2014.

In, H.-J.; O.-K. Kwon. External compensation of nonuniform electrical characteristics of thin-film transistors and degradation of OLED devices in AMOLED displays. 2009. *IEEE Electron Device Letters*. 30(4): 377–379.

Jeong, J.K. The status and perspectives of metal oxide thin-film transistors for active matrix flexible displays. 2011. *Semiconductor Science and Technology.* 26: 034008.

Kamiya, T.; K. Nomura; H. Hosono. Present status of amorphous In-Ga-Zn-O thin-film transistors. 2010. *Science and Technology of Advanced Materials.* 11: doi:10.1088/1468-6996/11/4/044305.

King, R. Expectations dim for OLED lighting. *IEEE Spectrum.* September 2011. 15–16.

Lee, J.-H.; J.-H. Kim; M.-K. Han. A new a-Si:H TFT pixel circuit compensating for the threshold voltage shift of a-Si:H TFT and OLED for active matrix OLED. 2005. *IEEE Electron Device Letters.* 26(12): 897–899.

Lee, S.-K.; H.Y. Jang; S.J. Jang; E.Y. Choi; B.H. Hong; J.C. Lee et al. All graphene-based thin film transistors on flexible plastic substrates. 2012. *Nano Letters.* 12: 3472–3476.

Lin, C.-L.; W.-Y. Chang; C.-C. Hung; C.-D. Tu. LTPS-TFT pixel circuit to compensate for OLED luminance degradation in three-dimensional AMOLED display. 2012. *IEEE Electron Device Letters.* 33(5): 700–702.

Lin, C.-L.; Y.-C. Chen. A novel LTPS-TFT pixel circuit compensating for TFT threshold-voltage shift and OLED degradation for AMOLED. 2007. *IEEE Electron Device Letters.* 28(2): 129–131.

Lu, C.-C.; Y.-C. Lin; C.-H. Yeh; J.-C. Huang; P.-W. Chiu. high mobility flexible graphene field-effect transistors with self-healing gate dielectrics. 2012. *ACS Nano.* 6(5): 4469–4474.

Markets. Available at: http://www.marketsandmarkets.com/PressReleases/global-OLED-market.asp. 2013. Accessed 10/13/2014.

Marketwatch. Available at: http://www.marketwatch.com/story/global-automotive-lighting-market-2014-2019-trend-profit-and-forecast-analysis-2014–07–30. 2014. Accessed 10/16/2014.

Miura, K.; T. Ueda; S. Nakano; N. Saito; Y. Hara; K. Sugi et al. Low temperature-processed IGZO TFTs for flexible AMOLED with integrated gate driver circuits. 2011. *SID Symposium Digest of Technical Papers.* 42(1): 21–24.

Nanomarkets. Available at: http://nanomarkets.net/market_reports/report/metal_oxide_thin_film_transistor_markets. 2012. Accessed 10/11/2014.

Nanomarkets. Available at: http://nanomarkets.net/market_reports/report/oled-automotive-lighting-2014. 2014a. Accessed 10/13/2014.

Nanomarkets. Available at: http://nanomarkets.net/market_reports/report/oled-lighting-markets-2014. 2014b. Accessed 10/13/2014.

Nanomarkets. Available at: http://nanomarkets.net/market_reports/report/oled-materials-markets-2014. 2014c. Accessed 10/13/2014.

Nathan, A.; A. Kumar; K. Sakariya; P. Servati; S. Sambandan; D. Striakhilev. Amorphous silicon thin film transistor circuit integration for organic LED displays on glass and plastic. 2004. *IEEE Journal of Solid-State Circuits.* 39(9): 1477–1486.

Navigant. Available at: http://www.navigantresearch.com/research/energy-efficient-lighting-for-commercial-markets. 2013. Accessed 10/10/2014.

Nomoto, K.; M. Noda; N. Kobayashi; M. Katsuhara; A. Yumoto; S.-I. Ushikara et al. Rollable OLED display driven by organic TFTs. 2012. *SID Symposium Digest of Technical Papers.* 42(1): 488–491.

Nomura, K.; H. Ohta; K. Ueda; T. Kamiya; M. Hirano; H. Hosono. Thin-film transistor fabricated in single-crystalline transparent oxide semiconductor. 2003. *Science.* 300: 1269–1272.

OLED-info. Available at: http://www.oled-info.com/cmel/displaybank_7_1m_amoleds_in_first_half_of_2008_over_10m_expected_in_second_half. 2008. Accessed 10/10/2014.

OLED-info. 2012: Available at: http://www.oled-info.com/samsung-track-scale-ltps-gen-8-amoled-production. Accessed 10/10/2014.

OLED-info. Available at: http://www.oled-info.com/flexible-oled-market-report. 2014a. Accessed 10/13/2014.

OLED-info. Available at: http://www.oled-info.com/handbook/. 2014b. Accessed 10/1/2014.

OLED-info. Available at: http://www.oled-info.com/oled-automotive-market-report. 2014c. Accessed 10/13/2014.

OLED-info. Available at: http://www.oled-info.com/oled-microdisplay-market-report. 2014d. Accessed 10/13/2014.

Park, J.-S. Oxide TFTs for AMOLED TVs. 2013. *Information Display.* 29(2): 16–19.

Park, J.-S.; T.-W. Kim; D. Stryakhilev; J.-S. Lee; S.-G. An; Y.-S. Pyo et al. Flexible full color organic light-emitting diode display on polyimide plastic substrate driven by amorphous indium gallium zinc oxide thin-film transistors. 2009. *Applied Physics Letters.* 95: 013503.

Park, J.S.; W.-J. Meng; H.-S. Kim; J.-S. Park. Review of recent developments in amorphous oxide semiconductor thin-film transistor devices. 2012. *Thin Solid Films.* 520: 1679–1693.

Chapter 17

Park, J.W.; J.H. Lee; Y.-Y. Noh. Optical and thermal properties of large-area OLED lightings with metallic grids. 2012. *Organic Electronics*. 13: 189–194.

Park, K.P.; T.N. Jackson; J.E. Anthony; D.A. Mourey. High mobility solution processed 6,13-bis(triisopropyl-silylethynyl)pentacene organic thin film transistors. 2007. *Applied Physics Letters*. 91: 063514.

Park, S.H.K.; M. Ryu; C.-S. Hwang; S. Yang; C. Byun; J.-I. Lee et al. Transparent ZnO thin film transistor for the application of high aperture ratio bottom emission AM-OLED display. 2009. *SID Symposium Digest of Technical Papers*. 39(1): 629–632.

Researchmoz. Available at: http://www.researchmoz.us/oled-lighting-markets-2014-report.html. 2014. Accessed 10/13/2014.

Sakanoue, T.; H. Sirringhaus. Band-like temperature dependence of mobility in a solution-processed organic semiconductor. 2010. *Nature Materials*. 9: 736–740.

Stewart, M.; R.S. Howell; L. Pires; M.K. Hatalis. Polysilicon TFT technology for active matrix OLED displays. 2001. *IEEE Transactions on Electron Devices*. 48(5): 845–851.

Street, R.A. Thin-film transistors. 2009. *Advanced Materials*. 21: 2007–2022.

Sony. Available at: http://www.sony.net/Products/SC-HP/cx_news/vol62/pdf/sideview62.pdf. 2010. Accessed 10/1/2014.

Su, C.-Y.; W.-H. Li; L.-Q. Shi; X.-W. Lv; K.-Y. Ko; Y.-W. Liu et al. A 31-in. FHD AM-OLED display using amorphous IGZO TFTs and RGB FMM. 2014. *SID Symposium Digest of Technical Papers*. 45(1): 846–848.

Tang, C.W.; S.A. VanSlyke. Organic electroluminescent diodes. 1987. *Applied Physics Letters*. 51: doi: 10.1063/1.98799.

Transparency. Available at: http://www.transparencymarketresearch.com/oled-displays.html. 2013. Accessed 10/13/2014. $5B in 2012 to $25B in 2018. Says that OLED TV will overtake mobile devices in 2015.

Tsai, C.-C.; Y.-J. Lee; J.-L. Wang; K.-F. Wei; I.-C. Lee; C.-C. Chen et al. High-performance top and bottom double-gate low-temperature poly-silicon thin film transistors fabricated by excimer laser crystallization. 2008. *Solid-State Electronics*. 52: 365–371.

Yang, C.-S.; L.L. Smit; C.B. Arthur; G.N. Parsons. Stability of low-temperature amorphous silicon thin film transistors formed on glass and transparent plastic substrates. 2000. *Journal of Vacuum Science and Technology B*. 18: 683–689. doi:10.1116/1.591259.

Yole. Available at: http://www.yole.fr/iso_upload/News/2014/PR_Printed%20Electronics_YOLE%20 DEVELOPPE MENT_ Februay%202014.pdf. 2014. Accessed 10/13/2014.

Zhou, L.; S. Park; B. Bai; J. Sun; S.-C. Wu; T.N. Jackson; S. Nelson; D. Freeman; Y. Hong. Pentacene TFT driven AMOLED displays. 2005. *IEEE Electron Device Letters*. 26(9): 640–642.

Index of Materials

Subject Index